HERIOT-WATT UNIVERSITY LIBRARY
WITHDRAWN

Synthetic Procedures in Nucleic Acid Chemistry

VOLUME 2

Synthetic Procedures in Nucleic Acid Chemistry

VOLUME 2

Physical and Physicochemical Aids in Characterization and in Determination of Structure

EDITED BY

THE LATE W. WERNER ZORBACH

DEPARTMENT OF CHEMISTRY AND CHEMICAL ENGINEERING
MICHIGAN TECHNOLOGICAL UNIVERSITY
HOUGHTON, MICHIGAN

AND

R. STUART TIPSON
FORMERLY OF

NATIONAL BUREAU OF STANDARDS
WASHINGTON, D.C.

WILEY — INTERSCIENCE
A DIVISION OF JOHN WILEY & SONS, INC.
NEW YORK · LONDON · SYDNEY · TORONTO

Library of Congress Catalog Card Number: 68-16058

ISBN 471 98418-3

Printed in the United States of America

10 9 8 7 6 5 4 3 2 1

Prologue

In the death of Professor W. W. Zorbach on June 28, 1970, the field of nucleic acid chemistry suffered an irreparable loss. The concept of initiating a series of volumes dealing with synthetic procedures in this field was entirely his, and it was he who undertook the onerous task of preparing a worldwide list of the leaders in the field and of extending invitations to them to contribute to the first volume. With minimal suggestions from Dr. Zorbach, each author was encouraged to select the topics he would discuss, and, because of the obvious need for the book, the response was enthusiastic. The wisdom of these policies is attested to by the high praise accorded Volume 1 by reviewers in journals published both in the United States and abroad.

For Volume 2, Dr. Zorbach decided that a collection of articles describing some of the most important of the physical and physicochemical aids in the characterization and determination of structure of nucleosides, nucleotides, and related compounds would constitute a highly useful adjunct to the synthetic procedures described in Volume 1. The edited manuscripts for Volume 2 were sent to the publishers a few months before his death.

Preface

In the present volume, nine chapters by experts in their fields briefly present the theory, and provide the most modern techniques, for important methods of characterization and determination of structure. The first two chapters, by A. Albert (Canberra), discuss the *ionization constants* and *ultraviolet absorption spectra* of pyrimidines and purines. A. E. Pierce (Rockford, Ill.), whose reagents are known to all who practice the technique, describes the *gas-phase analysis* of nucleic acid components as their trimethylsilyl derivatives. Complementary to this article is a chapter by D. C. DeJongh (Detroit, Mich.) that covers the *mass spectrometry* of nucleic acid components. The *optical rotatory dispersion* of nucleosides and nucleotides is treated by T. L. V. Ulbricht (London), and the *infrared spectroscopy* of nucleic acid components, by M. Tsuboi and Y. Kyogoku (Tokyo). Next discussed, by L. B. Townsend (Salt Lake City, Utah), is the use of *nuclear magnetic resonance spectroscopy* in the study of nucleic acid components and certain related derivatives. Professor Zorbach had realized that automation, coupled with the use of computers, has made *X-ray diffraction for determination of crystal structure* a procedure that might well become a routine laboratory tool in the near future. Hence a chapter, by S. T. Rao and M. Sundaralingam (Madison, Wis.), is devoted to this subject, which, only a few years ago, would not have received consideration as a method potentially having general utility. Finally, the exceedingly valuable technique of *chromatography* of nucleic acid components is discussed by S. Zadražil (Prague).

As with Volume 1, the present volume is intended primarily for the organic chemist, but it should also prove valuable to biochemists and medicinal chemists. Although the various techniques described are oriented toward application to compounds derived from nucleic acids, the general principles involved are equally applicable to many other groups of organic compounds. Consequently, the book should have wide usefulness both as a guide for laboratory work and as an advanced textbook that provides important insights into the use of techniques of wide applicability in biomedical, biochemical, and organic chemical research. The reception accorded Volumes 1 and 2 as companion volumes will decide whether this series will be continued.

R. Stuart Tipson

Kensington, Maryland
December 1970

Contents

Contents

Synthetic Procedures in Nucleic Acid Chemistry

VOLUME 2

CHAPTER 1

Ionization Constants of Pyrimidines and Purines

ADRIEN ALBERT

JOHN CURTIN SCHOOL OF MEDICAL RESEARCH,
AUSTRALIAN NATIONAL UNIVERSITY, CANBERRA, AUSTRALIA

I. INTRODUCTION

Ionization constants are those values that are used as a measure of strength of acids and bases. The term dissociation constant is a more general one as it includes other types of equilibria.

Ionization constants are determined because they reveal the proportions of the different ionic species into which a compound is divided at any chosen pH. Thus, the ionization constant of benzoic acid is 76 μM or 76 micromoles per liter (*i.e.*, 7.6×10^{-5} M in the style hitherto customary). From this figure, it can quickly be calculated from equation *1* that, in aqueous solution at pH 5,

$$\% \text{ Ionized (as acid)} = 100/[1 + \text{antilog } (\text{p}K_a\text{-pH})] \quad \ldots (1)$$

benzoic acid is 89% in the form of the anion (*1*) and 11% in the form of the neutral species (*2*); and also that, at

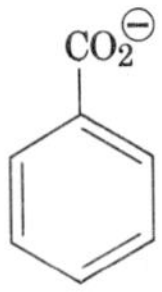

$CO_2^{\ominus}$ CO_2H

1 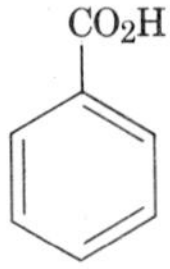 **2**

pH 4.12, the ratio of anion to neutral species is unity. For a base, equation *2* is used in the same way. To save

$$\% \text{ Ionized (as base)} = 100/[1 + \text{antilog } (\text{pH-p}K_a 0] \quad \ldots (2)$$

time, tables[1] may be used instead of these equations. In these equations, pK_a signifies the negative logarithm of the ionization constant, and pH has its usual meaning as a measure of hydrogen-ion activity. The use of pK_a values instead of ionization constants provides the convenience of small whole numbers.

Knowledge of the partition of a compound among its species is useful in many ways. For example, different ionic species have different ultraviolet spectra; hence, a spectrum has no meaning unless it is measured at a pH such that only one ionic species is present. Ionization constants, by defining the pH range in which a compound is least ionized, indicate the conditions under which it is least soluble in water, and most soluble in immiscible solvents. This knowledge has great value in preparative chemistry, and so has knowledge of the pK_a values of likely impurities. The adsorbability of a compound, and its ability to penetrate biological membranes, are both strongly influenced by its ionization constant.[2]

Equation *3* indicates the type of equilibrium to which ionization constants refer. It signifies that the dis-

$$CH_3\text{-}CO_2H \rightleftharpoons H^{\oplus} + CH_3\text{-}CO_2^{\ominus} \quad \ldots (3)$$

sociation of acetic acid in aqueous solution gives hydrogen cations and acetate anions, and that these form an equilibrium mixture. Thus, the product of the concentration of the ions formed by the ionization of acetic acid must bear a fixed ratio to the concentration of the non-ionized species (*i.e.*, the molecules). In other words, the product $[H^{\oplus}]\ [CH_3CO_2^{\ominus}]$ bears a fixed ratio to $[CH_3CO_2H]$, and this ratio is the acidic ionization constant $[K_a]$, shown in equation *4*. To define the ioniza-

$$K_a = \frac{[H^{\oplus}]\ [CH_3CO_2^{\ominus}]}{[CH_3CO_2H]} \qquad \ldots(4)$$

tion of any acid, equation *4* can be put in the general form shown in equation *5*, where $A^{\ominus}$ is any anion.

$$K_a = \frac{[H^{\oplus}]\ [A^{\ominus}]}{[HA]} \qquad \ldots(5)$$

Hence,

$$pK_a = pH + \log [HA] - \log [A^{\ominus}] \qquad \ldots(6)$$

Thus, the essential part of any description of ionization is the application of the law of mass action to describe the state of ionic equilibria.

The ionization of bases is most conveniently expressed on the same scale as the acidic ionization constant (K_a), and the use of a special scale based on K_b has gone out of use. The equilibrium described by K_a can conveniently be illustrated for ammonia, given in equation 7. This has been found by experiment to be 0.44 n$\underline{M}$ (0.44 nanomole

$$K_a = \frac{[H^{\oplus}]\ [NH_3]}{[NH_4^{\oplus}]} \qquad \ldots(7)$$

per liter) at 20°; whence, pK_a is 9.36.

The higher the pK_a of a base, the stronger the base, whereas the strongest acids are those having the lowest pK_a values.

Ionization constants can be determined in many ways.[1] However, they are most conveniently measured by one of two methods: (*a*) potentiometric titration, or (*b*) the measurement of ultraviolet spectra in a series of buffers. The first method is by far the more rapid and has the virtue of revealing impurities (even water) in the specimen, but it cannot be used for sparingly soluble substances. The second method suffers less from limitations of range and solubility, but it is useless if members of a pair of ionic species have the same spectrum, or no spectrum at all. Both methods require a small correction in order to convert the results into thermodynamic (*i.e.*, concentration-independent) values of pK_a. Thus, the pK_a of 2-naphthol was found, by potentiometric titration in 10m$\underline{M}$ solution in water, to be 9.58, and appropriate corrections[1] converted this into the thermodynamic pK_a of 9.63; a spectrometric determination gave a similar value (9.59). Most of the values given in the Tables in this Chapter have not been subjected to thermodynamic correction; nor have they been corrected for temperature,[3] apart from being chosen within a narrow range of temperature (20-25°). Nevertheless, they represent close approximations, and are valid data on which to base the discussion. A worked example of determination of an ionization constant is given in Section IV (see p. 37).

II. THE BASIC AND ACIDIC STRENGTHS OF PYRIMIDINES

1. General

The strength of pyrimidine as a base is about one ten-thousandth of that of pyridine. Whereas pyridine has a pK_a of 5.2 (and, hence, has about the same basic strength as aniline), pyrimidine has pK_a 1.31 for the acceptance of the first proton (to give the monocation) and pK_a -6.3 for the second proton. The low basic strength of pyrimidine arises from the inductive (-$\underline{I}$) and mesomeric (-$\underline{M}$) effects of the doubly bound nitrogen atom; these quantitatively resemble those of a nitro group. Hence, it is not surprising that the first pK_a of pyrimidine (1.31) is in the region of that of 3-nitropyridine (0.8). The exceedingly low pK_a for the addition of a second proton to pyrimidine is due to the operation of Coulomb's law, whereby one positive charge repels another; this means that the pyrimidine monocation can effectively repel hydrogen ions in its environment until an exceedingly high concentration of them is provided.

Apart from amino, hydroxyl, and thiol groups, most substituents exert on the basic strength of pyrimidine an effect similar to that exerted on the basic strength of aniline (see Table 8.11 in Ref. 1). Table I lists the pK_a values of many uncomplicated pyrimidines of this kind. Although many compounds suitable for inclusion in Table I have not yet been examined, the following situation emerges clearly. A methyl group is electron-releasing and raises the pK_a. A methoxyl and a methylthio group

TABLE I

The Ionization of Pyrimidines Bearing Uncomplicated Substituents[a]

Derivative of pyrimidine	pK_a	δ[b]	References
Unsubstituted	1.31[c]	0	4
5-Bromo-2-methoxy-	-0.77	-2.08	5
5-Bromo-4-methoxy-	1.35	0.04	5
2-Chloro-	<1	>-0.3	6
4,6-Dimethoxy-	1.49	0.18	7
4,6-Dimethyl-	2.7	1.4	6
2-Ethoxy-	1.27	-0.04	5
2-Methoxy-	1.05	-0.26	5
4-Methoxy-	2.5	1.2	8
2-Methoxy-4-methyl-	2.1	0.8	9
4-Methoxy-6-methyl-	3.65	2.34	9
4-Methoxy-2,6-dimethyl-	4.76	3.45	5
2-(Methoxycarbonyl)-	-0.68	-1.99	10
4-Methyl-	1.98	0.67	4
4-Methyl-2-(methylthio)-	1.95	0.64	9
4-Methyl-6-(methylthio)-	3.25	1.94	9
2-(Methylthio)-	0.59	-0.72	11
4-(Methylthio)-[d]	2.48	1.17	11

[a] (In water at 20-25°.) The decrease in basic strength for each degree rise in temperature is about 0.011 unit of pK_a for a base of pK_a 3.3, 0.017 unit for a base of pK_a 6.6, and 0.021 unit for a base of pK_a 10.0 (see Table 1.3 of Ref. 1). [b] This column records the difference in pK caused by the substituent(s). [c] Also, pK_a -6.3 for *di*-cation. [d] See Table I in Chapter 2 for pK_a values of several (methylsulfinyl)- and (methylsulfonyl)-pyrimidines.

are base-weakening in the 2-position (where their -I effect exceeds their +*M* effect), and base-strengthening in the 4-position (where +*M* exceeds -I). For the benzene series, a methylthio group does not show a +*M* effect so large as that often found when it occupies a γ-position on a heteroaromatic, nitrogen-containing ring. The relative closeness of the dipolar charges (compared to those of aniline) evoke an otherwise reluctant orbital change, to give tercovalent sulfur. The halogen and ester groups are base-weakening in the 2-position, and this effect is expected of them in other positions.

Many of the effects of these substituents are (roughly) additive. Thus, one 4-methyl group increases the basic strength by 0.67 unit of p*K*, and two of them, by 1.4. Again, a 4-methyl plus a 4-methylthio group increases the basic strength by 1.94 units, which is close to 1.84, the the sum of their individual effects. It should, however, be noted that 4,6-dimethoxypyrimidine is a weaker base than 4-methoxypyrimidine, because the doubly bonded oxygen atom (which the +*M* contribution permits any one methoxyl group to have at a given time, as in 3) leads to

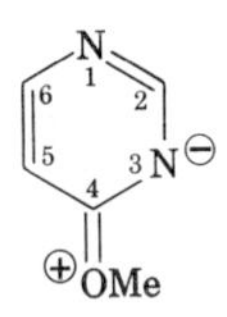

3

resonance in the neutral species (symmetrical), but not in the cation (nonsymmetrical).

2. Amino Substituents

The insertion of an amino group into a nitrogen heterocycle usually increases the basic strength by about 1 unit of p*K*, and the ring-nitrogen atom remains the basic center. The 3-, 5-, and 6-positions in quinoline provide examples of this normal effect, and the behavior of 5-aminopyrimidine seems to be similar (see Table II, A; p.9). However, when the insertion of an amino group permits more resonance in the cation than in the neutral species, a remarkable increase in the basic strength results; this was first demonstrated for the 2- and 4-positions in pyridine and quinoline,[4] and is often referred to as "4-aminopyridinium-type base-strengthening". For example, 4-aminopyridine is 4 units of p*K* stronger than pyridine (*i.e.*, 10,000 times as strong). The resonance situation in the aminopyrimidines will become clearer if that occurring with 4-aminopyridine (6) is first described.

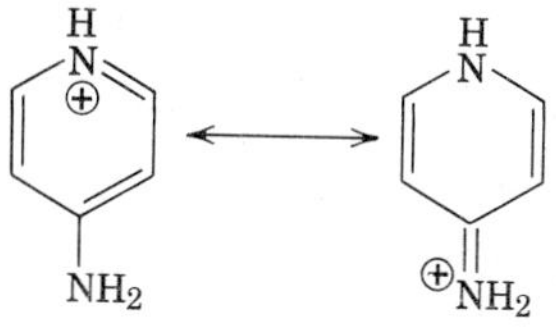

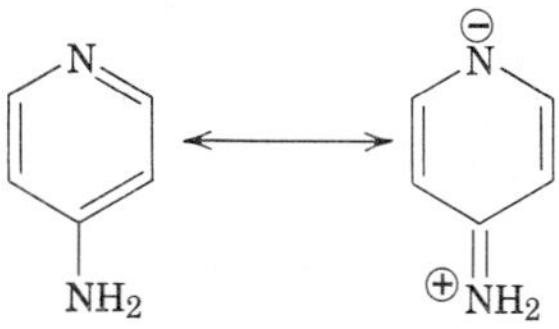

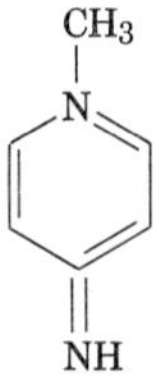

The main, canonical forms of the resonance hybrid of the 4-aminopyridine cation are 4 and 5. A resonance also exists in the neutral species, namely, between 6 and 7, but this is much smaller because of the large separation of charges in 7, an energy-demanding structure that is not present in the ion. The reality of this resonance in the neutral species is demonstrated by the high basic strength of the nuclear *N*-methyl derivative 8, which has $\mathrm{p}K_a$ <12.5 (*cf.*, 9.2 for 4-aminopyridine and 5.2 for pyridine) a difference which is interpreted as follows. The cation of the *N*-methyl derivative 8 is allowed complete resonance of the type 4 ⟷ 5, whereas no resonance of the type 6 ⟷ 7 is possible for the neutral species. In sum, 4-aminopyridine 6 is a surprisingly strong base, because more resonance is available in the cation than in the neutral species; but 1,4-dihydro-4-imino-1-methylpyridine (8) is an exceedingly strong base, because the resonance is possible only in the cation.

This base-strengthening effect, described for 4-aminopyridine, occurs also for 2-aminopyridine, but the enhancement is less (largely because the separation of charges in the neutral species is less, so that the resonance of this species is not so much lower than that of its cation). That this explanation is correct may be seen by comparing the pk_a of 2-aminopyridine (6.9) with that of its nuclear *N*-methyl derivative, namely, 12.2.

Turning now to the aminopyrimidines (Table II, A), it may be seen that the 2- and 4-amino derivatives show at least as much increase of basic strength as with the corresponding aminopyridines, and these increases may safely be attributed to similar resonance-effects. At first, it may seem surprising that 4-aminopyridine does not follow 2-aminopyridine (to which it has some structural similarity), but it is known that cations favor *para*- to *ortho*-quinonoidal forms where a choice exists, because the former (for example, 5) are energetically favored. That the primary aminopyrimidines do not exist to any appreciable extent in an imino form has been shown by comparison of the ionization constants and ultraviolet spectra with those of the various *N*-methylated derivatives.[8,33]

The introduction of a second, and a third, amino group into an aminopyrimidine always raises the basic strength, but never so much as to equal the sum of the separate increments. This situation is largely explained by the -I effect exerted on a neighboring ring-nitrogen atom by any amino group which is not (in a given canonical structure) involved in the *para*-quinonoidal resonance of a cation, as in 5.

Part B of Table II lists the basic strengths of aminopyrimidines that contain a further substituent having a nonionizing character. In general, these substituents have very much the same base-strengthening or base-weakening effects shown in Table I. The influence of many kinds of substituents (for example, $-NO_2$, or $-CN$) that were not available for inclusion in Table I can be traced here. Attention is also drawn to 4-amino-6-methoxypyrimidine and 4-amino-6-(methylthio)pyrimidine, each of which has two base-strengthening resonances that are structurally conflicting: as a result, these compounds achieve a basic strength that falls far short of the sum of the individual increments.

Many examples in Part B of Table II show that a methyl group attached to an exocyclic nitrogen atom, as in 2-(methylamino)pyrimidine, is slightly base-strengthening.

TABLE II

The Ionization of Aminopyrimidines[a]

Derivative of pyrimidine		pk_a	δ [b]	References
A.	Unsubstituted	1.31	0	4
	2-Amino-	3.54	2.23	4
	4-Amino-	5.71	4.40	4
	N-acetyl-	2.76	1.45	7
	5-Amino-	2.60	1.29	12
	2,4-Diamino-	7.26	5.95	4
	4,5-Diamino	6.03; <0	4.72	13
	N-5-formyl-	4.45	3.14	7
	4,6-Diamino-	6.01	4.70	14
	2,4,6-Triamino	7.63; 2.56	6.32	15
	2,4,6-Triamino-	6.84	5.53	4
	4,5,6-Triamino-	5.78; 1.47	4.47	15
B.	2-Amino-5-bromo-	1.95		16
	4-Amino-6-chloro-	2.10		14
	2-Amino-4,6-dimethyl-	4.99		17
	4-Amino-2,6-dimethyl-	6.98		17
	2-Amino-4-(dimethylamino)-	7.96		16
	4-Amino-2-(dimethylamino)-	7.64		18
	2-Amino-4-methoxy-	5.53		19
	4-Amino-2-methoxy-	5.3		20
	4-Amino-6-methoxy-	4.02		14
	2-Amino-4-methyl-	4.15		4

TABLE II (continued)

Derivative of pyrimidine	pK_a	δ^b	References
4-Amino-2-methyl-	6.53		21
4-Amino-6-methyl-	6.25		9
5-Amino-4-methyl-	3.15		9
4-Amino-2-(methylamino)-	7.55		22
4-Amino-6-(methylamino)-	6.32		23
4-Amino-6-(methylamino)-5-nitro-	2.75		23
4-Amino-5-(methylnitrosamino)-	3.69		24
2-Amino-4-(methylthio)-	4.75		19
4-Amino-2-(methylthio)-	4.91		19
4-Amino-6-(methylthio)-	3.94		25
2-Amino-5-nitro-	0.35		26
4-Amino-5-nitro-	1.98		27
2-(Benzylamino)-	3.56		28
4,6-Bis(dimethylamino)-	6.36		14
4,5-Bis(methylamino)-	6.03		29
4,6-Bis(methylamino)-	6.39		14
5-Bromo-2-(butylamino)-	2.21		5
5-Bromo-4-(butylamino)-	4.49		5
5-Carbamoyl-2-(methylamino)-	2.05		30
4-Chloro-6-(dimethylamino)-	2.42		14
2-Chloro-4-(methylamino)-	2.83		22
4-Chloro-2-(methylamino)-	2.63		22
4-Chloro-6-(methylamino)-	2.24		14
5-Chloro-2-(methylamino)-	2.04		31
5-Cyano-2-(methylamino)-	0.76		30
4,6-Diamino-5-bromo-	4.22		14
2,4-Diamino-6-chloro-[c]	3.57		22

	2,4-Diamino-6-methyl-	7.7	32
	2,4-Diamino-6-(methylthio)-	5.46	22
	4,5-Diamino-2-(methylthio)-	5.05	22
	2-(Dimethylamino)-	3.96	33
	4-(Dimethylamino)-	6.35	33
	2-(Dimethylamino)-4-methoxy-	5.87	19
	4-(Dimethylamino)-2-methoxy-	6.17	34
	4-(Dimethylamino)-6-methoxy-	4.29	14
	2-(Dimethylamino)-4-(methylthio)-	5.02	19
	4-(Dimethylamino)-2-(methylthio)-	5.73	19
	4-(Dimethylamino)-6-(methylthio)-	4.57	19
	2-(Ethylamino)-	4.03	16
	5-Formamido-4,6-bis(methylamino)-	5.00	35
	2-(Heptylamino)-	4.09	16
	2-Hydrazino-	4.55; -0.46	36
	4-Methoxy-6-(methylamino)-	4.23	14
	2-(Methylamino)-	3.82	33
	4-(Methylamino)-	6.12	33
	2-(Methylnitrosamino)-	<1.00	37
	4-(Methylnitrosamino)-	1.62	37
	4,5,6-Tris(methylamino)-	6.01	35
C.	4-Amino-1,6-dihydro-6-imino-1-methyl-	11.98	23
	1,2-Dihydro-2-imino-1-methyl-	10.75	8
	1,4-Dihydro-4-imino-1-methyl-	12.22	8
	1,2-Dihydro-1-methyl-2-(methylimino)-	11.74	28
	4-(Dimethylamino)-1,2-dihydro-2-imino-1-methyl-	13.68	16

[a]In water at 20-25°. Where two pK_a values are given, the first denotes equilibria for the addition of one proton, the second for two protons, respectively. [b]Increase in base strength caused by the substituent (recorded as the difference in pK_a). [c]Values for 70 examples of 5- and 6-substituted, 2,4-diaminopyrimidines are available (see Ref. 108).

An ethyl group exerts a slightly stronger effect, which remains essentially unaltered on further lengthening of the chain of the alkyl substituent.

Part C of Table II shows the large increments in base-strength caused by alkylating a nuclear nitrogen atom (as already described for the pyridine series). For these compounds, exposure to alkali usually causes the alkyl group to become transferred to the exocyclic nitrogen atom. This transfer, known as the Dimroth Rearrangement, takes place by opening and closing of the ring.[38] The pK_a of the rearranged product is much lower (often by four pK_a units), so that it has frequently been used for detecting an unexpected rearrangement of this kind.

When an *N*-methylated pyrimidine carries both an oxo and an amino substituent, tautomerism favors double-bonding of the former (base-weakening) over that of the latter (base-strengthening). Thus, the pk_a of 1-methylcytosine is only 4.6, whereas that of the 4-imino tautomer should be 10 or higher.

3. Substituents that Create Acidic Properties in Pyrimidines

When inserted into nitrogen heterocycles, hydroxyl groups are very apt to take part in tautomeric changes. These changes have been thoroughly discussed for the hydroxypyridines.[4,39] The following brief discussion of these compounds is intended to pave the way for understanding the ionization of the hydroxypyrimidines.

2-Pyrimidinol, for example, has both basic and acidic properties. The former are feeble (pK_a 0.75; *cf*., 5.2 for pyridine), but protonation has been shown to occur quite normally on the nitrogen atom, and the cation has the normal Kekulé type of structure (9). The acidic properties of 2-pyrimidinol are also feeble, for it has pk_a 11.6; that is, it is somewhat weaker than phenol (10.0), and much weaker than either 3-pyrimidinol (8.7) or a typical carboxylic acid, such as benzoic acid (4.1). The anion has been shown to have the normal Kekulé structure (10). Thus, some unusual structural feature must reside in the neutral species, and, in fact, this has been shown, by several different physical techniques, to have structure 11, which is that of a cyclic amide, stabilized, as all amides are, by some degree of resonance with a zwitterionic form (12). 4-Pyrimidinol has a similar structure, and even 3-pyrimidinol (which cannot, for reasons of valence, assume an amide form) exists in the zwitterionic form (13) to the extent of about 50% in cold water.

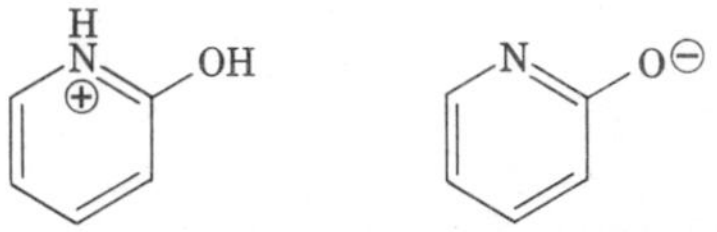

9 10

2- and 4-Pyrimidinol were shown[8,33] (by comparison of pK_a values and ultraviolet spectra with those of the corresponding *O*- and *N*-methylated derivatives, and by infrared spectra) not to exist to any extent in the hydroxyl form, such as 14, but in a stabilized, amide form corresponding to the pair of formulas 11 ⟷ 12.

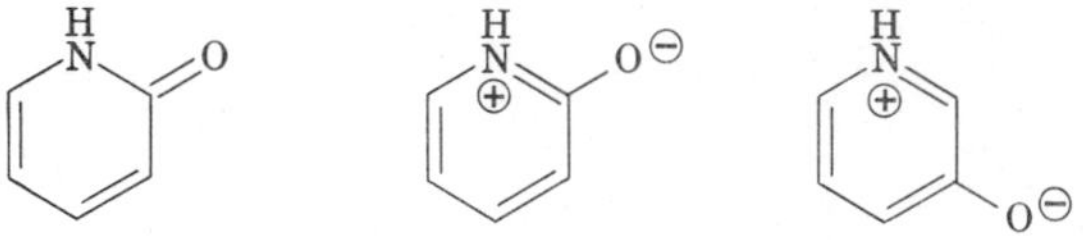

11 12 13

Closer investigation showed that, where both an *ortho*- and a *para*-quinonoid form of the amide could exist, as with 15 and 16, respectively, for 4-pyrimidinol, the former was strongly favored. This situation depends on

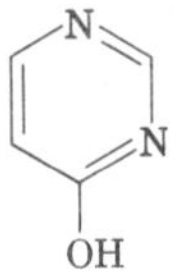

14

the smaller separation of charge (in the zwitterionic form that stabilizes the resonance) for an *ortho*, as compared to a *para*, form. This is exactly the opposite of the situation shown by the cations of amino compounds (see earlier); but the latter are ions, and hence have no charge separation.

A selection of pK_a values of hydroxypyrimidines is given in Table III. It is evident that the insertion of another nitrogen atom into the ring of the various hydroxypyridines (to give the corresponding hydroxypyrimidines) increases the acid strength by about 1.5 pk units. This is due to the $-I$ and $-M$ properties of the doubly bound nitrogen atom already discussed. Uracil (2,4-pyrimidinedione) is as weak an acid as 2-pyrimidinone.[a] However, barbituric acid (2,4,6-pyrimidinetrione) is considerably stronger, due to the highly resonant anion (formed by loss of one proton) whose canonical forms are (17) and similar formulas, having the negative charge on the other two oxygen atoms, in turn. The most favored

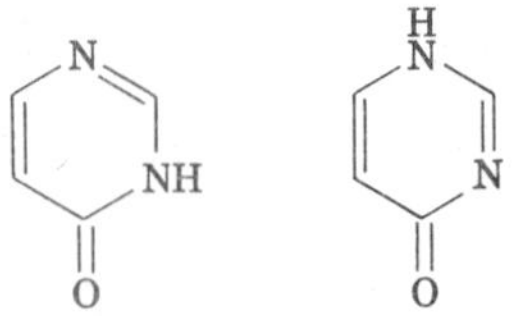

15 16

[a]The first site of ionization of uracil as an acid is not known, and has proved elusive to investigate. The negative charge is usually placed on O-2 in the anion. The pK_a for the second acidic ionization of uracil is 12 or higher, but is not known exactly.

TABLE III

The Ionization of Pyrimidines Bearing an Acidic Group[a]

Derivative of pyrimidine	Basic pK_a	Acidic[b] pK_a	References
A. Hydroxyl group(s) only			
2-Hydroxy-	2.24	9.17	40
4-Hydroxy-	1.85	8.59	8
5-Hydroxy-	1.87	6.78	41
2,4-Dihydroxy- (Uracil)	-3.4	9.38	42,43
4,5-Dihydroxy-	1.99	7.48, 11.61	43
4,6-Dihydroxy-	0.26	5.4	7.43
2,4,5-Trihydroxy-	—	8.11, 11.48	43
2,4,6-Trihydroxy- (Barbituric acid)	—	3.9, 12.5	43,44(a)
2,4,5,6-Tetrahydroxy-	—	2.83, 11	43
B. Hydroxyl group(s) plus nonionizing group(s)			
2-Benzyl-4,6-dihydroxy-	—	5.78	27
5-Bromo-2,4-dihydroxy-	-7.25	7.83	42
5-Bromo-4-hydroxy-	0.43	7.15	45
4-Chloro-2,6-dihydroxy-	—	5.67	46
5-Chloro-2,4-dihydroxy-	—	7.95	46
4-Chloro-6-hydroxy-	—	7.43	14

TABLE III (continued)

Derivative of pyrimidine	Basic pK_a	Acidic pK_a	References
5,5-Diethyl-2,4,6-trihydroxy- (Barbitone)[c]	—	7.89, 12.7	44
N^1-methyl-	—	8.45	44(a)
1,2-Dihydro-4-hydroxy-1-methyl-2-oxo- (1-Methyluracil)	-3.4	9.75	20
3,4-Dihydro-2-hydroxy-3-methyl-4-oxo- (3-Methyluracil)	—	9.95	20
1,2-Dihydro-4,6-dihydroxy-1-methyl-2-oxo- (N-Methylbarbituric acid)	—	4.2, 12.8	44
2,4-Dihydroxy-5-iodo-	—	8.25	46
2,4-Dihydroxy-5-methyl-(Thymine)	—	9.94	20
2,4-Dihydroxy-6-methyl-	—	9.64	20
4,6-Dihydroxy-2-methyl-	0.21	6.35	7
4,6-Dihydroxy-5-methyl-	-0.51	6.01	47
2,4-Dihydroxy-6-(methylsulfony)-	—	4.68	48
4,6-Dihydroxy-2-(methylthio)-	—	5.09	35
2,4-Dihydroxy-5-nitro-	—	5.56, 11.3	48
2,4-Dihydroxy-5-(trifluoromethyl)-	—	7.35	49
2-Ethoxy-4-hydroxy-	—	8.2	20
4-Ethoxy-2-hydroxy-	1.0	10.7	20
4-Fluoro-2,6-dihydroxy-	—	4.03	46
5-Fluoro-2,4-dihydroxy-	—	8.04	48,49
4-Hydroxy-5-methoxy-	1.75	8.60	43
4-Hydroxy-6-methoxy-	-0.22	8.47	7

	2-Hydroxy-4-methyl-	3.15	9.8	9
	4-Hydroxy-6-methyl-	2.15	9.0	9
	4-Hydroxy-6-(methylthio)-	-0.11	8.52	25
C.	Mercapto (thiol) group(s), alone or further substituted			
	2,4-Dimercapto-	—	6.46, 11.19	52
	4,6-Dimercapto-	-2.3	3.60, 9.70	53
	5-Bromo-2-mercapto-	-0.43	5.47	45
	5-Bromo-4-mercapto-	-0.46	5.60	45
	4,6-Dihydroxy-2-mercapto- ("Thiobarbituric acid")	-4.4	3.7, 7.89	53
	4-Hydroxy-2-mercapto- (2-Thiouracil)	—	7.74, 12.7	54
	1-Ethyl-	—	8.7	54
	3-Ethyl-	—	8.65	54
	4-Hydroxy-6-mercapto-	-1.7	4.33, 10.52	25
	4-Hydroxy-2-mercapto-6-methyl-	—	8.1	54
	4-Hydroxy-2-seleno	—	7.18	52
	2-Mercapto-	1.35	7.14	51
	4-Mercapto-	0.68	6.90	51
	4-Mercapto-2,6-dimethyl-	1.80	8.13	45
	4-Mercapto-6-methoxy-	-1.98	7.51	25
D.	Carboxylic acid group			
	5-Bromo-6-carboxy-2,4-dihydroxy-	—	2.38, 7.33	56
	2-Carboxy-	-1.13	2.85	10
	5-Carboxy-2,4-dihydroxy-	—	4.16, 8.89	56
	6-Carboxy-2,4-dihydroxy- (Orotic acid)	—	2.07, 9.45	56
	6-Carboxy-2,4-dihydroxy-5-nitro-	—	<1.5, 4.94	56

TABLE III (continued)

Derivative of pyrimidine	Basic pK_a	Acidic pK_a	References
5-Carboxy-2-(ethylthio)-4-hydroxy-	—	6.01, 10.52	56
E. Amino group(s) also present			
4-Amino-5-carboxy-2-methyl-	2.14	6.28	21,57
2-Amino-4,6-dihydroxy-	1.27	7.00	58
4-Amino-2,6-dihydroxy-	0.80	-	58
4-Amino-5-formamido-2-hydroxy-	3.55	10.87	59
2-Amino-4-hydroxy-	4.00	9.59	22
4-Amino-2-hydroxy- (Cytosine)	4.58	12.15	20,60
4-Amino-6-hydroxy-	1.36	10.05	19
2-Amino-4-mercapto-	2.86	8.03	19
4-Amino-2-mercapto-	3.33	10.58	19
4-Amino-6-mercapto-	-0.24	9.25	19
2,4-Diamino-6-hydroxy-	3.27	10.83	58,22
4,5-Diamino-2-hydroxy-	4.37	11.45	15
4,5-Diamino-6-hydroxy-	3.57	9.86	15
4,6-Diamino-2-hydroxy-	6.56	11.98	58
4,5-Diamino-2-mercapto-	2.96	10.39	15
4,6-Diamino-2-mercapto-	5.02	10.53	58
2-Hydroxy-4-(methylamino)-	4.55	>13	61
4-Hydroxy-2-(methylamino)-	3.93	9.82	22
4-Hydroxy-6-(methylamino)-	<1.7	10.47	14
2-Mercapto-4-(methylamino)-	3.09	11.10	61
4-Mercapto-6-(methylamino)-	-0.27	9.64	25

[a] In water at 20–25°. [b] The first of two values refers to the loss of the first proton (stronger acid), and the second to the loss of the second proton. [c] Values for other medicinal barbiturates are given in Ref. 44(b).

tautomer present in the neutral species of barbituric acid (in which this resonance is impossible) has been[44] established as 18. Because of the extra resonance

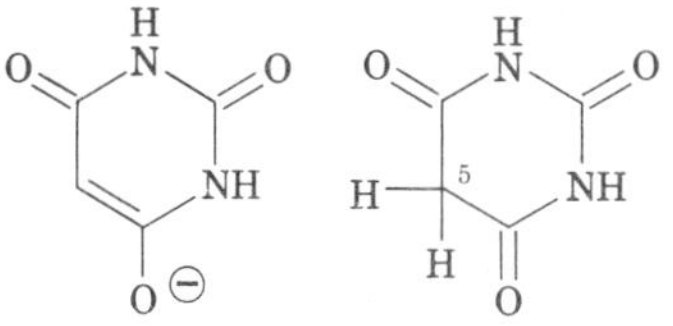

17

18
Barbituric acid
(preponderant tautomer)

possible in the mono-anion, barbituric acid (pK_a 3.9) is as strong as a carboxylic acid. On the other hand, the 5,5-dialkylbarbituric acids, used in medicine as depressants of the central nervous system, have to form the monoanion by loss of a proton from a nitrogen atom. Accordingly, they cannot form a highly resonant anion similar to 17 and are much weaker acids, with pK_a values between 7.5 and 8.0; however, this pK range is essential for their penetration, as neutral species, of the lipoidal membranes that surround parts of the nervous system. The tendency shown by barbituric acid to favor the tautomer having a methylene group is demonstrable also in 4,6-pyrimidinedione,[7] although this prototropy does not take place to the same extent (the exact proportions present at equilibrium are still controversial). No other hydroxypyrimidine appears to have this keto-methylene structure, although it is almost universal for the monhydroxy derivatives of 5-membered rings.[39]

5-Pyrimidinol has been shown, mainly by ultraviolet spectroscopy, to be almost free of a zwitterionic tautomer, similar to 13, which plays such an important part in the equilibria of 3-pyridinol.[50] The reason for this difference is that a second nuclear nitrogen atom decreases the proton-accepting property of the first nitrogen atom more than it enhances the proton-releasing property of the oxygen atom.

The basic strengths of 2- and 4-pyrimidinone are slightly higher than that of pyrimidine: the basic center is the nitrogen atom that is not involved in the amide tautomerism. 2-Pyrimidinone owes its slightly higher basic strength to stabilization of the symmetrical cation through the resonance 19 ⟷ 20. In uracil and thymine, in which both of the nitrogen atoms are thus in-

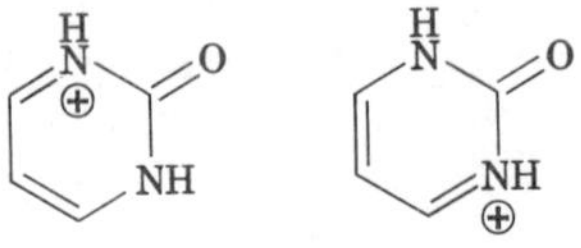

19 20

volved, very little basic strength remains.

When the acidic properties of a hydroxypyrimidine are eliminated by *N*-alkylation, the basic strength increases (but only slightly) because of electron release by the entering alkyl group. Thus, methylation of 2-pyrimidinone gives 1-methyl-2(1*H*)-pyrimidinone, whose basic pK_a value is 2.50; again, the two isomers produced by *N*-methylation of 4-pyrimidinone have[8,41] pK_a 2.02 and 1.84. Similarly, uracil and 1-methyluracil have the same (feebly basic) pK_a, namely, -3.4.

Table III,B lists a representative selection of hydroxypyrimidines bearing further substituents that cannot ionize. It may be seen that these substituents have an effect on the acidic properties qualitatively similar to that exerted on phenol (Ref. 1, Table 8.4) or benzoic acid (Ref. 1, Table 8.2). Quantitatively, the effect is somewhat greater. Alkoxyl and methylthio groups do not create disturbances like those already described for the aminopyrimidines. The 5-halogeno-uracils have a higher proportion (than in uracil) of a hydroxyl group, as in <u>14</u>, in equilibrium with the normal amide form: it is still a very small proportion, but it is of interest as a possible cause of mutations.[42]

Table III,C shows several mercaptopyrimidines that are considerably stronger than their oxygen analogs; thus 2,4-dithiouracil (pK_a 6.46) is almost three units stronger than uracil, whereas the value for 2-thiouracil lies between these values. *N*-Alkylation abolishes acidic properties, but affects the basic strength very little.[51]

The substitution of one hydrogen atom of the amino group of aminopyrimidines by the *p*-aminophenylsulfonyl group has been much used for preparing useful members of the sulfonamide series of antibacterial drugs. Such compounds have moderately strong acidic properties, because the propinquity of the $-SO_2-$ group causes the -NH- group to ionize. Thus, sulfadiazine (2-sulfanil-amidopyrimidine, <u>21</u>) has a pK_a of 6.5; sulfamethazine

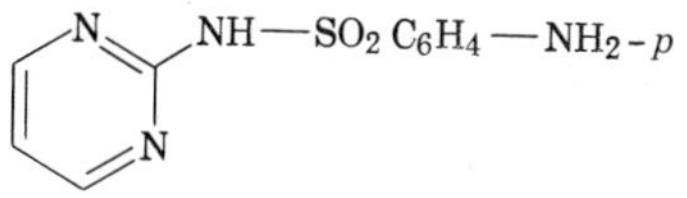

$$\text{2-pyrimidinyl}-NH-SO_2C_6H_4-NH_2\text{-}p$$

21
Sulfadiazine

(sulfadimidine), its 4,6-dimethyl derivative, is weaker, as would be expected, with[55] a pK_a of 7.4.

2-Pyrimidinecarboxylic acid is the only carboxypyrimidine for which the pK_a is known; this is 2.85, a value indicative of the acid-strengthening $-I$ effect of the neighboring, doubly bound, nitrogen atom. This value, and those for several pyrimidinecarboxylic acids bearing further substituents, constitute section D of Table III. It may be seen that the carboxyl group preserves a high acid strength, except where it is placed in the 5-position, where the $-I$ effect is lessened.

Section E of Table III lists examples that carry an amino group, as well as an acidic group. It may be seen that, in most cases, these groups do not interfere with one another; thus, 2-amino-4-pyrimidinone (isocytosine) has basic and acidic pK_a values of 4.00 and 9.59, respectively, which are similar to those of 2-aminopyrimidine (3.54) and 4-pyrimidinone (8.59). Cytosine (4-amino-2-pyrimidinone) is exceptional in having a surprisingly weal acidic pK, namely 12.15, whereas that of 2-pyrimidinone is 9.17 (*i.e.*, a thousand times stronger as an acid).

In 1961, the zwitterion structure <u>22</u> was proposed for cytosine,[62] but this appeared to be an improbable formulation, because the pK_a values of the component groups (*i.e.*, 5.71 for 4-aminopyrimidine and 9.17 for 2-pyrimidinone) lie too far apart to neutralize one another (such a tendency is usually just detectable if there are only 3 units, but is strong if there are less than 2 units, of separation). That the formula of cytosine is actually <u>23</u> was soon demonstrated by comparison of the ultraviolet spectra with those of all the methylated derivatives possible,[34] and by nuclear magnetic resonance studies.[63] Similarly, it was shown that protonation, to form the cation, takes place on N-3 of <u>23</u>. The most likely explanation of the weakness of cytosine as an acid is that the necessary removal of the proton from N-1 is hindered by a partial negative charge placed

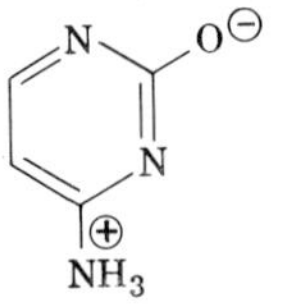

22

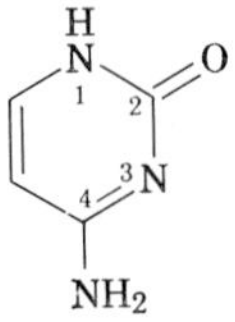

23
Cytosine

on that atom by the amino group, as already discussed for formula 7. In putting forward this explanation, it is only fair to point out that no such acid weakening exists in two analogously constituted compounds, namely, 4-amino-2-pteridinone[13] and 6-aminopurin-2-one (isoguanine).[64] Another example from the pyrimidine series, slightly more remarkable than cytosine, is 4-amino-2-pyrimidinethiol, the acidic pK_a (10.58) of which is much weaker than that of 2-pyrimidinethiol (7.14).

4. The Pyrimidine Nucleosides and Nucleotides

From Table IV, the pK_a values of pyrimidine nucleosides and nucleotides may be seen to correspond closely to those of the parent pyrimidines. D-Ribose has an acidic pK_a of 12.22, and 2-deoxy-D-*erythro*-pentose, of 12.67; this is the source of an extra ionization (about pK_a 12) in nucleosides. That it arises from ionization of a hydroxyl group in the sugar is shown by the entropy change, which is of a size that corresponds to the first ionization of a neutral molecule; this result indicates a large distance between the two ionizing groups.[60] Hydrogen bonding between oxygen atoms in the 2- and 2'-positions causes an ultraviolet shift when the 2'-hydroxyl group (in the D-ribosyl group) ionizes.[65] The sugar moieties are all inserted at the position shown for cytidine (24).

The transformation of nucleosides into nucleotides affects the pK_a values of the pyrimidines very little, but the phosphoric acid moiety ionizes in the ranges of 0.8 to 1.6 and 6.0 to 6.6. Consequently, the cytidylic acids are present entirely as zwitterions.

The titration curves obtained when an aqueous solution of 2'-deoxy-D-ribonucleic acid (DNA) (as its sodium salt at pH 6-7) is titrated to either 2.5 or 12 are not

TABLE IV

Values of pK_a for the Ionization of Pyrimidine Nucleosides and Nucleotides[a]

Compound	Pyrimidine ring	Sugar hydroxyl group	Phosphate group	References
Cytidine (1-β-D-Ribofuranosyl-cytosine) (24)	4.08	12.24	—	60
Cytosine, 1-(2-deoxy-β-D-ribo-furanosyl)-	4.25	—	—	65
Cytidine 2'-phosphate (2'-Cytidylic acid)[b]	4.30	—	0.8 6.19	65
Cytidine 3'-phosphate (3'-Cytidylic acid)[c]	4.16	—	0.8, 6.04	65
Cytidine 5'-phosphate (CMP)	4.5	—	6.3	66
Cytidine 5'-pyrophosphate (CDP)	4.6	—	6.4	66
Cytidine 5'-triphosphate (CTP)	4.8	—	6.6	66
Cytidine 2'-deoxy-, 5'-phosphate	4.44	—	—	65
Thymidine [1-(2-Deoxy-β-D-ribo-furanosyl)thymine]	9.79	12.85	—	60
Thymidine 5'-phosphate (5'-Thymidylic acid)	10.0	—	1.6,6.5	67
Uridine (1-β-D-Ribofuranosyl-uracil)[b]	9.30	12.59	—	60
Uracil, 1-(2-deoxy-β-D-ribofuran-osyl)-	9.3			68

TABLE IV (continued)

Values of pK_a for the Ionization of Pyrimidine Nucleosides and Nucleotides[a]

Compound	Pyrimidine ring	Sugar hydroxyl group	Phosphate group	References
Uracil, 5-β-D-ribofuranosyl- (Pseudouridine)	8.97[c]	—	—	69
Uridine 5'-phosphate (UMP)	9.5	—	6.4	66
Uridine 5'-pyrophosphate (UDP)	9.4	—	6.5	66
Uridine 5'-triphosphate (UTP)	9.5	—	6.6	66

[a]In water at 20–25°. [b]For the ionization constants of 5-bromo-, 5-chloro-, 5-fluoro-, and 5-iodo-uridine, see Ref. 70. [c]This value replaces the value of 9.6 found in the earlier literature.

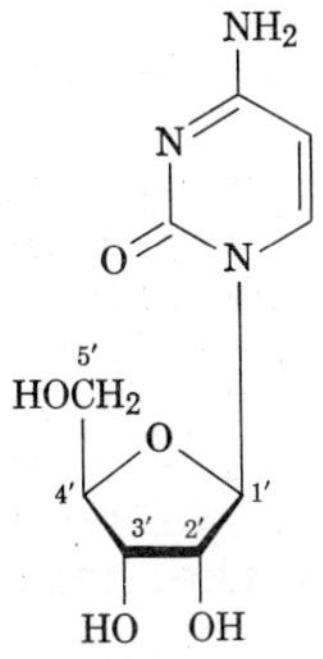

24
Cytidine

reproduced on back-titration.[71] This hysteresis effect is caused by the breaking of hydrogen bonds in the Watson-Crick spiral of the molecule. Thus, the basic group of the cytosine moiety, which has pK_a 4.40 in the forward titration, exhibits a pK_a of 4.85 in the back titration.

5. Conclusion

The preceding account of the ionization of pyrimidines has necessarily been brief, but it covers the common kinds of substituents and the major varieties of electronic effects on ionization. Many further values may be found in D. J. Brown's lists.[72] Attention given to pK_a values is constantly revealing new effects. One of the most recent of these is the demonstration that the cation of 5-nitropyrimidine is covalently hydrated, as in 25.

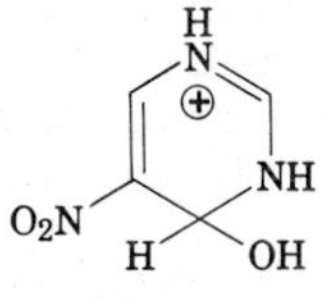

25

This situation was first suspected when the pK_a was found to be 0.7, whereas the value calculated was -2.0.

III. THE BASIC AND ACIDIC STRENGTHS OF PURINES

1. General

Purine (26) differs from pyrimidine in several ways. Firstly, it has acidic properties, which pyrimidine lacks. Loss of the proton from the 7-position[a] produces a resonant anion in which the negative charge is shared between N-7 and N-9. This ionization has a pK_a of 8.93, so that purine is a somewhat stronger acid than phenol. The corresponding value for benzimidazole (27) is 12.3, reflecting the neutral character of the

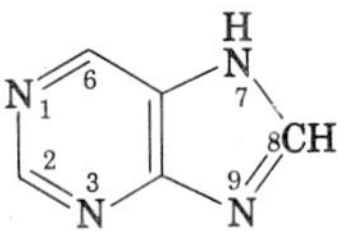

26
Purine

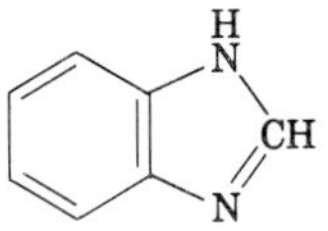

27
Benzimidazole

annelated benzene ring in 27, as compared with the strongly electron-attracting pyrimidine ring in purine (26).

Ascertaining the location of the basic center in purine (pK_a 2.39) is rather more difficult. Imidazole (28) and benzimidazole, which are much stronger bases (pK_a 7.0 and 5.5, respectively), obviously owe their basic strength to the addition of a proton to the doubly bonded nitrogen atom, thus permitting a base-strengthening resonance in which each nitrogen atom, in turn, carries the positive charge. Because the cation of 9-methylpurine has a spectrum somewhat different from that of 7-methylpurine, it has been suggested that

[a]X-Ray diffraction data show that, in the solid state, the mobile hydrogen atom is on the 7-position; however, the 9-position may be favored in solution, because the u.v. spectrum of purine (neutral species) rather more resembles that of the 9- than that of the 7-derivative.[75]

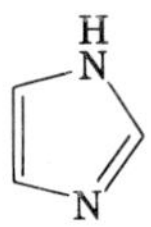

28
Imidazole

purine cannot possibly form its cation by protonation of the imidazole ring.

This view is supported by a comparison of the pK_a values of 6- and 8-(trifluoromethyl) purine (see Table V). The trifluoromethyl group exerts a pure $-I$ base-weakening effect. Because base strength is depressed much more by the 6- than by the 8-isomer, the view has been put forward that the basic center is in the pyrimidine ring.[76] This site was then assigned to N-1 from a comparison of the basic pK_a values of 2- and 6-(methylthio)purine (1.91 and 0, respectively), because the methylthio group is principally inductive ($-I$). At first, it was surprising that such a poorly basic ring as pyrimidine should capture a hydrogen ion in competition with such a highly basic ring as imidazole; however, it is evident that conjugation between the two rings in purine causes considerable delocalization of the electrons belonging to each ring, considered separately. Hence, the N^1-H⊕ cation just described may be stabilized by resonance with a small proportion of the canonical form 29. This is another example of the 4-aminopyridinium type of basic strengthening resonance, 4 ⟷ 5, already discussed.

29

TABLE V

The Ionization[a] of Purines Lacking Hydroxyl or Thiol Substituents

Derivative of purine	Basic pK_a	Acidic pK_a	References
Unsubstituted	2.39	8.93	64
2-Amino-	3.80, -0.28	9.93	64
6-Amino-	4.25, <1	9.83	79,64
8-Amino-	4.68	9.36	64
2-Amino-6,8-bis(trifluoromethyl)-	5.02	~0.3	80
2-Amino-8-(methylsulfonyl)-	2.08	5.61	81
2-Amino-8-(methylthio)-	4.40	8.48	81
2-Amino-8-phenyl-	3.98	9.20	64
2-Amino-6-(trifluoromethyl)-	1.85	8.87	80
2-Amino-8-(trifluoromethyl)-	2.59	6.14	81
2,8-Bis(methylthio)-	2.19	7.73	81
2-Chloro-	0.69	8.21	82
9-methyl-	0.65	—	82
6-Chloro-	0.45	7.88	82
9-methyl-	0.20	—	82
8-Chloro-	1.77	6.02	82
9-methyl-	2.00	—	82
6-Cyano-	~0.3	6.88	80
2,6-Diamino-	5.09, <1	10.77	64
2,6-Diamino-8-(trifluoromethyl)-	3.68	7.55	80

2,6-Dichloro-	-1.16	7.06	82
2-(Dimethylamino)-	4.02	10.22	64
6-(Dimethylamino)-	3.87, <1	10.5	64
8-(Dimethylamino)-	4.80, <1	9.73	64
2-Ethoxy-	2.46	9.47	83
9-methyl-	2.53	-	83
6-Ethoxy-	2.13	9.52	83
9-methyl-	1.90	-	83
8-Ethoxy-	-	-	-
9-methyl-	3.45	-	83
6-(Furfurylamino)- (Kinetin)	~4	~10	84
2-Methoxy-	2.44	9.2	64
6-Methoxy-	2.21	9.16	64
8-Methoxy-	3.14	7.73	78
6-(Methoxycarbonylamino)-	2.27	9.68	80
6-Methyl-	2.6	9.02	64
9-methyl-	3.2	-	85
7-Methyl-	2.29	-	75
8-methyl-	2.85	9.37	64
9-Methyl-	2.36	-	64
2-(Methylamino)-	4.01	10.32	35
6-(Methylamino)-	4.18, <1	9.99	64
9-methyl-	4.12	-	35
8-(Methylamino)-	4.78	9.56	64
8-(Methylsulfonyl)-	0.42	4.87	81
2-(Methylthio)-	1.91	8.91	64
6-(Methylthio)-	1.63	8.74	29,64
8-(Methylthio)-	2.95	7.67	64
9-methyl-	2.98	-	78
8-Phenyl-	2.68	8.09	64
2-Piperidino-	3.97[b]	10.24	83

TABLE V (continued)

The Ionization of Purines Lacking Hydroxyl or Thiol Substituents

Derivative of purine	Basic pK_a	Acidic pK_a	References
9-methyl-	4.10, -0.38	—	83
6-Piperidino-	4.33	10.04	83
9-methyl-	4.12, -1.54	—	83
2,6,8-Triamino-	6.23, 2.41	10.79	64
2,6,8-Trichloro-	-3.1	3.96	82
6-(Trifluoromethyl)-	<0	7.35	80
8-(Trifluoromethyl)-	~ 1.0	5.12	80
6-Ureido-	2.35	9.95	80

[a] In water at 20–25°. The first of two basic pK_a values concerns the first nitrogen atom to ionize as the solution of the neutral species is progressively acidified; the second value corresponds to a less basic nitrogen atom. The first of two acidic pK_a values concerns the loss of the first proton as the solution of the neutral species is made progressively alkaline; the second value corresponds to the ionization of a less acidic group. [b] These values are comparable to those of the corresponding (methylamino)purines.

Against this view, it has been argued that, because the pK_a values of 7-, 8-, and 9-methylpurine and of 8-aminopurine can be predicted (within 0.3 pH unit) from the values of the corresponding benzimidazoles, the site of protonation of purine is[77] N-7 or N-9. However, the same authors showed that, for other purines, protonation can occur on the pyrimidine ring because purin-8-ol is more basic than purine, whereas the corresponding hydroxybenzimidazole (2-OH) is weaker than benzimidazole by more than 7 pK units.

Because purine has acidic properties, it is not possible to arrange the Tables of pK_a values in exactly the way used for the pyrimidines. Table V contains the values for those purines that lack a hydroxyl or thiol substituent; this permits the response of the acidic (NH) group of purine to substituents to be evaluated before the complicating factor of acidity derived directly or indirectly from oxygen is introduced.

It may be seen from Table V that substituents on purine exert much the same effect on (*a*) basic ionization as they do for pyrimidine, and (*b*) acidic ionization as for phenol (see Table 8.4 of Ref. 1). An amino group in the 2-, 6-, or 8-position increases the basic strength. This result is brought about by a 2-(or 4-)aminopyridinium type of resonance, such as has been discussed in Section II, 2. In this way, the basic strength of 2,6,8-triaminopyrimidine reaches the relatively high pK_a of 6.23. That the three monoaminopurines have a true primary amino group has been shown by comparison with *N*-methylated derivatives (both nuclear and exonuclear), by use of pK_a values and ultraviolet and infrared spectra.[64,78]

2. Purines having Hydroxyl or Thiol Substituents

All three monohydroxypurines exist, preponderantly, as cyclic amides in which the mobile hydrogen atom is held by a neighboring nitrogen atom; and this situation causes a resonance-stabilized weakening of acidic properties, as with 2-pyridinone (11) ⟷ (12). This conclusion was reached from comparison of the pK_a values and ultraviolet and infrared spectra of *O*- and *N*-methylated derivatives.[64,78] It is noteworthy that substituents in the 8-position of purine do not constitute a special category. The pK_a values of purin-8-one and of purine-8-thiol are not remarkably different from those of their 2- and 6-isomers. However, each purinethiol is a stronger acid than the corresponding purinol, and the one purineselenol known is stronger still. The poly-hydroxypurines show a large increase in acid strength, because of the increased possibilities for resonance.

A selection of values for purines having oxygen- or sulfur-containing substituents is given in Table VI.

Studies of the ionization and of the ultraviolet spectra of xanthine (purine-2,6-dione) and its *N*-methylated derivatives showed that the mobile hydrogen atom is on N-7 in the neutral species, but on N-9 in the monanion.[89] The quaternary zwitterion 30 (3,7,9-trimethyl-2,6-dioxotetrahydropurinium betaine) obtained by methylation of 3,9-dimethylxanthine has pK_a values of 12.27 and 3.12 for the addition of 1 and 2 protons, respectively.[89]

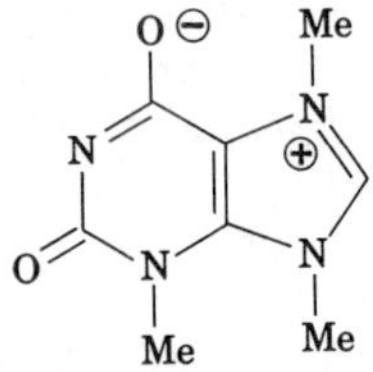

30

In guanine, the mobile proton is on N-1; when, to form the anion, this proton is lost, the negative charge is on the oxygen atom, as is usual for cyclic amides (see Section II, 3). Although, when forming the monocation, adenine accepts a proton on N-1, guanine accepts it[97] on N-7.

The pK_a values of several purine *N*-oxides (some of them in the tautomeric form of *N*-hydroxypurines) have been described for adenine,[98] 2,6-diaminopurine,[98] hypoxanthine,[99] xanthine, isoguanine,[100] and 1-methylxanthine.[101]

3. The Purine Nucleosides and Nucleotides

The pK_a values of purine nucleosides, for example, adenosine (31), are very similar to those of the parent purines, but there is an extra value in the range of 12 to 13 for the ionization of a hydroxyl group in the sugar (Table VII). The nucleotides exhibit similar pK_a values, together with two ionizations of the phosphoric acid moiety at 0.7–0.9 and 5.9–7.0, respectively (see Table VII). The adenylic acids are completely zwitterionic in aqueous solution, but the 5'-guanylic acid

TABLE VI

The Ionization[a] of Purines Having Hydroxyl or Thiol Substituents

Derivative of purine	Basic pK_a	Acidic pK_a	References
2-Amino-6-hydroxy-(Guanine)[b]	3.0	9.32, 12.6	87
1,7-dimethyl-	3.4	—	87
1,9-dimethyl-	3.3	—	87
1-methyl-	3.1	10.5	87
7-methyl-	3.5	10.0	87
9-methyl-	2.8	9.8	87
6-Amino-2-hydroxy-	4.5	9.0	64
8-Carboxy- (Purine-8-carboxylic acid)	2.91,~0	9.37	88
2,6-Dihydroxy- (Xanthine)[c]	—	7.7, 11.94	89
1,3 dimethyl- (Theophylline)	—	8.68	89
1,7-dimethyl-	—	8.65	89
1,9-dimethyl-	—	5.99	89
3,7-dimethyl- (Theobromine)	—	10.00	89
3,9 dimethyl-	—	10.14	89
1 methyl-	—	7.9, 12.2	89
3 methyl-	—	8.45, 11.92	89
7 methyl-	—	8.42, >13	89
9 methyl-	—	6.12, >13	89
1,3,7-trimethyl- (Caffeine)[d]	<1	—	91
2,8-Dihydroxy-	—	7.45	64
1-methyl-	—	7.94	61
3-methyl-	—	8.39	61

TABLE VI (continued)

The Ionization of Purines Having Hydroxyl or Thiol Substituents

Derivative of purine	Basic pK_a	Acidic pK_a	References
9-methyl-	—	8.74	61
6,8-Dihydroxy-	—	7.65, 9.87	64
1 methyl-	—	8.52, 11.83	14
9 methyl-	—	8.31, 11.74	14
6-Formyl- (Purine-6-aldehyde)	2.4	8.8	80
2-Hydroxy-	1.69	8.43, 11.90	64
9-methyl-	<1.5	9.19	78
6-Hydroxy- (Hypoxanthine)	1.98	8.94, 12.10	64
1,7-dimethyl-	2.16	—	78
7-methyl-	—	8.8	92
9-methyl-	1.86	9.32	78
8-Hydroxy-	2.58	8.24, >12	64
7,9-dimethyl-	2.8	—	78
7-methyl-	2.69	8.20	78
9-methyl-	2.80	9.05	78
2-Hydroxy-8-mercapto-	—	—	—
1-methyl-	—	6.61	61
9-methyl-	—	7.00, 11.21	61

2-Hydroxy-6-methyl-	2.29	8.87, 12.46	81
8-(Hydroxymethyl)-	2.62	8.79	93
6-[3-(Hydroxymethyl)-2-butenyl]amino- (Zeatin)	4.4	9.8	94
2-Hydroxy-8-(trifluoromethyl)-	—	5.35, 10.92	81
6-Hydroxy-2-(trifluoromethyl)-	~1.1	5.1, 11.2	80
6-Hydroxy-8-(trifluoromethyl)-	—	~5, 10.9	80
6-(Hydroxylamino)-	3.80	9.83	80
2-Mercapto-[e]	~0.5	7.15, 10.4	64
6-Mercapto-[e]	<0	7.77, 10.84	64
8-Mercapto-[e]	<0	6.64, 11.16	64
9-benzyl-	1.69	7.11	85
9-methyl-	<2.5	7.48	78
6-Seleno-	—	7.33	52
2,6,8-Trihydroxy- (Uric acid)[f]	—	5.4, 10.6	95,96
1,7-dimethyl-	—	5.7, 10.9	95
3-methyl-	—	6.2	95
7-methyl-	—	5.5, 10.6	95

[a]Defined in footnote *a* of Table V. [b]These values replace the approximate results given in Ref. 86. [c]These values for xanthine (and the methylated xanthines) replace the approximate results given in Ref. 90. [d]Caffeine has no acidic function. The basic properties are very weak,[91] and an older value (pK_a 2) is erroneous. [e]For *S*-methyl derivatives, see under "Methylthio" in Table V. [f]Only approximate values are available for uric acid and its *N*-methyl derivatives; it is not known for certain if uric acid has two (see Ref. 96) ionization steps near pK_a 5.4.

TABLE VII

The Ionization[a] of Purine Nucleosides and Nucleotides

Compound	Purine ring		Sugar hydroxyl group	Phosphate group	References
	Basic	Acidic			
Adenosine (9-β-D-Ribofuranosyladenine)	3.63	—	12.35	—	102,103
2'-phosphate	3.80	—	—	6.15	102
3'-phosphate	3.65	—	—	5.88	102
5'-phosphate (AMP)	3.74	—	13.06	6.05	102,103
5'-pyrophosphate (ADP)	4.20	—	—	7.00	104
5'-triphosphate (ATP)	4.00	—	—	6.48	102
Adenosine, 2'-deoxy-, 5'-phosphate	~4.4	—	—	6.4	67
Guanosine (9-β-D-Ribofuranosylguanine)	~1.6	9.33	~12.3	—	105,106
5'-phosphate (GMP)	~2.4	~9.4	—	~6.1	66
5'-pyrophosphate (GDP)	~2.9	~9.6	—	~6.3	66
5'-triphosphate (GTP)	~3.3	~9.3	—	~6.5	66
Inosine (9-β-D-Ribofuranosylhypoxanthine)	~1.5	8.82	—	—	105
Xanthosine (9-β-D-Ribofuranosylxanthine)	—	5.67	—	—	105

[a] pK_a values in water at 20–25°.

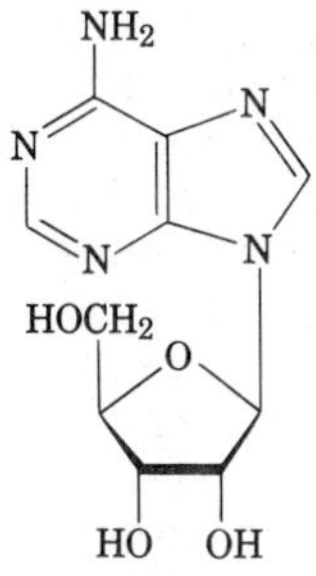

31
Adenosine

zwitterion is in equilibrium with about 12% of the neutral species.

When 2-deoxy-D-ribonucleic acid was titrated with acid and then back-titrated, little significant difference in pK_a was found for the basic group of the guanine residue.[71] This finding is in contrast to those for the adenine and cytosine residues, as discussed in Section II, 4 (see p. 22).

4. Conclusion

No comprehensive table of pK_a values exists for the purines. Consultation of References 64-106 will bring to light a few more values not relevant to the above discussion.

It has not yet been possible to demonstrate covalent hydration in the purine series,[81] but it is very common[107] for the "8-azapurines."

IV. EXAMPLE OF DETERMINATION OF AN IONIZATION CONSTANT (PRACTICAL DETAILS AND CALCULATIONS)

The relative merits of the two principal methods for determining ionization constants have been outlined in Section I (see p. 3). The *potentiometric method* will first be described. Where solubility[a] permits its use, this method has two great advantages: (*a*) high productivity, because three constants can readily be obtained in the course of an hour, and (*b*) its stoichiometric nature prevents a constant from being by-passed when there is no spectral change in the first of two consecutive ionizations. Extra advantages, derived from (*a*) and (*b*), are that the method can detect (and even circumvent) chemical change, and it can also detect any impurity.

Not every pH-measuring set is suitable for potentiometric titration, because the instrument has to hold its reference potential for at least 20 minutes (and, preferably, for one hour). This ability is readily ascertainable. The glass electrode selected should be tested for proportionality; that is, as soon as it has been set to read pH 4.00 when inserted in a buffer of pH 4.00 (by balancing the circuit), it must read pH 9.23 in pH 9.23 buffer *without any adjustment* of the circuit. Many electrodes retain proportionality for only the first three months of use; after this, they give erroneous readings during titrations (and, hence, incorrect pK_a values). The glass electrode should have a coaxially shielded lead, about 1 meter long. It is possible to obtain special electrodes for the region above pH 12, but these are usually injured by acid, and have a very short useful life. When not in use, all electrodes must be scrupulously cared for according to the manufacturer's instructions.

Titrations are conveniently conducted in tall beakers (*e.g.*, 50-ml), closed with a cork that has been bored with five holes: one each for the glass and calomel electrodes, one for a nitrogen-delivering capillary, one for the thermometer, and one to admit the tip of a buret or a micrometer syringe. The electrodes should fit loosely in the holes, and should be secured to a miniature retort-stand by spring-clips; the electrodes are then connected to the terminals of the pH set, and the retort stand is grounded by wire. Nitrogen is used for stirring the system; it should be freed of carbon dioxide and oxygen, and then be introduced, in a *slow* stream, under the surface of the liquid to be titrated. Alternatively, magnetic stirring may be employed. Stirring should be

[a]It is recommended that no solvent other than water be used in determining ionization constants. Mixtures of solvents lead to erroneous comparisons, because neutral molecules usually attract the more lipophilic solvent into their environment, thus displacing the equilibrium with the ionic species.[1]

stopped while a reading is being taken. The beaker is placed in a thermostated water-bath, to keep the temperature of the solution constant.

Titration is performed with hydrochloric acid or potassium hydroxide (0.1 or 1.0 *M*). The potassium hydroxide must be entirely free from carbonate. Sodium hydroxide is unsuitable, because glass electrodes are relatively permeable to sodium ions.

If the solubility of the compound permits, the usual practice is to titrate the unknown substance at 10m*M* concentration. At this concentration, the activity effects are small, and may be corrected for by the following equation:

$$pK_a^T = pK_a^M + 0.5 \sqrt{I_m} \text{ (for acids), or } -0.5 \sqrt{I_m} \text{ (for bases)} \quad \ldots(8)$$

where pK_a^T is the thermodynamic (*i.e.*, concentration independent) value, pK_a^M is the value determined by titration, and I_M is the ionic strength at the mid-point of the titration (*e.g.*, I_m is 0.005 for a 10m*M* titration). Potassium chloride cannot be added to secure "constant ionic strength," because its use invalidates the simplified calculation recommended here.

More concentrated solutions, up to 1.0*M*, may be titrated, but the activity correction becomes more complex. No titration is valid if the apparent pK_a obtained is less than the negative logarithm of the concentration. Thus, if an apparent pK_a of 1.8 is obtained on titrating a 10m*M* solution (for which the negative logarithm is obviously 2), the value 1.8 is spurious, and the true pK_a could be -5.1 (or any other value less than 2). A large number of such spurious, low values are to be found in the literature. For solutions more dilute than 1m*M*, a more sensitive apparatus, such as the vibrating-condenser electrometer, must be employed. Only pK_a values lying between 1.25 and 11.00 are accurately determinable with a glass electrode, but the (more troublesome) hydrogen electrode can be used up to pK_a 13.65.

Titration with a glass electrode is conducted as follows. After the electrode has been standardized both to phthalate <u>and</u> to borate buffer, and the solution has been brought to the temperature desired, the titrant is added from the buret in 9 equal portions, each capable of neutralizing one tenth of the unknown. The pH is read between each addition, and the results are set out as in Table VIII. "Stoichiometric concentrations" are those that would be present were each portion of acid to react with its equivalent of base; actually, some hydrolysis occurs, and this is compensated for by the term in

TABLE VIII

Determination of the Basic Ionization Constant of Adenosine[a]

1	2	3	4	5	6	7	8
Titrant (100m*M* HCl), ml	*p*H	Stoichiometric concentration [B]	[BH$^{\oplus}$]	$H^{\oplus}\}$	$\frac{[B]+\{H^{\oplus}\}}{[BH^{\oplus}]-\{H^{\oplus}\}}$	log of last column	p*k*$_a$ (= pH-column 7)
0.0	5.99	0.0050	0.0000				
0.5	4.58	0.0045	0.0005		45/5	0.95	3.63
1.0	4.26	0.0040	0.0010		40/10	0.60	3.66
1.5	4.05	0.0035	0.0015	0.0001	36/14	0.41	3.64
2.0	3.86	0.0030	0.0020	0.0001	31/19	0.21	3.65
2.5[b]	3.70	0.0025	0.0025	0.0002	27/23	0.07	3.63
3.0	3.57	0.0020	0.0030	0.0002	22/28	-0.10	3.67
3.5	3.43	0.0015	0.0035	0.0004	19/31	-0.21	3.64
4.0	3.28	0.0010	0.0040	0.0005	15/35	-0.37	3.65
4.5	3.14	0.0005	0.0045	0.0007	12/38	-0.50	3.64

Result: p*k*$_a$ = 3.65($\pm$0.02) at 5m*M* and 20°

= 3.62 (thermodynamic, at 20°)

[a]Titration of 100 ml of 5m*M* adenosine, in water at 20°. [b]Half an equivalent.

column 6. Column 5 gives the hydrogen-ion activity, obtained by subtracting the pH from 0, and taking the antilogarithm of the difference.

Results having a spread of more than ±0.06 are not valid, and their error is usually much greater than the apparent spread. Table VIII gives the results obtained on titrating the basic center of adenine.

Although the determination of ionization constants by *ultraviolet spectrophotometry* is more time-consuming than by potentiometry, the method is ideal when the pK_a lies outside the range of 1.25–11.00, and is mandatory if the compound is too insoluble for potentiometry. No corrections for hydrolysis are required, because direct, optical measurement gives the true concentration. Determinations are usually made at a dilution of 10 to 100μ*M*. Because buffer salts (usually 10m*M*) are present, equation *8* must be used if thermodynamic results are required.

The following five operations are needed in order to make spectrometric determination of a pK_a. (a) Prepare a stock solution, and dilutions of it in various buffers. (b) Search for pure spectra of the two ionic species involved in the equilibrium. (c) Choose a wavelength suitable for the determination, namely, one that has maximal absorption coupled with minimal interference from the other species (either species may be used for selecting this "analytical wavelength"). (d) Search for an approximate value of the pK_a, setting out the calculations in six columns, as follows.

1	2	3	4	5	6
pH	d	$d_I - d$	$d - d_N$	$\log \dfrac{(d-d_N)}{(d_I-d)}$	pK_a (= pH + column 5)

Where d is the density found at the analytical wavelength, d_I is the density of the pure ionic species, and d_N is the density of the pure neutral species.

(e) Accurately determine the pK_a by using a set of buffers that will produce a 10 to 90% ionization of the unknown in nine equal steps. Results having a spread greater than ±0.06 should be discarded, as in potentiometry.

For all values after the first one, whether determined potentiometrically or spectrometrically, compounds having more than one ionization constant of the same sign need rather complex activity corrections. For these, and for all other practical details of the determination of ionization constants, see Ref. 1.

REFERENCES

(1) A. Albert and E. P. Serjeant, *Ionization Constants of Acids and Bases*, Methuens, London, England (1962).

(2) A. Albert, *Selective Toxicity*, 4th edition, Methuens, London, England (1968).

(3) Phenols become stronger by 0.012 unit of pK for each degree rise in temperature, whereas the value for carboxylic acids does not change appreciably. For the effect of temperature on base strength, see footnote *a* to Table I.

(4) A. Albert, R. J. Goldacre, and J. N. Phillips, *J. Chem. Soc.*, (1948) 2240.

(5) D. J. Brown and R. V. Foster, *Aust. J. Chem.*, 19, 1487, 2321 (1966).

(6) V. Boarland and J. F. W. McOmie, *J. Chem. Soc.*, (1952) 3716, 3722.

(7) D. J. Brown and T. Teitei, *Aust. J. Chem.*, 17, 567 (1964); *cf.*, D. J. Brown, P. W. Ford, and K. H. Tratt, *J. Chem. Soc. (C)*, (1967) 1445.

(8) D. J. Brown, E. Hoerger, and S. F. Mason, *J. Chem. Soc.*, (1955) 211, 4035.

(9) J. R. Marshall and J. Walker, *J. Chem. Soc.*, (1951) 1004.

(10) S. F. Mason, *J. Chem. Soc.*, (1959) 1247.

(11) A. Albert and G. B. Barlin, *J. Chem. Soc.*, (1962) 3129.

(12) N. Whittaker, *J. Chem. Soc.*, (1951) 1565.

(13) A. Albert, D. J. Brown, and G. W. H. Cheeseman, *J. Chem. Soc.*, (1952) 1620, 4219.

(14) D. J. Brown and J. S. Harper, *J. Chem. Soc.*, (1961) 1298.

(15) S. F. Mason, *J. Chem. Soc.*, (1954) 2071.

(16) D. J. Brown and J. S. Harper, *J. Chem. Soc.*, (1963) 1276.

(17) D. J. Brown, B. T. England, and J. M. Lyall, *J. Chem. Soc. (C)*, (1966) 226.

(18) D. J. Brown and T. Teitei, *J. Chem. Soc.*, (1965) 755.

(19) D. J. Brown and T. Teitei, *Aust. J. Chem.*, 18, 559 (1965).

(20) D. Shugar and J. Fox, *Biochim. Biophys. Acta*, 9, 199 (1952).

(21) S. Mizukami and E. Hirai, *J. Org. Chem.*, 31, 1199 (1966).
(22) D. J. Brown and N. W. Jacobsen, *J. Chem. Soc.*, (1962) 3172.
(23) D. J. Brown and N. W. Jacobsen, *J. Chem. Soc.*, (1960) 1978.
(24) A. Albert, *J. Chem. Soc. (B)*, (1966) 427.
(25) D. J. Brown and T. Teitei, *J. Chem. Soc.*, (1963) 3535, 4333.
(26) D. J. Brown, personal communication.
(27) M. Biffin, D. J. Brown, and T.-C.Lee, *Aust. J. Chem.*, 20, 1041 (1967).
(28) D. J. Brown and J. S. Harper, *J. Chem. Soc.*, (1965) 5542.
(29) Recently determined in the Department of Medical Chemistry, Australian National University.
(30) D. J. Brown and M. Paddon-Row, *J. Chem. Soc. (C)*, (1966) 164.
(31) D. J. Brown and M. Paddon-Row, *J. Chem. Soc. (C)*, (1967) 903.
(32) J. Gage, *J. Chem. Soc.*, (1949) 469.
(33) D. J. Brown and L. N. Short, *J. Chem. Soc.*, (1953) 331.
(34) D. J. Brown and J. M. Lyall, *Aust. J. Chem.*, 15, 851 (1962).
(35) D. J. Brown and N. W. Jacobsen, *J. Chem. Soc.*, (1965) 3770.
(36) D. J. Brown and P. W. Ford, *J. Chem. Soc. (C)*, (1967) 568.
(37) E. Kalatzis, *J. Chem. Soc. (B)*, (1967) 273.
(38) D. J. Brown, in *Mechanisms of Molecular Migrations*, B. S. Thyagarajan (ed.), John Wiley and Sons, Inc., New York, N.Y. (1968), Vol. 1, pp. 209-245.
(39) A. Albert, *Heterocyclic Chemistry*, Athlone Press, London (Oxford University Press, New York), 2nd edition (1968).
(40) D. J. Brown, *Nature*, 165, 1010 (1950).
(41) S. F. Mason, *J. Chem. Soc.*, (1958) 674.
(42) A. R. Katritzky and A. J. Waring, *J. Chem. Soc.*, (1962) 1540.
(43) A. Albert and J. N. Phillips, *J. Chem. Soc.*, (1956) 1294.
(44) (a) J. J. Fox and D. Shugar, *Bull. Soc. Chim. Belges*, 61, 44 (1952); (b) A. Biggs, *J. Chem. Soc.*, (1956) 2485.
(45) D. J. Brown and T.-C.Lee, *Aust. J. Chem.*, 21, 243 (1968).
(46) I. Wempen and J. J. Fox, *J. Amer. Chem. Soc.*, 86, 2474 (1964).

(47) D. J. Brown and T. Teitei, *J. Chem. Soc.*, (1964) 3204.
(48) J. Jonas and J. Gut, *Collect. Czech. Chem. Commun.*, 27, 716 (1962).
(49) H. Gottschling and C. Heidelberger, *J. Mol. Biol.*, 7, 541 (1963).
(50) S. F. Mason, *J. Chem. Soc.*, (1957) 5010.
(51) A. Albert and G. B. Barlin, *J. Chem. Soc.*, (1962) 3129.
(52) H. Mautner, *J. Amer. Chem. Soc.*, 78, 5292 (1956).
(53) B. Stanovnik and M. Tisler, *Arzneim. Forsch.*, 14, 1004 (1964).
(54) D. Shugar and J. J. Fox, *Bull. Soc. Chim. Belges*, 61, 293 (1952).
(55) P. Bell and R. O. Roblin, *J. Amer. Chem. Soc.*, 64, 2905 (1942).
(56) E. R. Tucci, E. Doody, and N. C. Li, *J. Phys. Chem.*, 65, 1570 (1961).
(57) E. Hirai, *Chem. Pharm. Bull.* (Tokyo), 14, 861 (1966).
(58) T. Okano and S. Kojima, *Yakugaku Zasshi*, 86, 547 (1966).
(59) A. Albert, *J. Chem. Soc. (B)*, (1966) 438.
(60) J. J. Christensen, J. H. Rytting, and R. M. Izatt, *J. Phys. Chem.*, 71, 2700 (1967).
(61) D. J. Brown, *J. Appl. Chem.* (London), 9, 203 (1959).
(62) J. P. Kokko, J. H. Goldstein, and L. Mandell, *J. Amer. Chem. Soc.*, 83, 2909 (1961).
(63) A. R. Katritzky and A. J. Waring, *J. Chem. Soc.*, (1963) 3046.
(64) A. Albert and D. J. Brown, *J. Chem. Soc.*, (1954) 2060.
(65) J. J. Fox, L. F. Cavalieri, and N. Chang, *J. Amer. Chem. Soc.*, 75, 4315 (1953).
(66) R. M. Bock, N.-S. Ling, S. A. Morell, and S. H. Lipton, *Arch. Biochem. Biophys.*, 62, 253 (1956).
(67) R. O. Hurst, A. M. Marko, and G. C. Butler, *J. Biol. Chem.*, 204, 847 (1953).
(68) J. J. Fox and D. Shugar, *Biochim. Biophys. Acta*, 9, 369 (1952).
(69) J. Ofengand and H. Schaefer, *Biochemistry*, 4, 2832 (1965).
(70) K. Berens and D. Shugar, *Acta Biochim. Pol.*, 10, 25 (1963); *Chem. Abstr.*, 59, 7083 (1963).
(71) J. M. Gulland, D. O. Jordan, and H. F. W. Taylor, *J. Chem. Soc.*, (1947) 1131.
(72) D. J. Brown, *The Pyrimidines*, Interscience Publishers, Inc., New York, N.Y. (1962); see also, First Supplement (1970).
(73) M. E. C. Biffin, D. J. Brown, and T.-C.Lee,

J. Chem. Soc. (C), (1967) 573.
(74) G. D. Watson, R. M. Sweet, and R. E. Marsh, *Acta Crystallogr.*, 19, 573 (1965).
(75) A. Bendich, P. Russell, and J. J. Fox, *J. Amer. Chem. Soc.*, 76, 6073 (1954).
(76) A. Bendich, A. Giner-Sorolla, and J. J. Fox, *Ciba Found. Symp. Chem. Biol. Purines*, (1957) 11.
(77) J. Clark and D. D. Perrin, *Quart. Rev.* (London), 18, 295 (1964).
(78) D. J. Brown and S. F. Mason, *J. Chem. Soc.*, (1957) 682.
(79) A. Albert and E. P. Serjeant, *Biochem. J.*, 76, 621 (1960).
(80) A. Giner-Sorolla and A. Bendich, *J. Amer. Chem. Soc.*, 80, 3932, 5744 (1958); A. Giner-Sorolla, I. Zimmerman, and A. Bendich, *ibid*, 81, 2515 (1959).
(81) A. Albert, *J. Chem. Soc. (B)*, (1966) 438.
(82) G. B. Barlin and N. B. Chapman, *J. Chem. Soc.*, (1965) 3017.
(83) G. B. Barlin, *J. Chem. Soc. (B)*, (1967) 954.
(84) C. O. Miller, F. Skoog, F. Okumura, M. Von Saltza, and F. Strong, *J. Amer. Chem. Soc.*, 77, 2662 (1955).
(85) D. J. Brown, P. W. Ford, and K. H. Tratt, *J. Chem. Soc. (C)*, (1967) 1445.
(86) H. F. W. Taylor, *J. Chem. Soc.*, (1948) 765.
(87) W. Pfleiderer, *Ann.*, 647, 167 (1961).
(88) A. Albert, *J. Chem. Soc.*, (1960) 4705.
(89) W. Pfleiderer and G. Nübel, *Ann.*, 647, 155, 161 (1961).
(90) A. G. Ogston, *J. Chem. Soc.*, (1935) 1376.
(91) A. Turner and A. Osol, *J. Amer. Pharm. Assoc., Sci. Ed.*, 38, 158 (1949).
(92) A. G. Ogston, *J. Chem. Soc.*, (1936) 1713.
(93) A. Albert, *J. Chem. Soc.*, (1955) 2690.
(94) D. S. Letham, J. S. Shannon, and I. R. McDonald, *Proc. Chem. Soc.*, (1964) 230.
(95) E. Johnson, *Biochem.* J., 51, 133 (1952).
(96) A. Bernouilli and A. Loebenstein, *Helv. Chim. Acta*, 23, 245 (1940).
(97) C. A. Dekker, *Ann. Rev. Biochem.*, 29, 453 (1960).
(98) M. A. Stevens and G. B. Brown, *J. Amer. Chem. Soc.*, 80, 2755 (1958).
(99) J. C. Parham, J. Fissekis, and G. B. Brown, *J. Org. Chem.*, 31, 966 (1966).
(100) J. C. Parham, J. Fissekis, and G. B. Brown, *J. Org. Chem.*, 32, 1151 (1967).

(101) A. D. McNaught and G. B. Brown, *J. Org. Chem.*, 32, 3689 (1967).
(102) R. A. Alberty, R. M. Smith, and R. Bock, *J. Biol. Chem.*, 193, 425 (1951).
(103) R. M. Izatt, J. H. Rytting, L. D. Hansen, and J. J. Christensen, *J. Amer. Chem. Soc.*, 88, 2641 (1960).
(104) R. M. Izatt and J. J. Christensen, *J. Phys. Chem.*, 66, 359 (1962).
(105) A. Albert, *Biochem. J.*, 54, 646 (1953).
(106) P. A. Levene, H. S. Simms, and L. W. Bass, *J. Biol. Chem.*, 70, 243 (1926).
(107) A. Albert, *J. Chem. Soc. (B)*, (1966) 427; (*C*), (1968) 344.
(108) B. Roth and J. Strelitz, *J. Org. Chem.*, 34, 821 (1969).

CHAPTER 2

The Ultraviolet Spectra of Pyrimidines and Purines

ADRIEN ALBERT

JOHN CURTIN SCHOOL OF MEDICAL RESEARCH,
AUSTRALIAN NATIONAL UNIVERSITY, CANBERRA, AUSTRALIA

I. THE SPECTRA OF PYRIMIDINES

1. General

Because different ionic species have different ultraviolet spectra, it follows that a spectrum has no meaning unless it is measured at such a pH that only one ionic species can be present. This situation is usually achieved by performing the measurement in a buffer having a pH at least two units[a] removed from the pK_a.

[a]For the definition of pK_a, and the effect of pH upon ionization, see Chapter 1, Section I.

The practice of publishing only those spectra that represent a single ionic species did not begin,[1] for the pyrimidine series, until 1952. Since that time, there has been little departure from this principle, although the ultraviolet spectra of a few compounds have justifiably been measured in lipophilic solvents to elicit special effects.

The energy that an illuminated molecule absorbs from ultraviolet radiation raises the molecule from the ground state to an excited level. The wavelength of the absorbed light is determined by the energy of the transition, and the extinction, by the probability of the transition. These electronic transitions are always polarized directionally, either along the long axis of the molecule or along the short axis. Although spectra obtained for the gaseous state show a high degree of resolution, it is usually more convenient to measure spectra in solution, even though many sharp, separate peaks are certain to have coalesced into rounded envelopes. The resolution of spectra recorded for solutions is increased by lowering the temperature, or by employing hydrocarbon solvents, or both. However, for most purposes, it is convenient to use a temperature of 20–25°, and to use water as the solvent (in order to control the pH, as already discussed).

Above 180 nm, the ultraviolet spectra[a] of aromatic and heteroaromatic molecules usually show four principal bands (regions of strong absorption). These bands have been correlated with the fundamental principles of quantum mechanics through the pioneering studies of Sklar,[1a] of Mulliken,[1b] and of Förster.[1c] A reasonably good explanation of the spectra of N-heteroaromatic nuclei is based on the free-electron model of Platt[2] (1949), with which he evaluated the energy levels of π-electrons in aromatic systems and attempted to predict the spectra. His analysis agrees with the results of more recent attempts that use molecular orbital theory derived from LCAO (linear combination of atomic orbitals).

Of the four absorption bands mentioned as being present in the ultraviolet spectra of all aromatic molecules, two are of moderately high frequency (short wavelength) and were called B_a and B_b by Platt; the other two are of low frequency (longer wavelength) and are called L_a and L_b. The B bands are thought to be the response to perturbation by a single pair of opposite charges, and the

[a]The reading 180 nm is the same as 180 mμ or 1800 Å, which are no longer recommended as units.

L bands, to several such pairs. The present, simplified account will deal only with transitions to the "singlet" excited state, and, hence, use of state-prefixes (as in ^3B_a) will be unnecessary. The B_a and B_b bands of benzene lie close together, near 190 nm; the L_a band is seen as five or six diffuse peaks near 204 nm (log $\varepsilon_{max} \sim 3.8$); the L_b band occurs near 255 nm, and has about six weak peaks (log ε_{max} 2.35) which are sharply defined.

The principal key to the spectrum of any heteroaromatic nucleus is the close resemblance to the spectrum of the corresponding aromatic hydrocarbon. The resemblances are most readily studied for di- and tri-cyclic substances, for which the spectrum is spread over a wide range of wavelengths. If a heteroatom is present, the L_b peak is usually intensified, because the transitions in that region involve a charged, excited state[3]; whereas, for benzene, this state is absent, and transitions in this region are partly inhibited by the sheer symmetry of the molecule (so-called "forbidden transitions").

The spectrum of pyrimidine in the vapor phase shows many maxima. An intense maximum at 191 nm is thought to correspond[4] to the 204-nm maximum of benzene vapor, and lines at 233, 238, 243, and 248 nm correspond to the 255-nm region of benzene.[5] Other lines, at longer wavelengths (297, 305, 309, 312, 315, 316, and 322 nm)[5] resemble peaks seen in the low-intensity region of the ultraviolet spectrum of pyrimidine in a hydrocarbon solvent [Fig. 1 (A)]. This region is due to an $n \rightarrow \pi^*$ transition, namely, a transition from the lone pair of electrons (on the nitrogen atom) to an excited (π^*) orbital in the π-electron layer which covers the pyrimidine nucleus. In the spectrum of pyridine, this type of transition is obscured by the L_b band and, hence, is barely recognizable, but it shifts to longer wavelengths with increase in the number of doubly bound nitrogen atoms in the ring,[6] so that *s*-tetrazine ("1,2,4,5- tetra-azabenzene") has λ_{max} 542 nm because of this effect. The transition is, as would be expected, highly sensitive to change in solvent. Such hydrogen-bonding solvents as water or ethanol become strongly attached to the lone pair, and severely diminish this transition (compare curves A and B in Fig. 1). Because of the relative inconspicuousness of the $n \rightarrow \pi^*$ band of pyrimidine, and its virtual disappearance from the spectrum (a) of a solution in water, (b) in the cationic form, or (c) if a donor hydrogen-bonding substituent is present, the rest of this discussion will be confined to the more prominent bands of pyrimidine, which are $\pi \rightarrow \pi^*$ transitions, not sensitive to solvents.[7,8] Only spectra of

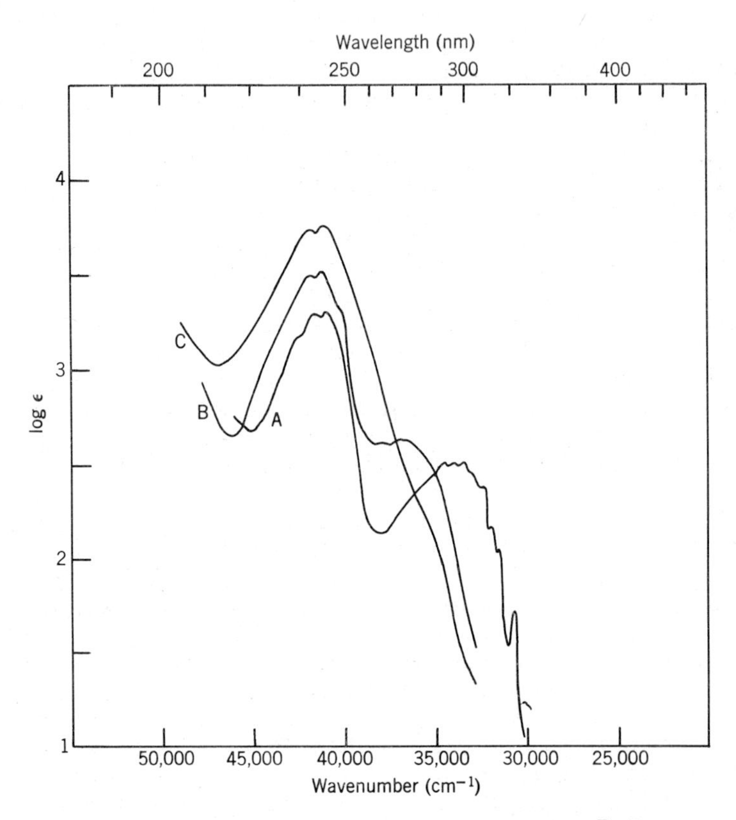

Fig. 1. Ultraviolet Spectra of Pyrimidine.[7,8] [A, Neutral species in cyclohexane; B, neutral species in water (buffered at pH 7); and C, cation in 2*M* sulfuric acid.] Note that these curves are linear as to frequency (to display details of resolution); the other curves shown in this Chapter are linear as to wavelength.

compounds in solution (usually in water) will be discussed. On protonation, the 243-nm band of pyrimidine gains in intensity, with almost no shift in wavelength, and barely any trace can be seen of the $n \rightarrow \pi^*$ band [see Fig. 1 (C)].

In general, substitution on pyrimidine moves the 243-nm band to a longer wavelength and, at the same time, increases the intensity. These changes are rather small for nonconjugating substituents.[9] When only one strongly conjugating substituent is present on a polysubstituted pyrimidine, the shifts in wavelength are roughly additive, and so are increases in intensity.[9] It was found that the following substituents show little evidence of conjugation with the ring: CH_3, Cl, $O^{\ominus}$, and NH_2, whereas C_6H_5, SCH_3, and $S^{\ominus}$ substituents show signs of extensive conjugation, as evidenced by abnormally high extinction coefficients.[9]

Closer study has shown that the effects of substituents on maximal wavelengths and absorptions follow much the same rules for the pyrimidine as for the benzene series. That is to say, wavelength and changes in intensity follow the usual, vectorial, addition rules based on transition dipole-moments. Electron-attracting substituents ($-CO_2H$, -Cl, -SOMe, $-SO_2Me$) produce only a small bathochromic shift when present on C-2 or C-5, and are more likely to lessen the intensity than to increase it. However, if present on C-4, they give rise to a larger bathochromic shift and an increase in intensity. Electron-releasing substituents ($-CH_3$, $-NH_2$), on the other hand, produce a far greater increase in both wavelength and intensity when present on C-2 or C-5 than they do on C-4.

It has already been mentioned that substituents that conjugate with the nucleus have the effect of increasing the intensity. If sufficiently conjugated, they can also split the $\pi \rightarrow \pi^*$ transition band (*i.e.*, the one near 240 nm in the spectrum of the parent pyrimidine), into two peaks, one of which is at considerably longer wavelengths and is almost as intense as the one that remains near 240 nm. Insensitivity to change of solvent, apart from the high intensity, has been used for showing that the components of the split peak are unrelated to the $n \rightarrow \pi^*$ transition; however, the $\pi \rightarrow \pi^*$ band is not necessarily split in the related cation. 2-Phenylpyrimidine (see Table I) and 2-aminopyrimidine (see Table II) furnish examples of these split $\pi \rightarrow \pi^*$ bands.[13a]

2. The Aminopyrimidines

The ultraviolet spectrum of 2-aminopyrimidine (1) is shown in Fig. 2, together with superimposed curves of 2-(methylamino)- and 2-(dimethylamino)-pyrimidine. These spectra constitute a family of curves in which each replacement of a hydrogen atom by a methyl group brings about a regular displacement of the whole curve to slightly longer wavelengths which also obtains in the series: aniline, *N*-methylaniline, *N*,*N*-dimethylaniline. The spectra of the cations of these compounds preserve this relationship (see Fig. 3), but the band

TABLE I

Ultraviolet Spectra of Some Monosubstituted Pyrimidines[a] (see, also, Table II)

Derivative of pyrimidine	pk_a	Species[b]	λ_{max} (nm)[c]	log ε_{max}	pH	References
Unsubstituted	1.31	0	238, 243, 272	3.48, 3.51 2.62	7.0	7,8
		+	242	3.70	-0.8	7,8
	-6.3	++	245	3.76	[d]	10
5-Bromo-	<1	0	217, 261	4.02, 3.46	7.0	9
2-Carboxy-	2.85[e]	-	246	3.42	7.0	6
	-1.13	0	240	3.43	0.8	6
4-Carboxy-	?	-	<u>253</u>	<u>3.51</u>	13.0	9
5-Carboxy-	?	-	<u>245</u>-<u>246</u>	<u>3.31</u>	13.0	9
2-Chloro-	<1	0	209, <u>251</u>	3.75, <u>3.43</u>	7.0	9
5-Chloro-	<1	0	211,<u>258</u>	3.85, <u>3.38</u>	7.0	9
2-Hydrazino-	4.55[f]	0	230, 297	4.12, 3.39	7.0	12
	-0.46	+	220, 276	4.03, 3.31	2.0	12

2-(Methoxy-carbonyl)-	−0.68	0	245	3.36	7.0	6
2-Methyl-	?	0	248	3.46	7.0	9
		+	251−252	3.87	0	9
4-Methyl-	1.98	0	244	3.53	7.0	11
		+	244	3.70	0	11
2-(Methylsul-finyl)-	< −3	0	247	3.57	5.0	12
4-(Methylsul-finyl)-	<0	0	254	3.63	5.0	12
5-(Methylsul-finyl)-	0.42	0	247	3.44	5.0	12
2-(Methylsul-fonyl)-	< −3	0	239, 243	321, 3.23	5.0	12
4-(Methylsul-fonyl)-	<0	0	229, 250, 277	3.90, 3.42 2.93	5.0	12
5-(Methylsul-fonyl)-	0.97	0	240, 243	3.06, 3.06	5.0	12
2-Phenyl-	?	0	251	4.18	7.0	9
		(+)	256-258, 287	4.06, 3.92	0	9
5-Phenyl-	?	0	256	4.08	E[g]	13

TABLE I (continued)

Ultraviolet Spectra of Some Monosubstituted Pyrimidines[a] (see, also, Table II)

Derivative of pyrimidine	pK_a	Species[b]	λ_{max} (nm)[c]	log ε_{max}	pH	References
2-(Phenylsulfinyl)-	< −3	0	235, <u>267</u>	4.04, <u>3.30</u>	5.0	12
2-(Phenylsulfonyl)-	< −3	0	226	4.11	5.0	12

[a]In water, at 20–25°, except where otherwise indicated. [b]++, di-cation; +, monocation; 0, neutral species; -, monoanion. [c]Inflections are underlined. [d]In 18M H_2SO_4 (H_o -10). [e]Ionization of carboxylic acid. [f]Ionization of hydrazino group. [g]In ethanol.

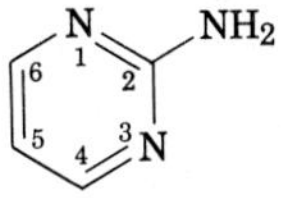

1

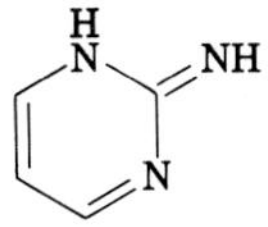

2

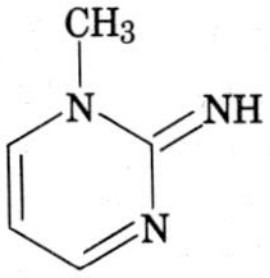

3

of longer wavelength is displaced farther, to still longer wavelengths, having the effect of further separating the two bands into which the $\pi \rightarrow \pi^*$ transition has been split by the conjugating nature of the amino group. That the formation of the cation did not cause the spectra to revert to that of the parent pyrimidine demonstrates that protonation takes place on a ring-nitrogen atom, whereas the spectrum of aniline reverts to that of benzene when the cation is formed.

The corresponding 4-aminopyrimidines (primary, secondary, and tertiary) show similar regularity in their spectra (see Figs. 4 and 5) for the neutral species and the cations, respectively. As might be expected for an

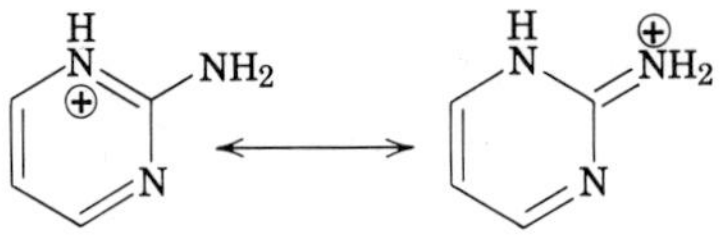

4 **5**

electron-releasing group on C-4, the spectra approach that of pyrimidine more closely than those of the 2-isomers. In Fig. 3, the two bands of the $\pi \rightarrow \pi^*$ transition are incompletely separated, and for the related cations (see Fig. 4), these bands have all but fused.[14]

That 2- and 4-aminopyrimidines are primary amines (*e.g.*, 1) and lack the isomeric iminodihydro structure (*e.g.*, 2) was shown in Chapter 1 by means of ionization constants, because true iminodihydro compounds, maintained in that strained state by alkylation of a nuclear nitrogen atom, are enormously stronger bases. A more direct proof of the primary amine structure has been obtained from the ultraviolet spectra.[15] For example, 2-(dimethylamino)pyrimidine, which must have a structure similar to (1), has a spectrum closely related to that of 2-aminopyrimidine by the expected bathochromic displacement; whereas the spectrum of 1,2-dihydro-2-imino-1-methylpyrimidine (3) is displaced toward much longer wavelengths, and differs in shape (Fig. 6). The same proof is offered for 4-aminopyrimidine (see Fig. 7). It should be noted that the cations of compounds having the amino structure and of those having the imino structure have almost identical spectra, because each structure

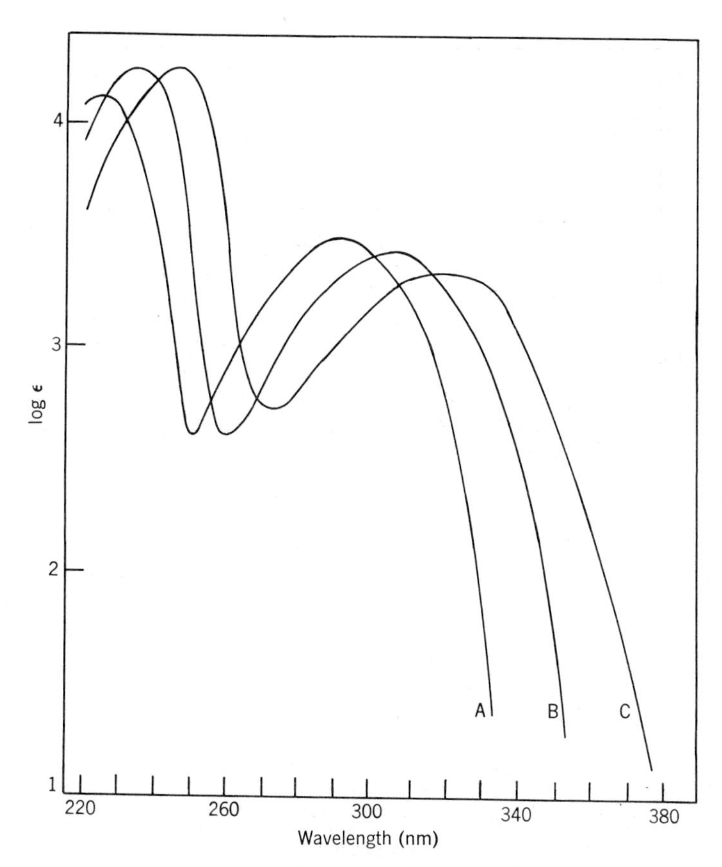

Fig. 2. Ultraviolet Spectra[14] of the Neutral Species, in Water at pH 7.0, of (A) 2-Aminopyrimidine, (B) 2-(Methylamino)pyrimidine, and (C) 2-(Dimethylamino)pyrimidine.

forms the same type of resonant cation, *e.g.*, 4 ⟷ 5; this resonance has already been discussed in Chapter 1 [Section II (2)].

The possibility cannot be excluded that the amino form (of the neutral species of 2- and 4-aminopyrimidine) is in equilibrium with a very small proportion of the imino form. By spectrometry, it is usually impossible to detect less than 1% of a minor component, but the ionization constants permit calculation of an (approximate) equilibrium ratio, designated K_T. This type of calcula-

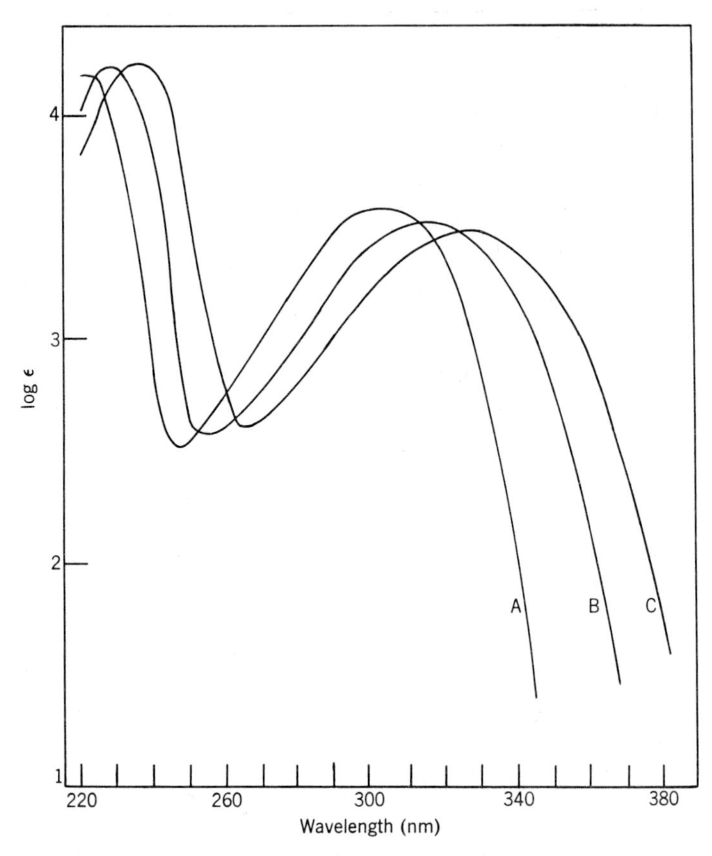

Fig. 3. Ultraviolet Spectra[14] of the Cationic Species, in Water at pH 1.0, of (A) 2-Aminopyrimidine, (B) 2-(Methylamino)pyrimidine, and (C) 2-(Dimethylamino)pyrimidine.

tion, introduced[18] for establishing the K_T of aminopyridines, is applicable when two tautomers have a common cation (usually a mesomeric cation, as here). The method makes use of the fact that replacement of a hydrogen atom by a methyl group does not alter a pK_a value or an ultraviolet spectrum very much. If pK_a^A is the pK of a primary amine, and pK_a^I is that of its (much more basic) imino tautomer, then

$$K_T = \text{antilog}\ (pK_a^A - pK_a^I) - 1.$$

On applying this equation to 2-aminopyrimidine, it may be seen that the approximate ratio of amino form to

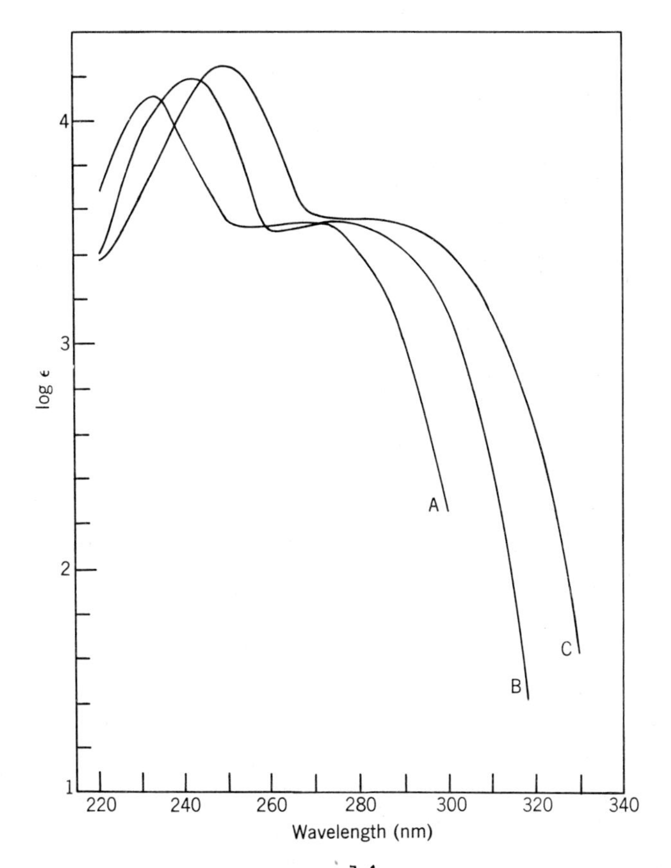

Fig. 4. Ultraviolet Spectra[14] of the Neutral Species of (A) 4-Aminopyrimidine (in Water at pH 13), (B) 4-(Methylamino)pyrimidine (in Water at pH 9), and (C) 4-(Dimethylamino)pyrimidine (in Water at pH 9).

imino form (in aqueous solution) is about a million to one; 4-aminopyrimidine, also, has a similar ratio of the two forms.

Numerical data on the spectra of the aminopyrimidines are given in Table II.

3. The Hydroxy and Thiol Derivatives of Pyrimidines

For the pyrimidinones, pyrimidinethiones, various

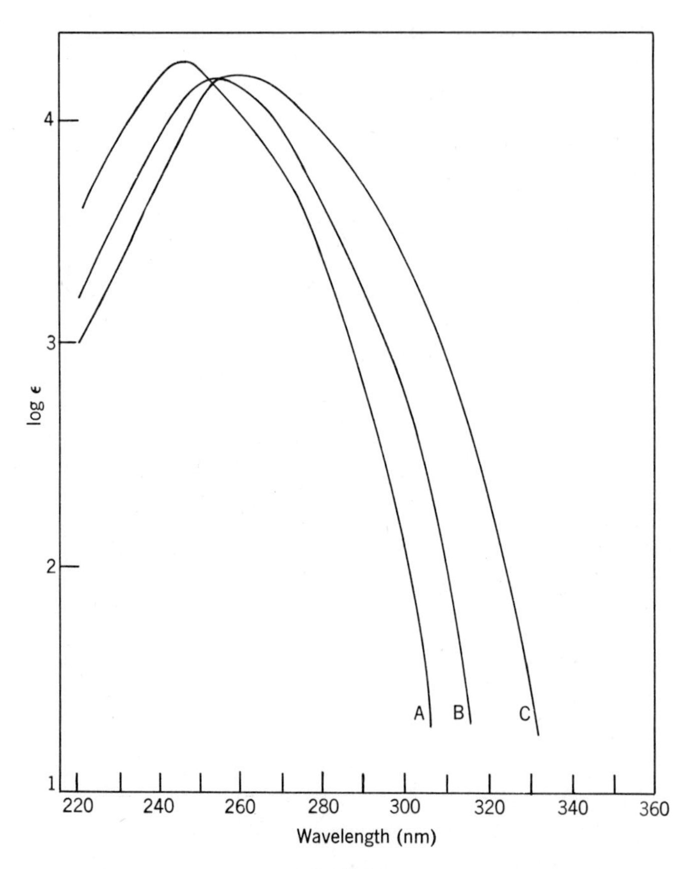

Fig. 5. Ultraviolet Spectra[14] of the Cationic Species of (A) 4-Aminopyrimidine (in Water at pH 0), (B) 4-(Methylamino)pyrimidine (in Water at pH 2), and (C) 4-(Dimethylamino)pyrimidine (in Water at pH 3).

tautomeric forms may be written. It has already been stated in Chapter 1 [Section II (3)] that the oxo (*e.g.*, 6) and thio forms of the 2- and 4-substituted pyrimidines are enormously favored over the hydroxy (*e.g.*, 8) and mercapto forms at equilibrium in aqueous solution. A proof of this statement, through comparison of the ultraviolet spectra of these derivatives, will now be given.

The ultraviolet spectra of the neutral species of 2- and 4-methoxypyrimidines (see Table III) are logically related to that of pyrimidine, following the rules already

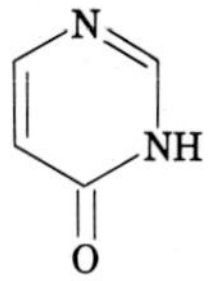

N
NH
O

6

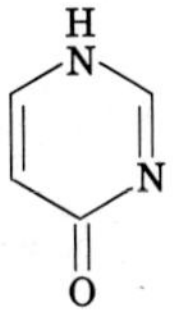

H
N
N
O

7

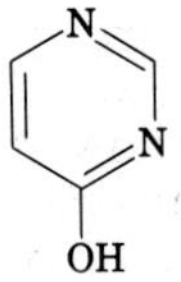

N
N
OH

8

CH_3
N
O
N $\oplus$ H

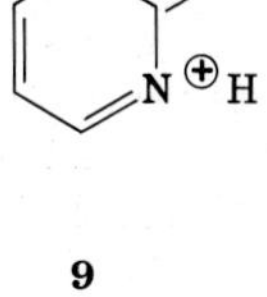

9

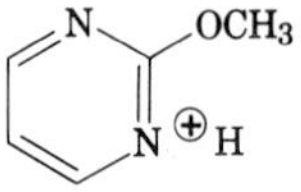

10

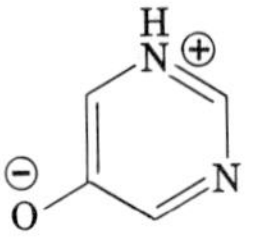

11

given. The spectra of the neutral species of 2- and 4-pyrimidinones (see Table III) are very different from those of their methyl ethers, but are extraordinarily like those of their *N*-methyl derivatives. It has, therefore, been concluded[14,15] that these hydroxypyrimidines exist mainly in the form of the oxo tautomers (*e.g.*, 6) when at equilibrium in aqueous solution, and that the proportion of hydroxy tautomer (*e.g.*, 8) is insignificant. Fig. 8 shows the spectral curves for 4-pyrimidinone and its three methyl derivatives. Here, the situation is more complex than for 2-pyrimidinone, because 4-pyrimidinone can give two different *N*-methyl derivatives. It is evident that the curve for 4-pyrimidinone could be (mainly) constructed from that of the 3-methyl derivative, plus a contribution from the 1-methyl derivative; from these curves, the ratio of structure 6 to 7 in a solution of 4-pyrimidinone has been estimated[7] as 5:2.

Because the *N*- and the *O*-derivatives of 2-pyrimidinol do not have a common cation (see the cationic spectra in Table III), the ratio of enol form to oxo form cannot be calculated by a method similar to that used for 2-aminopyrimidine. The ratios for 2- and 4-pyrimidinone are known to be approximately 1:340 and 1:2200, respectively,[22] and still larger ratios would be expected for 2- and 4-pyrimidinone. Careful study[7] of the spectra of

TABLE II

Ultraviolet Spectra of the Aminopyrimidines and Their *N*-Methyl Derivatives[a]

Derivative of pyrimidine	pK_a	Species	λ_{max} (nm)	log ε_{max}	pH	References
4-Acetamido-	2.76	0	232, 262	4.01, 4.01	5.0	
2-Amino- (1)	3.54	0	224, 292	4.13, 3.50	7.0	14
		+	221, 302-3	4.17, 3.60	1.0	14
4-Amino-	5.71	0	233, 268-9	4.26, 3.72	13.0	14
		+	246	4.27	0	14
5-Amino-	2.60	0	236, 298	4.04, 3.49	~5	16
		+	253, 332	4.16, 3.57	1.0	16
4,5-Diamino-	6.03	0	246, 289	3.89, 3.86	8.0	17
		+	284	3.94	3.2	17
N^5-formyl-	4.45	0	233, 279	3.97, 3.62	7.0	
1,2-Dihydro-2-imino-1-methyl- (3)	10.75	0	236, 345	4.19, 3.46	13.0	15
		+	222, 301	4.12, 3.63	7.0	15
1,4-Dihydro-4-imino-1-methyl-	12.2	0	253, 315	4.21, 2.79	13.0	15
		+	250	4.21	7.0	15
2-(Dimethylamino)-	3.96	0	243, 318	4.26, 3.35	7.0	14
		+	235, 324-5	4.24, 3.47	1.0	14

TABLE II (continued)

Ultraviolet Spectra of the Aminopyrimidines and Their *N*-Methyl Derivatives[a]

Derivative of pyrimidine	pK_a	Species	λ_{max} (nm)	log ε_{max}	pH	References
4-(Dimethyl-amino)-	6.35	0	250, 286	4.22, 3.56	9.3	14
		+	262	4.21	3.2	14
2-(Methylamino)-	3.82	0	234, 306-7	4.23, 3.43	7.0	15
		+	228, 315	4.23, 3.53	1.0	14
4-(Methylamino)-	6.12	0	242, 276-7	4.18, 3.54	9.0	14
		+	254	4.20	2.1	14
2,4,5-Triamino-	7.63	0	232-4, 303	3.92, 3.73	9.8	17
	2.56	+	228, <u>240</u>, 295	4.20, <u>4.07</u>, 3.62	5.2	17
		++	268	3.61	0	17
4,5,6-Triamino-	5.78	0	277, 380	3.89, 2.37	8.0	17
	1.47	+	287	4.01	3.6	17
		++	265	4.01	-0.8	17

[a]In water at 20–25°. The spectra of 2,6-diamino-[19], 4,6-diamino-[19], 2,4,6-triamino-[19], and 2,4,5,6-tetraamino-pyrimidine[20] are also available, but were measured at such pH values that the figures published do not represent a single ionic species.

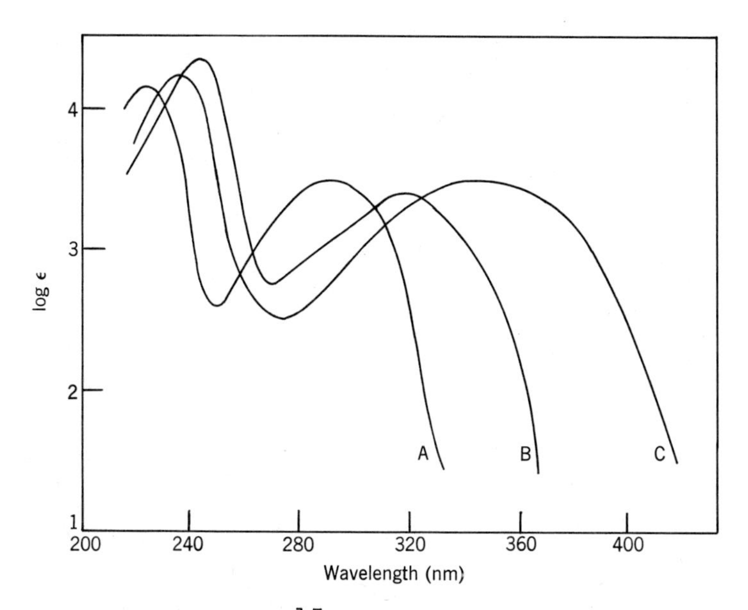

Fig. 6. Demonstration[15] that 2-Aminopyrimidine (A) has a Spectrum (and hence a Structure) like that of 2-(Dimethylamino)pyrimidine (B), and unlike that of 1,2-Dihydro-2-imino-1-methylpyrimidine (C) (in Water at pH 13; All Neutral Species).

the cations indicated that 1-methyl-2(1*H*)-pyrimidinone has the structure 9, whereas 2-methoxyprimidine has structure 10; this is the difference that prevents calculation of the tautomeric ratio. It may be seen from Fig. 9 that 4-pyrimidinone forms a cation that is like those of its two *N*-methyl derivatives, but quite unlike that of its *O*-methyl derivatives.

Somewhat similar spectroscopic studies[23] have shown that 5-pyrimidinol, which is mainly in the true phenolic form in aqueous solution, has about 2 per cent of the zwitterionic tautomer (11) in equilibrium with it.

Calculations of the energies required for electronic transitions in substituted pyrimidines have been used for predicting[7] that (for a particular ionic species and a given position of substitution) the wavelength of maximal absorption should increase in the following order: hydroxy < amino < mercapto (substituents). The results given in Table III confirm this prediction. The

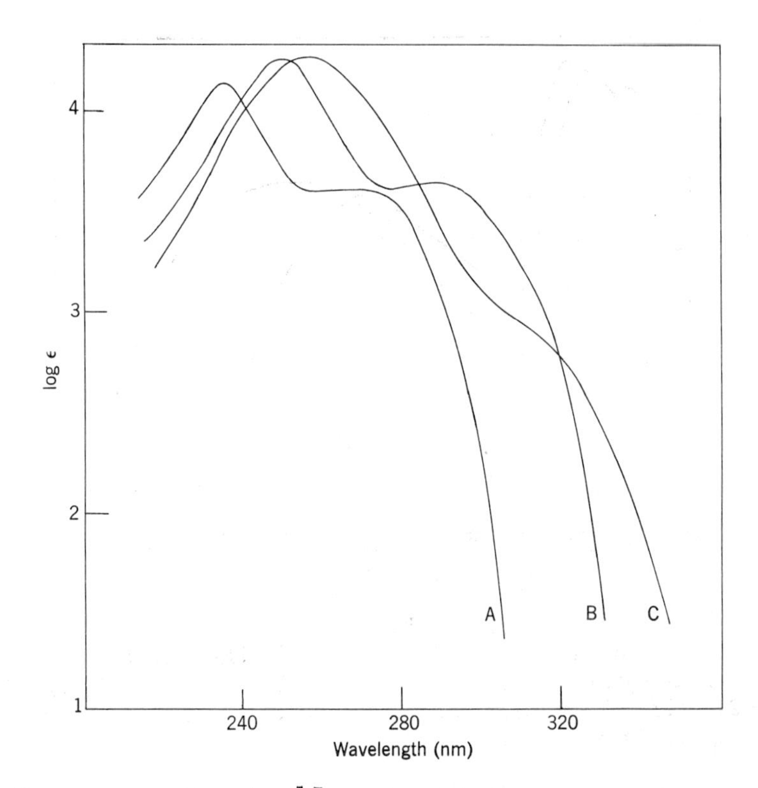

Fig. 7. Demonstration[15] that 4-Aminopyrimidine (A) has a Spectrum (and hence a Structure) like that of 4-(Dimethylamino) pyrimidine (B), and unlike that of 1,4-Dihydro-4-imino-1-methylpyrimidine (C) (in Water at pH 13; All Neutral Species).

well-known rule of Jones[24], namely, that the anion of a hydroxy derivative has an ultraviolet spectrum almost identical with that of the neutral species of the amino analog, is in harmony with the above calculation. In this connection, it is useful to note that a hydroxy compound, whether in the form of a cyclic amide or a true enol, usually increases in λ_{max} on conversion into the anion.

The spectra of 2- and 4-pyrimidinethione are summarized in Table III, along with those of their *S*- and *N*-methyl derivatives. The spectra show, even more clearly than the ionization constants, that tautomers having a hydro-

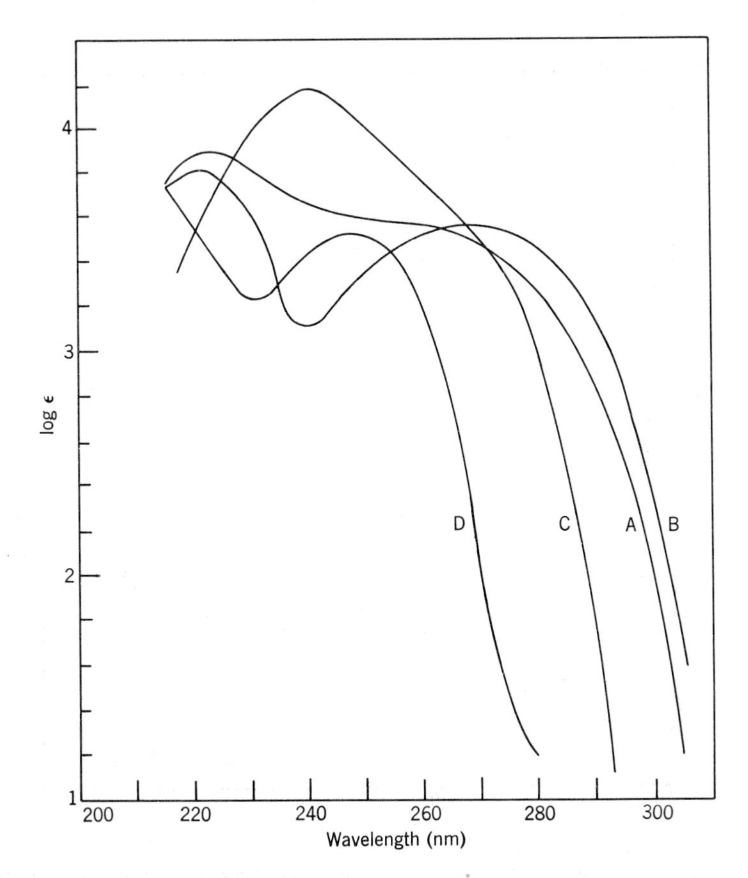

Fig. 8. Demonstration[15] That 4-Pyrimidinone (A) has a Spectrum Related to those of 3-Methyl-4(3*H*)-pyrimidinone (B) and 1-Methyl-4(1*H*)-pyrimidinone (C) in the ratio of 5:2, and Unrelated to that of 4-Methoxypyrimidine (D). (All Neutral Species in Water at the pH Values Given in Table III).

gen atom on nitrogen are favored over those having a hydrogen atom on sulfur. The mobile hydrogen atom of 4-pyrimidinethione is seen, from inspection of the actual curves,[21] to be situated almost entirely on N-3.

When more than one hydroxyl (or thiol) group is present on a pyrimidine, the assessment of the principal tautomeric forms becomes more difficult, because a larger number of models having a fixed structure must

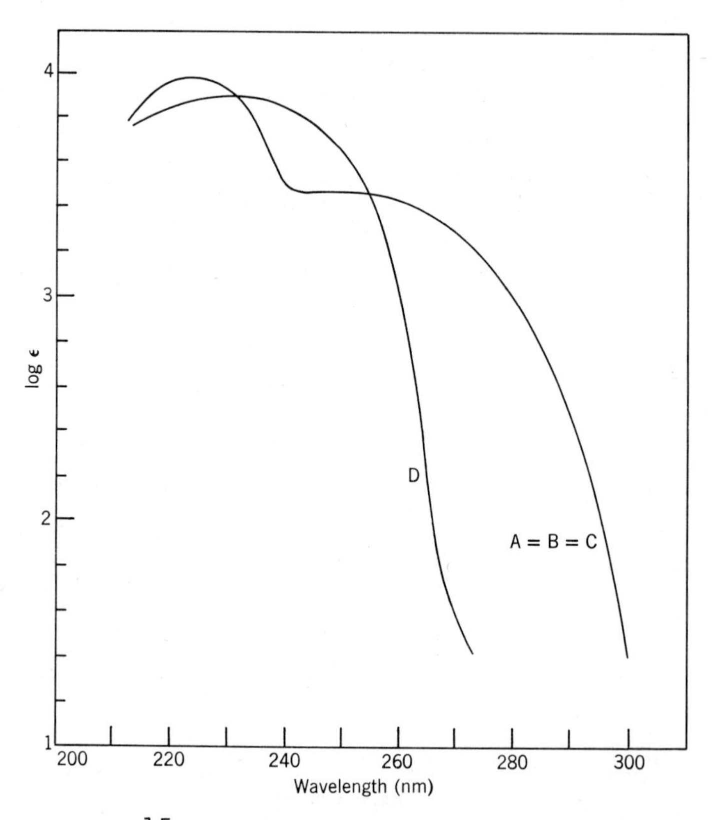

Fig. 9. Spectra[15] of the Cations of the Four Compounds, (A, B, C, and D, respectively) of Fig. 8.

be made and measured, and the results taken into consideration. Uracil (2,4-pyrimidinedione) has been investigated alongside many of its *N*-methyl and *O*-ethyl derivatives.[25,26] From the spectra of the four known monoalkyl (and two known dialkyl) derivatives, it has been provisionally concluded[25,26] that, in aqueous solution, the monoanion of uracil is a 1:1 tautomeric mixture of a form having one hydrogen atom still attached to N-1 and a form having one hydrogen atom attached to N-3. The proportions of the two anions differ greatly in the 5-halogenated uracils.[28] Table IV lists the spectra of such polyhydroxypyrimidines and their derivatives.

The nature of the principal tautomeric forms of bar-

TABLE III

Ultraviolet Spectra of the Monohydroxy- and Mercapto- pyrimidines and Their *N*-, *O*-, and *S*-Methyl Derivatives[a]

Derivative of pyrimidine	pK_a	Species	λ_{max} (nm)[b]	log ε_{max}	pH	References
A. Monohydroxy						
2-Hydroxy-	9.17	–	220, 292	4.06, 3.66	13.0	15
	2.24	0	212, 298	4.03, 3.67	6.2	16
		+	<215, 309	>3.2, 3.75	0	15
O-methyl-[c]	1.05	0	264	3.68	7.0	6
		+	273	3.70	-1.2	6
1-methyl-[d]	2.50	0	215, 302	4.0, 3.73	6.0	15
		+	<215, 313	>3.8, 3.85	0.3	15
4-Hydroxy-	8.59	–	227, 263	4.05, 3.56	13.0	14
	1.85	0	223, 260	3.86, 3.57	6.2	14
		+	224, 251	3.99, 3.47	-1.2	6,14
O-methyl-	2.5	0	248	3.53	7.0	6
		+	227, 240	3.89, 3.83	0	6
1-methyl-	2.02	0	240	4.16	6.0	15
		+	229, 250	4.01, 3.42	0	6
3-methyl-	1.84	0	221, 269	3.83, 3.59	5.0	15
		+	226, 258	3.96, 3.47	-0.5	15

TABLE III (continued)

Ultraviolet Spectra of the Monohydroxy- and Mercapto- pyrimidines and Their *N*-, *O*-, and *S*-Methyl Derivatives[a]

Derivative of pyrimidine	pK_a	Species	λ_{max} (nm)[b]	log ε_{max}	pH	References
5-Hydroxy-	6.78	–	238, 304	4.05, 3.65	9.5	6
	1.87	0	214, 271	3.99, 3.68	4.3	6
		+	223, 285	3.84, 3.67	-0.3	6
B. Mercapto						
2-Mercapto-	7.14	–	231, 270	3.69, 4.23	13.0	21
	1.35	0	278, 346	4.33, 3.42	4.9	21
		+	285, 378	4.51, 3.18	0	21
S-methyl-	0.59	0	250, 285	4.11, 3.25	7.0	21
		+	255, 313	4.10, 3.51	- 2.8	21
1-methyl-	1.66	0	279, 344	4.26, 3.45	9.0	21
		+	283, 378	4.34, 3.14	- 3.0	21
4-Mercapto-	6.90	–	292–4	4.04	13.0	21
	0.68	0	285, 327	4.03, 3.91	4.5	9
		+	305	4.22	- 2.3	21
S-methyl-	2.48	0	257, 279	3.85, 3.91	7.0	21
		+	221, 302	3.47, 4.23	0	21

1-methyl-	1.16	0	328	4.33	7.0	21
		+	304	4.24	-2.0	21
3-methyl-	0.56	0	287, 322	4.03, 3.91	7.2	21
		+	304	4.23	-3.0	21

[a]In water, at 20–25°. [b]Inflections are underlined. [c]2-Methoxypyrimidine. [d]1-Methyl-2(1*H*)-pyrimidinone.

TABLE IV

Ultraviolet Spectra of Polyhydroxypyrimidines and Hydroxypyrimidinethiols, and Their *N*-, *O*-, and *S*-Methyl Derivatives[a]

Derivative of pyrimidine	pK_a	Species	λ_{max} (nm)	log ε_{max}	pH	References
2,4-Dihydroxy- (Uracil)	9.38	—	284	3.79	12.0	25
	-3.4	0	203, 260	3.96, 3.91	7.0	25,26
5-bromo-	7.83	—[b]	289	3.81	12.0	27
		0	275	3.85	1.0	27
6-carboxy- (Orotic acid)	9.45	--	286	3.78	12.0	25
	2.07	—	207, 279	4.06, 3.88	7.2	25
5-chloro-	7.95	—[b]	287	3.83	12.0	27
		0	272	3.87	1.0	27
2,4-di-*o*-ethyl-	—	0	211. 259	3,86, 3.79	7.2	25
1,3-dimethyl-	—	0	266	3.95	7.0	25
2-*o*-ethyl-	8.2	—	221, 265	3.85, 3.83	12.0	25
		0	218, 260	3.82, 3.78	7.0	25
4-*o*-methyl-	10.7	—	220, 278	3.99, 3.81	13.0	25
	1.0	0	269	3.71	7.0	25
5-fluoro-	8.04	—[b]	282	3.60	12.0	27
		0	265	3.77	1.0	27

5-iodo-	8.25	—	305	—	11.0	28
		0	285	—	5.0	28
1-methyl-	9.75	—	265	3.85	12.0	25
	-3.4	0	268	3.99	7.0	25,26
3-methyl-1	9.95	-	283	4.03	12.0	25
		0	259	3.86	7.0	25,26
4-*o*-methyl		0	275	—	—	26
5-methyl- (Thymine)	9.94	—	291	3.74	12.0	25
		0	207.265	3.98, 3.90	7.0	25,27
6-methyl-	9.64	—	277	3.83	13.0	11
		0	261	4.00	4.6	11
5-nitro-	11.3	--	243, 362	3.81, 4.20	14.0	25
	5.56	—	231, 258, 342	3.78, 3.74, 4.20	7.2	25
		0	238, 300	3.86, 3.98	3.0	25
4,6-Dihydroxy-	5.4	—	252	3.88	7.8	29,30
	-0.26	0	253	3.98	2.0	29,31
		+	242	3.92	-2.8	29
4,6-di-*o*-methyl-	1.49	0	242	3.50	4.2	29,30
		+	247	3.85	-2.0	29
4-*o*-methyl-	8.47	—	230, 266	3.77, 3.30	10.8	29,30
	-0.22	0	237, 254	3.54, 3.56	2.0	29,31
		+	241	3.90	-2.4	29
1-methyl-4-*o*-methyl-	-0.44	0	225, 299	4.17, 4.24	0.4	29
		+	240	3.95	-2.5	29

TABLE IV (continued)

Ultraviolet Spectra of Polyhydroxypyrimidines and Hydroxypyrimidinethiols, and Their *N*-, *O*-, and *S*-Methyl Derivatives[a]

Derivative of pyrimidine	pk_a	Species	λ_{max} (nm)	log ε_{max}	pH	References
1-methyl-6-*O*-methyl-	-2.18	0	217, 253, 287	4.11, 3.34, 3.09	1.0	29,30
		+	248, <u>280</u>	3.44, <u>3.13</u>	-4.7	29
2,4,5-Trihydroxy-	11.48[c]	0	285	3.62	1.0	27
2,4,6-Trihydroxy-(Barbituric acid)	12.5	--	260	4.19	14.0	32
	3.9	—	258	4.33	7.0	32
		0	205, 258	4.02, 2.78	1.0	32
5,5-diethyl-(Barbitone)	12.7	--	255	3.90	14.0	32
	7.89	—	239	4.03	10.5	32
		0	213	3.96	5.0	32
1,3-dimethyl-	—	0	228	3.80	7.0	32
1-methyl-	8.45	—	245	3.94	11.0	32
		0	223	3.90	6.0	32

1,3-dimethyl-	4.6	—	260	4.28	7.2	32
		0	226	3.90	2.0	32
6-*o*-methyl-	—	0	228	3.80	7.0	32
1-methyl-	12.8	--	260	4.18	14.0	32
	4.2	—	259	4.31	7.2	32
		0	220	3.94	2.0	32
1,3,6,-tri-methyl-		0	257	4.08	7.0	32
2,4,6-tri-methyl-		0	248	3.89	7.0	32
4-Hydroxy-2-mercapto- (2-Thiouracil)[d]	12.7	--	259, 286	4.06, 3.89	14.0	36
	7.74	—	232, 259, 312	4.07, 4.03, 3.87	10.1	36
		0	213, 274	4.23, 4.15	5.8	36
1,3-dimethyl-	—	0	222,273	4.19, 4.08	7.0	36
1-ethyl-	8.7	—	237, 270	4.34, 4.20	12.0	36
		0	217, 270	4.25, 4.13	3.6	36
3-ethyl-	8.65	—	260, 313	3.85, 4.11	12.0	36
		0	218, 281	4.18, 4.16	5.8	36
S-ethyl-1-methyl-	0.9	0	234	4.43	5.8	36
S-ethyl-3-methyl-	—	0	292	4.00	2.0	36
4-Hydroxy-6-mercapto-	10.52	—	222, 294	4.16, 4.11	7.0	37
	4.33	0	229, 241, 304	3.96, 3.93, 4.09	0.5	37
	-1.7	+	231, 264	4.11, 3.97	-3.6	37
o-methyl-	7.51	0	300	4.22	3.0	37
S-methyl-	8.52	0	236, 278	4.19, 3.95	1.8	37

Footnotes to Table IV.

[a]In water, at 20–25°. [b]By selecting 12.0 instead of 11.0, these authors have apparently reported a mixture of the mono- and di-anion (see Ref. 28). For spectra of 1-methyl-5-bromo-(and 5-nitro)uracil, see Ref 15. [c]These authors also presented a spectrum recorded at pH 12, but this must inevitably have been for a mixture of ionic species. [d]For the spectrum of 2,4-dithiouracil in ethanol, see Ref. 38.

bituric acid (2,4,6-pyrimidinetriol) and barbitone (the 5,5-diethyl derivative of the latter) have been discussed in Chapter 1 [Section II (3)]. When an aqueous solution of barbituric acid is made increasingly alkaline, the first proton is lost from the methylene group at C-5, and the second is lost from one of the nitrogen atoms.[32] In the ionization of barbitone, all protons are necessarily lost from nitrogen atoms, apparently from N-4 and N-2, in that order.[32] Ultraviolet spectra of the medicinal 5,5-dialkylbarbiturates have been recorded; they do not differ greatly from one another.[33,34]

Spectra have been recorded[35] (without regard to ionic species) for dialuric acid (2,4,5,6-tetrahydropyrimidine) and its complete oxidation product, alloxan.

The spectra of cytosine (4-amino-2-pyrimidinone, 12) and its derivatives have shown conclusively that 12 is the correct structure for the neutral species.[39,40] Of the two methods of numbering the cytosine ring, that shown in 12 is used here. Table V lists the spectra of cytosine and many of its derivatives, as well as the spectra of other aminopyrimidinones and aminopyrimidinethiones.

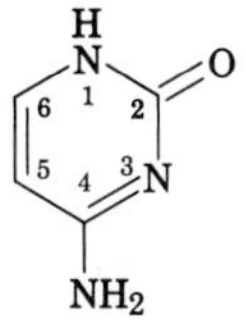

12

Cytosine

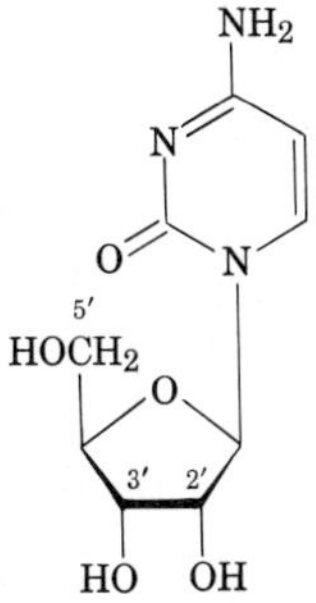

13

Cytidine

TABLE V

Ultraviolet Spectra of Aminohydroxypyrimidines and Their Derivatives[a]

Derivative of pyrimidine	pK_a	Species	λ_{max} (nm)[b]	log ε_{max}	pH	References
2-Amino-4-hydroxy-	9.59	0	280	3.64	7.0	42
	4.0					
1-methyl-	—	0	260	3.74	13.0	42
3-methyl-	—	0	225, 284	3.86, 3.96	9.8	42
o-methyl-	5.53	0	225, 277	4.10, 3.68	7.8	42
4-Amino-2-hydroxy- (Cytosine, 12)	12.5	—	282	3.90	14.0	25
	4.58	0	197, 267	4.35, 3.79	7.0	25
		+	210, 276	3.99, 4.00	2.0	25
1,3-dimethyl-	9.4	0	272	3.95	12.0	41
		+	281	4.01	4.0	41
1-methyl-	4.57	0	230, 274	3.89, 3.92	7.0	39,40
		+	213, 283	4.00, 4.09	2.0	39,40
3-methyl-	7.4	0	294	4.08	12.0	41
		+	274	3.97	4.0	41
5-methyl-	12.4	—	290	3.91	14.0	25
	4.6	0	211, 274	4.15, 3.79	7.0	25
		+	211, 284	4.08, 3.99	2.0	25

o-methyl-	5.3	0	225, 271	3.90, 3.86	7.2	25
		+	230, 261	3.95, 3.98	2.0	25
4-Amino-6-hydroxy-	10.05	–	213, 254	4.54, 3.60	12.5	42
	1.36	0	213, 258	4.42, 4.07	7.0	42
		+	217, 257	4.02, 4.07	-2.0	42
1-methyl-	0.98	0	216, 257	4.54, 3.80	4.8	42
o-methyl-	4.02	0	235	3.88	7.0	42
2-Amino-4-mercapto-	8.03	–	262, 312	3.77, 4.16	10.4	42
	2.86	0	236, 256, 342	3.67, 3.63, 4.11	5.4	42
		+	259, 325	3.72, 4.13	-0.2	42
3-methyl-	2.92	0	258, 337	3.68, 4.21	7.0	42
s-methyl-	4.75	0	234, 300	3.90, 4.02	7.0	42
4-Amino-2-mercapto-	10.58	–	222, 264, 298	4.32, 4.13, 3.80	13.3	42
	3.33	0	242, 269	4.26, 4.26	7.0	42
		+	224, 277, 315	4.15, 4.30, 3.74	0.2	42
s-methyl-	4.91	0	224, 250, 285	4.31, 4.00, 3.79	7.8	42
2,4-Diamino-6-hydroxy-	10.78	–	240, 263	3.61, 4.01	13.0	44
	3.33	0	211, 267	4.47, 4.19	7.0	44
		+	264	4.31	1.0	44
4,5-Diamino-2-hydroxy-	11.45	–	226, 303	3.91, 3.67	13.0	43

TABLE V (continued)

Ultraviolet Spectra of Aminohydroxypyrimidines and Their Derivatives[a]

Derivative of pyrimidine	pK_a	Species	λ_{max} (nm)[b]	log ε_{max}	pH	References
	4.37	0	292	3.58	7.0	43
		+	305	3.76	2.3	43
4,5-Diamino-6-hydroxy-	9.86	−	272, 370	3.87, 2.62	12.0	43
	3.57	0	278, 372	3.95, 2.44	6.7	43
	1.34	+	258	3.74	2.5	43
		++	257	3.85	-0.7	43
2-(Dimethylamino)-4-hydroxy-	9.89	−	231, 285	4.11, 3.73	13.3	42
	3.68	0	224, 297	4.30, 3.51	7.0	42
		+	222, 265	4.11, 3.80	0.2	42
4-(Dimethylamino)-6-hydroxy-	10.42	−	223, 260	4.47, 3.92	13.3	42
	1.22	0	224, 267	4.37, 4.11	7.0	42
		+	265	4.19	-1.3	42
1-methyl-	<1	0	228, 270	4.45, 4.11	7.0	42
o-methyl-	4.23	0	257	4.14	7.0	42

[a]In water, at 20–25°. [b]Inflections are underlined.

TABLE VI

Ultraviolet Spectra of Pyrimidine Nucleosides and Nucleotides[a]

Compound	pK_a	Species	λ_{max} (nm)[b]	log ε_{max}	pH	References
Cytidine (1-β-D-Ribofuranosylcytosine, 13)	12.24	−	273	3.96	14.0	45
	4.08	0	198, 230, 271	4.37, 3.92, 3.96	7.0	45,49
		+	213, 280	4.00, 4.13	2.0	45
2'-deoxy-	4.25	−	273	3.97	14.0	45
		0	198, 271	4.35, 3.95	7.0	45,49
		+	213, 280	4.01, 4.12	2.0	45
5'-phosphate	4.44	−	272	3.97	14.0	46
3-methyl-	8.7	0	266	3.95	12.0	41
		+	278	4.07	4.0	41
2'-phosphate (2'-Cytidylic acid)	4.30	0	270	3.92	6.2	46
		+	277	4.11	1.6	46
3'-phosphate (3'-Cytidylic acid)	4.16	0	273	3.92	6.2	46
		+	278	4.12	1.6	46
5'-phosphate (CMP)	6.3	−	280	4.12	9.0	48
	4.5	+	271	3.96	2.5	48

TABLE VI (continued)

Ultraviolet Spectra of Pyrimidine Nucleosides and Nucleotides[a]

Compound	pK_a	Species	λ_{max} (nm)[b]	log ε_{max}	pH	References
5'-pyrophosphate (CDP)	6.4	−	280	4.11	9.0	48
	4.6	+	271	3.96	2.5	48
5'-triphosphate (CTP)	6.6	−	280	4.11	9.0	48
	4.8	+	271	3.95	2.5	48
Thymidine [1-(2-Deoxy-β-D-*erythro*-pentofuranosyl)thymine]	9.79	--	268	3.88	14.0	45
	12.85	0	208, 267	3.98, 3.99	7.0	45,52
5'-phosphate (5'-Thymidylic acid)	10.0	0	207, 268	3.97, 3.98	7.1	49
Uracil, 5-β-D-Ribofuranosyl- (Pseudouridine)	8.97	−	285	?	11.8	50
		0	260	?	6.5	50
Uridine (1-β-D-Ribofuranosyluracil)	12.59	--	265	3.88	14.0	45

5-bromo-	—	0	279	3.99	1.0	27
5-chloro-	—	0	277	3.96	1.0	27
5-fluoro-	7.57	--	270	3.85	14.0	47
5-hydroxy-	—	0	280	3.91	1.0	27
5-iodo-	—	0	288	3.85	1.0	27
5'-phosphate (UMP)	9.5	--	261	3.89	11.0	48
5'-pyrophosphate (UDP)	6.4	-	262	4.00	2.0	48
	9.4	--	261	3.90	11.0	48
	6.5	—	262	4.00	2.0	48
5'-triphosphate (UTP)	9.5	--	261	3.91	11.0	48
	6.6	—	262	4.00	2.0	48
	9.30	0	205, 261	3.99, 4.00	7.3	49
2'-deoxy-	9.3	--	264	3.90	14.0	45
		0	262	4.01	7.0	45
5-bromo-	—	0	279	4.01	1.0	27
5-chloro-	—	0	277	3.96	1.0	27
5-fluoro-	7.66	0	268	3.95	1.0	27
		--	270	3.84	14.0	47
5-hydroxy-	—	0	280	3.85	1.0	27
5-methyl- (see Thymidine)						

[a]In water, at 20–25°. [b]Inflections are underlined.

4. The Pyrmidine Nucleosides and Nucleotides

Ultraviolet spectra of some pyrimidine nucleosides and nucleotides are given in Table VI.

As regards the pyrimidine nucleosides and nucleotides, it is helpful to remember that the replacement of an alkyl group by a sugar residue makes very little difference to the ultraviolet spectra. Only in the region of high alkalinity (above pH 11), where a sugar group begins to ionize, can a significant change in the spectrum be found. Differences in spectra between D-ribosides and the corresponding "2-deoxy-D-ribosides" can best be seen in this region.[45] Furanoid and pyranoid forms of nucleosides are distinguishable by comparison of their spectra.[45]

The ultraviolet spectra of the nucleic acids and polynucleotides show a well defined maximum at 259 nm that is due to the chromophoric pyrimidine and purine residues. The molecular extinction coefficient is much lower than for the constituent nucleosides, but, under hydrolytic conditions, increases with progressive denaturation,[51] even before hydrolysis begins. This hypochromism has been attributed[49] to dispersion-force interactions between the pyrimidine and purine chromophores as stacked in the helixes; this interaction leads to an interchange of intensity between various transitions, and can be hyperchromic at lower wavelengths.

II. THE SPECTRA OF PURINES

1. General

Purine (14) is much less volatile than pyrimidine, and, hence, its ultraviolet spectrum has been found unmeasurable in the vapor phase; however, such a measurement has proved just possible for 9-methylhypoxanthine.[52] Purine

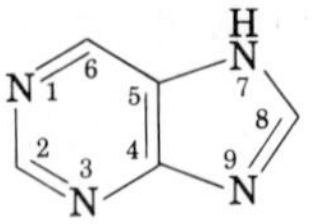

14

7*H*-Purine

is only slightly soluble in hydrocarbons, but its spectrum in methylcyclohexane has been recorded,[52] and has been found similar to that shown for the (more liposoluble) 9-methyl-9H-purine (see Fig. 10). The latter has, in the 310-nm region,[43] a low-intensity peak which is an n → π* transition similar to that discussed for pyrimidine. As with pyrimidine, this peak of 9-methyl-9*H*-purine almost disappears when the solvent is changed to one that forms hydrogen bonds [see Fig. 10 (A)], or when the compound is converted into the cation. Finally, no such peak is seen when intermolecular N—H···N hydrogen-bonding is possible, as in purine itself.

Early attempts[19,20] to equate the ultraviolet spectra of purines with those of the corresponding 4,5-diaminopyrimidines met with only partial success. At least, it established that, in both series, the curves of corresponding species have the same general outlines, in that an electron-releasing substituent (*e.g.*, $-NH_2$ and $-O^{\ominus}$) on C-2 increases the wavelength of the peak of maximum absorption (but sometimes diminishes its intensity), and that a similar substituent on C-6 decreases this wavelength, but increases its intensity. However 4,5-diaminopyrimidine has at 246 nm a peak that disappears when the neutral species is converted into the cation; no such peak occurs in the spectrum of purine. This phenomenon of the extra peak for 4,5-diaminopyrimidines is quite widespread, and may be due to conjugation of an amino group with the pyrimidine nucleus.[43] It is now realized that any account of the spectra of purines cannot exclude the contribution of the imidazole ring, which, in purines, permits a longer pathway of conjugation than is possible in pyrimidines.

Substituted purines have ultraviolet spectra consisting, usually, of two broad bands, *e.g.*, those at 188 and 263 nm in the spectra of unsubstituted purine (see Table VII). Mason[43] considered that the 263-nm peak was due to a longitudinal polarization (*i.e.*, between C-2 and C-8) of the purine nucleus; this judgment rested on (*a*) the effect of substituents, and (*b*) comparisons of the spectra with those of indene and indole. Similarly, the 188-nm peak was ascribed to a transverse polariza-

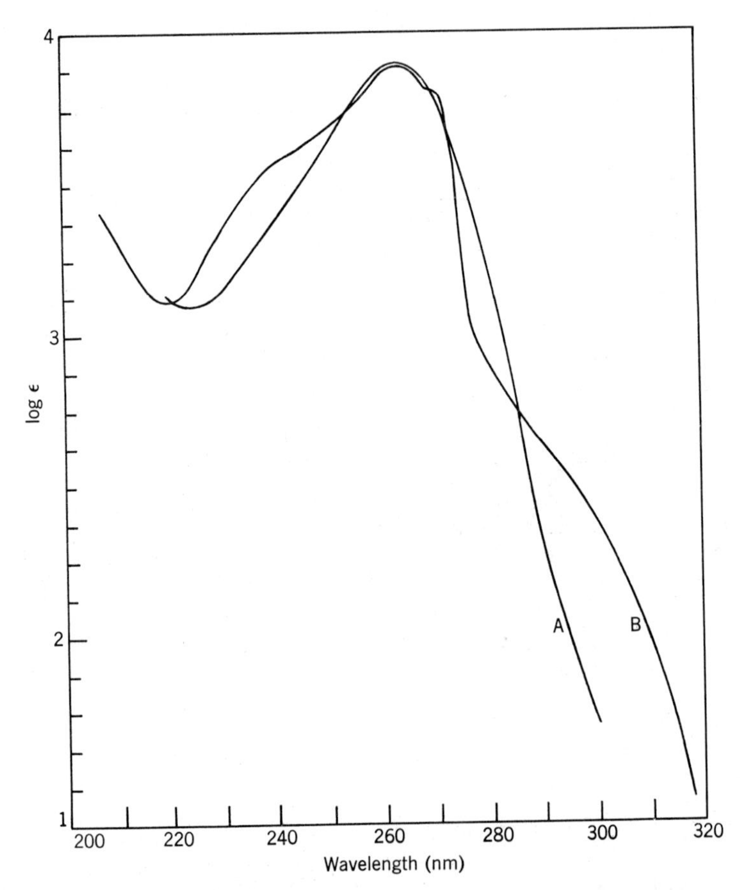

Fig. 10. Spectra[43] of 9-Methyl-9*H*-purine (Neutral Speci[illegible] in Water (A) and Cyclohexane (B).

tion of the molecule. Both Mason[43] and Clark and Tinoco regarded the 263-nm peak as being the first π → π* transition of purine and, hence, analogous to the 255-nm pea[illegible] for benzene (see Section I (1)). The latter authors dre[illegible] attention to a slight inflection at ~240 nm in the spect[illegible] of purine in methylcyclohexane, and suggested that it ma[illegible] be analogous to the peaks of benzene centered about 204[illegible] accordingly, they compared the 188-nm peak of purine to[illegible] 190-nm peak of benzene.

All of these assignments have been confirmed and extended by examination of magnetic circular dichroism.[52a][illegible]

In substituted purines, this technique separates a band (usually at 260–280 nm, which corresponds to the 263-nm peak in benzene) from a band of opposite magnetic sign (usually at 220-260 nm, which corresponds to the 204-nm peak in benzene). In ultraviolet spectroscopy, these two bands often overlap, but are partly resolved in guanine (245 and 274 nm) and in xanthine (225 and 266 nm). Magnetic dichroism does not resolve these bands so clearly in the pyrimidine series, but they occur at about the same wavelengths as in the purines, and have the same origins. (The peaks for pyrimidine nucleosides are well resolved.

The effect of substituents on the ultraviolet absorption of purine may be seen in Tables VII to IX. In general, substituents on C-2 shift the absorption peak to longer wavelengths (*i.e.*, are more bathochromic) than those on C-8, and those on C-8 more than those on C-6. The intensity of the absorption follows quite a different order: substituents on C-8 (followed closely by their 6-isomers) increase the intensity far more than do those on C-2. (For a tabulation and a discussion of these shifts, see Ref. 43.) This effect was attributed to enhancement, by electron-releasing substituents, of the following polarization of the purine nucleus[43]: 8 +→- 2. The order of bathochromy of common substituents changes from one position to another, and among various ionic species in any one position; but the mercapto derivatives always have the greatest bathochromic effect, and, in general, the order is very similar to that exerted by the same substituents on the 255-nm band of benzene. The magnitudes of the bathochromic shift arising from substituents on C-2 are somewhat greater (and on C-8 somewhat less) than the shifts caused by the same substituents on the benzene ring. Substituents on C-6 appear, in many cases, to fuse the two principal bands of purine into a single band; the effect (on the spectrum of purine) of substituents in this position is more difficult to predict than elsewhere. It is interesting that all methylthio derivatives absorb at wavelengths much longer than those for the corresponding methoxy derivatives (in fact, at about the same wavelengths as for the corresponding amines), and this strongly suggests conjugation of the sulfur atom with the nucleus.

The effects of more than one substituent are not additive in the way that obtains for pyrimidine derivatives (see Section I (1)). Instead, it would appear that substituents can conjugate with one another somewhat more freely than in the pyrimidine series.[43]

When purine, or a purine bearing an unconjugated (or only slightly conjugated) substitutent is converted into the anion, the long-wavelength peak in the spectrum almost invariably moves to a longer wavelength, often by as much as 8 nm. On the other hand, conversion into the cation usually has little effect on the spectrum of the neutral species; in the majority of examples, the long-wavelength peak is moved 2 or 3 nm to a shorter wavelength; in other cases, it remains stationary, or moves slightly in the opposite direction. The ionization of a purine often causes an inflection to appear where the neutral species has none, or a new peak may arise somewhere in the middle of the spectrum where the neutral species has only an inflection.

Figure 11 presents a typical example of the effects of ionization on spectra of members of the purine series. It is very difficult to use such small changes diagnostically. The bathochromic shift upon formation of the cation of 2-aminopurine is larger (by 9 nm) than usual, and it

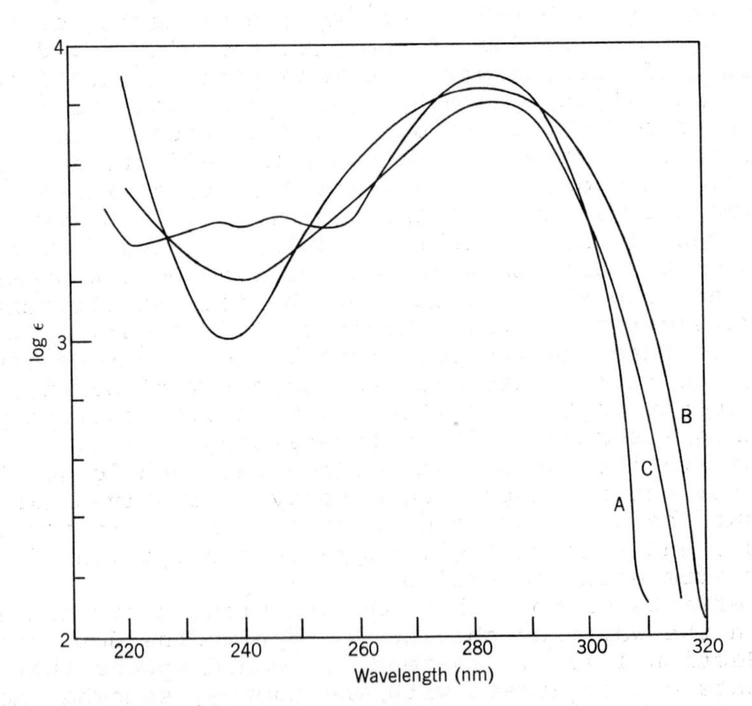

Fig. 11. Spectra[43] of 2-Methoxypurine in Water: the Neutral Species (A) at pH 6, the Anion (B) at pH 11, and the Cation (C) at pH 0.

TABLE VII

Ultraviolet Spectra of Purines Lacking Hydroxyl (or Thiol) Substituents[a]

Derivative of purine	pK_a	Species	λ_{max} (nm)	log ε_{max}	pH	References
Unsubstituted	8.93	−	219, 271	3.92, 3.88	11.0	43
	2.39	0	188, 263	4.37, 3.90	5.8	43,49
		+	201, 260	4.36, 3.79	0.4	43,49
Purine 2-amino-	9.93	−	<u>276</u>, 303	<u>3.61</u>, 3.76	12.0	43
	3.80	0	<u>236</u>, 305	<u>3.70</u>, 3.78	7.0	43
	-0.28	+	<u>237</u>, 314	<u>3.62</u>, 3.60	1.8	43
		++	<u>235</u>, 325	<u>3.81</u>, 3.62	-3.5	43
6,8-bis(trifluoromethyl)-	5.02	0	328	3.86	2.1	61
	~0.3	+	334	3.53	-0.9	61
8-(methylsulfonyl)-	5.61	−	223, 273, 318	4.43, 3.68, 3.98	8.0	64
	2.08	0	223, 281, 322	4.45, 3.68, 3.90	3.9	64
		+	225, 326	4.60, 3.74	0	64
8-(methylthio)-	8.48	−	231, <u>277</u>, 317	4.31, <u>3.60</u>, 4.15	11.0	64

TABLE VII (continued)

Ultraviolet Spectra of Purines Lacking Hydroxyl (or Thiol) Substituents[a]

Derivative of purine	pK_a	Species	λ_{max} (nm)	log ε_{max}	pH	References
	4.40	0	224, 319	4.25, 4.12	6.5	64
		+	222, 243, 288, 326	4.25, 4.26, 3.87, 3.98	2.0	64
8-phenyl-	9.20	—	239, 330	4.31, 4.28	11.4	43
	3.98	0	238, 329	4.22, 4.28	6.5	43
		+	257, 332	4.38, 4.08	1.0	43
6-(trifluoromethyl)-	8.87	—	283, 323	3.68, 3.70	12.0	61
	1.85	0	323	3.78	5.1	61
		+	323	3.78	0.2	61
8-(trifluoromethyl)-	6.14	—	220, 270, 308	4.37, 3.58, 3.84	9.0	64
	2.59	0	217, 281, 312	4.40, 3.48, 3.80	4.4	64
		+	221, <u>241</u>, 318	4.59, <u>3.65</u>, 3.64	0	64
6-amino- (Adenine)	9.83	—	267	4.08	12.0	43
	4.25	0	260	4.13	7.0	43
		+	262	4.12	2.1	43

1-methyl-	11.0	−	239, 270	? , 4.16	13.0	57
	7.2	+	228, 259	? , 4.07	4.0	57
3-methyl-[b]	5.7	0	244, 273	? , 4.11	13.0[c]	57, 58
		+	235, 274	? , 4.20	2.0	57
7-methyl-	3.6	0	270	4.02	11.0	59
		+	272	4.18	1.0	59
9-methyl-	3.25	0	262	4.08	11.0	59
		+	261	4.16	1.0	59
N-oxide	9.0	0	231, 263	4.62, 3.91	7.0	60
	2.6	+	259	4.06	1.0	60
8-amino-	9.36	−	230, 290	3.98, 4.07	12.0	43
	4.68	0	241, 283	3.51, 4.16	7.1	43
		+	288	4.20	2.4	43
2,8-bis(me-thylthio)-	7.33	−	220, 248, 318	4.22, 4.36, 4.21	10.0	64
	2.19	0	224, 248, 319	4.17, 4.24, 4.20	5.0	64
		+	240, 264, 326	4.04, 4.38, 4.03	-0.2	64
2-chloro-	8.21	−	213, 278	4.33, 3.89	10.5	65
	0.69	0	207, 273	4.31, 3.87	5.0	65
		+	221, 271	3.52, 3.91	-1.4	65
9-methyl-	0.65	0	272	3.89	2.0	65
		+	271, 278	3.94, 3.72	-1.6	65
6-chloro-	7.88	−	210, 274	4.33, 3.92	10.0	65
	0.45	0	245, 265	3.64, 3.95	4.0	65,53
		+	233, 262, 269	3.65, 3.97, 3.85	-1.9	65

TABLE VII (continued)

Ultraviolet Spectra of Purines Lacking Hydroxyl (or Thiol) Substituents[a]

Derivative of purine	pK_a	Species	λ_{max} (nm)	log ε_{max}	pH	References
9-methyl-	0.20	0	266	4.05	2.4	65
		+	244, 263, 270	3.72, 4.01, 3.88	-2.0	65
8-chloro-	6.02	—	275	4.02	9.0	65
	1.77	0	216, 269	4.15, 3.98	4.0	65
		+	269	3.97	-0.3	65
9-methyl-	2.00	0	267	4.03	4.2	65
		+	270	3.90	-0.2	65
6-cyano-	6.88	—	292	3.88	9.1	61
	~0.3	0	288	3.96	4.9	61
		+	247, 284	3.58, 3.96	-1.7	61
2,6-diamino-[d]	10.77	0	246, 279	3.85, 3.95	7.5	43
	5.09	+	241, 282	3.96, 4.02	3.0	43
	<1.0	++[e]	247, 296	4.01, 3.88	-1.2	43
N-oxide	12.0		228, 295	4.43, 3.90	13.0	60
	9.7	?	230, 290	4.50, 3.85	5.5	60
	3.7		248, 290	3.93, 3.94	1.0	60
	1.0					
8-(trifluoromethyl)-	7.55	—	245, 288	3.42, 3.87	14.0	61
	3.68	0	284	4.03	5.9	61

2,6-dichloro-	7.06	—	219, 280	4.33, 3.93	9.5	65
	-1.16	0	211, 249, 275	4.35, 3.55, 3.95	4.0	65
		+	235, 242 273	3.63, 3.49 3.98	-2.3	65
2-(dimethyl-amino)-	10.22	—	232, 327	4.39, 3.67	12.7	43
	4.02	0	223, 248 332	4.41, 4.02 3.71	7.0	43
		+	228, 248 340	4.52, 3.97 3.48	1.7	43
6-(dimethyl-amino)-	10.5	—	221, 281	4.21, 4.25	13.0	43
	3.87	0	275	4.25	7.0	43
		+	276	4.19	1.7	43
3-methyl-	5.8	+	290	?	1.0	58
8-(dimethyl-amino)-	9.73	—	230, 306	3.91, 4.22	12.0	43
	4.80	0	250, 296	3.47, 4.27	7.3	43
		+	230, 305	4.04, 4.29	2.7	43
2-ethoxy-	9.47	—	211, 283, 290	4.25, 3.89, 3.85	12.0	68
	2.46	0	246, 284, 290	3.37, 3.88 3.83	6.0	68
		+	285, 294	3.80, 3.72	-0.4	68
9-methyl-	2.53	0	209, 237, 282, 294	4.32, 3.26, 3.91, 3.67	7.0	68
		+	209, 265, 285, 295	4.34, 3.58 3.78, 3.67	0	68
6-ethoxy-	9.52	—	211, 262	4.30, 4.01	12.0	68
	2.13	0	248, 253, 260	3.99, 4.02, 3.92	6.0	68

TABLE VII (continued)

Ultraviolet Spectra of Purines Lacking Hydroxyl (or Thiol) Substituents[a]

Derivative of purine	pK_a	Species	λ_{max} (nm)	log ε_{max}	pH	References
9-methyl-	1.90	+	249, 255, 261	3.98, 4.03, 3.96	-0.4	68
		0	249, 254, 262	4.03, 4.04, 3.83	5.0	68
		+	251, 254, 263	4.01, 4.01, 3.83	-0.4	68
8-ethoxy-9-methyl-	3.45	0	272	3.97	6.0	68
		+	212, 273	4.42, 3.89	1.0	68
1-ethyl-	5.08	0	219, 274	4.02, 3.79	7.0	56
		+	268	3.82	1.0	56
7-ethyl-	2.67	0	267	3.75	7.0	56
		+	258	3.66	1.0	56
9-ethyl-	2.67	0	264	3.86	7.0	56
		+	263	3.73	1.0	56
6-(furfuryl-amino)-(Kinetin)	~10	—	274	4.25	14.0	69
	~4	0	267	4.27	6.4	69
		+	274	4.23	0	69

2-hydrazino-[f]						59,67
6-iodo-[g]						70
2-methoxy-	9.2	—	283	3.88	11.4	43
	2.44	0	246, 283	3.41, 3.91	6.0	43
		+	284	3.83	0	43
6-methoxy-	9.16	—	261	3.99	11.3	43
	2.21	0	252	3.99	5.6	43
		+	254	4.01	0.2	43
N-methyl-[h]						59
8-methoxy	7.73	—	279	3.98	10.0	71
	3.14	0	271	4.03	5.4	71
		+	271	4.05	1.0	71
6-(methoxycar-bonylamino)-[i]	9.68	0	274	4.13	7.1	61
	2.27	+	275	4.19	o.1	61
6-methyl-[j]	9.02	—	271	3.93	11.5	43
	2.6	0	261	3.92	5.9	43
		+	265	3.88	0	43
9-methyl-	3.2	0	262	3.89	6.0	73
		+	265	3.81	0	73
7-methyl-[k]	2.29	0	267	3.91	9.2	53
		+	258	3.83	0.2	53
8-methyl-	9.37	—	274	3.92	12.0	43
	2.85	0	266	4.01	5.9	43
		+	264	3.92	0	43
9-methyl-	2.36	0	264	3.90	8.5	53,55
		+	263	3.77	0.6	53
2-(methyl-amino)-	10.32	—	226, 272, 316	4.37, 3.48, 3.72	12.5	74

TABLE VII (continued)

Ultraviolet Spectra of Purines Lacking Hydroxyl (or Thiol) Substituents[a]

Derivative of purine	pK_a	Species	λ_{max} (nm)	log ε_{max}	pH	References
	4.01	0	219, 240, 319	4.42, 3.88, 3.74	7.0	74
		+	223, <u>244</u>, 327	4.55, <u>3.81</u>, 3.57	1.8	74
6-(methyl-amino)-	9.99	–	273	4.20	12.0	43
	4.18	0	266	4.21	7.1	43
		+	267	4.18	2.0	43
9-methyl-	4.12	0	209, 267	4.27, 4.20	6.5	74
		+	209, 264, <u>271</u>	4.25, 4.23, <u>4.17</u>	1.9	74
8-(methyl-amino)-	9.56	–	<u>230</u>, 298	<u>4.00</u>, 4.16	12.0	43
	4.78	0	<u>245</u>, 290	<u>3.50</u>, 4.22	7.2	43
		+	<u>230</u>, 296	<u>3.90</u>, 4.24	2.7	43
6-*N*-methylcar-boxamide[l]	8.9	–	292	3.86	12.0	61
	~1.0	(0)	287	4.00	7.7	61
		(+)	276	4.03	0.1	61

8-(methylsulfonyl)-	4.87	—	279	4.13	7.5	64
	0.42	0	215, 271	4.39, 4.01	2.7	64
2-(methylthio)-	8.91	—	240, 300-2	4.28, 3.79	11.0	43
	1.91	0	232, 250, 305	4.22, 3.93, 3.78	5.9	43
		+	241, 250, 314	4.13, 4.09, 3.64	0	43
6-(methylthio)-	8.74	—	222, 290	4.27, 4.31	11.2	43
	1.63	0	255, 290	3.66, 4.35	5.8	43
		+	222, 313	4.08, 4.41	-3.5	43
8-(methylthio)-	7.67	—	220, 296	4.23, 4.27	9.9	43
	2.95	0	246, 290	3.59, 4.30	5.1	43
		+	232, 305	4.04, 4.32	0	43
9-benzyl-	2.84	0	215, 254, 289	4.27, 3.58, 4.23	6.0	73
		+	237, 302	4.14, 4.14	0	73
9-methyl-	2.98	0	211, 255, 289	4.10, 3.57, 4.24	7.0	71
		+	238, 301	4.25, 4.23	0.3	71
8-phenyl-	8.09	—	233, 304	4.24, 4.42	10.3	43
	2.68	0	231, 298	4.06, 4.42	5.4	43
		+	237, 304	4.13, 4.41	0	43
2-piperidino-	10.24	—	234, 317	4.36, 3.61	12.5	68
	3.97	0	225, 253, 332	4.39, 4.10, 3.65	7.0	68
		+	231, 253, 343	4.52, 4.14, 3.46	1.7	68

TABLE VII (continued)

Ultraviolet Spectra of Purines Lacking Hydroxyl (or Thiol) Substituents[a]

Derivative of purine	pK_a	Species	λ_{max} (nm)	log ε_{max}	pH	References
6-piperidino-	10.04	−	221, 284, 292	4.22, 4.30, 4.22	12.5	68
	4.33	0	215, 281	4.19, 4.29	7.0	68
		+	208, 280	4.14, 4.25	2.0	68
9-methyl-	4.12	0	216, 282	4.20, 4.30	6.5	68
	-1.54	+	213, 275	4.20, 4.27	1.3	68
		++	211, 280	4.17, 4.18	-3.9	68
6-succinamido-[m]						75
2,6,8-triamino-	10.79	−	226, 261, 295	4.31, 3.69 4.08	13.0	43
	6.23	0	249, 293	3.80, 4.08	8.5	43
	2.41	+	221, 250, 299	4.31, 3.69, 4.24	4.3	43
		++	248, 305	4.12, 4.11	0.3	43
2,6,8-trichloro-	3.96	−	220, 285	4.34, 4.08	7.0	65
	-3.1	0	213, 248, 280, 288	4.43, 3.65, 4.11, 4.00	1.0	65
		+	210, 245, 279, 287	4.45, 3.54, 4.11, 4.09	-5.1	65

6-(trifluoro-methyl)-	7.35	—	275	3.87	10.3	61
	<0	0	270	3.91	3.2	61
8-(trifluoro-methyl)-	5.12	—	271	3.94	8.0	61
	1.0	0	264	3.89	3.0	61
2,6,8-trimethyl-[j]						72
6-ureido-[n]	9.95	0	267	4.17	8.2	61
	2.35					

[a]In water, at 20–25°. Inflections are underlined. [b]For u.v. spectra of 2- and 8-methyladenine (without differentiation of species), see Refs. 62 and 63, respectively. [c]There is no change in the spectrum between pH 7 and 13, and the compound fails to migrate as an anion on applying paper electrophoresis[38] at pH 12. [d]For spectra (undifferentiated) of 6,8-diaminopurine and many other 6,8-disubstituted purines, see Ref. 66. [e]At least partly ++ (mixed with +). [f]For spectra (undifferentiated) of 2- and 6-hydrazinopurine and many of their derivatives, see Refs. 59 and 67. [g]For spectrum (undifferentiated) see Ref. 70. [h]For spectra of *N*-methyl derivatives, see Ref. 59 (species not differentiated). [i]Ethoxy analog is given in the same reference. [j]For u.v. spectra of 2-methyl-, 2,6-, 2,8-, and 6,8-dimethyl-, and 2,6,8-trimethyl-purine (species not differentiated), see Ref. 72. [k]For spectra of 3-methylpurine (without differentiation of ionic species), see Ref. 54. [l]For the dimethylamide, see the same paper.[61] [m]For spectra (undifferentiated), see Ref. 75. [n]The same reference also contains u.v. spectra at pH values that are too close (numerically) to the pK_a values for the spectra to have significance.

is only rarely that lower-wavelength bands differ so much after a change in species, as in Fig. 13.

2. The Aminopurines

The spectra of all three monoaminopurines resemble those of the same ionic species of the corresponding dimethylamino derivatives, although the spectra of the latter are naturally displaced to a longer wavelength. An example is shown in Fig. 12, and Table VII lists all of the relevant maxima; this type of evidence suggests that the aminopurines are true primary amines, and crystalline adenine hydrochloride has been shown, by X-ray crystallography,[76] to have this structure. The magnitude of the increase in basic strength caused by attaching a primary amino group to the purine molecule confirms the primary amine character of the products (see Chapter 1). However, very little work has been done towards completing the evidence by comparison with authentic imino compounds, as has been done so well for the pyrimidines [see Section I (2)]. This situation can be traced to a reluctance to synthesize the large number of N-methyl derivatives that would be needed for a thorough comparison. Because of the anion-forming NH group in purine, no less than two hydrogen atoms must be replaced by *N*-methyl groups, to give analogs of such pyrimidines as 3 ; the large number of isomers of these dihydroiminopurines necessarily arises from the number of nitrogen atoms in purine.

Aminopurines form the cation by addition of a proton to a ring-nitrogen atom. For very few heterocycles is the proton added to a primary amino group, but when this does occur, the bathochromic effect of the amino group in the neutral species is completely lost. This effect is found for aniline, the cation of which gives a spectrum little different from that of benzene. In the one example where the spectrum of the cation of an aminopurine (namely, adenine) resembles that of the neutral species of purine, X-ray crystallography[76] has shown that the cationic proton is firmly attached to N-1, but is hydrogen-bonded to the primary amino group.

The unusual properties of 1- and 3-methyladenine indicate that a more thorough exploration of (nuclear) N-methylated aminopurines would be rewarding. In 3-methyladenine, the main absorption peak of adenine is split, and the longer-wavelength portion is shifted bathochromically (see Table VII). The complete lack of acidic properties, and the similarity in shape of all its spectra to those of 6-(dimethylamino)-3-methyl-3*H*-

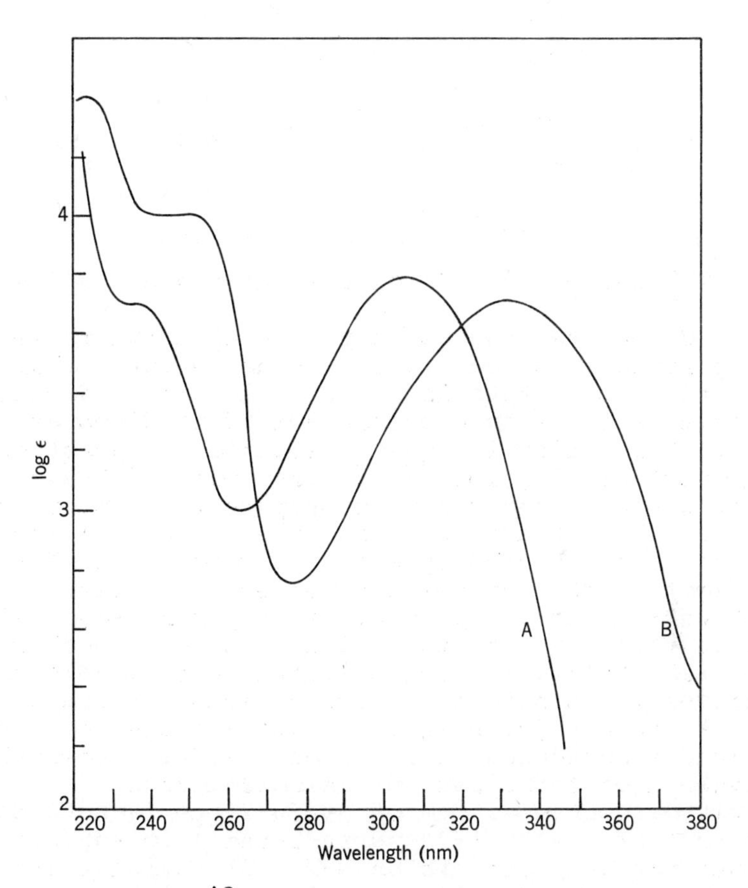

Fig. 12 Spectra[43] of Both 2-Aminopurine (A) and 2-(Dimethylamino)purine (B), as Neutral Species in Water at pH 7.0.

purine, have led to acceptance[58] of structure 15 for this compound. 1-Methyladenine shows a similar split of the main adenine peak, but here the cation is not bathochromic compared to that of adenine. This isomer forms an anion. 1-Methyladenine is about 1000 times as strong a base as adenine. It is generally considered[57,58] that these properties all indicate that 1-methyladenine has structure 16.

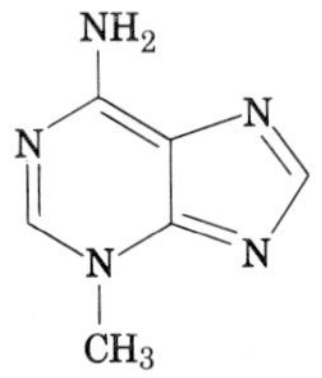

15 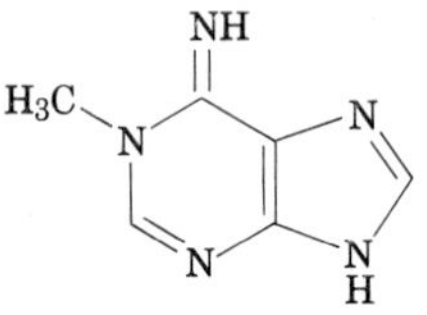 **16**

3. The Hydroxy and Thiol Derivatives of Purines

Infrared spectra conclusively show that the three monohydroxypurines exist mainly in the amide form (*e.g.*, 17) both in the solid state and in solution in chloroform.[71] Ultraviolet spectra have been examined to discover where the equilibrium lies in aqueous solution, a question of great importance in biological work. To reach a conclusion, it would be necessary to prepare the compounds in which the mobile hydrogen atom, in each possible position, has been replaced by a methyl group; therefore, eight dimethyl derivatives are needed. However, it has been shown[43,71] that the spectrum of purin-2-one (Fig. 13, Table VIII) is not at all like that of 2-methoxypurine (Table VII), so that only the four *N*,*N*-dimethyl derivatives would need to be synthesized for comparison. This work has not yet been done, but infrared spectrometry (under nonaqueous conditions) clearly shows that the oxygen atom forms part of an amide group.

The ultraviolet spectrum of purin-2-one is sufficiently unlike that of 9-methyl-9*H*-purin-2-one[71] to suggest that the other mobile proton is on N-7. Ultraviolet spectra have been examined for 1-methyl-1*H*-purin-2-one, 3-methyl-3*H*-purin-2-one, and 3,7-dimethyl-3*H*-purin-2-one[72a], but only at pH 8, where a mixture of ionic species probably exists; hence, no information can be educed from these figures. Here is a promising topic for further research.

Purin-6-one(hypoxanthine) presents a more difficult case, because it, 6-methoxypurine, and 1,7-dimethyl-1*H*-purin-6-one have closely similar ultraviolet spectra. The infrared spectra strongly indicate[71] that the mobile hydrogen atom derived from the 6-hydroxyl group of hypoxanthine is linked mainly to N-1. The ultraviolet spectra of all *N*-methyl derivatives and of one *N*,*N*-dimethyl derivative of hypoxanthine are assembled in

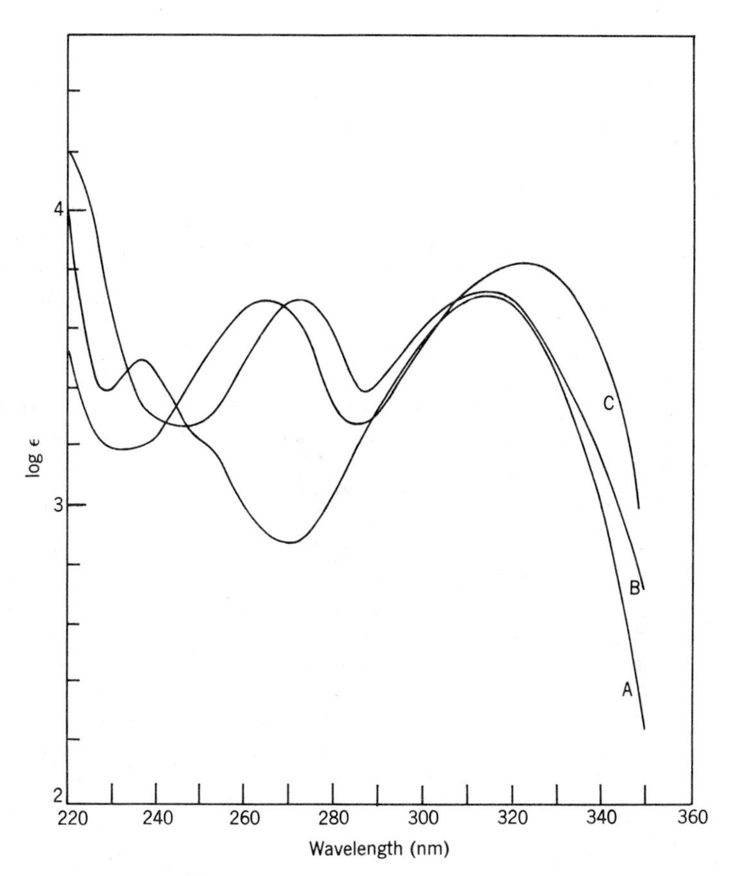

Fig. 13 Spectra[43] of Purin-2-one in Water: the Neutral Species (A) at pH 6.0, the Anion (B) at pH 10.2, and the Cation (C) at H_o -0.75.

Table VIII, and that of 6-methoxypurine is given in Table VII. The ionization constants and ultraviolet spectra for all are very similar. The pk_a values and spectra of 3-methyl-3*H*-purin-6-one (and the pk_a values of the 9-methyl isomer) do not seem related in the normal way to those of hypoxanthine, but the values for the N-1 and N-7 isomers are compatible, and so are those of 1,7-dimethyl-1*H*-purin-6-one. This evidence is inconclusive, but, as the differences involved are likely to be small, the synthesis of further dimethyl derivatives may not seem an attractive topic to pursue.

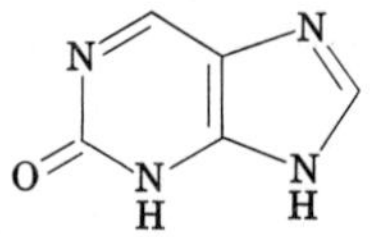

17

The ultraviolet evidence suggests that 18 is the main structure for purin-8-one in water. The spectra of 7-methylpurin-8-one, 9-methyl-9*H*-purin-8-one, and 7,9-dimethyl-9*H*-purin-8-one are closer to that of purin-8-one (see Table VIII) than is that of 8-methoxypurine (see Table VII).

So far, this discussion has concerned the neutral species of hydroxypurines. The spectrum of the anion of purin-6-one closely resembles that of the neutral species of 6-aminopurine (λ_{max} 258 and 260 nm, respectively) and this indicates that, on basification, the first proton is lost from N-1, thus following the pattern found in the 2- and 4-pyrimidinones (see Section I, 3). In this region of the molecule, unusual opportunities for hydrogen-bonding in the anion (19) must blur any possibility of differentiation of tautomers by spectral means. The spectra of the anions of purin-2-one and purin-8-one do no resemble those of the neutral species of the corresponding aminopurines. This departure from the usual Jones rule[24] indicates that the loss of the first proton (upon basification) occurs from the imidazole ring.[43] Moreover, the spectrum of the anion of purin-8-one is more like that of the anion of its 7-methyl than of its 9-methyl derivative; this suggests that the first proton is lost from N-9.[71]

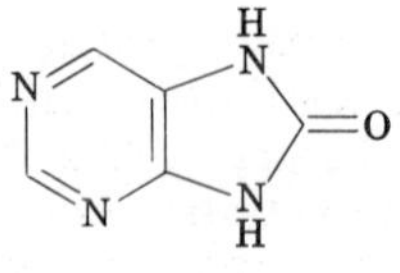

18 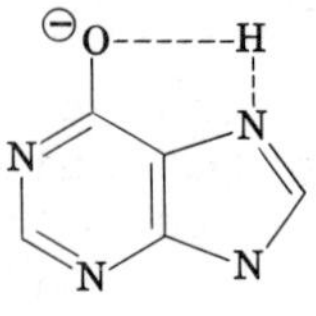 **19**

TABLE VIII

Ultraviolet Spectra of Purines Having Hydroxyl or Thiol Substituents[a]

Derivative of purine	pk_a	Species	λ_{max} (nm)	log ε_{max}	pH	References
Aminodihydroxy-[b]						19,20
2-Amino-6-hydroxy- (Guanine)	12.6	--	273	3.98	14.0	77
	9.32	–	243, 273	3.93, 4.00	11.0	77
	3.0	0	245, 274	4.04, 3.92	6.0	77
		+	248, 270	4.06, 3.86	0	77
1,7-dimethyl-	3.4	0	250, 283	3.75, 3.87	6.0	77
		+	252, 273	4.01, 3.83	0	77
1,9-dimethyl-	3.2	0	255, 269	4.09, 4.00	6.0	77
		+	254, 279	4.05, 3.88	0	77
1-methyl-	10.5	–	262, 277	3.90, 3.91	13.0	77
	3.1	0	249, 273	4.01, 3.91	6.0	77
		+	250, 270	4.03, 3.85	0	77
7-methyl-	10.0	–	240, 280	3.81, 3.89	13.0	77
	3.5	0	248, 283	3.79, 3.89	6.0	77
		+	250, 270	4.04, 3.85	0	77
9-methyl-	9.8	–	258, 268	4.01, 4.05	13.0	77
	2.8	0	252, 270	4.10, 3.97	6.0	77
		+	251, 276	4.08, 3.88	0	77

TABLE VIII (continued)

Ultraviolet Spectra of Purines Having Hydroxyl or Thiol Substituents[a]

Derivative of purine	pk_a	Species	λ_{max} (nm)	log ε_{max}	pH	References
6-Amino-2-hydroxy-[c]	9.0	—	235, 284	3.72, 4.09	11.1	43
	4.5	0	240, 286	3.89, 3.90	7.0	43
		+	230, 284	3.69, 4.03	2.0	43
9-methyl-[d]						78
1-oxide	11.48	--	225, 260, 300	4.32, 3.68, 3.91	13.1	79
	6.41	—	228, 251, 295	4.34, 3.72, 3.86	9.0	79
	3.64	0	243, 288	3.72, 3.88	5.0	79
		+	283	4.00	1.7	79
Aminomercapto-[e]						
8-Carboxy- (Purine-8-carboxylic acid)[f]	9.37 2.91	—	275	4.09	6.5	82
Diaminohydroxypurines[b]						19,20
2,6-Dihydroxy- (Xanthine)	11.94	--	242, 283	3.65, 3.97	14.0	84

	7.7	—	241, 276	3.95, 3.97	10.0	84
		0	225, 266	3.49, 4.03	5.0	84
1,3-dimethyl- (Theophylline)	8.68	—	274	4.09	11.0	84
		0	270	4.02	6.0	84
1,7-dimethyl-	8.65	—	233, 288	3.70, 3.94	11.1	84
		0	268	4.02	6.0	84
1,9-dimethyl-	5.99	—	248, 276	3.94, 3.95	9.0	84
		0	238, 263	3.82, 3.98	3.0	84
3,7-dimethyl- (Theobromine)	10.00	—	234, 273	3.85, 4.01	13.0	84
		0	271	4.01	7.0	84
3,9-dimethyl-	10.14	—	240, 269	3.82, 4.01	13.0	84
		0	235, 268	3.91, 3.99	7.0	84
1-methyl-	12.2	--	242, 283	3.69, 4.00	14.0	84
	7.9	—	242, 276	3.93, 3.98	10.0	84
		0	223, 267	3.52, 4.04	5.0	84
3-methyl-	11.92	--	232, 274	3.77, 4.12	14.0	84
	8.45	—	274	4.10	10.0	84
		0	270	4.05	6.0	84
7-methyl-	>13	—	231, 287	3.67, 3.93	11.0	84
	8.42	0	268	4.01	6.0	84
9-methyl-	>13	—	245, 277	3.98, 3.97	10.0	84
	6.12	0	234, 265	3.92, 4.01	3.0	84
1,3,7-trimethyl- (Caffeine)	—	0	272	4.02	6.0	84
1,3,9-trimethyl-	—	0	237, 268	3.99, 4.00	6.0	84

TABLE VIII (continued)

Ultraviolet Spectra of Purines Having Hydroxyl or Thiol Substituents[a]

Derivative of purine	pK_a	Species	λ_{max} (nm)	log ε_{max}	pH	References
1-oxide	12.4	---	226, 287	4.40, 3.90	14.0	79
	9.94	--	225, 283	4.30, 3.85	11.2	79
	6.54	-	242, 276	3.83, 3.88	8.5	79
	6.85	0	267	3.98	3.0	79
		+	261	3.93	-1.0	79
2,8-Dihydroxy-	>12	–	262, 306	3.98, 3.87	10.0	43
	7.45	0	230, 310	3.90, 3.70	5.1	43
1-methyl-	~12	–	222, 264, 310	4.28, 4.03, 4.02	9.9	85
	7.94	0	215, 313	4.44, 3.89	5.8	85
3-methyl-	8.39	–	223, 325	4.18, 4.18	10.5	85
		0	220, 251, 310	4.19, 3.71, 4.04	6.0	85
9-methyl-	~11.5	–	224, 314, 318	4.46, 3.71, 3.72	10.1	85
	8.74	0	218, 307	4.49, 3.85	6.5	85
6,8-Dihydroxy-	9.87	--	271	4.14	12.0	43
	7.65	–	265	4.04	8.7	43
		0	257, 280	4.08, 3.76	5.1	43
1-methyl-	11.83	–	274	4.11	10.2	86
	8.52	0	259	4.10	4.7	86

9-methyl-	11.74	—	264	4.12	9.9	86
	8.31	0	259	4.12	4.7	86
Dimercaptopur-ines[g]						66,87
6-Formyl- (Pur-ine-6-alde-hyde)	8.8	—	277	3.80	11.9	88
	2.4	0	267	3.93	5.5	88
		+	264	3.88	1.6	88
2-Hydroxy-	11.90	--	219, 265, 312	4.29, 3.60 3.83	13.0	43
	8.43	—	271, 313	3.68, 3.68	10.2	43
	1.69	0	238, 315	3.46, 3.69	6.1	43
		+	264, 322	3.67, 3.81	-0.8	43
9-methyl-[h]	9.19	—	240, 308	3.65, 3.91	11.2	71
	<1.5	0	218, 246, 259, 316	4.48, 3.41, 3.38, 3.68	6.5	71
6-Hydroxy- (Hy-poxanthine)	12.0	--	262	4.04	13.0	43
	8.94	—	258	4.05	10.4	43
	1.98	0	249	4.02	5.2	43
		+	248	4.02	-0.8	43
1,7-dimethyl-	2.16	0	255	3.94	7.0	71
		+	250	4.00	0	71
1-methyl-	8.85	—	260	3.99	11.1	89
		0	251	3.97	5.2	89
		+	249	3.97	0	89
3-methyl-	8.34	—	265	4.04	11.1	89
	2.61	0	264	4.15	5.1	89
		+	253	4.04	0	89

TABLE VIII (continued)

Ultraviolet Spectra of Purines Having Hydroxyl or Thiol Substituents[a]

Derivative of purine	pk_a	Species	λ_{max} (nm)	log ε_{max}	pH	References
7-methyl-	9.00	−	262	4.03	11.1	89
	2.12	0	256	3.98	5.1	89
		+	250	4.01	0	89
9-methyl-	9.32	−	254	4.11	12.0	71
	1.86	0	250	4.08	5.5	71
		+	250	3.99	-0.3	71
1-oxide	10.10	--	229, 264	4.64, 3.81	13.0	90
	5.65	−	228, 259	4.56, 3.75	8.0	90
		0	250	3.92	3.0	90
8-Hydroxy-	>12	−	285	4.11	10.1	43
	8.24	0	235, 277	3.51, 4.05	5.4	43
	2.58	+	280	4.02	0	43
7,9 dimethyl-	2.8	0	241, 279	3.56, 4.01	7.0	71
		+	285	4.03	0.3	71
7-methyl-	8.20	−	286	4.06	12.0	71
	2.69	0	240, 278	3.60, 3.97	5.5	71
		+	284	3.90	0.3	71
9-methyl-	9.05	−	<u>257</u>, 289	<u>3.63</u>, 3.97	12.0	71
	2.80	0	<u>235</u>, 277	<u>3.45</u>, 4.01	5.5	71
		+	<u>281</u>	<u>4.07</u>	0.3	71

6-(Hydroxyamidino)-	9.4	—	326	3.93	11.7	61
	~2.0	0	272, 305	3.90, 3.83	7.3	61
		+	274, 285	3.85, 3.83	0.1	61
6-(Hydroxylamino)-	9.83	0	268	4.07	6.7	61
	3.80	+	271	4.12	1.2	61
2-Hydroxy-8-mercapto- ^{i}						
1-methyl-	6.61	—	218, 295, 341	4.15, 3.87, 4.27	8.7	85
		0	213, 255, 344	4.00, 4.19, 4.32	4.5	85
9-methyl-	11.21	--	243, 336	4.25, 4.22	13.0	85
	7.00	—	255, 345	4.34, 4.09	8.6	85
		0	254, 279, 342	4.22, 4.00, 4.24	3.0	85
8-(Hydroxymethyl)-	8.79	—	275	4.01	11.4	91
	2.62	0	266	4.01	6.1	91
		+	265	3.94	0	91
2-Hydroxy-6-methyl- ^{j}	12.46	—	215, 269, 312	4.38, 3.68, 3.78	10.7	64
	8.87	0	213, 236, 312	4.53, 3.56, 3.82	5.6	64
	2.29	+	259, 318	3.64, 3.91	0	64
6-[3-(Hydroxymethyl)-2-butenyl]amino- (Zeatin)	9.8	—	220, 276	4.20, 4.16	13.0	92

TABLE VIII (continued)

Ultraviolet Spectra of Purines Having Hydroxyl or Thiol Substituents[a]

Derivative of purine	pK_a	Species	λ_{max} (nm)	log ε_{max}	pH	References
	4.4	0	212, 270	4.23, 4.20	7.2	92
		+	207, 275	4.16, 4.17	1.0	92
2-Hydroxy-8-(methylthio)-	12.38	--	228, 266, 323	4.33, 3.73 4.15	14.4	64
	7.20	–	220, 234, 284, 326	4.23, 4.23, 4.06, 3.92	10.0	64
	2.30	0	217, 240, 287, 326	4.23, 4.19, 3.66, 4.00	5.0	64
		+	212, 282, 342	4.18, 3.84 4.31	0.2	64
2-Hydroxy-8-(trifluoromethyl)-	10.92	--	218, 263, 316	4.39, 3.64, 3.92	13.0	64
	5.35	–	215, 271, 322	4.42, 3.78, 3.73	8.0	64
		0	213, 281, 316	4.31, 3.77, 3.75	2.0	64
6-Hydroxy-2-(trifluoromethyl)-	11.2	--	263	3.97	14.0	61
	5.1	–	258	3.96	9.1	61
	~1.1	0	253	3.91	3.2	61
		(+)	247	3.96	0	61

6-Hydroxy-8-(trifluoromethyl)-	10.9	--	268	4.08	14.0	61
	~5.0	−	261	4.09	8.6	61
2-Mercapto-	10.4	−	235, 263, 328[k]	4.12, 4.19, 3.49	8.8	43
	7.15	0	241, 285, 345-8	4.10, 4.25, 3.18	5.0	43
	~0.5	+	227-232, 287, 382	3.87, 4.27, 3.26	-1.2	43
6-Mercapto-[l]	10.84	−	228, 312	3.98, 4.16	9.3	43
	7.77	0	225, 325	3.87, 4.27	5.1	43
	<0	(+)	225, 324	3.82, 4.17	-3.5	43
1-methyl-	8.64	−	237, 321	4.03, 4.38	11.1	89
	<3	0	234, 320	3.97, 4.32	5.1	89
3-methyl-	7.51	−	245, 332	4.05, 4.48	11.1	89
	1.69	0	245, 338	3.96, 4.51	5.1	89
		+	244, 334	3.74, 4.47	0	89
7-methyl-	7.92	−	232, 316	3.94, 4.24	11.1	89
		0	329	4.31	5.1	89
9-methyl-	7.96	−	234, 309	4.11, 4.33	11.1	89
	1.42	0	229, 321	4.10, 4.41	5.1	89
		+	326	4.27	0	89
8-Mercapto-	11.16	--	230, 315	4.18, 4.31	13.0	43
	6.64	−	228, 313	4.13, 4.37	8.9	43
		0	231, 310	4.01, 4.48	4.5	43
9-methyl-	7.48	−	233, 311	4.18, 4.35	10.0	71
	<2.5	0	232, 309	4.09, 4.43	5.0	71
2,6,8-Trihydroxy- (Uric Acid)	10.6	--	295	4.13	12	94
	5.4	−	292	4.09	8.3	94
		0	284	4.08	3.0	94

TABLE VIII (continued)

Ultraviolet Spectra of Purines Having Hydroxyl or Thiol Substituents[a]

Derivative of purine	pK_a	Species	λ_{max} (nm)	log ε_{max}	pH	References
1,7-dimethyl-	10.9	--	296	4.10	12	94
	5.7	0	286	4.04	3.0	94
3-methyl-[m]	>11	(--)	293	4.17	12	94
	6.2	0	287	4.04	3.0	94
7-methyl-	10.6	--	297	4.11	12	94
	5.5	—	293	4.09	8.3	94
		0	286	4.06	3.0	94

[a]In water, at 20–25°. Inflections are underlined. [b]The u.v. spectra of all of the aminodihydroxypurines (and diaminohydroxypurines) have been reported,[19,20] but without differentiation of ionic species. [c]For u.v. spectra (ionically undifferentiated) of 6-aminopurin-8-one and 8-aminopurin-6-one, see Ref. 66. [d]Ultraviolet Spectra of 6-amino-9-methylpurin-2-one (ionically undifferentiated) are given in Ref. 78. [e]For u.v. spectra (undifferentiated) of 2-aminopurine-6-thiol, 6-aminopurine-2-thiol, 6-aminopurine-8-thiol, and 8-aminopurine-6-thiol, see Refs. 80, 81, 66, and 66 respectively. [f]Ultraviolet spectra (undifferentiated) of purine-6-carboxylic acid are in Ref. 83. [g]For u.v. spectra (undifferentiated) of 2,6-and 6,8-purinedithiols, see Refs. 66 and 87. [h]For other *N*-methylpurin-2-one spectra (undifferentiated), see Ref. 72a. [i]For u.v. spectra (undifferentiated) of several hydroxypurinethiols, see Refs. 66 and 87. [j]Ultraviolet spectra (undifferentiated) of

8-methylpurin-6-one are given in Ref. 63. [k]Very broad band. [l]Ultraviolet spectra (undifferentiated) of purine-6-selenol are given in Ref. 93. [m]Also, undifferentiated spectra of 1-methyl- and 1,3-dimethyl-uric acid are given in the same reference. For 9-methyluric acid, see Ref. 95, which has depictions of the spectra of all four of the monomethyluric acids.

The following considerations of ultraviolet spectra and ionization constants have indicated[84] that, in xanthine (purine-2,6-dione, see Table VIII), there is a mobile hydrogen atom on N-9 in the monoanion, although it is mainly on N-7 in the neutral species. 9-Methyl- and 1,9-dimethylxanthine are more acidic than xanthine or its 1-methyl, 7-methyl, and 1,7-dimethyl derivatives. Again, only 9-substituted xanthines have the extra absorption maximum at 245–250 nm. Spectra of the monoanions of xanthine and its 1- and 9-methyl and 1,9-dimethyl derivatives are almost identical, whereas those of the 7-methyl- and 1,7-dimethyl analogs are completely different.[84] With increasing alkalinity, the sequence of loss of protons appears to be: first, from N-3; then, from N-7; and finally, from N-1. This places negative charges on O-2; on O-2 and N-7; and on O-2, N-7, and O-6, for the mono-, di-, and tri-anion, respectively.

Similar considerations[77] have led to the conclusion that the neutral species of guanine (2-aminopurin-6-one, see Table VIII) is an equilibrium mixture of two forms in comparable proportions, one having a mobile proton on N-7 and one having it on N-9. With increasing basification,[77] a mobile proton is lost from N-1, and then one from N-7.

The ultraviolet spectra of all of the purinethiols studied differ sufficiently from those of the corresponding (methylthio)purines to suggest that, at equilibrium in water, they exist mainly in thione form (-NH-C:S) In the purine-6-thiol series, the 1-methyl derivative is a weaker acid than purine-6-thiol or its 3-, 7-, or 9-methyl derivatives, which has been taken to indicate[89] that the first proton lost on basification of purine-6-thiol comes from N-1. The spectral resemblance between purine-6-thiol and its 1-methyl derivative is even more striking than in the hypoxanthine series. The high λ_{max} and high extinction coefficients of 3-methyl-3*H*-purine-6-thiol led to the suggestion[89] that it has an electronic structure different from those of its isomers.

4. The Purine Nucleosides and Nucleotides

As with the pyrimidine nucleosides, the replacement of an alkyl group by a sugar residue on the purine nucleus, as in (20), makes little difference to the spectrum. However, below 200 nm, the hydroxyl group-absorption can be found, and, in alkaline solutions (above pH 11), the ionization of such groups should be detectable, although it has so far been little sought. The absorption spectra of the nucleotides are remarkably independent of the number of phosphoric residues and the state of their ionization (see Table IX).

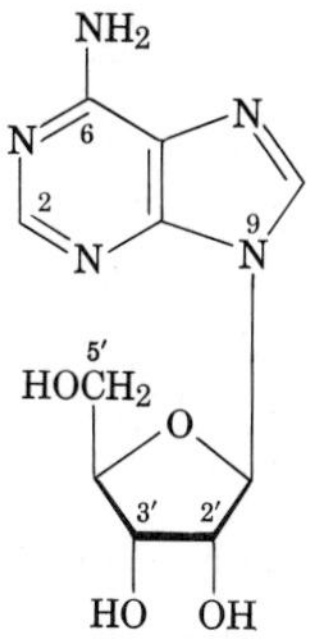

20

Poly(adenylic acid) shows considerably greater absorption (at all peaks)[49] at 74° than at 24°. 2'-Deoxy-D-ribonucleic acid also shows this effect, and the lower absorbance does not return after the solution has been cooled. Other spectral data on the nucleic acids are given in Section I, 4.

5. Solid-state Spectra

Solid-state, ultraviolet absorption spectra have been recorded and measured for crystals of 9-methyladenine, 1-methylthymine, and the dimer formed by association of one molecular proportion of each. Polarized, ultraviolet light was passed, in turn, through two axes of the crystals, permitting the first and second $\pi \rightarrow \pi^*$ transitions to be separated and allocated.[98]

TABLE IX

Ultraviolet Spectra of Purine Nucleosides and Nucleotides[a]

Compound	pk_a	Species	λ_{max} (nm)	log ε_{max}	pH	References
Adenosine (<u>20</u>)	12.35	0	190, 206, 260	4.30, 4.33, 4.17	7.9	49
	3.63	+	190, 205, 257	4.27, 4.33 4.16	1.5	49
2'-deoxy-	?	0	189, 208, 260	4.31, 4.32, 4.17	7.9	49
		+	188, 204, 257	4.28, 4.35, 4.16	1.5	49
5'-phosphate	6.4	?	185, 208, 260	4.32, 4.32, 4.18	7.9	49
	~4.4		188, 205, 258	4.28, 4.34, 4.16	2.2	49
5'-phosphate (AMP)	13.06	?	259	4.19	11.0	96
	6.05		259	4.19	7.0	96
	3.74		257	4.18	2.0	96
5'-pyrophosphate (ADP)	7.00	?	259	4.19	11.0	96
	4.20		259	4.19	7.0	96
			257	4.18	2.0	96
5'-triphosphate (ATP)	6.48	?	259	4.19	11.0	96

	4.00		259	4.19	7.0	96
			257	4.17	2.0	96
Guanosine	9.33	0	188, 253	4.43, 4.14	5.5	49
	~1.6					
2'-deoxy-, 5'-phosphate	?	?	188, 253	4.42, 4.13	7.7	49
			188, 253	4.42, 4.13	4.7	49
			193, 249	4.44, 4.15	1.4	49
5'-phosphate (GMP)	~9.4	?	258	4.06	11.0	96
	~6.1		252	4.14	7.0	96
	~2.4		256	4.09	1.0	96
5'-pyrophosphate (GDP)	~9.6	?	257	4.07	11.0	96
	~6.3		253	4.14	7.0	96
	~2.9		256	4.09	1.0	96
5'-triphosphate (GTP)	~9.3	?	257	4.08	11.0	96
	~6.5		253	4.14	7.0	96
	~3.3		256	4.09	1.0	96
Inosine	8.82	—	253	4.12	11.2	97
	~1.5	0	249	4.09	6.0	97
		+	251	4.23	-0.6	97
1-oxide	5.46	—	229, 256, 294	4.48, 3.79, 3.60	9.0	90
		0	251	3.96	3.0	90
Xanthosine	~13	--	276	3.97	14.0	97
	5.67	—	278	3.95	8.1	97
	~0	0	263	3.95	3.0	97

[a] In water, at 20–25°.

REFERENCES

(1) J. J. Fox and D. Shugar, in various contributions beginning in 1952 (see Refs. 25, 32, 36, and 45).
(1a) A. L. Sklar, *J. Chem. Phys.*, 10, 135 (1942).
(1b) T. Förster, *Z. Naturforsch.* 2a, 149 (1947).
(1c) R. S. Mulliken, *J. Chim. Phys.*, 46, 497, 675 (1949)
(2) J. R. Platt, *J. Chem. Phys.*, 17, 484 (1949).
(3) J. R. Platt, *J. Chem. Phys.*, 19, 263 (1951).
(4) M. A. El Sayed, *J. Chem. Phys.*, 36, 552 (1962).
(5) F. M. Uber, *J. Chem. Phys.*, 9, 777 (1941); F. M. Uber and R. Winters, *J. Amer. Chem. Soc.*, 63, 137 (1941).
(6) S. F. Mason, *J. Chem. Soc.*, (1959) 1247, 1253.
(7) D. J. Brown, *The Pyrimidines*, Interscience Publishers, New York, N. Y. (1962), pp. 477-498.
(8) S. F. Mason, in *DMS UV Atlas of Organic Compounds*, Vol. 2, Section G6, Butterworths, London, England (1966), pp. 2–4.
(9) V. Boarland and J. F. W. McOmie, *J. Chem. Soc.*, (1952) 3716, 3722.
(10) M. Iwaizumi and H. Azumi, *Nippon Kagaku Zasshi*, 84, 694 (1963).
(11) J. R. Marshall and J. Walker, *J. Chem. Soc.*, (1951) 1004.
(12) D. J. Brown and P. W. Ford, *J. Chem. Soc. (C)*, (1967) 568.
(13) A. Maggiolo and P. B. Russell, *J. Chem. Soc.*, (1951) 3297.
(13a) Optical circular dichroism has recently shown that each of the two bands in the ultraviolet spectrum of a simple pyrimidine consists of two bands of opposite sign and different origin. [D. W. Miles, M. J. Robins, R. K. Robins, M. W. Winkley, and H. Eyring, *J. Amer. Chem. Soc.*, 91, 824, 831 (1969).]
(14) D. J. Brown and L. N. Short, *J. Chem. Soc.*, (1953) 331.
(15) D. J. Brown, E. Hoerger, and S. F. Mason, *J. Chem. Soc.*, (1955) 211, 4035.
(16) N. Whittaker, *J. Chem. Soc.*, (1951) 1565.
(17) S. F. Mason, *J. Chem. Soc.*, (1954) 2071.
(18) S. J. Angyal and C. L. Angyal, *J. Chem. Soc.*, (1952) 1461.
(19) L. F. Cavalieri and A. Bendich, *J. Amer. Chem. Soc.*, 72, 2587 (1950).
(20) L. F. Cavalieri, A. Bendich, J. F. Tinker, and G. B. Brown, *J. Amer. Chem. Soc.*, 70, 3875 (1948).
(21) A. Albert and G. B. Barlin, *J. Chem. Soc.*, (1962) 3129.
(22) A. Albert and J. N. Phillips, *J. Chem. Soc.*, (1956) 1294.

(23) S. F. Mason, *J. Chem. Soc.*, (1958) 674.
(24) R. N. Jones, *J. Amer. Chem. Soc.*, 67, 2127 (1945).
(25) D. Shugar and J. J. Fox, *Biochim. Biophys. Acta*, 9, 199 (1952).
(26) K. Nakanishi, N. Suzuki, and F. Yamazaki, *Bull. Chem. Soc. Jap.*, 34, 53 (1961).
(27) E. R. Garrett, J. K. Seydel, and A. J. Sharpen, *J. Org. Chem.*, 31, 2219 (1966).
(28) I. Wempen and J. J. Fox, *J. Amer. Chem. Soc.*, 86, 2474 (1964).
(29) D. J. Brown and T. Teitei, *Aust. J. Chem.*, 17, 567 (1964).
(30) A. R. Katritzky, F. D. Popp, and A. J. Waring, *J. Chem. Soc. (B)*, (1966) 565.
(31) G. M. Kheifets and N. V. Khromov-Borisov, *Zh. Org. Khim.*, 2, 1511 (1966).
(32) J. J. Fox and D. Shugar, *Bull. Soc. Chim. Belges*, 61, 44 (1952).
(33) R. E. Stuckey, *J. Pharm. Pharmacol.*, 15, 370 (1942).
(34) S. Goldschmidt, W. Lamprecht, and E. Helmreich, *Z. Physiol. Chem.*, 292, 125 (1953).
(35) J. W. Patterson, A. Lazarow, F. J. Lemm, and S. Levey, *J. Biol. Chem.*, 177, 187 (1949); R. S. Tipson and L. H. Cretcher, *J. Amer. Pharm. Assoc., Sci. Ed.*, 40, 399, 440 (1951).
(36) D. Shugar and J. J. Fox, *Bull. Soc. Chim. Belges*, 61, 293 (1952).
(37) D. J. Brown and T. Teitel, *J. Chem. Soc.*, (1963) 4333.
(38) H. G. Mautner, *J. Amer. Chem. Soc.*, 78, 5292 (1956).
(39) D. J. Brown and J. M. Lyall, *Aust. J. Chem.*, 15, 851 (1962).
(40) A. R. Katritzky and A. J. Waring, *J. Chem. Soc.*, (1963) 3046.
(41) P. Brookes and P. D. Lawley, *J. Chem. Soc.*, (1962) 1349.
(42) D. J. Brown and T. Teitei, *Aust. J. Chem.*, 18, 559 (1965).
(43) S. F. Mason, *J. Chem. Soc.*, (1954) 2071.
(44) D. J. Brown and N. W. Jacobsen, *J. Chem. Soc.*, (1962) 3172.
(45) J. J. Fox and D. Shugar, *Biochim. Biophys. Acta*, 9, 369 (1952).
(46) J. J. Fox, L. F. Cavalieri, and N. Chang, *J. Amer. Chem. Soc.*, 75, 4315 (1953).
(47) I. Wempen, R. Duschinsky, L. Kaplan, and J. J. Fox, *J. Amer. Chem. Soc.*, 83, 4755 (1961).
(48) R. N. Bock, N.-S. Ling, S. A. Morell, and S. H.

Lipton, *Arch. Biochem. Biophys.*, 62, 253 (1956).
(49) D. Voet, W. B. Gratzer, R. A. Cox, and P. Doty, *Biopolymers*, 1, 193 (1963).
(50) J. Ofengand and H. Schaefer, *Biochemistry*, 4, 2832 (1965).
(51) D. O. Jordan, *The Chemistry of Nucleic Acids*, Butterworths, London, England (1960), p. 221.
(52) L. B. Clark and I. Tinoco, *J. Amer. Chem. Soc.*, 87, 11 (1965).
(52a) W. Voelta, R. Records, E. Bunnenberg, and C. Djerassi, *J. Amer. Chem. Soc.*, 90, 6163 (1968).
(53) A. Bendich, P. J. Russell, and J. J. Fox, *J. Amer. Chem. Soc.*, 76, 6073 (1954).
(54) L. B. Townsend and R. K. Robins, *J. Heterocycl. Chem.*, 3, 241 (1966).
(55) A. Albert and D. J. Brown, *J. Chem. Soc.*, (1954) 2060.
(56) R. W. Balsiger, A. L. Fikes, T. P. Johnston, and J. A. Montgomery, *J. Org. Chem.*, 26, 3446 (1961).
(57) P. Brookes and P. D. Lawley, *J. Chem. Soc.*, (1960) 539.
(58) B. C. Pal and C. A. Horton, *J. Chem. Soc.*, (1964) 400.
(59) (a) R. N. Prasad and R. K. Robins, *J. Amer. Chem. Soc.*, 79, 6401 (1957); (b) R. K. Robins and H. H. Lin, *ibid.*, 79, 490 (1957); (c) N. J. Leonard and J. A. Deyrup, *ibid.*, 84, 2148 (1962).
(60) M. A. Stevens, D. I. Magrath, H. W. Smith, and G. B. Brown, *J. Amer. Chem. Soc.*, 80, 2755 (1958).
(61) A. Giner-Sorolla and A. Bendich, *J. Amer. Chem. Soc.*, 80, 3932, 5744 (1958).
(62) J. Baddiley, B. Lythgoe, and A. R. Todd, *J. Chem. Soc.*, (1944) 318.
(63) H. C. Koppel and R. K. Robins, *J. Org. Chem.*, 23, 1457 (1958).
(64) A. Albert, *J. Chem. Soc. (B)*, (1966) 438.
(65) G. B. Barlin and N. B. Chapman, *J. Chem. Soc.*, (1965) 3017.
(66) R. K. Robins, *J. Amer. Chem. Soc.*, 80, 6671 (1958).
(67) J. A. Montgomery and L. B. Holum, *J. Amer. Chem. Soc.*, 79, 2185 (1957).
(68) G. B. Barlin, *J. Chem. Soc.(B)*, (1967) 954.
(69) C. O. Miller, F. Skoog, F. S. Okumura, M. H. von Saltza, and F. M. Strong, *J. Amer. Chem. Soc.*, 78, 1375 (1956).
(70) G. B. Elion and G. Hitchings, *J. Amer. Chem. Soc.*, 78, 3508 (1956).
(71) D. J. Brown and S. F. Mason, *J. Chem. Soc.*, (1957) 682.
(72) R. N. Prasad, C. W. Noell, and R. K. Robins,

J. Amer. Chem. Soc., 81, 193 (1959).
(72a) F. Bergmann, H. Kwietny, G. Levin, and D. J. Brown, *J. Amer. Chem. Soc.*, 82, 598 (1960); A. Kalmus and F. Bergmann, *J. Chem. Soc.*, (1960) 3679.
(73) D. J. Brown, P. W. Ford, and K. H. Tratt, *J. Chem. Soc. (C)*, (1967) 1445.
(74) D. J. Brown and N. W. Jacobsen, *J. Chem. Soc.*, (1965) 3770.
(75) J. Baddiley, J. G. Buchanan, F. J. Hawker, and J. E. Stephenson, *J. Chem. Soc.*, (1956) 4659.
(76) W. Cochran, *Acta Cryst.*, 4, 81 (1951).
(77) W. Pfleiderer, *Ann.*, 647, 167 (1961).
(78) R. Falconer, J. M. Gulland, and L. F. Story, *J. Chem. Soc.*, (1939) 1784.
(79) J. C. Parham, J. Fissekis, and G. B. Brown, *J. Org. Chem.*, 32, 1151 (1967).
(80) G. B. Elion and G. H. Hitchings, *J. Amer. Chem. Soc.*, 77, 1676 (1955).
(81) A. Bendich, J. F. Tinker, and G. B. Brown, *J. Amer. Chem. Soc.*, 70, 3109 (1948).
(82) A. Albert, *J. Chem. Soc.*, (1960) 4705.
(83) L. B. Mackay and G. H. Hitchings, *J. Amer. Chem. Soc.*, 78, 3511 (1956).
(84) W. Pfleiderer and G. Nübel, *Ann.*, 647, 155 (1961).
(85) D. J. Brown, *J. Appl. Chem.* (London), 9, 203 (1959).
(86) D. J. Brown and J. S. Harper, *J. Chem. Soc.*, (1961) 1298.
(87) A. G. Beaman, *J. Amer. Chem. Soc.*, 76, 5633 (1954).
(88) A. Giner-Sorolla, I. Zimmerman, and A. Bendich, *J. Amer. Chem. Soc.*, 81, 2515 (1959).
(89) G. B. Elion, *J. Org. Chem.*, 27, 2478 (1962).
(90) J. C. Parham, J. Fissekis, and G. B. Brown, *J. Org. Chem.*, 31, 966 (1966).
(91) A. Albert, *J. Chem. Soc.*, (1955) 2690.
(92) G. Shaw, B. M. Smallwood, and D. V. Wilson, *J. Chem. Soc. (C)*, (1966) 921.
(93) H. Mautner, *J. Amer. Chem. Soc.*, 78, 5292 (1956).
(94) E. A. Johnson, *Biochem. J.*, 51, 133 (1952).
(95) R. Falconer and J. M. Gulland, *J. Chem. Soc.*, (1939) 1369.
(96) R. M. Bock, N.–S. Ling, S. A. Morell, and S. H. Lipton, *Arch. Biochem. Biophys.*, 62, 253 (1956).
(97) G. H. Beaven, in E. Chargaff and J. N. Davidson (Eds.), *The Nucleic Acids*, Vol. 1, Academic Press, Inc., New York, N.Y. (1955), p. 493.
(98) R. F. Stewart and N. Davidson, *J. Chem. Phys.*, 39, 255 (1963).

CHAPTER 3

Gas-Phase Analysis of Nucleic Acid Components as Their Trimethylsilyl Derivatives

ALAN E. PIERCE

PIERCE CHEMICAL COMPANY, ROCKFORD, ILLINOIS

I. INTRODUCTION

The rapidly increasing use of gas-liquid chromatography (g.l.c.) and mass spectrometry has sharpened the search for ways of preparing compounds that are sufficiently volatile to be analyzable by these methods of gas-phase analysis. One important means is trimethylsilylation, namely, the introduction of the trimethylsilyl (TMS) group, $(CH_3)_3$ Si-, into a molecule, replacing active hydrogen atoms and yielding derivatives having a volatility suitable for gas-phase analysis. Many reagents for accomplishing this are available, and the procedures are simple.

During the past half-dozen years, trimethylsilylation procedures have been applied to over 2,500 compounds, including alcohols, phenols, acids, and amines, and most of the products have been purified or analyzed by gas-phase methods. This method of derivatization has been particularly useful with compounds of biological interest, such as amino acids, catecholamines, glycerides, phenolic acids, purine and pyrimidine compounds, steroids, and sugars. This Chapter will consider, through 1968, the literature relating to nucleic acid components. A more comprehensive review of trimethylsilylation has already appeared.[1]

The substitution of the TMS group for active hydrogen usually decreases the polarity and hydrogen bonding in the molecule, thus increasing the volatility and heat-stability as compared with those of the parent compound. The increase in volatility is spectacular for polyhydroxy compounds; for example, octakis(trimethylsilyl)sucrose and *N,O*-bis(trimethylsilyladenine are distillable, and have been analyzed by g.l.c. Even where the sample compound is itself volatile enough for subjection to g.l.c., the use of trimethylsilyl derivatives may increase the accuracy by improving the resolution and the peak symmetry, or by decreasing the absorption of the sample on the column.

Although the main analytical application of trimethylsilylated compounds has been in g.l.c., there has been an increase in interest in their mass spectrometry; here, as with g.l.c., the volatility of these derivatives has made possible the analysis of many compounds not previously considered suitable for examination by mass spectrometry. Determination of molecular weight and the formation, from trimethylsilyl derivatives, of characteristic fragments have often aided in identification of the parent compound. Mass spectrometry of deuterium-labeled trimethylsilyl compounds has, in some instances, clarified the course and mechanism of fragmentation.[2] The combination of g.l.c. and mass spectrometry makes an extremely powerful tool for identification; with coupled instruments, the chromatographic effluent may be continuously monitored by a fast-scan spectrometer, providing identification of single (and even of overlapping) chromatogram peaks.

II. TRIMETHYLSILYLATION METHODS

Of the numerous reagents and procedures for trimethylsilylation that have been applied to compounds containing active hydrogen, only those that have been employed for the analysis of nucleic acid components by g.l.c. will be discussed here. Where the main objective was the isolation of the derivative, other methods[2] have been used for the trimethylsilylation of these components; some of these methods could undoubtedly be adapted for use in g.l.c., where isolation of the derivatives is not necessary.

1. Hexamethyldisilazane and Chlorotrimethylsilane

Either hexamethyldisilazane (HMDS) or chlorotrimethylsilane (CTMS), alone, may effect trimethylsilylation of compounds containing active hydrogen compounds, as shown in reactions *1* and *2*.

$$2\ ROH + \underset{(HMDS)}{Me_3SiNHSiMe_3} \longrightarrow 2\ ROSiMe_3 + NH_3 \qquad ...(1)$$

$$ROH + \underset{(CTMS)}{Me_3SiCl} \longrightarrow ROSiMe_3 + HCl \qquad ...(2)$$

Both reactions are performed under reflux, and require a fairly long time for completion. However, reaction *1* is expedited by such acid catalysts as ammonium sulfate or CTMS, and, in many cases, reaction *2* becomes rapid at room temperature if an acid acceptor (such as pyridine or triethylamine) is present. Sweeley *et al.*[3] developed the use of a solution of 2:1 (v/v) HMDS—CTMS in pyridine, as in reaction *3*. The CTMS catalyzes the HMDS reaction, and

$$3\ ROH + Me_3SiNHSiMe_3 + Me_3SiCl \longrightarrow 3\ ROSiMe_3 + NH_4Cl \ ...(3)$$

the ammonia from the HMDS reaction is the acid acceptor for the CTMS reaction.

Trimethylsilylation reaction *3* is rapid, many compounds reacting completely within a few minutes at room temperature. Some of the ammonium chloride is precipitated, but it does not interfere with the analysis by g.l.c., which is performed directly on the resulting mixture. The solvent and the excess reagents pass quickly through the column, giving no interference with the peaks of the derivatives.

This HMDS—CTMS—pyridine reagent, especially if heated, trimethylsilylates pyrimidines, purines, and sugar phosphate moieties, forming trimethylsilyl ethers of hydroxyl groups of sugar residues and of enolized carbonyl groups, (trimethylsilyl)amines from amino groups, and trimethylsilyl esters from hydroxyl groups of phosphates. The behavior of 5'-guanylic acid is illustrative.

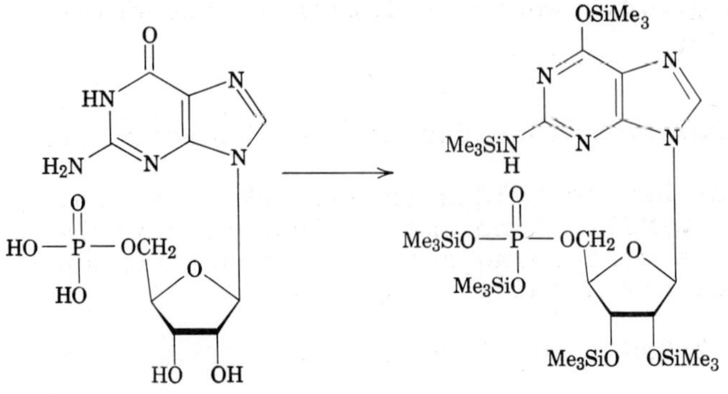

5

5′-Guanylic acid

2. *N*,*O*-Bis(trimethylsilyl)acetamide Compounds

One of the most powerful trimethylsilylating agents is N,O-bis(trimethylsilyl)acetamide (BSA),[4] which is a liquid at room temperature and may be used alone,[5] or in a solvent (such as acetonitrile[5,6] or pyridine), and the reaction may be catalyzed by CTMS.[2,6] The reaction mix-

$$ROH + CH_3C\begin{matrix}\diagup OSiMe_3 \\ \diagdown\!\!\diagdown NSiMe_3\end{matrix} \rightarrow ROSiMe_3 + CH_3C\begin{matrix}\diagup\!\!\diagup O \\ \diagdown NHSiMe_3\end{matrix}$$

ture may be analyzed directly by g.l.c.; the excess of the reagent and the *N*-(trimethylsilyl)acetamide formed are eluted quickly, well ahead of the trimethylsilylated sample.

Trifluorobis(trimethylsilyl)acetamide (TFBSA) is very similar in activity[6] to BSA. In comparison with BSA, it has the advantages of (*a*) being miscible with acetonitrile, (*b*) being more reactive, and (*c*) yielding chromatograms having fewer extraneous peaks.[6]

III. TRIMETHYLSILYLATION RESULTS

1. Pyrimidine and Purine Bases

Of these bases, the most difficult to trimethylsilylate completely is guanine,[6,7] and the conditions found necessary for this compound have, therefore, been used for treating all purine and pyrimidine samples prior to g.l.c. Sasaki and Hashizume[7] used 2:1 (v/v) HMDS–CTMS in pyridine, with refluxing for one hour, whereas Gehrke *et al.*[6] employed BSA in acetonitrile in sealed tubes, at 150° for 45 minutes. The former procedure yielded one peak from cytosine; the latter, two peaks each from cytosine and 5-methylcytosine. Other bases gave a single peak by either method. When trimethylsilylation by HMDS—CTMS was incomplete, cytosine, adenine, and guanine gave double peaks.[7] This behavior is explained by the greater difficulty of trimethylsilylating the amino groups in comparison with the enolizable carbonyl groups. Hence, if the reaction is incomplete, a mixture of *O*-(trimethylsilyl)ated and *N,O*-(trimethylsilyl)ated products results.

Gehrke *et al.*[6] found that, for most of these bases, the yield of trimethylsilylation product obtained with BSA is substantially higher than with HMDS—CTMS. It is possible that the double peaks from cytosine and BSA are due to *N*- and *N,N*-bis(trimethylsilyl) derivatives. [*O*-Trimethylsilyl)ation is assumed to take place first.] Furthermore, the addition of a moderate proportion of ammonium chloride to the BSA reagent was found to result in the formation of only one derivative. The use of ammonium salts as effective catalysts, particularly in the trimethylsilylation of amino groups, is well documented.[1] It would appear, therefore, that, when catalyzed by ammonium salts, BSA produces only the *N,N*-bis(trimethylsilyl) derivative from cytosine.

With BSA and ammonium salts, adenine and guanine also showed a higher yield of product than with BSA alone. The quenching effect of an excessive proportion of ammonium salts on the yields from cytosine, adenine, or guanine with BSA has not, as yet, been explained. As would be expected, the yields from the treatment of uracil and thymine with BSA were not affected by the addition of ammonium salts, as these bases are devoid of amino groups.

The purine and pyrimidine bases obtained from yeast ribonucleic acid (RNA) by hydrolysis with perchloric acid followed by ion-exchange purification were analyzed[6] by g.l.c., after trimethylsilylation with BSA. For each of the various bases, manipulative losses were shown to amount to 18 $\pm$1.5%.

The trimethylsilylation of all of the bases reported in Table I is assumed to be complete, considering the vigorous reaction-conditions employed, although the only derivative whose isolation has as yet been reported is the uracil derivative,[7,8] which, by elemental[7] and i.r.[7,8] analysis, was shown to be 2,4-bis(trimethylsiloxy)pyrimidine. Hence, adenine, cytosine, hypoxanthine, thymine, and uracil yield bis(trimethylsilyl) derivatives; guanine and xanthine give tris(trimethylsilyl) derivatives.

The trimethylsilylated bases are distillable; they are liquids, or crystalline solids having low melting points, and are readily soluble in nonpolar solvents.[8] They are readily decomposed by water or alcohol, to regenerate the original base. Because of their sensitivity to moisture, they should be stored in the reaction mixture, where the excess of trimethylsilylating reagent furnishes protection.

2. Nucleosides and Nucleotides

As with the pyrimidines and purines, the amino groups attached to the heterocyclic rings of the nucleosides and nucleotides require vigorous conditions for trimethylsilylation, and the silazane link so formed is readily cleaved. When trimethylsilylation was conducted with HMDS—CTMS in pyridine at room temperature, adenosine, cyidine, 2'-deoxyadenosine, and 2'-deoxyguanosine exhibited double peaks;[9] this indicates that incomplete trimethylsilylation had occurred, as single peaks were obtained after the reaction mixture had been heated.[7] Thus, Sasaki and Hashizume[7] trimethylsilylated adenosine with refluxing HMDS—CTMS—pyridine, and showed, by elemental and i.r. analysis, that the product initially formed was 2',3',5'-tris-_O_-(trimethylsilyl)adenosine; when the latter compound was further trimethylsilylated, the _N_-(TMS)-2',3',5'-_tris_-_O_-(TMS) derivative was obtained, and this showed greater retention in g.l.c. with DC-430. With the fully trimethylsilylated amino compounds, it was found necessary to work under strictly anhydrous conditions (to avoid double peaks in g.l.c., resulting from partial hydrolysis by atmospheric moisture).[7] Nucleosides are readily separated from their trimethylsilyl derivatives by thin-layer chromatography in _p_-dioxane, as the latter move with the solvent front.[10]

Because phenylphosphoric acid is trimethylsilylated by hot HMDS—CTMS, it was assumed that the phosphoric acid group of nucleotides is similarly esterified.[11] Even nicotinamide adenine dinucleotide and 5'-cytidylic acid may be trimethylsilylated, as shown by paper chromatography,[12] although the former derivative is not eluted in g.l.c.[11] and the results obtained with the latter derivative have thus far been equivocal.[11,12]

Jacobson *et al.*[5] enzymically hydrolyzed RNA to the nucleosides, and trimethylsilylated the dried digestion mixture with BSA at 120° (it was not found necessary to remove salt or protein prior to derivatization). An aliquot of the reaction mixture that contained 10 μg (or less) of total nucleoside was analyzed by g.l.c. on 4% OV-17. For several ribonucleases used, the quantitative results were in excellent agreement with other values in the literature. The peak for the guanosine derivative showed a minor, trailing shoulder which was absent when the trimethylsilylation was conducted at a higher temperature.

The dimethylsilyl derivative of adenosine[11] has been prepared from chlorodimethylsilane–tetramethyldisilazane in pyridine.Its retention time is about half that of the corresponding trimethylsilyl derivative (on g.l.c. with SE-30), suggesting that, for compounds of very low volatility, dimethylsilyl derivatives should be prepared.

IV. GAS-PHASE ANALYSIS

For g.l.c., columns of diatomaceous earth must be thoroughly deactivated[13,14] to diminish the possibility of adsorption or decomposition of the trimethylsilylated sample. This deactivation is accomplished by washing with acid followed by treatment with dichlorodimethylsilane or HMDS, which convert active sites (silanol groups) to the (less polar) silyl ethers. Various supports for the g.l.c. of trimethylsilylated bases have been compared.[6]

The stationary phases that have mainly been used for g.l.c. of trimethylsilyl derivatives of nucleic acid components are nonselective methyl silicones, namely, SE-30, DC-430, and OV-17. The elution time obtained by using these phases depends on the molecular weight of the sample and on the percentage of the stationary phase on the support.[12] With a more polar, selective phase, such as XE-60 (a cyanoethyl methyl silicone), retention on the basis of the electric charge of the compound may be more important than its molecular weight.[12] For example, when an XE-60 column is used, the order of elution of the two adenosine derivatives is the opposite of that observed with SE-30 columns.[10]

Synthetic mixtures of seven trimethylsilylated bases, or of four trimethylsilylated bases and four trimethylsilylated D-ribonucleosides were each well resolved on 5% DC-430, with programming, from 140°, at 6°/minute.[7] Trimethylsilyl derivatives of up to five nucleosides from the RNA of yeast or *E. coli* were resolved on 4% OV-17, with programming, from 160°, at 2°/min. The separation of these derivatives of adenosine, cytidine, guanosine, and uridine from each other was good, but those of pseudouridine and 1-β-D-ribofuranosylthymine each appeared as a single peak. Elution temperatures are given in the Tables. The elution temperatures for the individual trimethylsilylated bases on 4% SE-30, with programming, from 95°, at 6.7°/min, are also listed.[6]

The quantification of g.l.c. results has been investigated by Sasaki and Hashizume[6] by use of thermal-conductivity detectors, and by Gehrke *et al.*[6] by using flame-ionization detectors. The peak-area response with these detectors, relative to that of phenanthrene, was determined for various trimethylsilylated bases[7] and nucleosides,[7] and the results showed satisfactory accuracy and reproducibility. Jacobson *et al.*[5] used an argon ionization detector, and observed a linear response for trimethylsilylated nucleosides from RNA relative to trimethylsilylated adenosine.

The mass spectrometry of TMS and TMS-d_9 (prepared from BSA-d_{18}-CTMS-d_9) derivatives of single nucleosides and nucleotides was investigated, following their elution from a gas chromatograph having a 1% SE-30 or 1% OV-17 column.[2] From the mass spectra, the following information was obtained: (*a*) the molecular weight of the TMS derivative, (*b*) the identity of the base and some information as to its substitution, and (*c*) the identity of the sugar as being a ribose or a 2'- or 3'- deoxyribose (a 2'- or 3'-deoxy-*erythro*-pentose).

V. EXPERIMENTAL DIRECTIONS

The following two procedures are recommended for the preparation of a trimethylsilylated sample for g.l.c.:

A. About 10 mg of the dry base, nucleoside, or nucleotide is dissolved or suspended in 0.5 ml of dry pyridine, and 0.2 ml of HMDS and 0.1 ml of CTMS are added. The mixture is refluxed for 1 hr. under anhydrous conditions, and cooled, and an aliquot (1 to 10 µl) of the reaction mixture is injected into the chromatograph.[7,11]

B. About 5 mg of the sample is weighed into a screw-capped, culture tube, containing a Teflon-coated, magnetic stirring-bar. A 100-mole excess of BSA (1–2 ml) and 3 volumes of acetonitrile are added, and the tube is sealed with a Teflon-lined cap. The tube is heated and the contents stirred for 45 min at 150° (oil bath). The mixture is cooled, and analyzed directly.[5] Alternatively, the acetonitrile is omitted, and heating is maintained for 2 hr at 120°.[5]

For quantitative results in either procedure, phenanthrene (5–10 mg) may be added to the sample as an internal standard for comparison of peak areas.[7]

VI. TABLES

Tables I and II list the components (of nucleic acids) whose trimethylsilyl derivative(s) have been reported for use in g.l.c. or mass spectrometry.

1. Trimethylsilylation

Where substantiated in the original paper, the number of trimethylsilyl groups introduced is indicated (see Table II). In all cases where two degrees of trimethylsilylation of nucleosides are shown, the less completely silylated product has a free amino group on the base moiety, whereas the fully trimethylsilylated derivative has a trimethylsilylated amino group.

2. Method

The trimethylsilylation methods are designated as follows. A, Hexamethyldisilazane (HMDS)—chlorotrimethylsilane (CTMS) [2:1 (v/v)] in pyridine at room temperature[9,10,12]; B, HMDS-CTMS [2:1 (v/v)], refluxed for 1 hr[7,11]; C, *N,O*-Bis(trimethylsilyl)acetamide (BSA) in acetonitrile at 150° for 45 min,[6] or alone at 120° for 2 hr[5]; and D, BSA—CTMS, heated.[2,6]

3. Stationary Phase

The materials SE-30 (General Electric), DC-430 (Dow-Corning), and OV-1 (Ohio Valley Specialty Chemical Co.) are methyl silicones. Material XE-60 (General Electric) is a cyanoethyl methyl silicone containing 25% of cyanoethyl groups.

4. Column Temperature

Undesignated temperatures are for constant-temperature g.l.c. columns. Elution temperatures for programmed runs are designated "pr."

5. Retention Data and Physical Constants

Boiling point (b) and melting point (m) are indicated. Where the relative retention value for g.l.c. is given, the internal standard is shown after the shilling mark (Ado = adenosine; phen = phenanthrene). Where no relative retention value is given, the retention time (in min) is listed, if known. Where the analysis was made primarily by mass spectrometry, this is indicated by the symbol "m.s."

TABLE I

Trimethylsilylated Purines and Pyrimidines

Compound trimethyl-silylated	Formula	Method	Stationary phase	Column temp., °C.	Retention time (min); phys. constants of product	References
Adenine	$C_5H_5N_5$	B	5% DC-430	180	5.45	7
				230	1.10	
				193 pr		
		C	4% SE-30	185 pr		6
Cytosine	$C_4H_5N_3O$	B	5% DC-430	180	0.28/Ado	7
				168 pr		
		C	4% SE-30	148 pr		6
5-methyl-	$C_5H_7N_3O$	C	4% SE-30	152 pr		6
				165 pr		
Guanine	$C_5H_5N_5O$	B	5% DC-430	180	2.87/Ado	7
				230	2.18/Ado	
				214 pr		
		C	4% SE-30	211 pr		6
Hypoxanthine	$C_5H_4N_4O$	B	5% DC-430	180	0.84/Ado	7
				189 pr		
		C	4% SE-30	176 pr		6
Purine	$C_5H_4N_4$	C	4% SE-30	140 pr		6
Thymine	$C_5H_6N_2O_2$	B	5% DC-430	161 pr		7
				180	0.19/Ado	
Uracil	$C_4H_4N_2O_2$	B	5% DC-430	180	0.16/Ado (b$_9$ 98-99°)	7

				157 pr	(m 31.5-33°)	
		C	4% SE-30	120 pr		5
Xanthine	$C_5H_4N_4O_2$	B	5% DC-430	180	1.94/Ado	6
				230	1.77/Ado	
				206 pr		
		C	4% SE-30	202 pr		5

TABLE II

Trimethylsilylated Nucleosides and Nucleotides

Compound trimethyl-silylated	Formula	No. of TMS groups intro-duced	Method	Stationary phase	Column temp., °C.	Retention time (min); phys. con-stants	Refer-ences
Adenosine	$C_{10}H_{13}N_5O_4$		A	0.2% SE-30	250	2.0	12
				0.75% SE-30	253	2.2	
				1.2% SE-30	254	5.7	
				5% SE-30	258	5.2	
		3	A	1.2% SE-30	209	21.6	9
					242	6.0	
					254	5.4	
					228	12.9	10
					249	6.2	
				4% XE-60	225	10.2	
					240	7.9	
		3	B	5% DC-430	238	7.5 (m 36–37)	7
		4	A	1.2% SE-30	228	14.7	10
					249	6.7	
				4% XE-60	225	6.8	
					240	4.5	
		4	B	5% DC-430	246 pr		7
					238	8.9	
					230	11.0	
					250	5.2	

			C	4% SE-30	260 pr		6
			C	4% OV-17	213 pr		5
		4	D	1% SE-30		m.s.	2
2'-deoxy-	$C_{10}H_{13}N_5O_3$	3	A	1.2% SE-30	242	5.1	9
					254	3.6	
			B	5% DC-430	220	8.8	7
					230	0.84/Ado	
					250	0.87/Ado	
		3	D	1% SE-30		m.s.	2
5'-phosphate	$C_{10}H_{14}N_5O_6P$		A	1.2% SE-30	249	6.2	10
		4	D	1% SE-30		m.s.	2
3'-deoxy-	$C_{10}H_{13}N_5O_3$	3	D	1% SE-30		m.s.	2
N,N-dimethyl-	$C_{12}H_{17}N_5O_4$	4	D	1% SE-30		m.s.	2
1-methyl-	$C_{11}H_{15}N_5O_4$	4	D	1% SE-30		m.s.	2
5'-S-methyl-5'-thio-	$C_{11}H_{15}N_5O_3S$		A	1.2% SE-30	231	13.3	10
					239	8.0	
5'-propionate	$C_{13}H_{17}N_5O_5$		A	1.2% SE-30	249	9.0	10
3',5'-cyclic phosphate	$C_{10}H_{12}N_5O_6P$	3	D	1% SE-30		m.s.	2
S-Adenosyl-L-methionine	$C_{13}H_{23}N_6O_5S$		A	1.2% SE-30	249	8.3	10
2'-Adenylic acid	$C_{10}H_{14}N_5O_7P$		A	1.2% SE-30	231	6.6	10
2'(3')-Adenylic acid			B	5% DC-430	250	9.05	11
					265	5.6	
		5	D	1% SE-30		m.s.	2

TABLE II (continued)

Trimethylsilylated Nucleosides and Nucleotides

Compound trimethylsilylated	Formula	No. of TMS groups introduced	Method	Stationary phase	Column temp., °C.	Retention time (min); phys. constants	References
3'-Adenylic acid			A	1.2% SE-30	231	7.0	10
5'-Adenylic acid			A	0.75% SE-30	255	3.2	12
			A	1.2% SE-30	228	17.0	10
					249	8.2	
			B	5% DC-430	250	11.8	11
					265	7.1	
			C	4% SE-30	276 pr		6
		5	D	1% SE-30		m.s.	2
Cytidine	$C_9H_{13}N_3O_5$		A	1.2% SE-30	209	3.9	9
			A	1.5% SE-30	255	16.0	12
					265	7.0	
		4	B	5% DC-430	230	1.91/Ado	7
					250	1.69/Ado	
					255 pr		
			C	4% OV-17	237 pr		5
		4	D	1% SE-30		m.s.	2
2'-deoxy-	$C_9H_{13}N_3O_4$	4	D	1% SE-30		m.s.	2
5'-phosphate	$C_9H_{14}N_3O_7P$	4	D	1% SE-30		m.s.	2

Compound	Formula			Column			Value	Ref.
5'-Cytidylic acid	$C_9H_{14}N_3O_8P$		A	0.75% SE-30	255			12
		4	D	SE-30			m.s.	2
Cytosine, 1-β-D-arabinofuranosyl-	$C_9H_{13}N_3O_5$	4	D	1% SE-30			m.s.	2
Guanosine	$C_{10}H_{13}N_5O_5$		A	0.75% SE-30	255		3.9	12
			A	1.2% SE-30	209		38.4	9
					242		10.5	
			B	5% DC-430	254	pr		7
					230		1.80/Ado; 9.7	
					250		1.55/Ado	
			C	4% SE-30	265	pr		6
			C	4% OV-17	225	pr		5
					228	pr		
		5	D	1% SE-30			m.s.	2
2'-deoxy-	$C_{10}H_{13}N_5O_3$		A	1.2% SE-30	242		9.9	9
					254		6.6	
		4	B	5% DC-430	250		1.47/Ado	7
		4	D	1% SE-30			m.s.	2
5'-phosphate	$C_{10}H_{14}N_5O_7P$	5	D	1% SE-30			m.s.	2
2'(3')-Guanylic acid	$C_{10}H_{14}N_5O_8P$		B	5% DC-430	250		12.0	11
					265		7.3	
5'-Guanylic acid			A	0.75% SE-30	255			12
			B	5% DC-430	250		14.3	11
					265		8.65	
		6	D	1% SE-30			m.s.	2

TABLE II (continued)

Trimethylsilylated Nucleosides and Nucleotides

Compound trimethylsilylated	Formula	No. of TMS groups introduced	Method	Stationary phase	Column temp., °C.	Retention time (min); phys. constants	References
Hydrouracil, 1-β-D-ribofuranosyl-	$C_9H_{14}N_2O_6$	4	D	1% SE-30		m.s.	2
Inosine	$C_{10}H_{12}N_4O_5$		A	1.2% SE-30	242	6.0	9
		3	B	5% DC-430	220	(m 265–266°)	7
		4	B	5% DC-430	220	8.45	7
					230	0.95/Ado	
					250	0.89/Ado	
		4	D	1% SE-30		m.s.	2
2'-deoxy-	$C_{10}H_{12}N_4O_4$	3	D	1% SE-30		m.s.	2
5'-Inosinic acid	$C_{10}H_{13}N_4O_8P$		B	5% DC-430	250	10.0	11
					265	6.0	
Nicotinamide mononucleotide [3-Carbamoyl-1-(5-*o*-phosphono-β-D-ribofuranosyl)-	$C_{11}H_{15}N_2O_8P$		A	1.5% SE-30	180	3.3	12

pyridinium hydroxide, inner salt]							
Thymidine	$C_{10}H_{14}N_2O_5$		A	1.2% SE-30	214 254	9.6 2.6	9
		2	D	1% SE-30		m.s.	2
		2	B	5% DC-430	210	(m 108–109°)	7
		3	B	5% DC-430	210 230 250	6.8 0.54/Ado 0.55/Ado	7
			C	4% SE-30	230 pr		6
5'-Thymidylic acid	$C_{10}H_{15}N_2O_9P$		B	5% DC-430	250 265	6.3 4.05	11
		4	D	1% SE-30		m.s.	2
Thymine, 1-β-D-ribofuranosyl-	$C_{10}H_{14}N_2O_6$		C	4% OV-17	204 pr		5
			C	4% OV-17	190 pr		5
Uracil, 5-β-D-ribofuranosyl-(Pseudouridine)	$C_9H_{12}N_2O_6$	5	D	1% SE-30		m.s.	2
Uridine	$C_9H_{12}N_2O_6$		A	0.75% SE-30	255	1.8	12
			A	1.2% SE-30	209 214 242 254	12.9 12.0 4.5 3.3	9
			A	4% SE-30	234 pr		6
		3	B	5% DC-430	210	8.7 (m 39–44°)	7
		4	B	5% DC-430	238 pr		7

TABLE II (continued)

Trimethylsilylated Nucleosides and Nucleotides

Compound trimethylsilylated	Formula	No. of TMS groups introduced	Method	Stationary phase	Column temp., °C.	Retention time (min); phys. constants	References
					230	0.65/Ado	
					250	0.66/Ado	
			C	4% OV-17	204 pr		5
		4	D	1% SE-30		m.s.	2
2'-deoxy-	$C_9H_{12}N_2O_5$		A	1.2% SE-30	214	8.7	9
					254	2.1	
		2	B	5% DC-430	230	0.46/Ado	7
					250	0.50/Ado	
		2	D	1% SE-30		m.s.	2
5-fluoro-	$C_9H_{11}FN_2O_5$		A	1.2% SE-30	214	8.7	9
5-fluoro-	$C_9H_{11}FN_2O_6$		A	1.2% SE-30	214	12.0	9
5-hydroxy-	$C_9H_{12}N_2O_7$	5	D	1% SE-30		m.s.	2
2'(3')-Uridylic acid	$C_9H_{13}N_2O_9P$		B	5% DC-430	250	5.5	11
					265	3.5	
5'-Uridylic acid			A	0.75% SE-30	255	3.2	12
			B	5% DC-430	250	7.3	11
					265	4.4	
			C	4% SE-30	270 pr		6
		5	D	1% SE-30		m.s.	2
Xanthosine	$C_{10}H_{23}N_4O_6$		A	1.2% SE-30	242	8.4	9
		5	B	5% DC-430	230	1.33/Ado	7
					250	1.20/Ado	
					220	12.2	
			D	1% SE-30		m.s.	

REFERENCES

(1) A. E. Pierce, *Silylation of Organic Compounds*, Pierce Chemical Company, Rockford, Illinois (1968).

(2) J. A. McCloskey, A. M. Lawson, K. Tsuboyama, P. M. Krueger, and R. M. Stillwell, *J. Amer. Chem. Soc.*, 90, 4182 (1968).

(3) C. C. Sweeley, R. Bentley, M. Makita, and W. W. Wells, *J. Amer. Chem. Soc.*, 85, 2497 (1963).

(4) J. F. Klebe, H. Finkbeiner, and D. M. White, *J. Amer. Chem. Soc.*, 88, 3390 (1966).

(5) M. Jacobson, J. F. O'Brien, and C. Hedcoth, *Anal. Biochem.*, 25, 363 (1968).

(6) C. W. Gehrke and C. R. Ruyle, *J. Chromatogr.*, 38, 473 (1968). For a preliminary report, see C. W. Gehrke, D. L. Stalling, and C. D. Ruyle, *Biochem. Biophys. Res. Commun.*, 28, 869 (1967).

(7) Y. Sasaki and T. Hashizume, *Anal. Biochem.*, 16, 1 (1966).

(8) T. Nishimura and I. Iwai, *Chem. Pharm. Bull.* (Tokyo), 12, 352 (1964).

(9) R. L. Hancock and C. L. Coleman, *Anal. Biochem.*, 10, 365 (1965).

(10) R. L. Hancock, *J. Gas Chromatogr.*, 4, 363 (1966).

(11) T. Hashizume and Y. Sasaki, *Anal. Biochem.*, 15, 199 (1966).

(12) R. L. Hancock, *J. Gas Chromatogr.*, 6, 431 (1968).

(13) E. C. Horning, K. C. Maddock, K. V. Anthony, and W. J. A. VandenHeuvel, *Anal. Chem.*, 35, 526 (1963).

(14) D. M. Ottenstein, *J. Gas Chromatogr.*, 1, 11 (1963).

CHAPTER 4

Mass Spectrometry of Nucleic Acid Components

DON C. DEJONGH

DEPARTMENT OF CHEMISTRY, WAYNE STATE UNIVERSITY, DETROIT, MICHIGAN

I. INTRODUCTION

A requirement particularly relevant to the study of nucleic acid components by mass spectrometry is the volatilization of the sample into the ion source. If the sample is thermally stable at the temperature required to volatilize it at 1–100 nm Hg, the operating pressure of the ion source, it is possible to obtain its mass spectrum.[1–3] Purines and pyrimidines have sufficient volatility[4–6] at 130–200°. Nucleosides require higher temperatures,[6,7] although pyrolysis of the nucleoside in the inlet system may occur under operating conditions that are too severe.[8] Among the common nucleosides, guanosine and cytidine are too polar and not sufficiently volatile for mass spectrometry, but stable, volatile derivatives can be prepared. Similarly, the technique cannot be applied directly to nucleotides.[7]

The trimethylsilyl (TMS) derivatives[7] of the nucleosides and nucleotides have greater volatility than the underivatized compounds. Mass spectra have thus far been obtained for the TMS derivatives of the four major nucleosides (and 5'-monophosphates) of RNA (D-ribonucleic acid) and DNA (2'-deoxy-D-ribonucleic acid), and of an additional eleven nucleosides and two nucleotides.

For introducing the purines and pyrimidines into the ion source, a heated glass-inlet system[9] or a direct-introduction probe[4-6] may be used. However, use of a probe type of inlet is necessary for the nucleosides.[6,10] The TMS derivatives are sufficiently volatile to be used with a gas-chromatographic inlet-system, from which mass spectra of the eluted components may be obtained directly. A sample size in the microgram region is adequate for routine operation, and much smaller quantities (for example, nanograms) are sufficient, if necessary, although not on a routine basis.

Electron bombardment[1-3] has been used for ionization of the components of nucleic acids. Most of the mass spectra have thus far been reported at 70 eV, but a substantial simplification of the spectra of the purines and pyrimidines occurs if the energy of the electron beam is lowered to 12–20 eV.[4,5] Field-ionization and photoionization mass spectra have not yet been reported; spectra obtained in this way might be simpler than those obtained by electron impact, and the relative intensity of the molecular-ion region might be enhanced, which could be particularly important when mixtures are studied.

The fragmentation pathways of the ions observed in the mass spectra of purines and pyrimidines have been examined by using exact-mass measurements, metastable peaks and specific substitution with deuterium. Elemental compositions of selected ions, or of all of the ions, are determinable from exact masses measured on high resolution, mass spectrometers.[1-3,11] Not only are these measurements useful for interpreting the mass spectra, but they are valuable for studying impure samples. Shoul the impurity or impurities have heteroatoms not contained in the compound of interest, or *vice versa*, or should the impurity have a different carbon-to-hydrogen ratio, elimination from consideration of ions attributable to the impurity is possible. In the absence of exact-mass data, this is not possible unless the spectrum of the impurity is known and can, perhaps, be subtracted from the mass spectrum of the impure sample.

The technique most commonly employed for introducing deuterium into purines and pyrimidines is the replacement of exchangeable protons; it is applicable if the sample can be dissolved in warm, neutral or basic D_2O, and is then followed by evaporation of the excess D_2O. Amino, imidazole, and hydroxyl protons are exchanged by brief warming. Prolonged refluxing causes replacement of the C-8 proton, also,[4] in purine. Trimethylsilyl derivatives may be prepared with trimethyl-d_9-silyl-*N*-(trimethyl-d_9-silyl)acetimidate or chlorotrimethyl-d_9-silane.[7]

Metastable peaks (m^*) result if an ion of mass m_1 fragments to an ion of mass m_2 plus a neutral fragment in the field-free region of the instrument.[1-3] These peaks are recorded as low-intensity, broad peaks at $m^* \cong (m_2)^2/m_1$. The presence of such a peak for a transition is evidence that m_2 originates from m_1. The absence of a metastable peak does not indicate that such a transition does not occur, because some fragmentations do not give rise to metastable peaks. Transitions supported by the presence of a metastable peak will be indicated by m^* placed above the arrow in the following discussions.

The mass spectra of purines, pyrimidines, nucleosides, and TMS derivatives of nucleotides will be discussed in the following Sections. In a few cases, mass spectrometry has played a role in elucidating the structures of previously unreported compounds. Most of the spectra are, however, obtained from commonly available model compounds of known structure. Thus, mass spectrometry may aid in the characterization and identification of newly discovered components of nucleic acids, as well as in the recognition of known compounds. The technique is also useful for determining the location and number of biologically or synthetically incorporated D (^{2}H), ^{18}O, and ^{13}C.

II. MASS SPECTRA OF PURINES

The mass spectrum of[12,13] purine (1) is dominated by the molecular ion ($M^{+\cdot}$) at *m/e* 120. Two molecules of

$$[\text{purine}]^{+\cdot} \xrightarrow{m^*} m/e\ 93 + \text{HCN} \xrightarrow{m^*} m/e\ 66 + \text{HCN}$$

m/e 120
(from 1)

hydrogen cyanide (HCN) are eliminated from $M^{+\cdot}$, and metastable peaks are present for each transition. In molecules in which deuterium is placed on specific carbon and nitrogen atoms of the heterocyclic ring, HCN and DCN are lost in all possible sequences from the molecular ion.[12] Thus, many paths of fragmentation contribute to the mass spectrum. For example, almost 80% of the HCN lost from *m/e* 120 involves[12] C-2—N-3 and C-6—N-1.

The 70-eV, mass spectrum[4,14] of adenine (2) is given

NH_2 (adenine ring, $[\;]^{+\cdot}$)

m/e 135 (from **2**) $\xrightarrow{m^*}$ *m/e* 108 + HCN $\xrightarrow{m^*}$ *m/e* 81 + HCN $\xrightarrow{m^*}$ *m/e* 54 + HCN

in Fig. 1. The peaks at *m/e* 108, 81, and 54 are due to three successive expulsions of one molecule of HCN. The

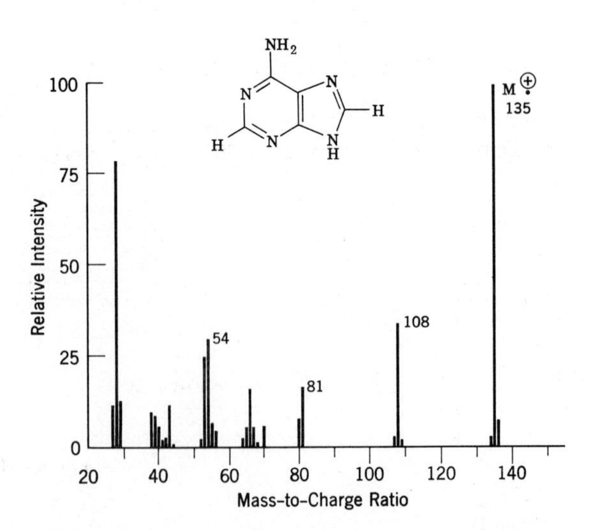

Fig. 1 The 70-eV mass spectrum of adenine (2).

molecular ions of adenine-*N,N,9*-d_3 and of adenine -*N,N,8,9*-d_4 lose HCN and DCN from many sites in different sequences, as does purine (1).

N-Methyladenine (3) initially eliminates CH_2N or $CH_3N\cdot$ from the methylamino group of the molecular ion.[4,14]

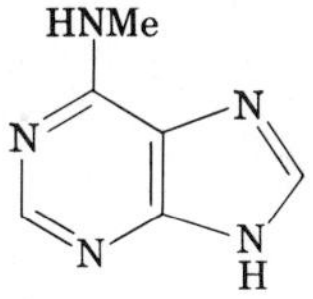

3

This fragmentation is followed by three successive losses of one molecule of HCN. It is interesting that the molecular ion of *N,N*-dimethyladenine also eliminates $CH_3N\cdot$; one methyl group apparently migrates from N-6 to a nitrogen atom of the heterocyclic system.[15] With larger substituents on the 6-amino group, the main fragmentations take place along the side chain.[14]

The structure of zeatin (4), a factor inducing cell division, was found, with the aid of mass spectrometry,[14,16]

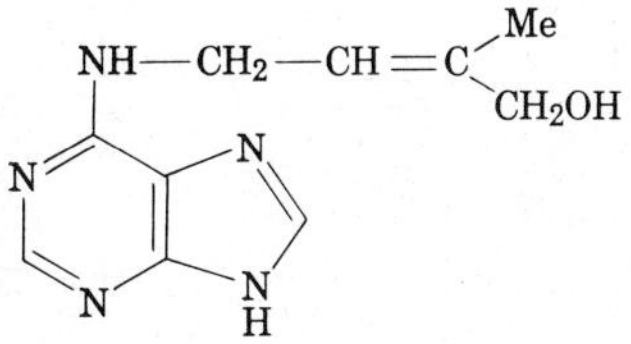

4

to be 6-[3-(hydroxymethyl)-*trans*-2-butenyl]aminopurine. The molecular ion at *m/e* 219 shifts to *m/e* 222 after exchange in D_2O, indicating the presence of three readily exchangeable protons. Prominent peaks are found at *m/e* 202 and 188, due to losses of $HO\cdot$ and $HOCH_2\cdot$ from $M^{\oplus\cdot}$, respectively. Peaks at *m/e* 135 and *m/e* 136 are characteristic of adenine (see Fig. 1) and of protonated adenine ions, formed by loss of the N^6 side-chain, with hydrogen atoms rearranged from the side chain to the amino group or to the nitrogen atoms of the ring.

The molecular ion is the most intense peak in the 70-eV mass spectrum (see Fig. 2) of guanine[4] (5). The peak at *m/e* 109 is formed by loss of N-1, C-2, and the amino nitrogen atom from $M^{\oplus\cdot}$. This ion (*m/e* 109) ejects CO and

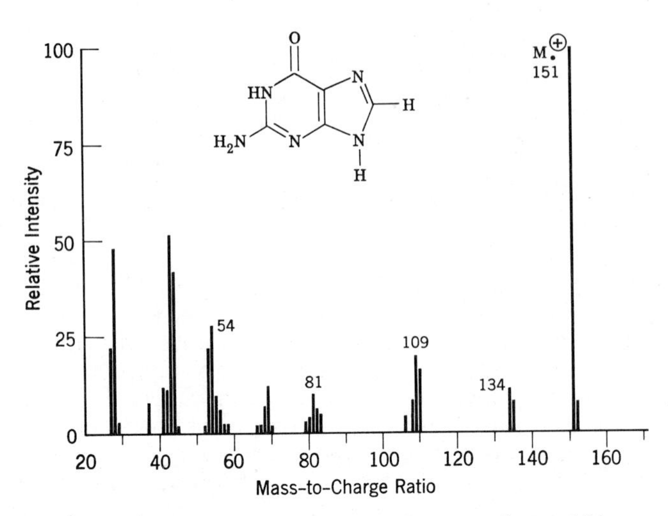

Fig. 2 The 70-eV mass spectrum of guanine (5).

m/e 151 (from **5**) $\xrightarrow{m^*}$ m/e 109 + CH_2N_2

m/e 109 → m/e 81 + CO

m/e 81 → m/e 54 + HCN

HCN successively, producing the peaks at m/e 81 and 54. The fragment at m/e 110 is formed when $CHN_2\cdot$ is eliminated from $M^{\oplus\cdot}$. Other peaks are found at m/e 135 (M-$NH_2\cdot$), m/e 134 (M-NH_3), and m/e 108 (M-HNCO). These assignments are supported by measured elemental compositions and by peak shifts in the mass spectra of D_2O-exchanged samples. The mass spectra of the 1-, 3-, and 7-methyl derivatives of guanine have also been reported.[4]

Fragmentation of hypoxanthine (6) parallels, to a large extent, that of guanine.[4] The major route begins with elimination of HCN from $M^{\oplus\cdot}$, involving N-1 and C-2. This fragment-ion subsequently eliminates CO and HCN, repre-

$$[\text{hypoxanthine}]^{+\cdot} \xrightarrow{m^*} m/e\ 109 + HCN$$
$$m/e\ 109 \xrightarrow{m^*} m/e\ 81 + CO$$
$$m/e\ 81 \xrightarrow{m^*} m/e\ 54 + HCN$$

m/e 136
(from **6**)

senting a continuation of the guanine fragmentation pattern.

III. MASS SPECTRA OF PYRIMIDINES

The 70- and 20-eV mass spectra of uracil (7) are shown[5] in Fig. 3. A substantial simplification of the spectrum

$$[\text{uracil}]^{+\cdot} \xrightarrow{m^*} m/e\ 69 + HNCO$$
$$m/e\ 69 \longrightarrow m/e\ 42 + HCN$$
$$m/e\ 69 \xrightarrow{m^*} m/e\ 41 + CO$$
$$m/e\ 69 \longrightarrow m/e\ 28 + HC_2O\cdot$$

m/e 112
(from **7**)

is achieved by lowering the energy of the electron beam to 20 eV, due to the elimination of the high-energy fragmentation paths. The molecular ion at *m/e* 112 expels HNCO to produce a peak at *m/e* 69 (C_3H_3NO); this fragment can lose CO, producing an ion at *m/e* 41 (C_2H_3N), or HCN, producing an ion at *m/e* 42 (C_2H_2O), or it may form $CH_2N^{+\cdot}$ at *m/e* 28.

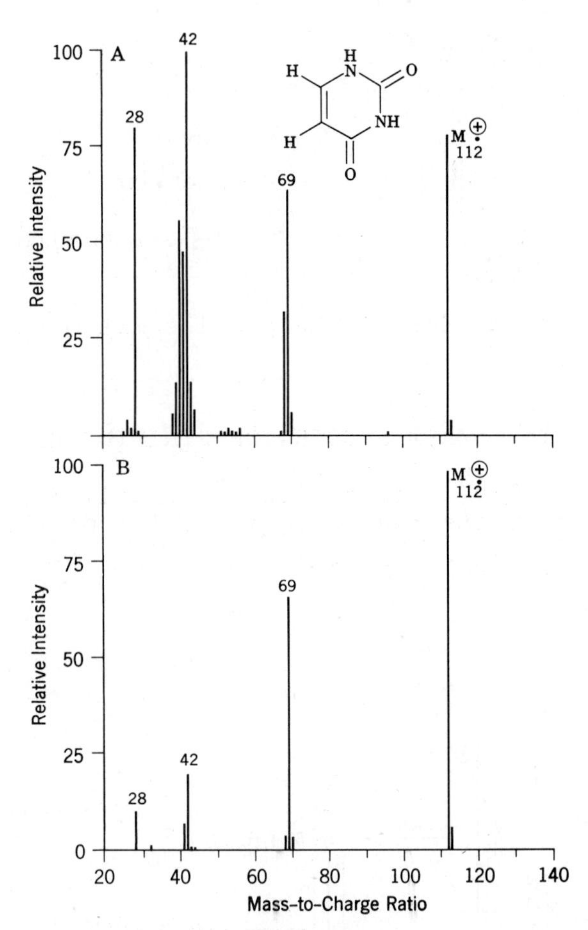

Fig. 3 The mass spectra of uracil (7) at (a) 70 eV and (b) 20 eV.

6-Methyluracil (8, Fig. 4) also loses HNCO from its molecular ion, followed by loss of the methyl group.[5] The peak at m/e 42 corresponds to $C_2H_4N^{\oplus}$ and $C_2H_2O^{\oplus}$ in the ratio of 3:1.

The 20-eV mass spectrum[5] of thymine (9) is also included in Fig. 4, in order to illustrate how structural changes may alter the appearance of the mass spectra. Once again, HNCO is expelled from the molecular ion, but the methyl group is not subsequently lost. The presence of the methyl group at C-5 favors decarboxylation of the

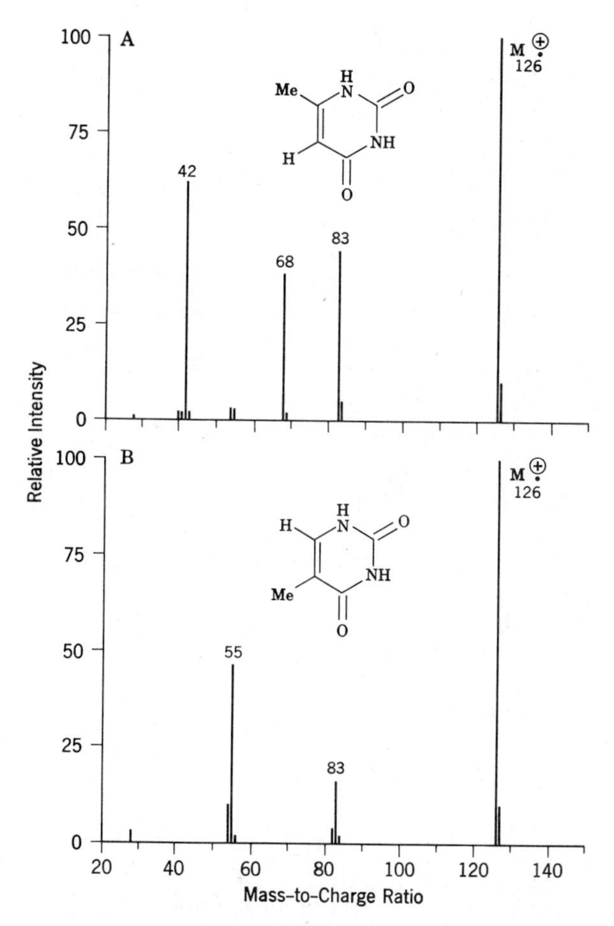

Fig. 4 The 20-eV mass spectra of (a) 6-methyluracil (8) and (b) thymine (9).

ion at *m/e* 83, instead of loss of HCN.

The mass spectra of hydrouracil and 5-methylhydrouracil were recorded at low temperatures of the ion-source.[5] Behavior typical of thermal decomposition of the sample is observed on increasing the temperature.

The 12-eV mass spectrum of cytosine (10) contains the molecular ion at *m/e* 111 and three fragment-ion peaks, *m/e* 83, 69, and 68. The 70-ev mass spectrum, which is more complex than that observed at 12 eV, is given in Fig. 5. The molecular ion loses $H_2N\cdot$ (in a process not observed at 12 eV), CO, NCO·, and HNCO. The molecular

$$\left[\text{6-methyluracil}\right]^{+\cdot} \longrightarrow m/e\ 83 + \text{HNCO}$$

m/e 126
(from **8**)

m/e 83 $\xrightarrow{m^*}$ m/e 68 + $CH_3\cdot$

m/e 83 $\longrightarrow$ m/e 42 + $HC_2O\cdot$

m/e 83 $\longrightarrow$ m/e 42 + CH_3CN

$$\left[\text{thymine}\right]^{+\cdot} \longrightarrow m/e\ 83 + \text{HNCO}$$

m/e 126
(from **9**)

m/e 83 $\xrightarrow{m^*}$ m/e 55 + CO

$$\left[\text{cytosine}\right]^{+\cdot}$$

m/e 111
(from **10**)

$\longrightarrow$ m/e 95 + $NH_2\cdot$

$\longrightarrow$ m/e 83 + CO

$\longrightarrow$ m/e 69 + $NCO\cdot$

$\longrightarrow$ m/e 68 + HNCO

m/e 68 $\longrightarrow$ m/e 41 + HCN

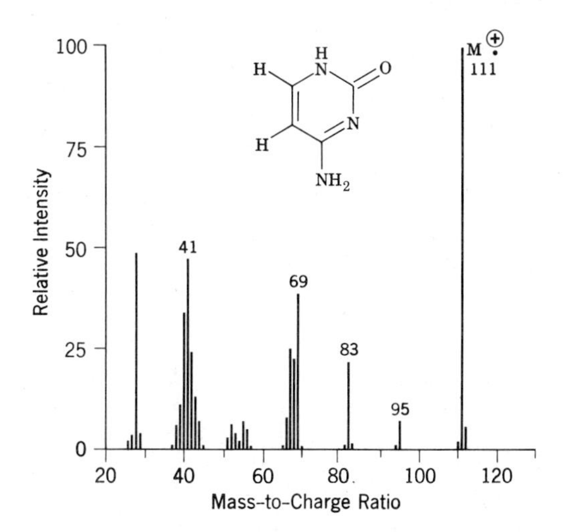

Fig. 5 The 70-eV mass spectrum of cytosine (10).

ion of 5-methylcytosine also eliminates $NH_2\cdot$, CO, NCO·, and HNCO;[5] there is a metastable peak for the loss of $H_2NCO\cdot$ from the molecular ion.

The major fragmentation path of the molecular ion of pyrimidine is the sequential loss of two HCN molecules.[5] The most important processes for the fragmentation of the molecular ion of 2-aminopyrimidine involve loss of HCN, followed either by loss of C_2H_2 or of a second molecule of HCN. The mass spectra of 22 additional simple pyrimidines have been published.[9,17] In most cases, the amino group directs the fragmentation. If functional groups more complex than 2-amino are attached to the ring, as in 2-piperidinopyrimidine and 4-(2-pyrimidinyl)morpholine, fragmentations triggered by these groups may compete effectively with those mentioned in connection with 2-aminopyrmidines.

The mass spectra of photo-cycloaddition products of thymines and uracils and of a photo-coupled product of 5-bromouracil have been obtained.[18] The most characteristic fragmentation of these dimers, which are joined through a cyclobutane ring, is scission of the cyclobutane ring, generating ions of the same compositions as those from the corresponding monomer. The lower halves of the spectra contain all of the peaks observed in the spectra of the corresponding monomers. The mass spectra must be obtained under optimal conditions, in order that molecular ions may be observed and the possibility of contributions from thermal decompositions may be lessened. Characteristic fragmentations have also been documented for dimeric photoproducts derived from oxetane and azetidine linkages and for a coupled product linked through a 5,5'-bond.[18]

IV. MASS SPECTRA OF COMMON NUCLEOSIDES

The mass spectra of many free nucleosides can be obtained by subliming them directly into the ionizing electron beam of the mass spectrometer. From the mass spectra of the nucleoside and its *N*,*O*-perdeutero analog, the base may be identified and considerable information about the sugar may be obtained.

The pyrimidine or purine moiety, for example, B in 11, can be recognized from intense peaks at B plus one hydrogen (B+*H*, 12), having the same mass-to-charge ratio as the free base, or B plus two hydrogen atoms (B+2*H*). The hydrogen atoms are rearranged to B from the hydroxyl groups of the sugar moiety (S), with cleavage of the C'-1-N bond.[3,10] The peak at B+2*H* is more prominent in the mass spectra of D-ribosyl than of 2'-deoxy-D-ribosyl (2-deoxy-D-*erythro*-pentosyl) derivatives, and more promi-

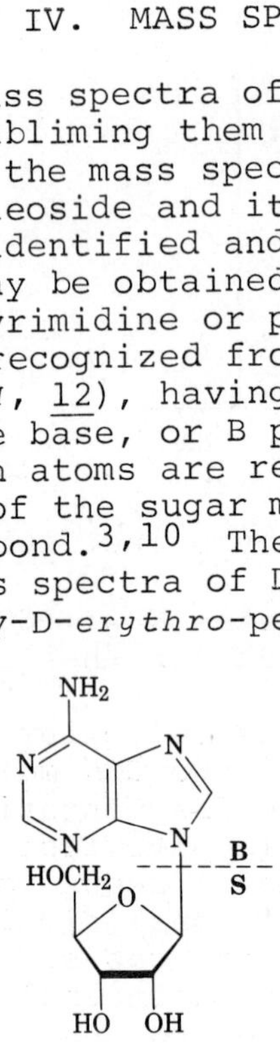

11
$M^{+\cdot}$, *m/e* 267

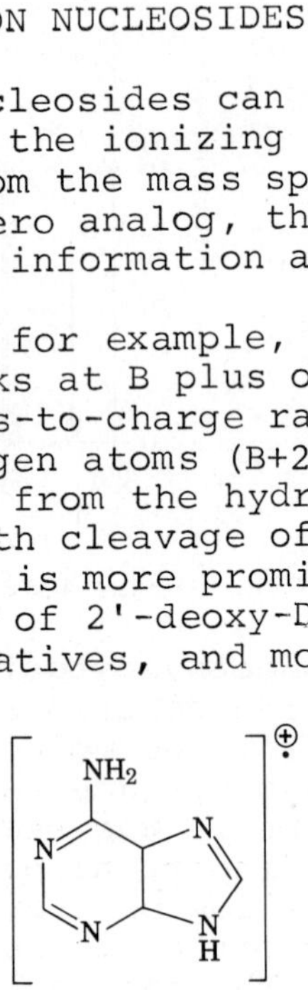

12, *m/e* 135 (**B** + *H*)

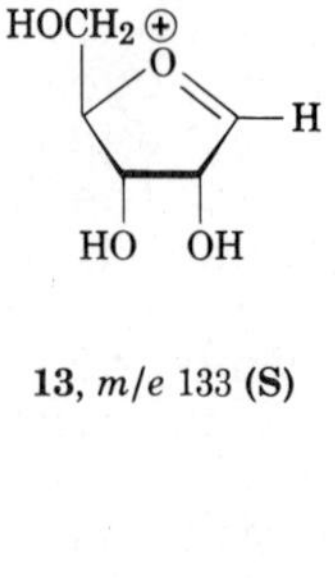

13, *m/e* 133 (**S**)

nent in the mass spectra of nucleosides of pyrimidine bases than of those of purine bases.

A peak at *m/e* 133 (13) is characteristic of D-ribose (S); it shifts to *m/e* 117 in the mass spectra of nucleosides containing 2-deoxy- D-ribose because of the difference of one oxygen atom.[3,10] The intensity of the peak due to S is much lower if the base is a purine than if it is a pyrimidine. The D-ribose fragment is generally less abundant than the 2-deoxy-D-ribose fragment. There are marked variations in the intensities of certain peaks in the mass spectra of epimers; however, the fragments formed are the same.

Commonly, a peak is found at a mass-to-charge ratio corresponding to 30 mass units higher[10] than the base (B+30). Deuterium-labeling studies show that this ion retains the base moiety (B), C'-1, the ring-oxygen atom, and a rearranged hydrogen atom, as shown in (14). The

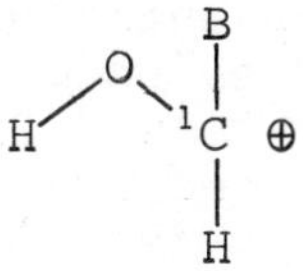

14, B+30

rearranged hydrogen atom (shown on the ring-oxygen atom in 14) appears to be abstracted preferentially from the 2'-OH group; B+30 is of low intensity in the mass spectra of nucleosides of 2-deoxy-D-ribose.

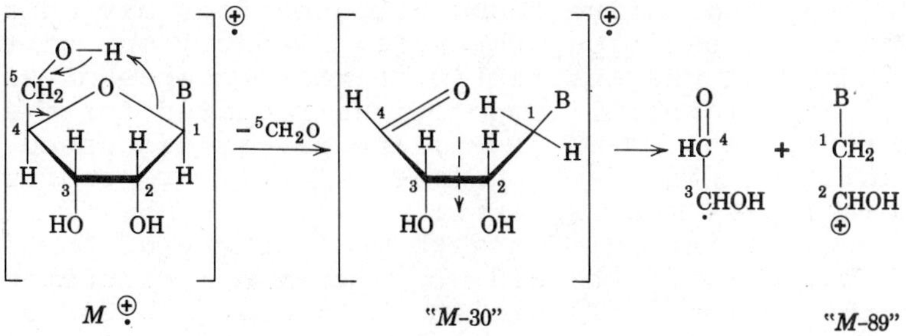

$-{}^{5}CH_2O$

M ⊕

"*M*-30"

"*M*-89"

Scheme 1

Another important peak for recognizing structural features is one that retains[3,10] the base (B), C'-1, and C'-2. A route suggested for its formation is shown in

Scheme 1, in which a total of 89 mass units is lost from the D-ribose portion of the molecular ion. The peak-shift in the spectrum of the D_2O-exchanged compound confirms that the base and two hydroxyl hydrogen atoms are present in the M-89 ion. In the mass spectrum of 1-(5-deoxy-β-D-lyxofuranosyl)uracil, no peak corresponding to this fragmentation is present, indicating that a 5'-OH group is necessary.

Such characteristics of the mass spectra of nucleosides may be recognized in the mass spectra of adenosine (11, Fig. 6), 2'-deoxyadenosine (15, Fig. 7) and uridine (16, Fig. 8).[3,10] The $M^{\oplus}$, *M*-30, *M*-89, B+*30*, B+2*H*, and B+H peaks are prominent in the mass spectrum of 11, whereas the peak at *m/e* 133 for the sugar moiety (S) is of low intensity. In the mass spectrum of 2'-deoxyadenosine (see Fig. 7), B+30 and B+2*H*, as well as *m/e* 117 from the deoxy sugar moiety, are of low intensity, or lower than in Fig. Thus, deoxy sugars in nucleosides may be recognized by their effect on the fragmentation, in addition to their mass difference of 16 mass units. In the mass spectrum of uridine (see Fig. 8), B+2*H* and *m/e* 133 (S) are more intense than in the mass spectrum of adenosine (see Fig. 6). The B+2*H* peak is of lower intensity than the B+*H* peak in the mass spectrum of 2'-deoxyuridine.[3,10]

The complete, high-resolution, mass spectrum of adenosine, presented as an "element map" in which the peaks are ordered according to heteroatom content, has been published.[11,19] The molecular ion is found to correspond to $C_{10}H_{13}N_5O_4$. All ions that contain one to four oxygen atoms and no nitrogen atoms are found with no more than five carbon atoms; this indicates that the four oxygen atoms are part of a C_5 moiety. All five nitrogen atoms are found with as few as five carbon atoms; because the total number of carbon atoms is ten, these five must be other than those five found with the four oxygen atoms. An ion of composition $C_5H_5N_5$ is, in fact, one of the most abundant ions characteristic of the presence of an adenine base. The elemental compositions required for *m/e* 133 (S), B+*H*, B+2*H*, B+30, *M*-89, *M*-30, and $M^{\oplus}$, as discussed in the beginning of this Section (see p. 156), are confirmed by the exact-mass measurements.

The fragmentation of the thymine moiety of thymidine has been studied with the aid of exact-mass measurements at high resolution.[5] The molecular ion at *m/e* 242 does not undergo fragmentation processes characteristic of thymine. However, the B+*H* and B+2*H* ions at *m/e* 126 and *m/e* 127 decompose by the same paths as does a 2,4-dideoxy pyrimidine molecular ion.

Derivatives of guanosine and cytidine must be prepared, in order to increase their volatility to the point at whic mass spectra can be obtained[7] (see Section VI, p. 170).

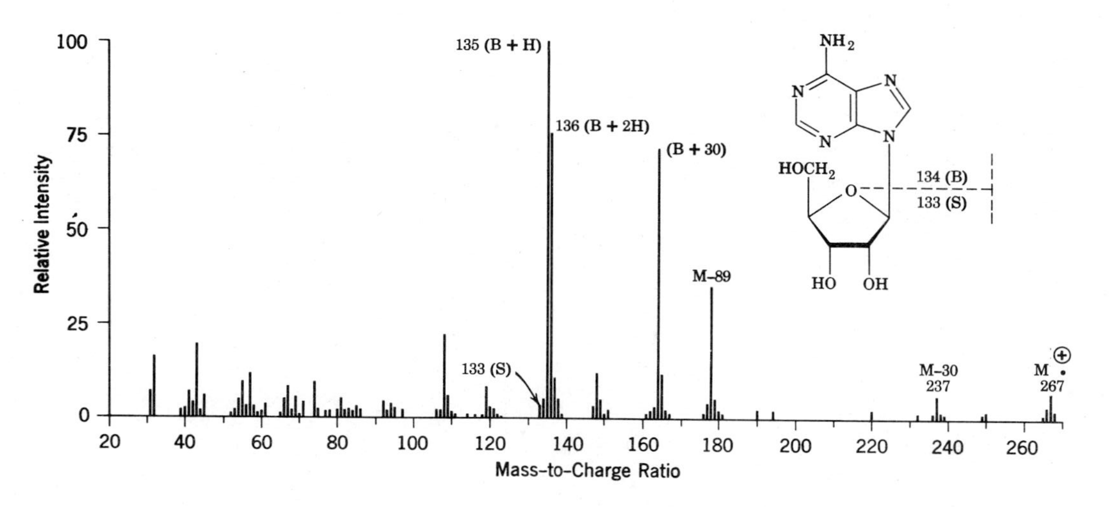

Fig. 6 The 70-eV mass spectrum of adenosine (11).

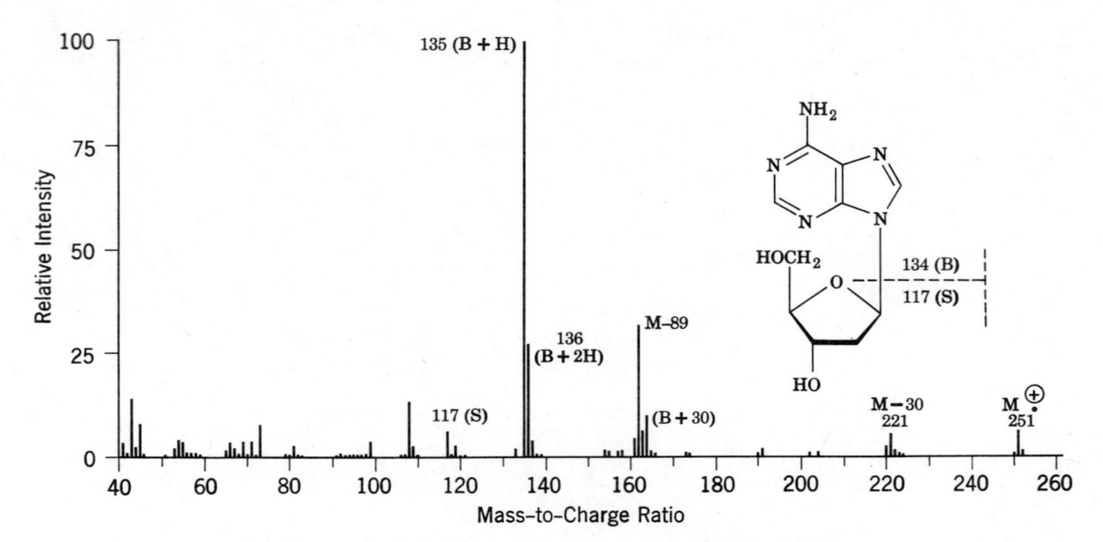

Fig. 7 The 70-eV mass spectrum of 2'-deoxyadenosine (15).

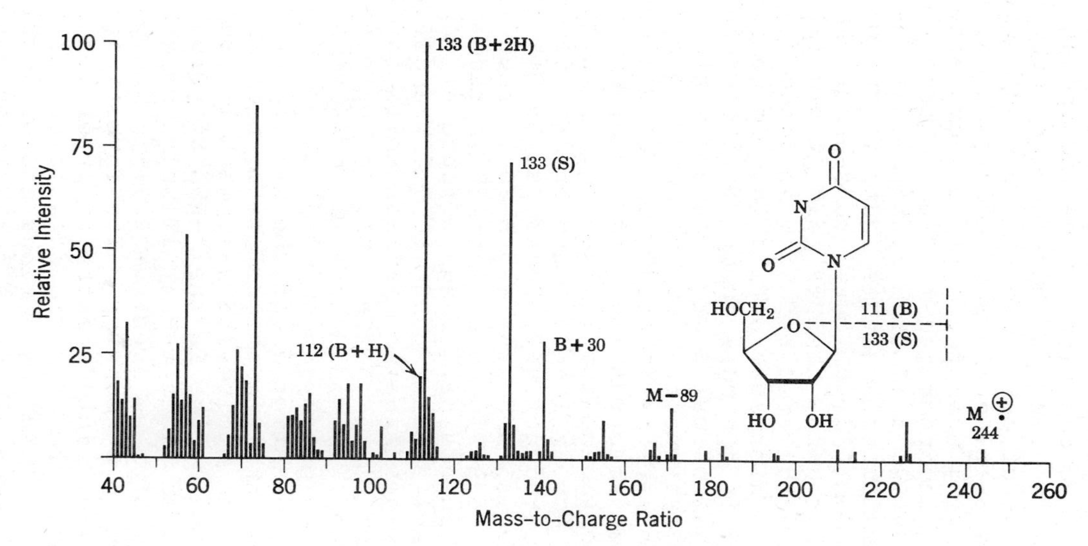

Fig. 8 The 70-eV mass spectrum of uridine (16).

V. APPLICATIONS OF MASS SPECTROMETRY IN STUDIES OF NUCLEIC ACID COMPONENTS

Mass spectrometry has played an important role in studies on the mechanism of the biological methylation of nucleic acids. The formation of methylated bases in transfer D-ribonucleic acid (t-RNA) by the use of methionine having three deuterium atoms in the γ-methylthio group (*i.e.*, CD_3-S-) has been investigated, and the number and position of the incorporated deuterium atoms have been analyzed by mass spectrometry.[8] In each case, three atoms of deuterium were present, with transmethylation to the 6-amino group of adenosine and to C-5 of uridine, involving transfer of an intact methyl group. No deuterium was incorporated into the unmethylated adenosine or uridine isolated from deuterated t-RNA, demonstrating that the deuterium atoms found in the methylated bases are not randomly distributed throughout the nucleoside.[8] On the other hand, the B+*H* peak at *m/e* 126 in the mass spectrum of methylated uridine, isolated from t-RNA of cells grown on nonisotopic methionine, is completely shifted to *m/e* 129 in the mass spectrum of the nucleoside isolated from t-RNA of cells grown on methionine-methyl-d_3.

Mass spectrometry helped[20] to confirm the structure of the antibiotic cordycepin as 3'-deoxyadenosine (17). The mass spectrum (see Fig. 9) of synthetically prepared 3'-deoxyadenosine is identical with that of cordycepin. The molecular ion is found at *m/e* 251, corresponding to a deoxyadenosine. Peaks due to B+*H* and B+2*H* are found at *m/e* 135 and 136, and B+30 is found at *m/e* 164. The peak at *M*−89, found at *m/e* 178 in the mass spectrum (see Fig. 6 of adenosine, is again found at *m/e* 178 in Fig. 9; here, *m/e* 178 corresponds to *M*−73, a difference (from the *M*−89 of adenosine) of 16 mass units for the oxygen atom absent from C'-3. A fragmentation path for 17, analogous to Scheme 1, would have $M^{\oplus}$ at *m/e* 251, *M*−30 at *m/e* 221, and *M*−73 (no oxygen on C'-3) at *m/e* 178.

A description of the use of a time-shared computer to help analyze the complete, high-resolution, mass spectrum of the nucleoside *N*-(3-methyl-2-butenyl)adenosine (18), which occurs in the t-RNA of yeast and mammalian tissue and in plant t-RNA, illustrates a highly sophisticated use of mass spectrometry.[11] The mass-spectral data were fed to a computer. The investigator, at an electric typewriter linked to the computer containing the high-resolution data, arrived at a molecular formula of $C_{15}N_5O_4$ having a total of 8 rings and/or double bonds (the number of hydrogen atoms was not printed out in this program) by commanding the computer to search the data. The informa-

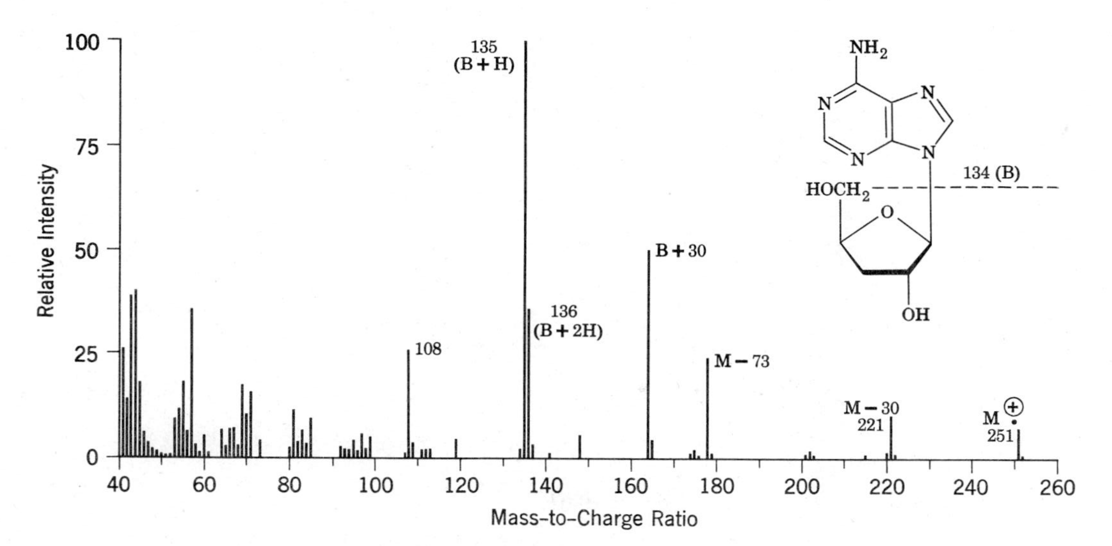

Fig. 9 The 70-eV mass spectrum of cordycepin (17).

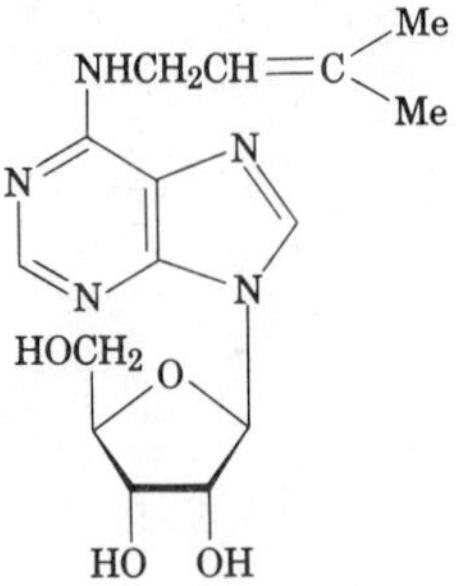

18

tion stored in the computer program also contained a list of masses of ions that are known to be characteristic of certain structural features of molecules of known structure. Thus, when the investigator further commanded the computer to search the intense ions in the spectrum of the nucleoside and to compare these ions with those in its list, "133.05080 pentoside" and "135.05530 adenine derivative" were printed out among the results. These peaks are *m/e* 135 (B+*H*) and *m/e* 133 (S) in Fig. 6.

At this point, the investigator knew that the unknown nucleoside was both a pentosyl and an adenine derivative, of elemental composition $C_{15}N_5O_4$, having 8 rings and/or double bonds.[11] Because adenine is C_5N_5 (with two rings and four double bonds) and a pentose residue is C_5O_4, with one ring, a total of C_5 and one ring or double bond was unaccounted for. The computer replied that $C_{10}O_4$ ions were absent from the spectrum, but that $C_{10}N_5$ and C_5N ions were present. Thus, the five carbon atoms unaccounted for were present as a substituent, attached to a nitrogen atom of the adenine moiety that is readily lost, specifically, from N-6. The position of the double bond in the C_5 side-chain was deduced from the n.m.r. spectrum. The complete "element map" of 18 has been published.[19]

The structure of 18 was reported independently by another group with the aid of mass spectrometry, but with no high-resolution data.[22] The mass spectrum of 18 showed a molecular-ion peak at *m/e* 335, and a peak for B+*H* at *m/e* 203; there were also prominent peaks at *m/e* 188, 160, 148, 136, and 135. This general fragmentation pattern is the same as that found in the mass spectrum of zeatin (4) after its $M^{\oplus}$ has lost the oxygen in a fragmentation process.

With the aid of mass spectrometry,[23] a hydroxylated derivative (19) of 18 has also been identified as a constituent of plant t-RNA.

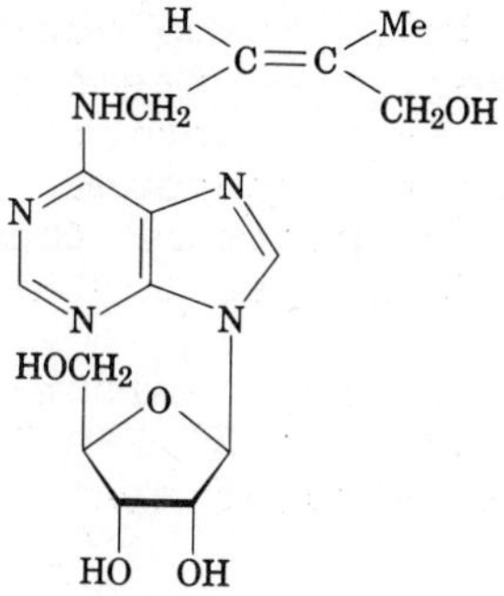

19

A sulfur-containing nucleoside from yeast t-RNA has been assigned the structure methyl 2-thio-5-uridineacetate (methyl 1,2,3,4-tetrahydro-4-oxo-2-thiopyrimidine-acetate, 20) on the basis of high-resolution mass-spec-

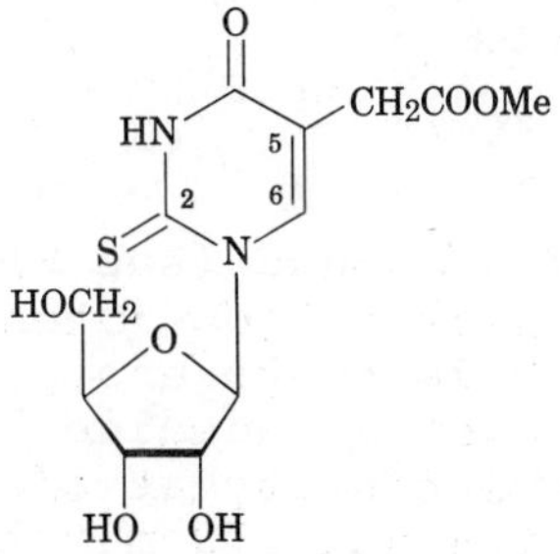

20

trometry, chemical properties, and ultraviolet spectra.[24] The isomeric 6-substituted 2-thiouridine structure was not completely ruled out, however. Due to the scarcity of the isolated material and the presence of impurities, it was necessary to obtain the elemental composition of all of the ions from the high-resolution mass-spectrum by computer. The ions due to 20 were recognized from their characteristic heteroatom content, which differed from the heteroatom content of the impurities. Peaks at B+*H* and B+2*H* were present at $C_7H_8N_2O_3S$ and $C_7H_9N_2O_3S$, indicating a two nitrogen pyrimidine base, not a five-nitrogen purine base; the presence of ions of composition $C_5H_9O_4$ corresponds to a pentosyl moiety. No molecular ion is present for $C_{12}H_{16}N_2O_7S$; however, the presence of ions corresponding to M-$CH_3O\cdot$ and M-H_2S confirms the elemental composition postulated for the compound. More-detailed interpretation of the mass spectrum, together with other physical and chemical data, led to the tentative assignment 20.

Another naturally occurring, sulfur-containing nucleoside has been identified[25] as 5'-*S*-methyl-5'-thioadenosine (21). The mass spectra of synthetically prepared

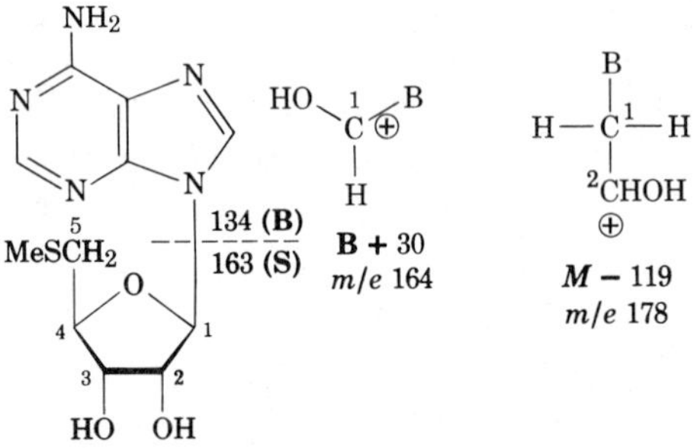

NH2
MeSCH2
HO OH
134 (B)
163 (S)
HO C⊕ B
H
B + 30
m/e 164
H—C—H
B
CHOH
⊕
***M* − 119**
m/e 178

21

21 and of the isolated compound (see Fig. 10) are identical. A molecular-ion peak of low intensity is found at *m/e* 297. The B+*H* and B+2*H* peaks are present at *m/e* 135 and 136, characteristic of an adenine base. The sugar fragment (S) is of low intensity at *m/e* 163. The prominent fragments at *m/e* 164 and 178 may be assigned to B+30 and *M*-119, in analogy with assignments made for the mass spectrum of adenosine.[3,10]

The behavior of puromycin nucleoside (22), and some derivatives, upon electron impact has been investigated.[26] The base peak in the mass spectrum (see Fig. 11) of 22 occurs at *m/e* 164, corresponding to B+2H. A metastable peak is present for the formation of *m/e* 134 ($C_6H_6N_4$)

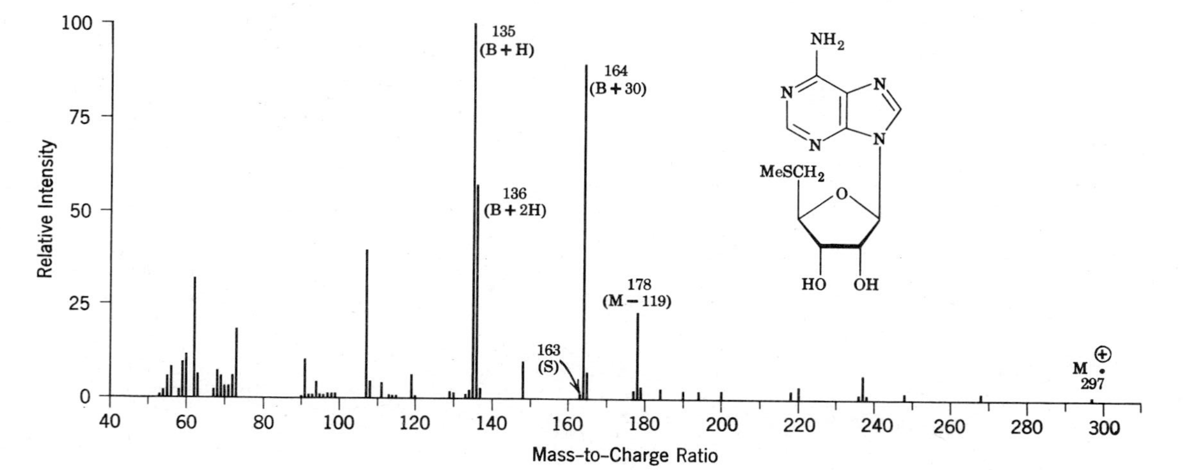

Fig. 10 The 70-eV mass spectrum of 5'-_S_-methyl-5'-thioadenosine (_21_).

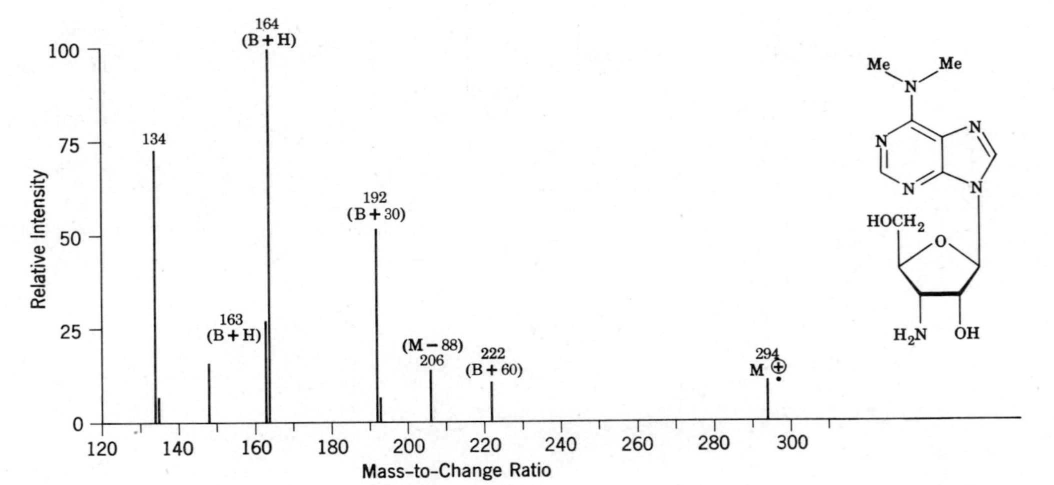

Fig. 11 The 70-eV mass spectrum of puromycin nucleoside (22).

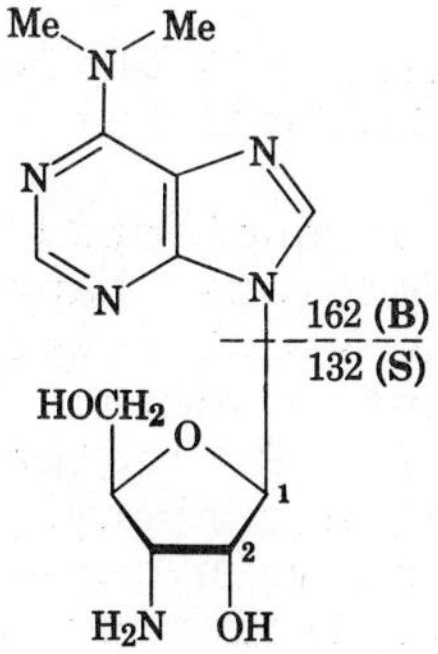

22

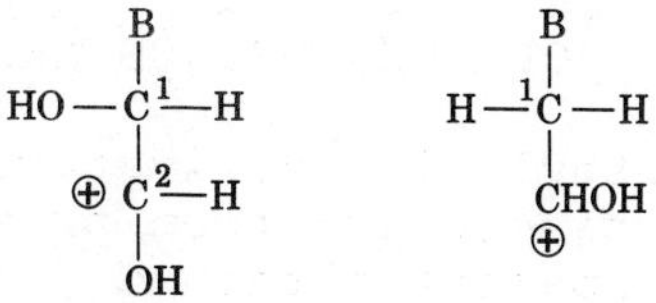

B + 60
m/e 222

M − 88
m/e 206

23

from B+H (m/e 163) with the loss of $CH_3N\cdot$; loss of $CH_3N\cdot$ also occurs from the molecular ion of N,N-dimethyladenine,[15] as discussed in Section II (see p. 149). The peak at m/e 132, corresponding to the sugar moiety (S), is negligible. The B+30 peak is present at m/e 192, and M-88 is found at m/e 206. The peak at m/e 222 has been interpreted as being B+60 (23); B+60 peaks have been observed as low-intensity ions in the mass spectra of other nucleosides.[6]

The mass spectrum of puromycin (24) is summarized in

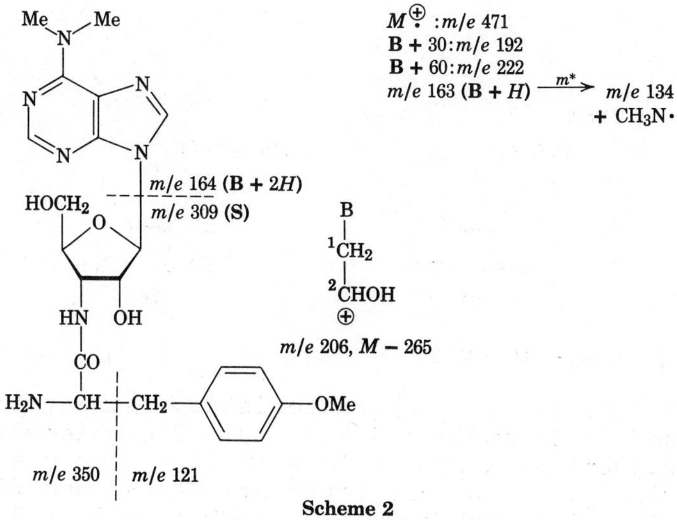

$M^{+\cdot}$: m/e 471
B + 30: m/e 192
B + 60: m/e 222
m/e 163 (B + H) $\xrightarrow{m^*}$ m/e 134 + $CH_3N\cdot$

m/e 164 (B + 2H)
m/e 309 (S)
m/e 206, M − 265
m/e 350 | m/e 121

Scheme 2

Scheme 2.[26] Cleavages are present that are characteristic of the amino acid that is attached to the sugar, as well as fragments characteristic of the puromycin nucleoside.

VI. MASS SPECTRA OF TRIMETHYLSILYL DERIVATIVES OF NUCLEOSIDES AND NUCLEOTIDES

The trimethylsilyl derivatives of nucleotides, nucleosides, and bases have been prepared (see Chapter 3) and their mass spectra obtained.[7] The spectra were interpreted with the aid of high-resolution data, metastable peaks, and deuterium labeling [$(CD_3)_3Si$ instead of $(CH_3)_3Si$]. Amino groups, enolizable carbonyl groups, and sugar and phosphate hydroxyl groups were generally mono-(trimethylsilyl)ated; trimethylsilylation of the bases in thymidine, 2'-deoxyuridine, and cytidine 5'-phosphate occurred to only a small extent, however. The mass spectra may be obtained directly (as the derivatives are eluted from a gas chromatograph) or after introduction through a direct-inlet probe.[7] The mass spectrum of the pentakis(trimethylsilyl) derivative (<u>25</u>) of guanosine is

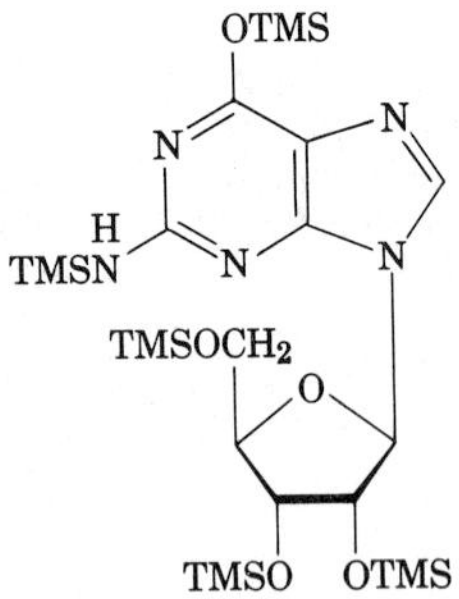

25 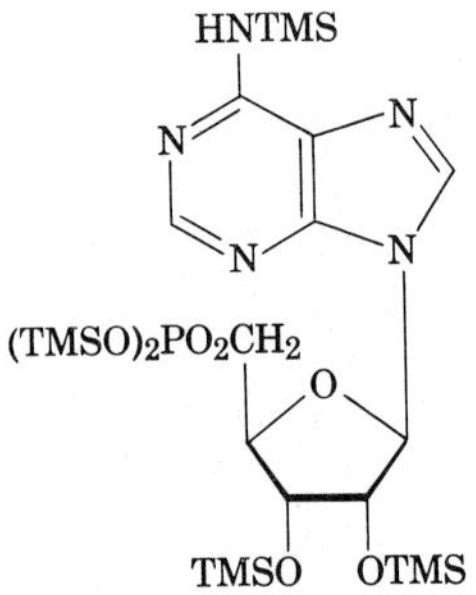 **26**

where TMS = Me_3Si.

given in Fig. 12. The characteristic fragmentations are summarized in Scheme 3 and Fig. 12. The molecular ion and a fragment peak due to loss of $CH_3\cdot$ from one of the trimethylsilyl groups are found at *m/e* 643 and 628, respectively. The ions at *m/e* 73 [$(CH_3)_3Si^{\oplus}$], 147 [$CH_3)_3SiOSi(CH_3)_2$], and 217 [$C_3H_4(OTMS)_2^{\oplus}$] are frequentl

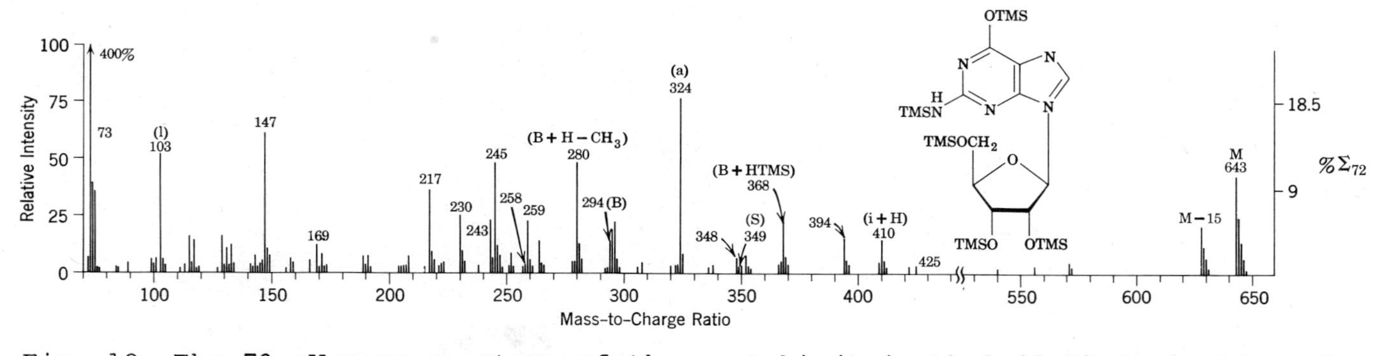

Fig. 12 The 70-eV mass spectrum of the pentakis(trimethylsilyl) derivative of guanosine (25). [Reproduced, by permission, from *J. Amer. Chem. Soc.*, 90, 4182 (1968).]

observed in the mass spectra of (trimethylsilyl) ated compounds.[27,28] D-(or L)-Ribose derivatives generally give[7] an abundant ion at m/e 230 [$C_4H_4(OTMS)_2^{\cdot\oplus}$].

m
R—CH₂ B (TMS)
B (+H, 2H or HTMS)
S
1
R = OTMS,
OPO(OTMS)₂
O
a $\xrightarrow[m^*]{-CO}$ B + 2H
TMSO OTMS
i(+H) $\xrightarrow[m^*]{CH_3\cdot}$ j
k(+H)

c $\xrightarrow{-TMSOH}$ d $\xrightarrow[m^*]{-CH_3\cdot}$ e
–H (S → c)
S $\xrightarrow{-(CH_2R + H)}$ f
–TMSOH (S → g)
g $\xrightarrow{-(R + H)}$ h

Scheme 3

The characteristic B+H peak is found at m/e 280 (see Fig. 12), after loss of $CH_3\cdot$ from a TMS group. A small peak at m/e 349 is present for the sugar moiety (S) and its three TMS groups, although a large peak at m/e 259 is formed when trimethylsilanol (TMSOH) is eliminated from m/e 349. The B+30 fragment is prominent at m/e 324. The peak corresponding to M-89 in the mass spectrum (see Fig. 6, p. 159) of adenosine (11) is found at m/e 410 (se Fig. 12).

Thus, the mass spectra of the TMS derivatives of nucleosides are similar to the mass spectra of the free nucleosides, except for the shifts caused by retention of the TMS groups, and for fragmentations and ions characteristic of the TMS groups. Determination of molecular weight and elemental composition, identification of the base, and confirmation that the sugar is a pentose or a 2- or 3-deoxypentose are possible. The addition of TMS groups greatly increases the molecular weight, as well as the volatility, but makes the mass spectra more complex and therefore more difficult to interpret.

In Fig. 13 is shown the mass spectrum of the pentakis-(trimethylsilyl) derivative of adenosine 5'-phosphate (26). Fragmentations are summarized in Scheme 3, and, in this case, peaks found in the mass spectra of nucleo-

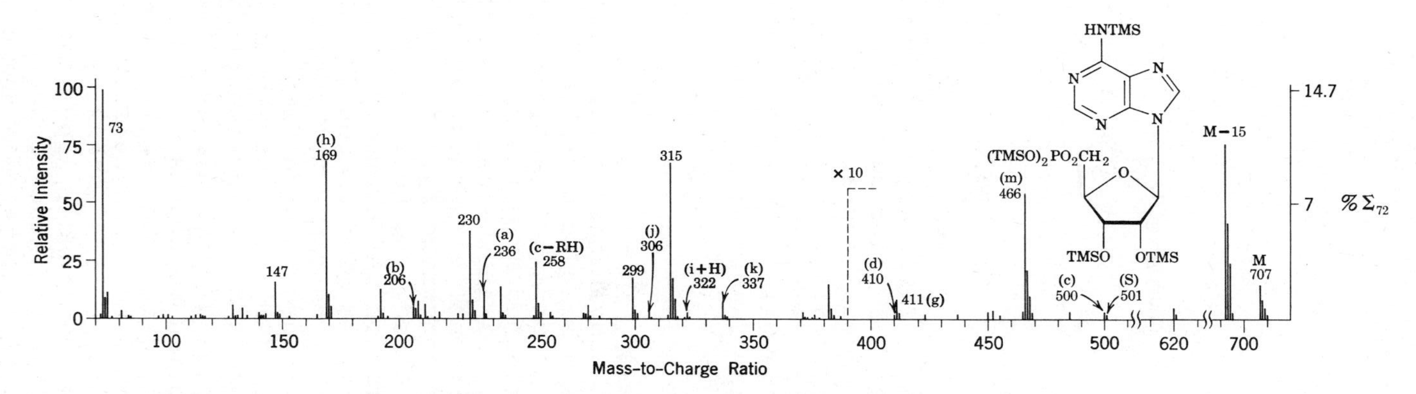

Fig. 13 The 70-eV mass spectrum of the pentakis(trimethylsilyl) derivative of adenosine 5'-phosphate (26). [Above *m/e* 390, the peaks are shown ten times enlarged (x 10). Reproduced, by permission, from *J. Amer. Chem. Soc.*, 90, 4182 (1968).]

$(TMSO)_2PO_2CH_2$ ... $O^{\oplus}$ ... H $\longrightarrow$ *m/e* 411 + TMSOH

TMSO OTMS $\qquad$ $\longrightarrow$ *m/e* 169 + $HO_2P(OTMS)_2$

m/e 501

sides are relatively unimportant. Loss of the 5'-substituent, $(TMSO)_2PO_2\cdot$, leads to the peak at *m/e* 466. Intramolecular rearrangement of a TMS group is responsible for the unusual ion at *m/e* 315, namely, $HO\text{-}\overset{\oplus}{P}\text{-}(OTMS)_3$ (that is, $C_9H_{28}O_4PSi_3$); this ion loses CH_4, forming *m/e* 299. The sugar moiety (S) at *m/e* 501 eliminates TMSOH, followed by $HO_2P(OTMS)_2$. The mass spectrum thus permits elucidation of the fundamental structural features.

VII. SUMMARY

The advantages of mass spectrometry are: the small sample size required, the exact molecular weights that can be obtained, and the structural information that may be gleaned from the fragmentation processes. The disadvantages are: loss of sample, the high investment in equipment, personnel, and maintenance of a capability in mass spectrometry, the need for volatility of the sample, and often, the need for model compounds before the fragmentation patterns may be analyzed in terms of structure.

Samples of a compound whose volatility is too low for use may be converted into more volatile derivatives or may be degraded to simpler molecules, which can then be studied by mass spectrometry. A gas chromatograph, coupled to a mass spectrometer via an appropriate "interface," affords a particularly efficient purification and identification apparatus.

The data obtained from the mass spectra of purines, pyrimidines, and nucleosides of known structure represent a detailed and extensive background of information on these compounds. Applications of this information to newly discovered compounds, as discussed in Section V (p. 162), illustrate the potentiality of mass spectrometry for nucleic acid research. Mass spectrometry has supplied significant, and, at times, decisive, data for identification of nucleic acid components; many more examples of its application in this area will undoubtedly appear in the years ahead.

REFERENCES

(1) J. H. Beynon, R. A. Saunders, and A. E. Williams, "The Mass Spectra of Organic Molecules," Elsevier Publishing Company, Amsterdam (1968).

(2) R. W. Kiser, "Introduction to Mass Spectrometry and Its Applications," Prentice-Hall, Inc., Englewood Cliffs, N. J. (1965).

(3) K. Biemann, "Mass Spectrometry," McGraw-Hill Book Company, Inc., New York (1962).

(4) J. M. Rice and G. O. Dudek, *J. Amer. Chem. Soc.*, 89, 2719 (1967).

(5) J. M. Rice, G. O. Dudek, and M. Barber, *J. Amer. Chem. Soc.*, 87, 4569 (1965).

(6) J. A. McCloskey, Ph.D. Thesis, Massachusetts Institute of Technology (1963).

(7) J. A. McCloskey, A. M. Lawson, K. Tsuboyama, P. M. Krueger, and R. N. Stillwell, *J. Amer. Chem. Soc.*, 90, 4182 (1968).

(8) B. E. Tropp, J. H. Law, and J. M. Hayes, *Biochemistry*, 3, 1837 (1964).

(9) T. Nishiwaki, *Tetrahedron*, 22, 3117 (1966).

(10) K. Biemann and J. A. McCloskey, *J. Amer. Chem. Soc.*, 84, 2005 (1962).

(11) K. Biemann and P. V. Fennessey, *Chimia*, 21, 226 (1967).

(12) A. Tatematsu, T. Goto and S. Matsuura, *Nippon Kagaku Zasshi*, 87, 71 (1966).

(13) T. Goto, A. Tatematsu, and S. Matsuura, *J. Org. Chem.*, 30, 1844 (1965).

(14) J. S. Shannon and D. S. Letham, *N. Z. J. Sci.*, 9, 833 (1966).

(15) Y. Rahamim, J. Sharvit, A. Mandelbaum, and M. Sprecher, *J. Org. Chem.*, 32, 3856 (1967).

(16) D. S. Letham, J. S. Shannon, and I. R. McDonald, *Proc. Chem. Soc.*, (1964) 230.

(17) T. Nishiwaki, *Tetrahedron*, 23, 1153 (1967).

(18) C. Fenselau and S. Y. Wang, *Tetrahedron*, 25, 2853 (1969).

(19) S. Tsunakawa, *Shitsuryo Bunseki*, 15, 143 (1967).

(20) S. Hanessian, D. C. DeJongh, and J. A. McCloskey, *Biochim. Biophys. Acta*, 117, 480 (1966).

(21) K. Biemann, S. Tsunakawa, J. Sonnenbichler, H. Feldmann, D. Dütting, and H. G. Zachau, *Angew. Chem. Intern. Ed. Engl.*, 5, 590 (1966).

(22) R. H. Hall, M. J. Robins, L. Stasiuk, and R. Thedford, *J. Amer. Chem. Soc.*, 88, 2614 (1966).
(23) R. H. Hall, L. Csonka, H. David, and B. McLennan, *Science*, 156, 69 (1967).
(24) L. Baczynskyj, K. Biemann, and R. H. Hall, *Science*, 159, 1481 (1968).
(25) T. M. Chu, M. F. Mallette, and R. O. Mumma, *Biochemistry*, 7, 1399 (1968).
(26) S. H. Eggers, S. I. Biedron, and A. O. Hawtrey, *Tetrahedron Lett.*, 3271 (1966).
(27) J. A. McCloskey, R. N. Stillwell, and A. M. Lawson, *Anal. Chem.*, 40, 233 (1968).
(28) D. C. DeJongh, T. Radford, J. D. Hribar, S. Hanessian, M. Bieber, G. Dawson, and C. C. Sweeley, *J. Amer. Chem. Soc.*, 91, 1728 (1969).

CHAPTER 5

Optical Rotatory Dispersion of Nucleosides and Nucleotides*

T. L. V. ULBRICHT**

TWYFORD LABORATORIES LTD., LONDON, ENGLAND

*The author thanks Mr. G. T. Rogers for his helpful criticism, and for drawing the Figures.

**Present address: Planning Section, Agricultural Research Council, 160 Great Portland Street, London, England.

I. INTRODUCTION

1. Optical Activity

To be asymmetric in *n*-dimensional space, the shape of an object has to be defined by not less than $n + 1$ properties. For example, in 3-dimensional space, four properties are required. These properties may be expressed as a polar vector and an axial vector (each consisting of two factors) which define the sense of a screw (*i.e.*, whether left- or right-handed): a left- and a right-handed screw are related in the same way as an object and its nonsuperposable mirror image. In stereochemical terms, a molecule will be optically active if it lacks a center of inversion, a plane of symmetry, and an alternating rotation-reflection axis of symmetry. A compound containing a carbon atom to which are attached four different groups is the simplest example.

The experimentally observable phenomenon of optical activity consists in the rotation of the plane of linearly polarized light by an optically active medium (in practice, a solution of the substance under investigation in a suitable, optically inactive solvent). Linearly polarized light may be considered as consisting of equal components of left- and right-circularly polarized light of identical frequency. These two components have unequal velocities in an optically active medium, and the resultant phase-shift rotates the plane of polarization of the emergent light through an angle, α, the optical rotation. This rotation is called dextrorotatory (+) if, like right-circularly polarized light, it is clockwise, and levorotatory(-), if, like left-circularly polarized light, it is counterclockwise.

2. Optical Rotatory Dispersion and Circular Dichroism

The specific rotation, $[\alpha]$, is defined by equation *1*

$$[\alpha]_{\lambda}^{t} = \alpha \times 100/\underline{l} \times \underline{c}, \qquad \ldots(1)$$

where α is the observed rotation, in circular degrees, t is the temperature, λ is the wavelength at which the rotation is measured, $\underline{l}$ is the light path, in decimeters, and $\underline{c}$ is the concentration in grams per 100 milliliters of solution. The molecular rotation $[M]$ is defined by equation *2*.

$$[M]_{\lambda} = [\alpha]_{\lambda} \times M/100, \qquad \ldots(2)$$

where M is the molecular weight of the solute.

For many years, the specific rotation was usually measured at a single wavelength, the D-line (at 589 nm) of sodium; this specific rotation is known as $[\alpha]_D$. However, the specific rotation varies with the wavelength, and this variation or spectrum of specific rotation *versus* wavelength is known as the optical rotatory dispersion (o.r.d.).

If the optically active medium absorbs left- and right-circularly polarized light unequally, the emergent light will also be elliptically polarized. This unequal absorption is called circular dichroism (c.d.). The phenomenon of the unequal velocity of transmission and the unequal absorption of left- and right-circularly polarized light is called the Cotton effect, after its discoverer.

Fig. 1 shows the relationship between the u.v. absorption spectrum, o.r.d., and c.d. The example is an ideal-

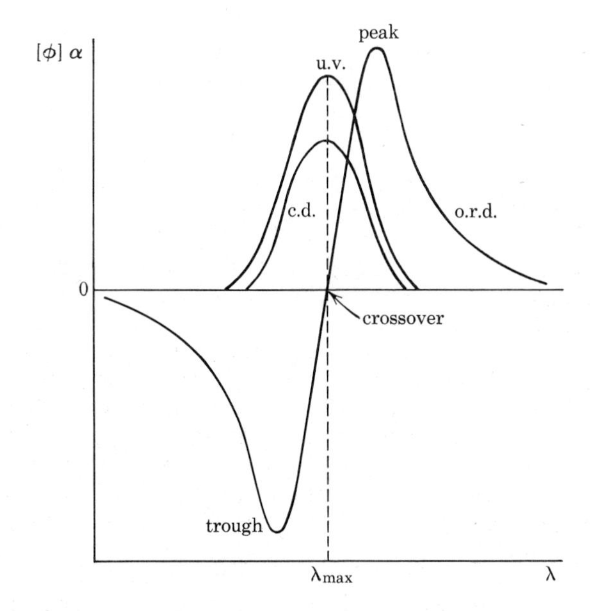

Fig. 1. Relationship Between U.V. Absorption, C.D., and O.R.D. (Positive Cotton-effect).

ized one, with a u.v. spectrum due to a single (optically active) transition, giving a symmetrical curve. The maxima and minima in o.r.d. and c.d. are called extrema. The c.d. extremum coincides with λ_{max} of the absorption spectrum, but the c.d. may be positive (see Fig. 1) or negative (see Fig. 2), depending on whether the left- or

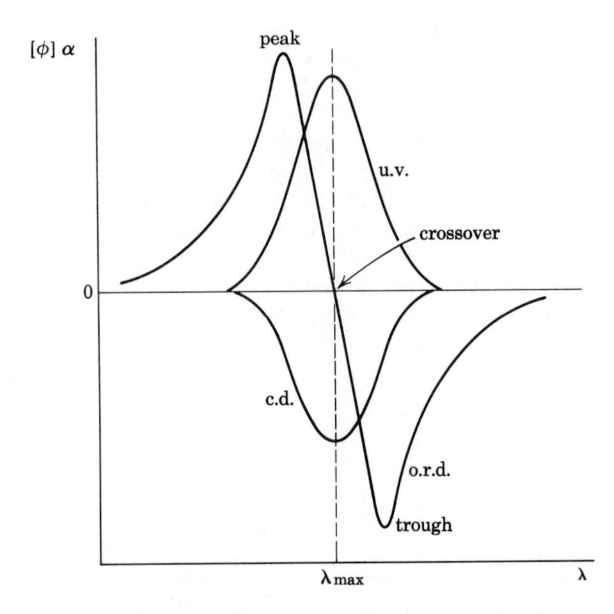

Fig. 2. Relationship Between U.V. Absorption, C.D., and O.R.D. (Negative Cotton-effect).

the right-circularly polarized component is the more strongly absorbed. The Cotton effect is called positive if the peak (maximum) of the o.r.d. is on the long-wavelength side of λ_{max} (see Fig. 1), and negative if the trough (minimum) of the o.r.d. is on the long-wavelength side (see Fig. 2). It will be noted that, in the idealized case, there is no rotation at λ_{max}; this is the "crossover" point. The amplitude of a Cotton effect is equal to the molecular rotation [*M*] at the first extremum minus the molecular rotation at the second extremum, divided by 100.

The curves shown in Figs. 1 and 2 are due to a single, optically active band; should there be more than one such band, there will be multiple Cotton effects, giving much more complicated spectra. It is obvious from these idealized curves, and confirmed in practice, that c.d. curves tend to be less complicated than o.r.d. curves. In cases of multiple Cotton-effects, overlapping in the o.r.d. spectrum may make interpretation very difficult, and the c.d. spectrum is often clearer, because it is more highly resolved. Nevertheless, even if instruments of comparable sensitivity are used, the peak and trough of the o.r.d. may often be more readily measured than the c.d. extremum and, ideally, both measurements should be made on the same sample.

II. OPTICAL ROTATORY DISPERSION OF NUCLEOSIDES AND NUCLEOTIDES

1. Studies Made Prior to 1963

In nucleic acid chemistry, as in most other fields of chemistry, the specific rotation at the D-line of sodium has been measured as a physical constant characteristic of the particular compound, like the melting point. The difference between the two is that the melting point measures, in degrees, something quite definite (namely, a phase transition), whereas the $[\alpha]_D$ is an arbitrarily chosen value. Just how arbitrary it is may be appreciated by imagining what the situation would be had, for a period of many decades, measurements of the u.v. and i.r. absorption of chemical compounds been made at but a single wavelength. Under the circumstances, it is remarkable that measurements of the $[\alpha]_D$ have proved as useful as they have.

If measured accurately, under specified conditions of concentration, solvent, and temperature, the $[\alpha]_D$ is a constant, and it has been used (in the same way as the melting point) to check the purity of a compound, and also to make sure that it is the same optical isomer as a reference sample. This technique was important, for example, in comparing synthetic nucleosides with the naturally occurring ones, which had been proved to be either β-D-ribosyl or "2-deoxy-β-D-ribosyl" (2-deoxy-β-D-<u>erythro</u>-pentosyl) derivatives. When synthetic methods yielded both the α-D and the β-D anomers as products, the the β-D anomer could be identified, because it had the same specific rotation as the natural product. Difficulties in the assignment of anomeric configuration arose when analogs began to be synthesized; such compounds are not constituents of nucleic acids, and are, therefore, not known in Nature. This problem first became acute in connection with the synthesis of "2-deoxy-D-ribosyl" derivatives. Normally, acylated derivatives of the sugars are used in the synthesis of nucleosides, and the C-2 acyloxy group participates in the reaction, exercising stereochemical control in favor of the β-D anomer.[2] In the absence of an acyloxy group at C-2, there is no stereochemical control, and syntheses yield mixtures of the two anomers, in unpredictable proportions.

Attempts were made to apply Hudson's rules of isorotation to nucleosides. These rules state that, of a pair of anomeric D-glycosides, the more dextrorotatory is the α-D anomer.[3] With nucleosides, exceptions to these rules were soon found; o.r.d. studies later showed why these rules are sometimes obeyed, and sometimes not, as will be explained later in this Chapter (see p. 203).

Prior to 1963, a few attempts were made to measure the o.r.d. of nucleic acid derivatives, but, unfortunately, these measurements were valueless. It was later realized that the instruments then available were insufficiently sensitive, and, also, that it is possible to obtain rotatory artifacts unless very dilute solutions are used. Papers thereon published before the end of 1963 should, therefore, be ignored.

2. Information Revealed by Optical Rotatory Dispersion and Circular Dichroism Studies of Nucleosides and Nucleotides

There are two main reasons why o.r.d. and c.d. studies of nucleosides and nucleotides are undertaken. The first is to establish the anomeric configuration of the compound under investigation. This subject will be discussed in detail later (see p. 184), and it will be explained that such studies constitute the method of choice for furanosyl derivatives of pyrimidines and purines (and their phosphates), which thus far are the most thoroughly investigated groups.

The only other physical methods available for the determination of anomeric configuration are X-ray crystallography, impracticable as a routine method, and n.m.r. spectroscopy. Even with a 100-MHz n.m.r. spectrometer linked to a computer, much larger quantities of material are required for n.m.r. than for o.r.d. studies; moreover the assignment made may be uncertain, unless both anomers are available and can be compared, which is not a requirement for o.r.d. In some cases, a chemical method, such a anhydronucleoside formation, may be used to determine the anomeric configuration, but this procedure is wasteful of material.

Secondly, these procedures are valuable in conformation studies. As will be shown in the following Section, attempts to understand the o.r.d. of nucleosides and nucleotides have led to the realization that a major factor in determining the sign of the Cotton effect is the conformation of the molecule. Consequently, study of the o.r.d. of such compounds, under different conditions, provides information about the conformation. Optical rotatory dispersion has played a major part in meaningful studies of the conformation of oligo- and poly-nucleo tides.

3. Pyrimidine Nucleosides and Nucleotides

Since 1963, sensitive o.r.d. instruments have been available that record the o.r.d. in the u.v., as well as in the visible, region. In nucleosides and their derivatives, the chromophore (the heterocyclic base residue) is adjacent to the (optically active) sugar moiety, and, as a result, these compounds show Cotton effects.[7] Apart from certain exceptions, which will be discussed subsequently (see p. 193), pyrimidine β-D-nucleosides give positive Cotton-effects, whereas pyrimidine α-D-nucleosides give negative Cotton-effects. Changes in stereochemistry other than at the anomeric center (C-1 of the sugar residue) merely change the magnitude, but not the sign, of the Cotton effect.

Figure 3 shows the o.r.d. of thymidine (1), a β-D com-

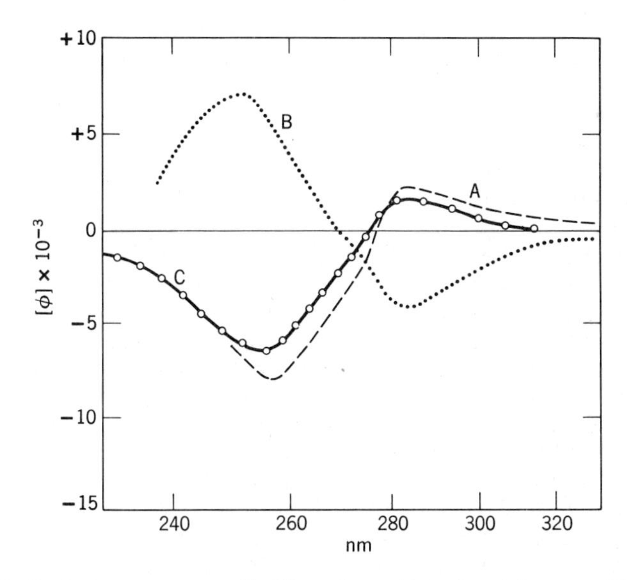

Fig. 3. O.R.D. curves of Thymine Derivatives (in Water). (A) (----) thymidine (1); (B) (.....) α-anomeric form of thymidine (2); (C) (o—o—o) 3'-thymidylic acid.

pound which has a peak at 282 nm and a trough at 255 nm. The crossover corresponds to the λ_{max}, at 267 nm. The α-D anomer (2) gives a negative Cotton-effect. Figure 3 also shows the o.r.d. of 5'-thymidylic acid, a β-D compound, and it will be noted that the phosphate group makes no significant difference to the o.r.d.; this is generally true for other nucleotides. Figure 4 shows a pair of

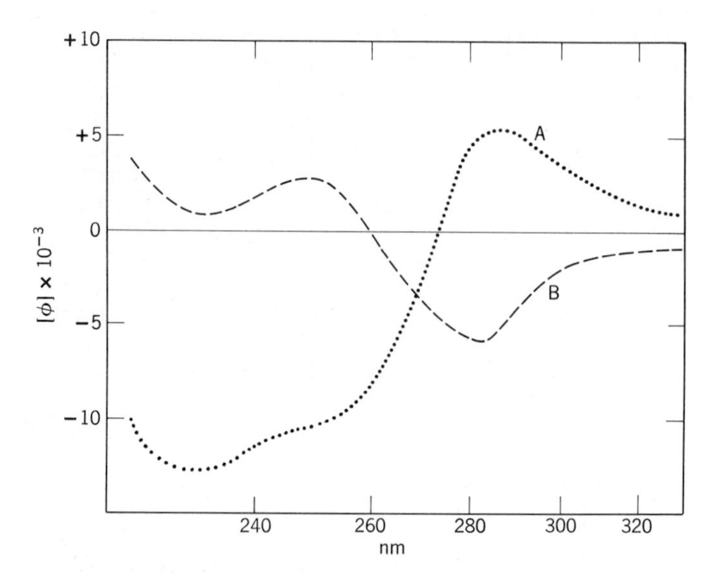

Fig. 4. O.R.D. Curves of Cytosine Nucleotides (in Water). (A) (.....) 3'-cytidylic acid (3); (B) (----) α-anomeric form of 3'-cytidylic acid (4).

anomeric ribonucleotides, 3'-cytidylic acid (3), a β-D compound that gives a positive Cotton-effect, and the corresponding α-D anomer (4), which gives a negative Cotton-effect. Many more such anomeric pairs have been measured, and similar results have been obtained.

Figure 5 shows a number of uridine (β-D) derivatives, all of which give positive Cotton-effects. There is little difference between uridine (5) and its 2',3',5'-triacetate (6), indicating that hydrogen bonding involving the sugar hydroxyl groups cannot be important insofar as the sign and magnitude of the Cotton effect are concerned. In contradistinction, both the C-2' epimer, 1-β-D-arabinofuranosyluracil (spongouridine, 7), and the 2-O-ethyl-2',3'-O-isopropylidene derivative (8) of uridine, give positive Cotton-effects of larger magnitude.

An intercomparison of cytidine (9), 2'-deoxycytidine (10), and 1-β-D-arabinofuranosylcytosine (11) (see Fig. 6) shows that replacement of the 2'-hydroxyl group by a hydrogen atom has little effect; however, as with the uracil compounds, the C-2' epimer (a D-arabinosyl derivative

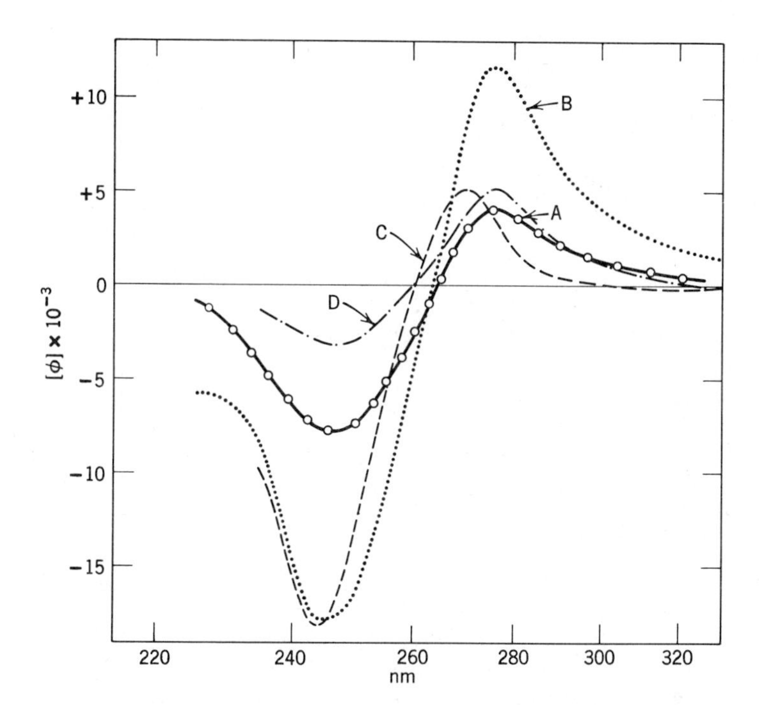

Fig. 5. O.r.d. Curves of Uracil Nucleosides. (A) (o—o—o) uridine (5, in water); (B) (.....) spongouridine (1-β-D-arabinofuranoxyluracil) (7, in water); (C) (----) 2-*O*-ethyl-2',3'-*O*-isopropylideneuridine (8, in methanol); (D) (-.-.-.) 2',3',5'-tri-*O*-acetyluridine (6, in methanol).

gives a significantly larger Cotton-effect. This result has been confirmed by an extensive intercomparison of D-ribosyl nucleosides with those containing D-arabinosyl and D-lyxosyl groups. However, the stereochemistry at C-3' has little effect on the sign and magnitude of the Cotton effect, and the Cotton effects for D-xylosylnucleosides do not differ significantly from those containing D-ribosyl groups.

2,2'-Anhydro-1-β-D-arabinofuranosyluracil (12) also gives a positive Cotton-effect, of much higher amplitude than that of uridine (5) (see Fig. 7), whereas 2,3'-anhydro-1-β-D-xylofuranosyluracil (13) gives a small, positive Cotton-effect. For 2,5'-anhydronucleosides, such as 2,5'-anhydro-5-methyluridine (14) (see Fig. 7), the sign of the Cotton effect is negative, and the amplitude is very large. On the other hand, 5',6-anhydro-6-hydr-

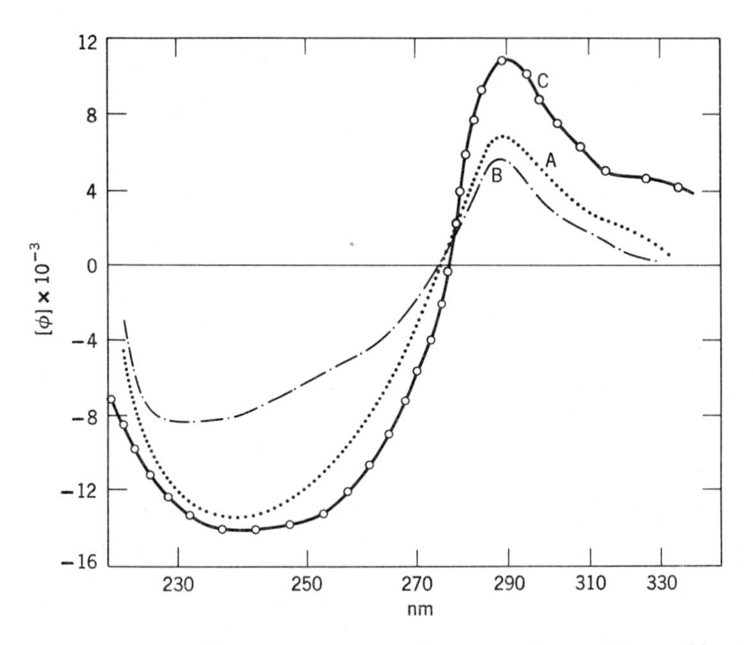

Fig. 6. O.R.D. Curves of Cytosine Nucleosides (in water).
(A) (....) cytidine (9); (B) (-.-.-.) 2'-deoxycytidine (10)
(c) (o-o-o-) 1-β-D-arabinofuranosylcytosine (11).

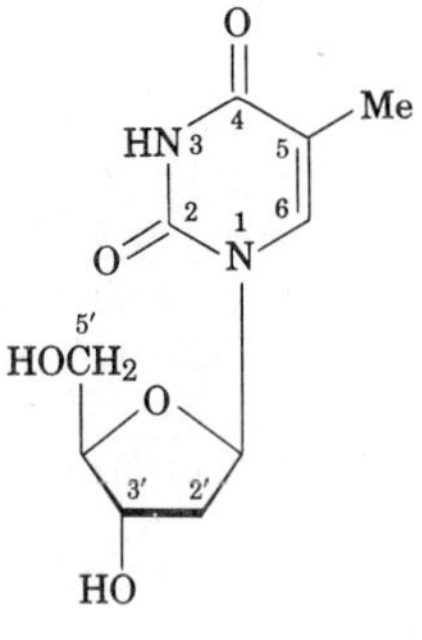

1 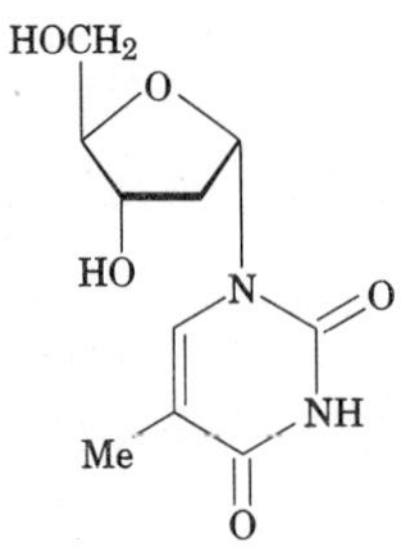 **2**

3

4

5

6

7

8

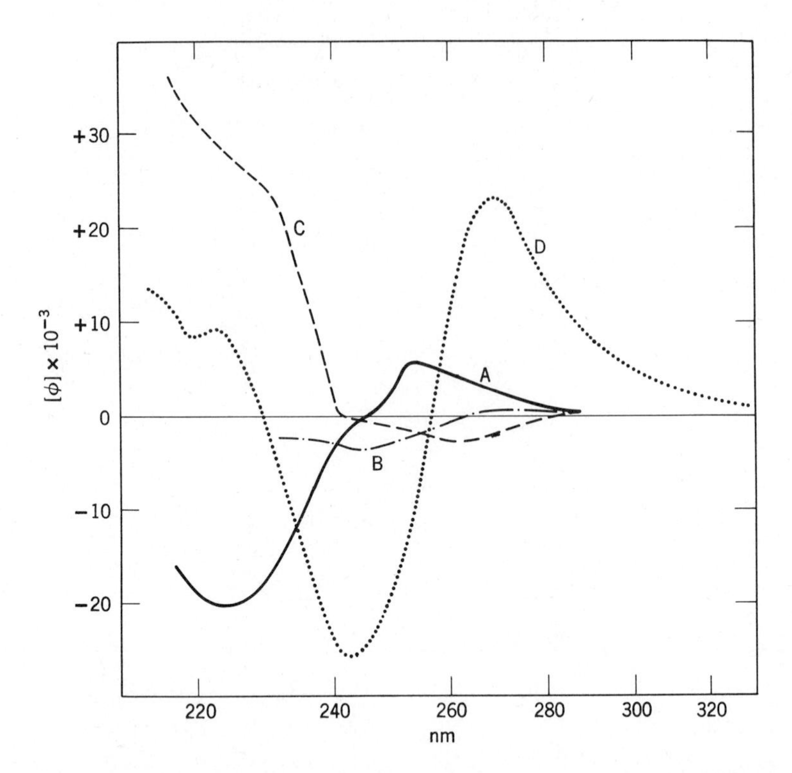

Fig. 7. O.R.D. Curves of Pyrimidine Anhydronucleosides (in water). (A) (———) 2,2'-anhydro-1-β-D-arabinofuranosylfuranosyluracil (12); (B) (-.-.-.) 2,3'-anhydro-1-β-D-xylofuranosyluracil (13); (C) (----) 2,5'-anhydrothymidine (14); (D) (....) 5',6-anhydro-6-hydroxyuridine (15).

NH_2 N O N $HOCH_2$ O HO R

9, R = OH
10, R = H

NH_2 N O N $HOCH_2$ O HO HO

11

oxyuridine (15) (see Fig. 7) shows a very large, positive Cotton-effect.

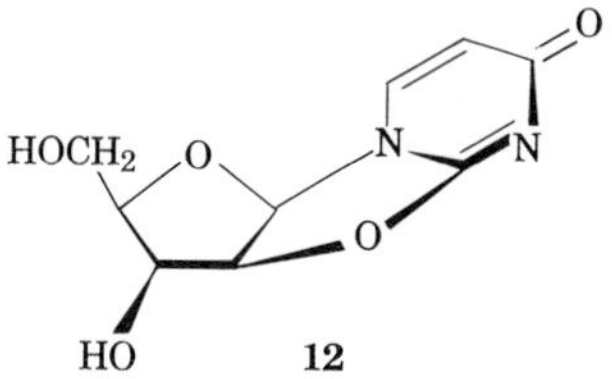

HOCH2 O N N O O HO 12

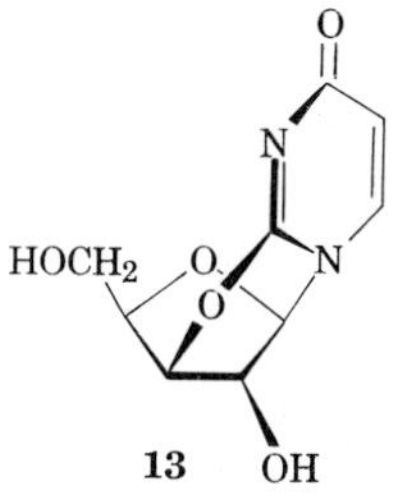

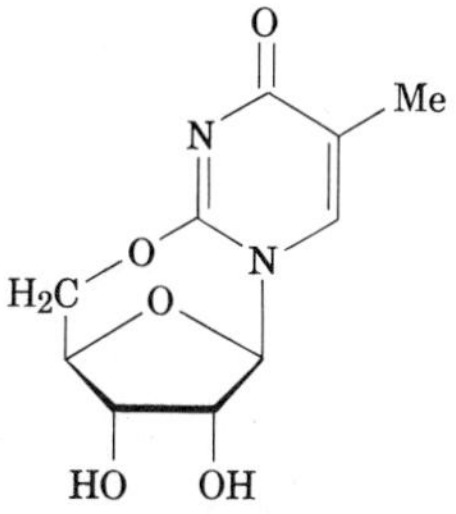

HOCH2 O N O N O O OH 13

O Me N O N H2C O HO OH 14

O NH O N O H2C O HO OH 15

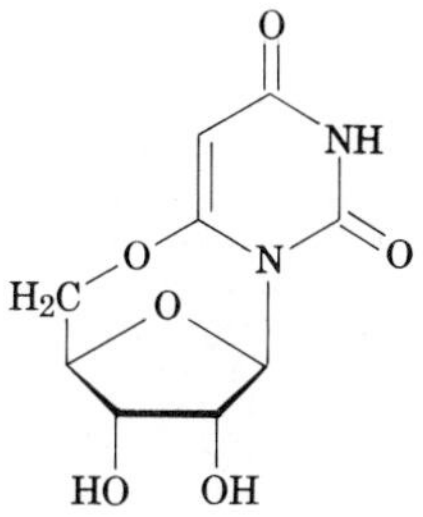

The above group of uracil-containing anhydronucleosides constitutes an important series, because the third ring present imparts to these molecules a fixed conformation. In each compound, the relative positions of the pyrimidine and D-ribose rings are different. The change in the sign and magnitude of the Cotton effect has been interpreted in terms of the changed conformation of the planar pyrimidine moiety with reference to the sugar ring, that is, the position of the chromophore relative to the asymmetric centers in the glycosyl group.

In simple pyrimidine nucleosides (not containing a third ring of the types present in compounds 12 through 15), rotation about the glycosyl-pyrimidine (C-1'-N) bond is theoretically possible. A study of models shows, however, that, in nucleosides of pyrimidines containing a carbonyl group at C-2, such as uracil, thymine, and cytosine, the proximity of this carbonyl group to the groups on C-2' and C-5' of the sugar interferes with free rotation about the nucleoside (glycosyl) bond. As a consequence, the pyrimidine ring, although not rigidly held, has a favored conformation, such that the carbonyl group at C-2 is directed away from the furanoid ring, as shown for uridine (5); this is known as the anti conformation of the molecule.

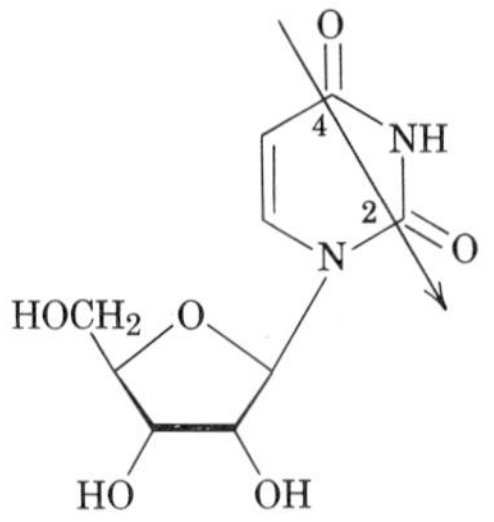

5

The effect of inversion at C-2' in a β-D-ribofuranosylpyrimidine, giving, for example, 1-β-D-arabinofuranosyluracil (7), is to restrict rotation about the nucleoside bond even more, and, as noted earlier, the magnitude of the Cotton effect in such compounds is increased. This effect also obtains with 2-*O*-ethyl-2',3'-*O*-isopropylideneuridine (8) (see Fig. 5), in which there is virtually no free rotation about the nucleoside bond owing to the large size of the 2-*O*-ethyl group, which must be directed away from the furanoid ring. The 2,2'-anhydroarabinofuranosylpyrimidines give Cotton effects of the same sign as, and similar magnitude to, those of the D-arabinofuranosylpyrimidines themselves (see Figs. 5 and 7).

5',6-Anhydro-6-hydroxyuridine has a fixed conformation (15), like that proposed for uridine, and it is noteworthy that the positive Cotton-effect of 15 is the largest of any measured; this supports the hypothesis that it is the anti conformation that is associated with a positive Cotton-effect. In contradistinction, the 2,5'-anhydrouridines (such as 14) give negative Cotton effects; these have the opposite (syn) conformation [compared with uridine (5) and with 15], as the pyrimidine ring has been rotated 180° about the nucleoside bond. 2,3'-Anhydro-1-β-D-lyxofuranosyluracil (13), which has a conformation intermediate between that of the 2,2'- and 2,5'-anhydronucleosides, has a small, positive Cotton-effect.

An attempt to illustrate the relative stereochemistry of the anhydronucleosides 12–15 is shown in Figure 8, which is a projection of a series of furanosyl derivatives of uracil. The nucleoside bond is at right angles to the plane of the page, at the center of the circle.

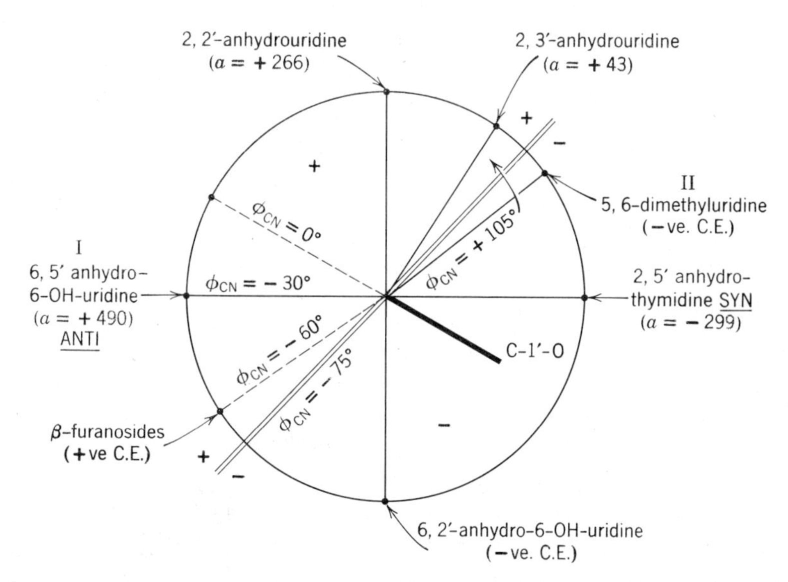

Fig. 8. O.R.D. of Pyrimidine Nucleosides. Relationship Between Sign and Magnitude of Cotton Effect and the Sugar-Base Torsion Angle, ϕ_{CN} (see p. 192).

The pyrimidine ring lies above the plane of the page. The sugar ring lies below the plane of the page, at an angle to it, and is held in such a position that the projection of the C-1'—ring-oxygen bond onto the plane of the page is in the position shown (heavy line). If the pyrimidine ring is now rotated while the sugar ring is held fixed, the oxygen atom at C-2 of the pyrimidine describes a circle. Different points on this circle represent different nucleoside conformations, as shown. If the conformation of the nucleoside is such that the same oxygen atom is in one half of the circle, the Cotton effect is positive; if the conformation of the nucleoside is such that the oxygen atom at C-2 is in the other half of the circle, the Cotton effect is negative.

Rule for Predicting the Sign of the Cotton Effect

On the basis of these and other results,[4-8] a rule has been formulated for predicting the sign of the Cotton effect for D-pentofuranosylnucleosides containing uracil, thymine, and cytosine. The rule is restated here in a slightly modified form.[8a] The sign of the Cotton effect will be positive if the following conditions are fulfilled:

1. the nucleoside has a favored conformation due to restricted rotation about the glycosidic bond such that the sugar-base torsion angle (ϕ_{CN}) is in the range -75° through 0° to +105°, and
2. a line from the C-4=O (or C-4—NH_2) group through the C-2=O group passes from above to below the plane of the furanoid ring.

The rule is illustrated for β-D compounds in the projection diagram given in Fig. 8.

The sugar-base torsion angle, ϕ_{CN}, is the angle formed by the trace of the plane of the base with the projection of the C-1'-O bond.[9] The line dividing the circle into positive and negative halves corresponds to ϕ_{CN} = -75° and +105°.

In Fig. 8, the amplitudes of the Cotton effects of the various compounds are given. The value ϕ_{CN} = -60° for ordinary β-D-nucleosides[9a] corresponds to a positive Cotton effect, as shown.

Examples of Applications of the Rule

All of the anhydronucleosides already discussed fulfil the conditions of the rule, as do compounds 12, 13, and 15; whereas 14 does not fulfil condition 1.

Pyrimidine α-D-nucleosides and pseudouridine (β-D) (16) do not fulfil condition 2; the line in these compounds passes from below to above the plane of the furanoid ring, and the Cotton effects are negative. For 3-β-D-ribofuranosylnucleosides, for example, 17, there is no favored conformation, because both oxo groups are in an "*ortho*"

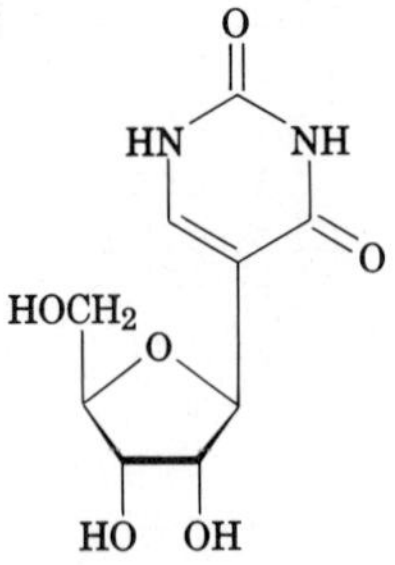

16

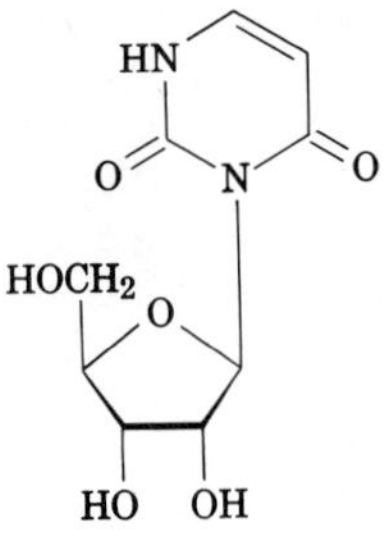

17

position to N-1. The first condition is not fulfilled, and the Cotton effect is negative.

The same condition is also not fulfilled in such 2',3'-unsaturated compounds as 1-(2,3-dideoxy-β-D-*glycero*-pent-2-enofuranosyl)uracil (18). The hydrogen atoms on C-2' and C-3' are in the plane of the sugar ring; there is virtually free rotation about the nucleoside bond, and the Cotton effect is negative. (The fact that two asymmetric centers have been replaced by a new π-electron system that is α,β to the anomeric center may also be an important factor.)

The compound 2,2'-anhydro-6-hydroxy-5-iodo-1-β-D-arabinofuranosyluracil (19) has ϕ_{CN}= +120°, and so does not fulfil condition 1; it has a negative Cotton effect.[10] Compound 19 has the same relationship to 12 as 15 has to 14; in each case, a rotation of 180° about the nucleoside

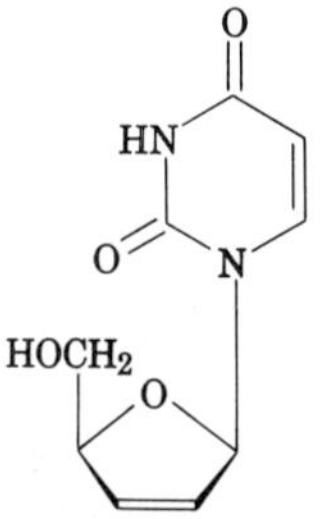

18

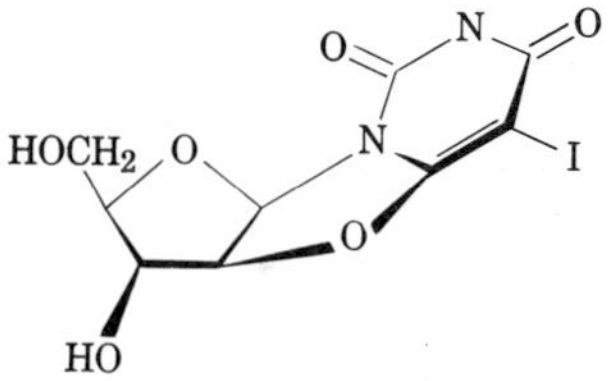

19

bond changes the sign of the Cotton effect.

The largest group of compounds studied consists of α-D- and β-D-furanosylpyrimidines that lack any special or unusual structural features; these all give positive and negative Cotton-effects, respectively, as expected.

It was assumed originally[6] that, although the chromophores of nucleosides and anhydronucleosides may differ somewhat (as their u.v. data indicate), this factor does not in itself lead to a drastic alteration of their o.r.d. behavior. Later c.d. studies[10a] then showed that the long-wavelength Cotton-effect observed in the o.r.d. spectra of pyrimidine anhydronucleosides is due to two separate Cotton-effects. For example, corresponding to the positive Cotton-effect centered at 245 nm in the o.r.d. spectrum of 2,2'-anhydrouridine (12), there is a small, negative Cotton-effect at 270 nm and a large, positive one at 240 nm. Robins and coworkers[10a] assumed that the small, long-wavelength Cotton-effect is due to a B_{2u} transition, but no evidence of any kind was presented in support of this opinion. Were this assignment correct, the rule would be invalid.

As a result of further investigations, it has been shown that the small, long-wavelength Cotton-effect is due to an $n \rightarrow \pi^*$ transition.[10b] Characteristics of $n \rightarrow \pi^*$ transitions are: (1) low intensity of the u.v. absorption, which may be difficult to detect; (2) a blue shift on going from a nonpolar to a polar solvent, with the biggest shift on going to water and a diminution in ε_{max}; (3) in aqueous acid (nitrogen protonation), there is a further blue-shift, or the band may disappear altogether.[1]

The c.d. spectrum of 2,2'-anhydrouridine diacetate shows a small, negative Cotton-effect at 281 nm (see Fig. 8a). In acetonitrile, this shows a small blue-shift, and in water, it disappears altogether. It is obvious from the ultraviolet spectrum (Fig. 8a) that this small, negative Cotton-effect is associated with a transition of very low intensity. All the above evidence suggests that the Cotton effect at 280 nm is due to an $n \rightarrow \pi^*$ transition. On going from acetonitrile to water as the solvent, a blue shift of 15 nm would be expected, and the resultant Cotton effect at 263 nm would, because of its low intensity, be hidden by the large, positive Cotton-effect. (The latter is probably due to an overlap of the B_{1u} and B_{2u} Cotton-effects.)

Similarly, it was shown that, in 2,5'-anhydro-uridine derivatives, there is a small, positive, long-wave-length Cotton-effect that is also an $n \rightarrow \pi^*$ transition. This work, therefore, confirms that the B_{2u} Cotton-effect in 2,2'-anahydrouridines is positive and in 2,5'-anhydrouridines is negative, in accord with the rule for predicting the sign of this Cotton effect in pyrimidine necleosides.[8a]

Compounds not Covered by the Rule

a. Azapyrimidine (as-Triazine) Nucleosides

Early in these studies, it was discovered that the sign of the Cotton effect in ordinary azapyrimidine nucleosides is reversed. The β-D anomers (for example, 6-azauridine) give negative Cotton-effects, whereas the corresponding α-D anomers give positive Cotton-effects.[6] There is no obvious reason why the conformation of 6-azauridine (<u>20</u>) should differ from that of uridine (<u>5</u>); hence, it might be that the change in the sign of the Cotton effect is due to a difference in the direction of the transition moment (because condition 2 of the rule governing the sign of the Cotton effect is related to this). To investigate this point, it was necessary to study appropriate anhydronucleosides. However, azapyrimidine anhydronucleosides give complicated o.r.d. curves. The c.d.

spectrum of 6-azauridine[10d] in water at pH 9 (anionic species) shows a small, positive Cotton-effect at 297 nm that does not correspond to any observed u.v. maximum (λ_{max} 264 nm; see Fig. 9). There are also negative and positive Cotton-effects at 259 and 215 nm, respectively. At pH 4 (protonated species), the latter Cotton-effects are still found, but that at 297 nm has disappeared (see Fig. 9). Almost identical results have been obtained with 5-methyl-6-azauridine (D-ribofuranosylazathymine). L-6-Azauridine gives similar results, but, as expected, the Cotton effects (C.E.) are of opposite sign (at pH 9: neg. C.E.

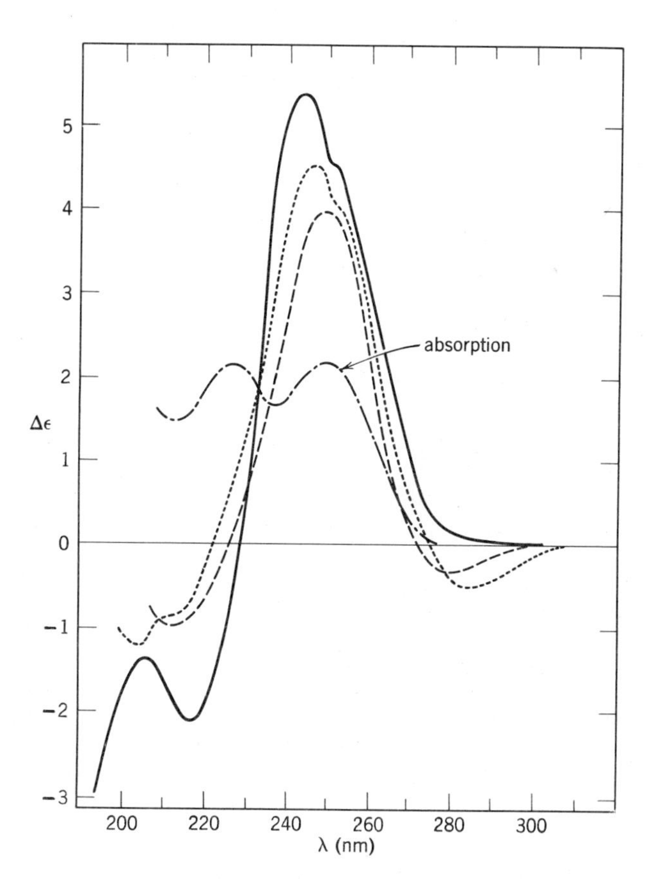

Fig. 8a. U.V. Absorption (-.-.-.) and C.D. Spectra of 3',5'-Di-*O*-acetyl-2,2'-anhydrouridine [in water at pH 7.0 (____), in p-dioxane (......), and in acetonitrile (---)].

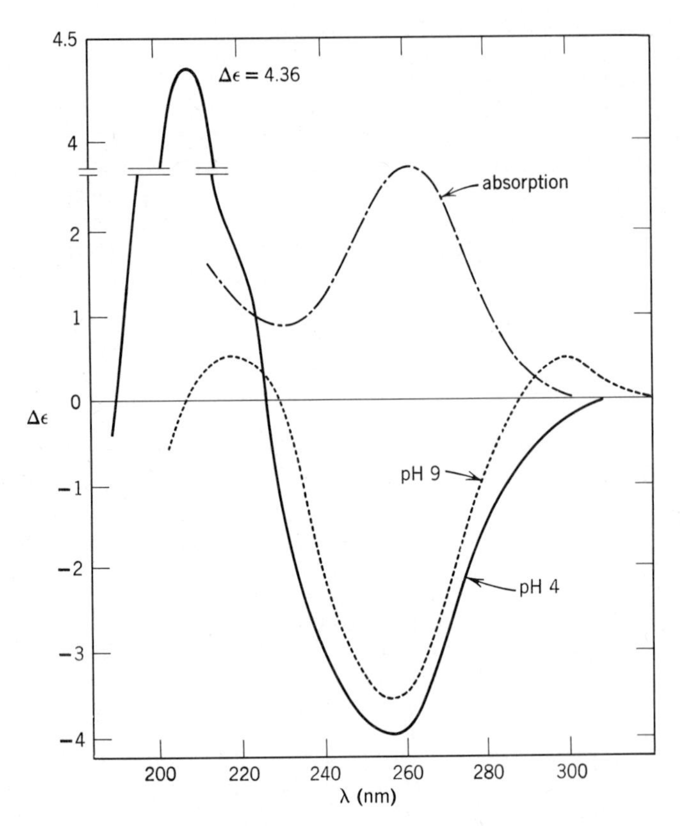

Fig. 9. U.V. Absorption and C.D. Spectra of 6-Azauridine.

at 298, pos. C.E. at 260, and neg. C.E. at 214 nm; at pH 4, the C.E. at 298 nm vanishes).

Unfortunately, 6-azauridine is too insoluble in non-polar solvents for studies to be carried out in them, but such experiments are possible with 2,2'-anhydroazauridine (21). The c.d. spectrum of 21 in *p*-dioxane (see Fig. 9a) shows several Cotton-effects. The small Cotton-effect at 312 nm shows a small blue-shift (3 nm) on going to a more polar solvent, namely, acetonitrile, and a further, larger blue-shift (15 nm) on going to water as the solvent. It is interesting that, in the u.v. spectrum in water, it is just possible to see a shoulder in the expected region (see Fig. 9a, arrow).

These c.d. studies clearly indicate that the small Cotton-effect in the 300-nm region is due to an $n \rightarrow \pi^*$ transition. These results, therefore, show that the B_{2u} Cotton-effect for 6-azauridine is opposite in sign to that

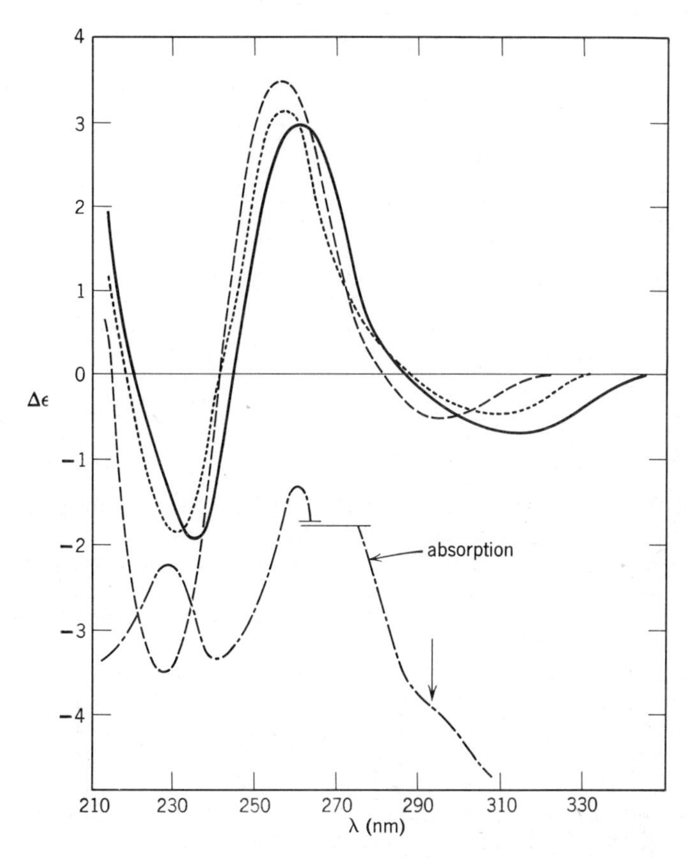

Fig. 9a. U.V. Absorption (-.-.-) and C.D. Spectra of 2,2'-Anhydro-6-azauridine [in *p*-dioxane (____), acetonitrile (....), and water at pH 7.0 (----)].

for uridine, whereas, for the corresponding anhydronucleosides (in which the nucleoside conformation is fixed), the sign is the same. This observation might suggest that the difference in sign of the Cotton effect for uridine and 6-azauridine is due to a difference in conformation. However, it seems more probable that it is due to a difference in the direction of the transition moment. It is possible to construct a projection diagram for β-D-azapyrimidine nucleosides, but data on too few compounds are available to permit fixing with any precision the line dividing the circle into positive and negative halves. However, the position of the line must be similar to its position in the purine diagram (see Fig. 12; p. 202).

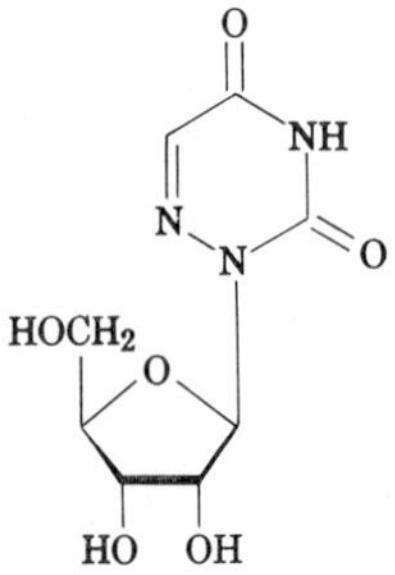

20 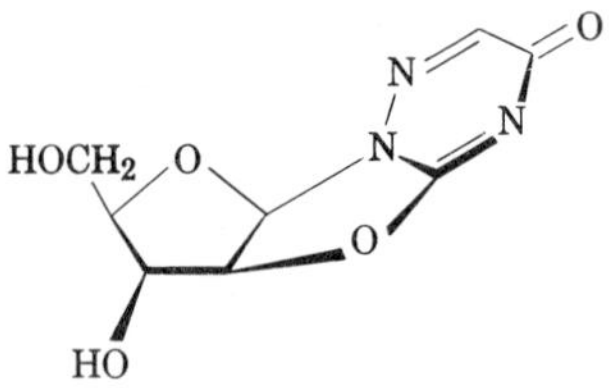 **21**

b. Compounds Containing Additional Chromophores.

An interaction between two chromophores in the same molecule gives rise to Cotton effects that may be of very high amplitude; in any case, the optical rotatory behavior will be different from that of a compound containing only one of the chromophores.

Nucleosides containing a 2',3'-double bond have already been mentioned (see p. 193). A more common example in nucleoside chemistry is a compound in which the sugar hydroxyl groups are protected with aromatic groups, such as benzyl, benzoyl, or *p*-toluoyl. The o.r.d. of very few such derivatives of pyrimidine nucleosides has been measured so far, but it is known from studies of purine nucleosides that the presence of such groups may change the sign of the Cotton effect.

c. Pyranosides.

If the glycosyl group is pyranoid, the conformational situation in a pyrimidine nucleoside is changed in two ways. Firstly, there are two chair conformations (1*C* and *C*1) of the D-pyranoid part. For the anomers of "2-deoxy-D-ribopyranosylthymine" [(2-deoxy-D-*erythro*-pentopyranosyl)-thymine], for example, it was shown that the conformation of the glycosyl group in the α-D anomer differs from that in the β-D anomer. If this situation obtains, it complicates the interpretation of o.r.d. results, and it would be necessary to study a series of pyranosyl nucleosides having known conformations of the glycosyl group in order to understand their o.r.d. behavior. Secondly, because the molecule is more flexible as a whole than that of a furanosyl nucleoside, rotation about the nucleoside bond is less restricted and, consequently, Cotton effects tend to be smaller.

Despite the foregoing, nearly all pyranosyl pyrimidines whose anomeric configurations are known, and whose o.r.d. have been measured, have obeyed the rule formulated for the furanosyl nucleosides. An exception is α-D-glucopyranosylthymine, which gives a positive Cotton-effect.

d. Summary.

The great majority of pyrimidine nucleosides thus far prepared synthetically are either β-D- or α-D-furanosyl derivatives not containing special structural features, such as an additional ring, and these compounds, without exception, give positive and negative B_{2u} Cotton-effects, respectively. Hence, measurement of the o.r.d. or c.d. constitutes an excellent method for determining the anomeric configuration of such compounds.

For predicting the sign of the Cotton effect in pyrimidine nucleosides, a rule has been formulated that covers the aforementioned compounds and, in addition, compounds in which the glycosyl group is attached at C-5 or N-3 (instead of N-1), anhydronucleosides, and other derivatives. This rule relates the sign and magnitude of the Cotton effect to the conformation, particularly as regards the orientation of the nucleoside bond; it does not yet cover azapyrimidine derivatives, pyranosyl nucleosides, or compounds containing additional chromophores, although some information concerning the o.r.d. and c.d. of these is already available.

4. Purine Nucleosides and Nucleotides

Because the o.r.d. and c.d. of pyrimidine nucleosides and nucleotides have been considered in some detail, that of the related purine compounds may be treated more briefly. The amplitude of the Cotton effect for purine nucleosides is considerably lower than for pyrimidine derivatives; this situation is undoubtedly related to the fact that rotation about the nucleoside bond in glycosylpurines is virtually unrestricted.[9] Consequently, it is to be expected that the o.r.d. of purine nucleosides may be more sensitive to changes in structure, including substitution on the heterocyclic ring, than with pyrimidine derivatives.

X-Ray diffraction studies on nucleosides and nucleotides in the solid state have shown that, with the exception of 2'-deoxyguanosine in one crystal complex, they all exist in the *anti* conformation.[9a] This configuration is also that assumed by all of the nucleotides in helical DNA and RNA. The *anti* conformation of pyrimidine nucleosides and nucleotides has been unequivocally established by o.r.d., as described; however, attempts to establish the conformations of purine nucleosides have been beset by difficulties, owing to one of the facts mentioned at the beginning of this Chapter, namely, that changes in the substitution on the heterocyclic ring can significantly affect the o.r.d.

Recent n.m.r. studies have not only confirmed the anti conformation of pyrimidine derivatives, but have also shown that a conformation approaching the anti preponderates in the ordinary purine nucleosides and nucleotides[11, 12, 12a] at neutral pH. Because the β-D anomers give negative Cotton-effects and the α-D anomers give positive ones,[13] a conformation approaching anti can definitely be related with a negative Cotton-effect for a purine β-D-nucleoside. This holds true for a large number of furanosyl derivatives of adenine, 8-azaadenine (7-amino-1*H*-*v*-triazolo[4,5-*d*]pyrimidine), 6-chloropurine, guanine, hypoxanthine, 6-methoxypurine, purine, uric acid, and xanthine, in which the sugar residue may be D-ribose, "2-deoxy-D-ribose" (2-deoxy-D-*erythro*-pentose), D-xylose, D-arabinose, D-lyxose, or D-glucose.

Fig. 10 shows the o.r.d. curves given by adenosine, gua-

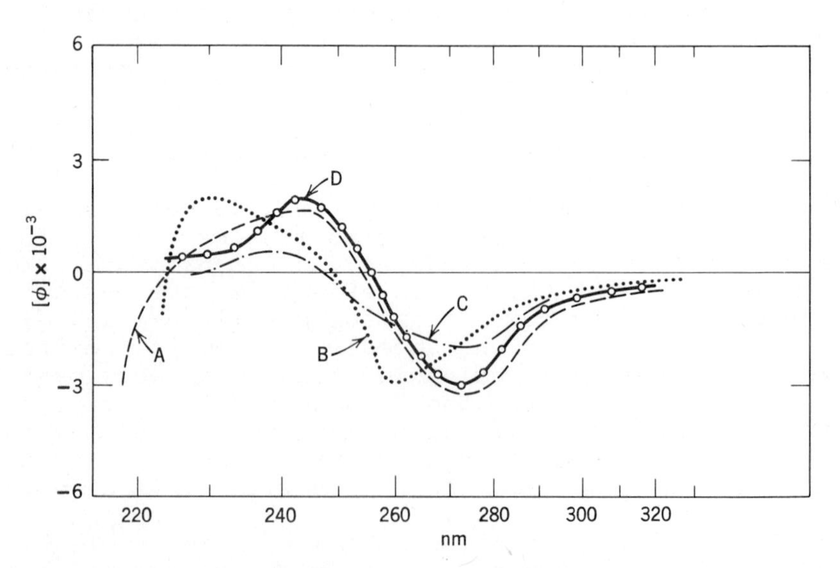

Fig. 10. O.R.D. Curves of Purine Nucleosides and Nucleotides (in Water). (A) (-----) adenosine; (B) (........) guanosine; (C) (-.-.-.) 3'-guanylic acid; (D) (-o-o-o) 5'-adenylic acid.

nosine, and the corresponding 5'-phosphates. As with the pyrimidine derivatives, there is no significant difference between the curve for a nucleoside and that of its nucleotide; all of these compounds give small, negative Cotton-effects. Two pairs of anomers are shown in Fig. 11: the D-xylofuranosyladenines and the 6-chloro-9-(2-

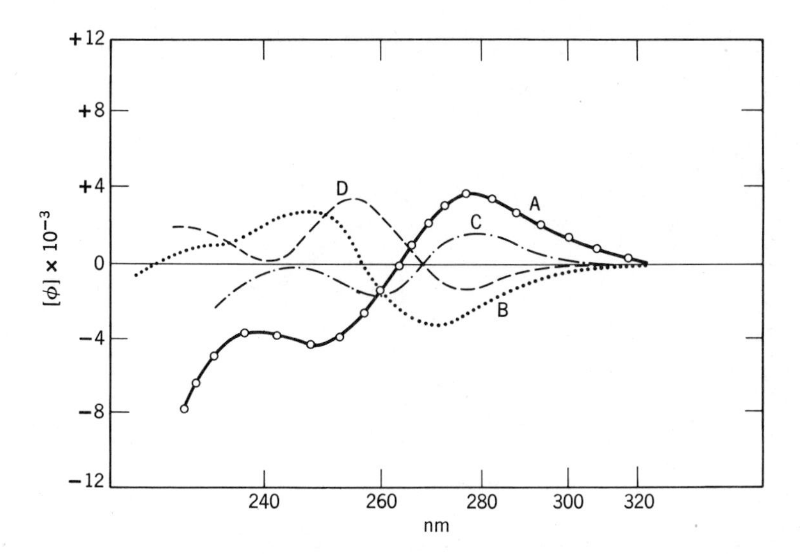

Fig. 11. O.R. D. Curves of Anomeric Purine Nucleosides (in Water) (A) (0-0-0-0) 9-α-D-xylofuranosyladenine; (B) (.....) 9-β-D-xylofuranosyladenine; (C) (.-.-.-.) 6-chloro-9-(2-deoxy-α-D-*erythro*-pentofuranosyl)-9*H*-purine; (D) (---) 6-chloro-9-(2-deoxy-β-D-*erythro*-pentofuranosyl)-9*H*-purine.

deoxy-D-*erythro*-pentofuranosyl)-9*H*-purines. In each case, the β-D anomers give negative Cotton-effects, and the α-D anomers, positive ones.

The spectra of a number of purine anhydronucleosides have been measured. The cyclic derivative (22) of adenosine gives a negative Cotton-effect,[14] and the anhydro-derivatives (23–25) of 8-hydroxyadenosine, in which the oxygen atom at C-8 is linked either to C-2, C-3, or C-5 of the sugar residue, respectively, all give positive Cotton effects.[15] From all of the foregoing results, a projection diagram[15a] (see Fig. 12) similar to that already shown for the pyrimidine nucleosides can be constructed. The line dividing the circle into positive and negative halves appears, in this case, to be in line with the C-1'–oxygen bond. The conformation of purine β-D-nucleosides is closer to anti than to syn, and this

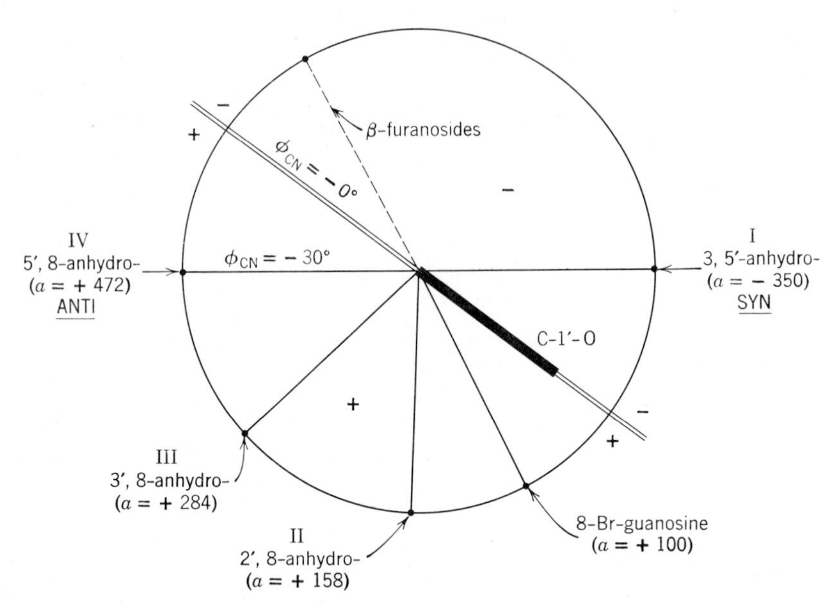

Fig. 12. Relationship Between the Sugar-base Torsion Angle, ϕ_{CN}, and the Sign and Magnitude of the Cotton Effect in Anhydroadenosines and other Purine Nucleosides.

would permit hydrogen bonding between the 2'-hydroxyl group and N-3, as has been suggested for certain D-ribose derivatives.[11]

22 **23**

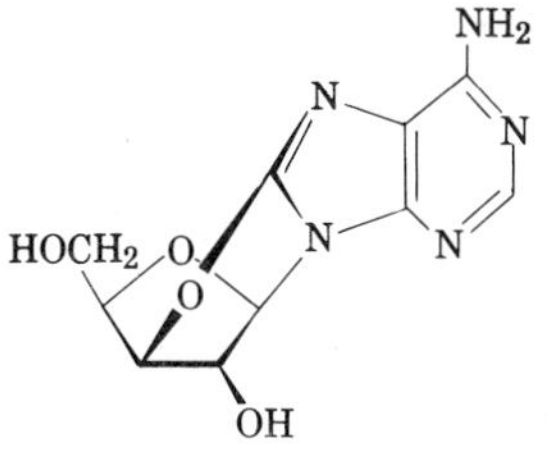

24 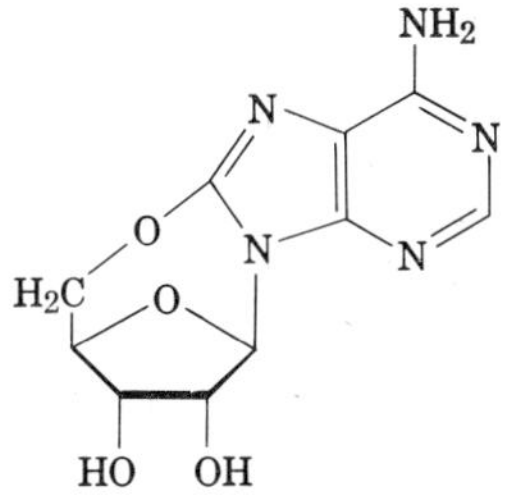 25

Such a conformation in a dinucleotide would enable the phosphate group to occupy a position close to H-8 of the purine, and would account for the deshielding observed in n.m.r. studies of these compounds.[12a]

β-D-Nucleosides and nucleotides of guanine are unique, in that the sign of the Cotton effect is reversed in acid at pH values at which guanosines are protonated. It was suggested[15b] that this is due to a change of conformation, *i.e.*, that the conformation in the protonated compound is more syn. Conceivably, however, it could be due to the fact that protonation of guanosine leads to a change in the direction of the transition moment.

8-Bromoguanosine gives a positive Cotton-effect.[15c] Because of the size of the 8-substituent, it cannot have the anti conformation, and, in the solid state, it has the syn conformation[15d] with $\phi_{CN} = +127°$. However, in aqueous solution, the conformation of 8-bromoguanosine is probably opposite to that suggested for normal purine β-D-nucleosides, *i.e.*, with $\phi_{CN} = 150°$ (see Fig. 12).

a. Hudson's Rules of Isorotation. Because purine β-D-nucleosides give negative Cotton-effects and the corresponding α-D anomers give positive ones, the α-D anomer will be more dextrorotatory than the β-D anomer at the D line of sodium, unless the two o.r.d. curves cross each other between the first extremum and the D line of sodium; this is rarely the case, and, hence, purine nucleosides obey Hudson's rules. The α and β anomers of D-arabinopyranosyladenine do not obey Hudson's rules, and, in this instance, the o.r.d. curves do cross each other in the long-wavelength region;[16] however, these compounds are pyranoid. In contradistinction, pyrimidine β-D-nucleosides give positive Cotton-effects, and the α-D anomers give negative ones. Hence, for pyrimidine nucleosides, Hudson's rules are reversed.[5]

b. 8-Azapurines. Furanosyl-8-azapurine (v-triazolo-[4,5-d]pyrimidine) nucleosides behave like ordinary purine nucleosides. For example, 8-azaadenosine (7-amino-3-β-D-ribofuranosyl-3*H*-*v*-triazolo [4,5-d] pyrimidine) gives a negative Cotton-effect (see Fig. 13); α-D anomers give positive Cotton-effects.

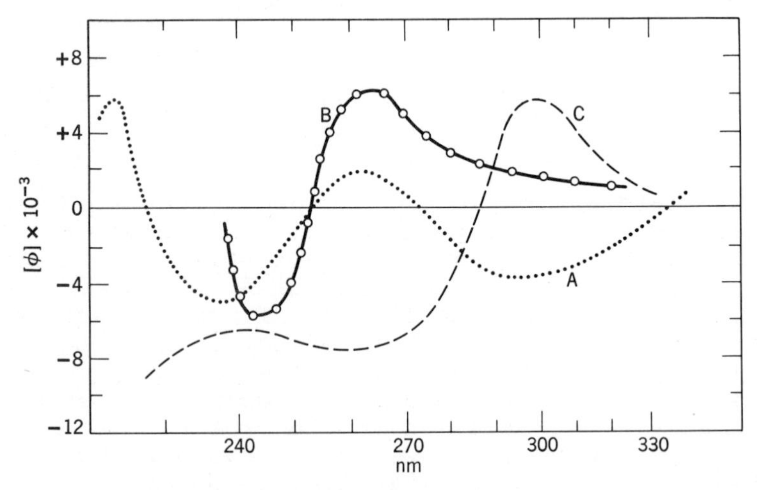

Fig. 13. O.R.D. Curves of Purine Nucleosides (in Water). (A) (.....) 8-Aza-adenosine; (B) (o—o—o) 9-β-L-ribofuranosyladenosine; (C) (---) 3-β-D-ribofuranosyluric acid (26).

c. Derivatives of L Sugars. It is obvious that the anomeric configuration determines the sign of the Cotton effect for pyrimidine and purine nucleosides and nucleotides. Because the stereochemistry at C-1' is the same for both β-L- and α-D-glycosyl derivatives, it is to be expected that such a compound as 9-β-L-ribofuranosyladenine, for example, should give a positive Cotton-effect, and it does (see Fig. 13). Similarly, the signs of the Cotton effects in β-L-azauridine are opposite to those in β-D-azauridine.[10d]

d. 3-Glycosylpurines. Such compounds are interesting in view of their structural similarity to pyrimidine nucleosides. 3-β-D-Ribofuranosyluric acid (26), for example, is closely related to uridine (5). As with uridine, there is a carbonyl group at C-2 of the heterocyclic ring, and rotation about the glycosyl bond must be restricted. It would therefore be expected that the conformation would be *anti* with respect to the pyrimidine ring, and that compound 26 would show a positive Cotton-effect; it does, in fact (see Fig. 13), as does the corresponding α-D-glucopyranosyl derivative.

On the other hand, 3-isoadenosine (3-β-D-ribofuranosyladenine, 27), which does not contain a carbonyl group at C-2, and in which rotation about the glycosyl bond is not restricted at all, gives a small, negative Cotton-effect. If these compounds are compared with xanthosine (which also has two carbonyl groups on the pyrimidine ring), it is seen that, if this compound had the purine *syn* conformation (as in 28) the steric relationship of the 4,5 double bond and the D-ribose ring would be almost the same as in uridine (5) and 3-β-D-ribofuranosyluric acid (26). However, as already stated (see p. 200),

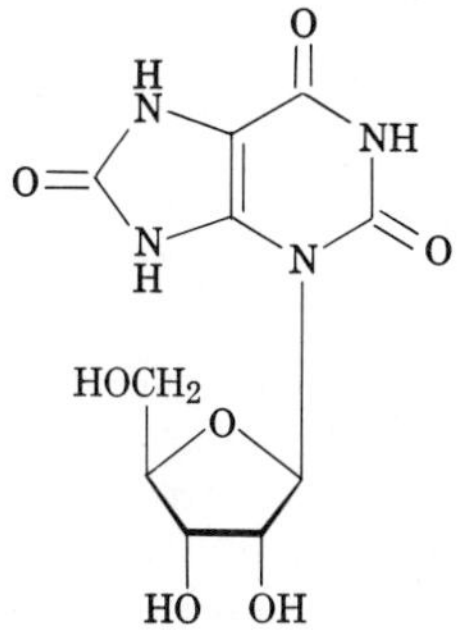

26

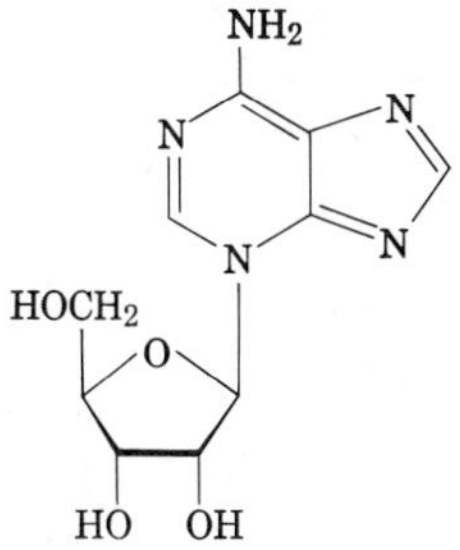

27

9-β-D-glycofuranosyl-9*H*-purines are known to have the *anti* conformation (as in 29), and, because the steric relationship is different from that of 5 and 26, these compounds give negative Cotton-effects, whereas 5 and 26 give positive ones. Although this explanation is oversimplified, it reasonably accounts for the difference in the sign of the Cotton effect between pyrimidine and purine derivatives.

e. Compounds Containing Additional Chromophores. Purine nucleosides having a sulfur-containing group at C-5' give positive Cotton-effects. It was argued that, because of the large size of the group at C-5', such compounds were certain to have the *anti* conformation, and that "normal" purine nucleosides, which give negative Cotton-effects, were *syn*.[17] It was assumed that, because there was no

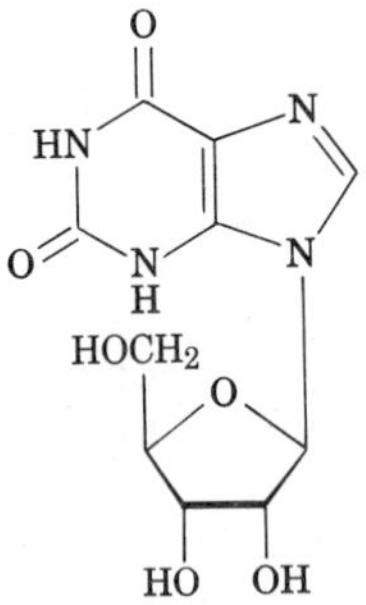

28

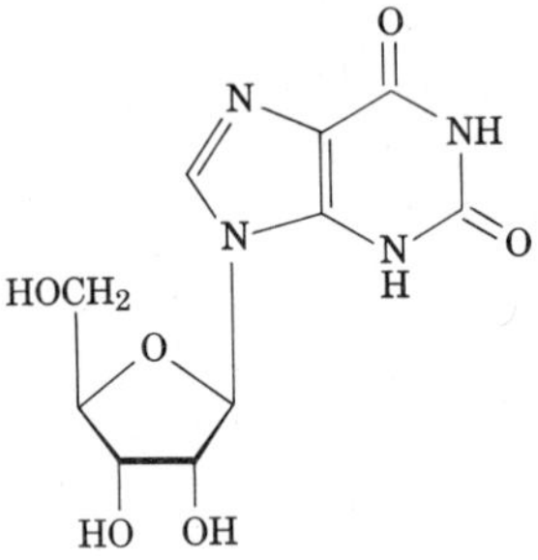

29

significant difference in the u.v. spectra of adenosine and its 5' sulfur-containing derivatives, there was no interaction between the heterocyclic ring and the sulfur-containing group. These arguments are incorrect for three reasons: (1) other purine nucleoside derivatives with such large groups on C-5' as acetoxyl, behave normally (negative Cotton-effect); (2) adenosine and adenylyl-(3' → 5')-adenosine have very similar u.v. spectra, and yet give totally different o.r.d. curves, and (3), as already mentioned (see p. 200), use of an independent technique (n.m.r.) has established that purine derivatives have a conformation approaching anti. It must therefore be concluded that the sulfur-containing groups do interact with the chromophore, leading to a change either in the conformation or in the electronic transitions in compounds containing such groups.

Chromophore interaction is also found in purine nucleosides containing aromatic protecting groups, such as *p*-toluoyl, on the sugar residue. For example, the β-D-nucleoside 30 gives a positive Cotton-effect, and the corresponding α-D anomer, a negative one. Hence, this pair of purine nucleoside derivatives does not obey Hudson's Rules.

This Chapter is concerned only with experimental results directly relating to the major topics, namely, anomeric configuration and molecular conformation. Other work on the o.r.d. and c.d. of nucleosides and nucleotides is discussed in a comprehensive review by Yang and Samejima.[18]

5. Other Nucleosides

No systematic work on the o.r.d. of nucleosides or nu-

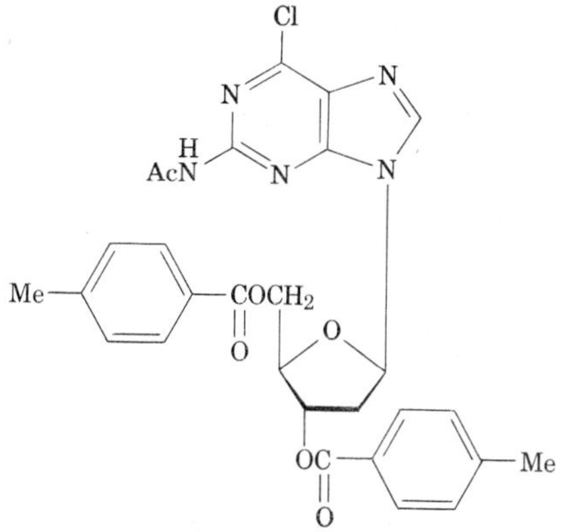

30

cleotides derived from other heterocyclic derivatives has as yet been published, and little appears to have been done. A preliminary study of pteridine nucleoside analogs, many of which were glycosides, did not reveal any obvious pattern, but this may have been because the compounds were pyranosides.[16]

An interesting pair of compounds are the β-D-ribofuranosyl derivatives (31 and 32) of 2-pyridone and 2-pyridazone, respectively. The pyridine derivative (31) gives a positive Cotton-effect, as does the corresponding pyrimidine derivative, whereas 32 gives a negative Cotton-effect,[16] as does the corresponding 6-azapyrimidine derivative. This result again suggests that a ring-nitro-

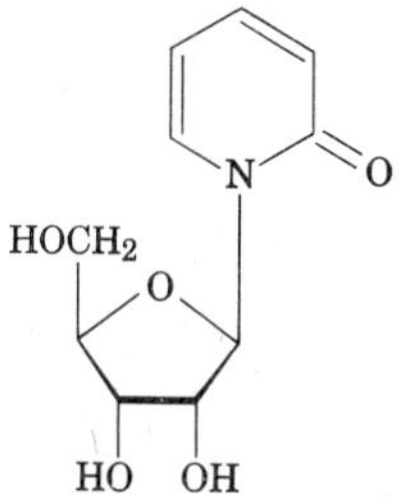

31 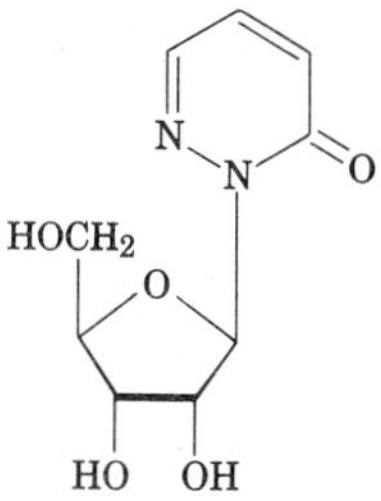 32

gen atom adjacent to that bearing the glycosyl group exerts a special effect on the o.r.d.

6. Oligonucleotides

The o.r.d. of di- and other oligo-nucleotides is strikingly different from that of the monomers, and is much more like that of polynucleotides.[19] For example, the o.r.d. of adenylyl-(3'→5')-adenosine ["diadenylic acid"] (ApA) resembles that of poly(adenylic acid) (see Fig. 14)

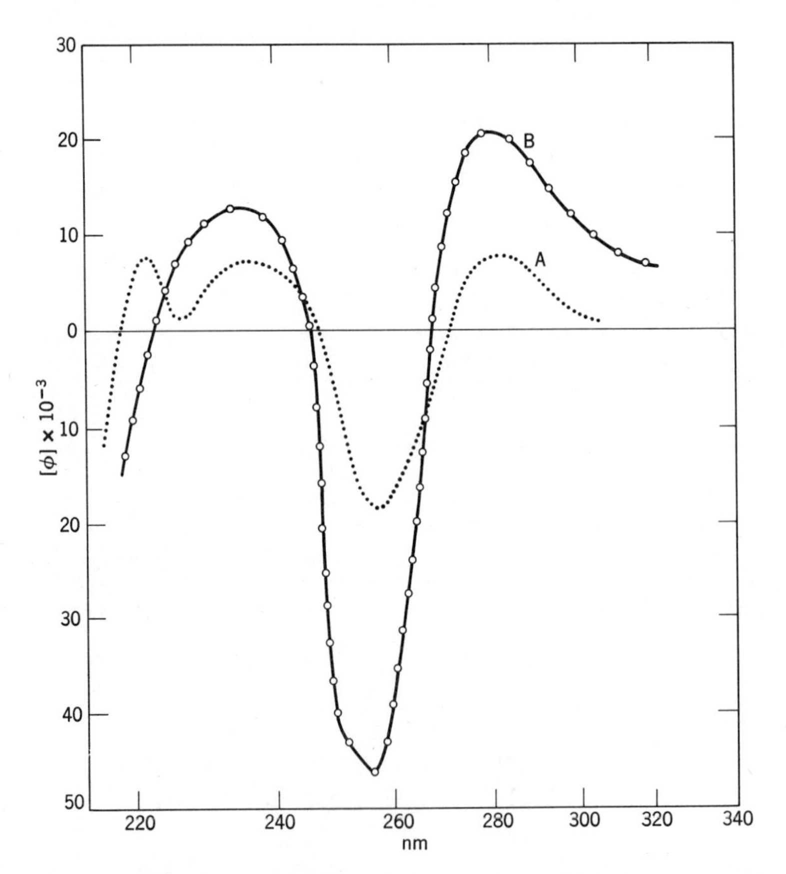

Fig. 14. O.R. D. Curves of Adenylic Acid Polymers (in 0.15 *M* Potassium Fluoride). (A) (....) adenylyl-(3'→5')-adenosine ("diadenylic acid"); (B) (O-O-O-O) poly(adenylic acid).

and is quite unlike that of 5'-adenylic acid. A convincing explanation for this difference has been given by

Tinoco *et al.*[20] According to these authors, "diadenylic acid" is a stacked molecule, that is, the planes of the adenine rings are parallel to each other and overlap. The resultant interaction between the two π-electron systems leads to a split in the u.v. maximum; because the two peaks are very close, only a broadening is observed experimentally. Hence, there are two Cotton effects, one positive and one negative, and the overlapping of the negative parts of the two curves is responsible for a large trough near λ_{max}.

The o.r.d. of all 16 dinucleoside phosphates has been measured; each compound gives a distinctive curve.[21] Measurement of the o.r.d. may therefore be used for identifying these compounds, including sequence isomers. In addition, valuable information may be obtained about the relative tendency of the various bases to stack.

The o.r.d. of oligo- and poly-nucleotides has been reviewed.[18]

III. A NOTE ON THE USE OF CIRCULAR DICHROISM

Although c.d. measurements of oligo- and poly-nucleotides have been made for some time, instruments sensitive enough to measure the c.d. of the monomers did not become available until 1967. A paper on the c.d. of some purine derivatives has appeared,[22] and very recent work on the c.d. of pyrimidine and azapyrimidine nucleosides has been described in the appropriate Sections (see p. 194). As explained in the Introduction (see p. 180), c.d. is potentially a very valuable technique, and it is to be hoped that its wider application to nucleosides and nucleotides will help clarify problems to which the answers are still obscure.

IV. INSTRUMENTATION AND EXPERIMENTAL METHODS

Reviews of the several commercial o.r.d. instruments now available have been published.[23,24] In deciding on the instrument to be used for a particular study, the investigator should give particular attention to the wavelength range. For a routine measurement of the o.r.d. of nucleosides and mononucleotides, a range down to 220–230 nm is sufficient. On the other hand, if rotatory effects associated with the absorption maximum at ~200 nm are to be studied, an instrument giving reproducible results down to ~190 nm is required.

A comprehensive discussion of experimental methods and related topics has been published.[24] For measuring the o.r.d. of nucleosides and simple nucleotides, the light transmission, concentration, stray light, and rotatory artifacts are the most important, and are considered together, as they are interrelated.

1. Light and Concentration

Clearly, sufficient light must pass through the solution and reach the photomultiplier in order to give a satisfactory signal-to-noise ratio. In practice, it is not advisable to allow the transmission to fall below 20%, which, at λ_{max}, gives an optical absorbance of ~1.0 nm in a 1-cm cell. To obtain the most meaningful results, more than one concentration should be used over the wavelength range. For example, for compounds available in adequate quantity, about 1 mg, accurately weighed, can routinely be dissolved in 5 ml of solvent, and the o.r.d. can be recorded into the ultraviolet region until the transmission falls below 20%. Various dilutions (1:5 is often optimal) may then be tried. Cells having a smaller path-length (0.1 or 0.2 cm) may be used to advantage, and permit use of more concentrated solutions and, therefore, of less material. The Cotton effect may, if necessary, be measured with as little as 25 µg. Any undissolved material or dust should be removed by filtration or centifugation prior to the determination.

If the absorbance of the solution is too great, not only does the noise level become too high, but stray light becomes a problem as it can give rise to rotatory artifacts.[25] To avoid this difficulty, the optical absorbance should be kept <2.0 over the whole wavelength-range. In addition, the measurement should be repeated to check reproducibility; and also, in case of any doubt, at a lower concentration.

2. Solvents

The o.r.d. of compounds that are soluble in organic solvents (isopropylidene and acyl derivatives, for example) may be measured in any organic solvent that has suitable optical properties; spectroscopic-grade methanol appears to have been used almost exclusively.

Most unsubstituted nucleosides may be dissolved in distilled water. Nucleotides, and nucleosides that have a pK value near neutrality, should be measured in a dilute buffer (not more concentrated than 0.15 *M*) of suitable pH. (A change in pH does not change the sign of the Cotton effect of most nucleosides and nucleotides, but guanine derivatives give positive Cotton-effects at pH values on the acid side below the pK value, probably because of a change in conformation.[26] The o.r.d. of oligo- and poly-nucleotides may change profoundly with pH, and it is essential to use appropriate buffers. Recommended buffers are Tris [2-amino-2-(hydroxymethyl)-1,3-propanediol] (down to 220 nm) and fluoride (down to 185 nm). A useful neutral buffer, which can also be used for oligo- and polynucleotides, is 0.15 *M* potassium fluoride.

3. Temperature

The o.r.d. of monomeric compounds is not very sensitive to small fluctuations in temperature. Jacketed cells are available for studies on temperature-dependence.

4. Polarimeter Cells

Only fused-quartz cells having integral end-plates should be used. The cell should fit snugly into the sample compartment, and be so placed that it always occupies the same position. It is best, in fact, to leave it in position, and to employ syringes for changing samples and making dilutions (for modified sample-holders, see Ref. 24). After the solution has been washed out with solvent (water, and methanol, respectively) the cell may be blown dry with nitrogen. The optical surfaces should not be touched; they may be washed with methanol and gently dried with lens paper (in a circular motion).

5. Measurements and Results

A blank (cell with solvent) should be determined both before and after the solution has been measured. In the presentation of results, solvent, concentration, maximum optical absorbance, minimum light transmission, path length (if critical), and temperature should be stated. The rotation measured should be calculated as the molecular rotation $[M]$, and it is useful to calculate the amplitude, a, of the Cotton effect (see p. 180). For oligo- and poly-nucleotides, it is necessary to calculate the mean residue rotation, $[m]$, defined by equation *3*,

$$[m]_\lambda = \alpha M'/100 \qquad \ldots(3)$$

where α is the observed rotation, in degrees, and M' is the mean residue molecular weight of the polymer (321 for RNA and 309 for DNA, assuming equimolar base composition).

REFERENCES

(1) T. L. V. Ulbricht, *Quart. Rev.* (London), 13, 48 (1959).

(2) T. L. V. Ulbricht, *Introduction to the Chemistry of Nucleic Acids and Related Natural Products*, Oldbourne Press (1966), p. 32.

(3) C. S. Hudson, *J. Amer. Chem. Soc.*, 31, 66 (1909).

(4) T. L. V. Ulbricht, J. P. Jennings, P. M. Scopes, and W. Klyne, *Tetrahedron Lett.*, (1964) 695.

(5) T. R. Emerson and T. L. V. Ulbricht, *Chem. Ind.* (London), (1964) 2129.

(6) T. L. V. Ulbricht, T. R. Emerson, and R. J. Swan, *Biochem. Biophys. Res. Commun.*, 19, 643 (1965).

(7) T. L. V. Ulbricht, T. R. Emerson, and R. J. Swan, *Tetrahedron Lett.*, (1966) 1561.

(8) T. R. Emerson, R. J. Swan, and T. L. V. Ulbricht, *Biochemistry*, 6, 843 (1967).

(8a) G. T. Rogers and T. L. V. Ulbricht, *Biochem. Biophys. Res. Commun.*, 39, 414 (1970).

(9) J. Donahue and K. N. Trueblood, *J. Mol. Biol.*, 2, 363 (1960).

(9a) A. E. V. Haschemeyer and A. Rich, *J. Mol. Biol.*, 27, 369 (1967).

(10) M. Honjo, Y. Furukawa, M. Nishikawa, K. Kamiya, and Y. Yoshioka, *Chem. Pharm. Bull.* (Tokyo), 15, 1076 (1967).

(10a) D. W. Miles, M. J. Robins, R. K. Robins, M. W. Winkley, and E. J. Eyring, *J. Amer. Chem. Soc.*, 91, 824, 831 (1969).

(10b) G. T. Rogers and T. L. V. Ulbricht, *FEBS Lett.*, 7, 337 (1970).

(10c) S. F. Mason, *J. Chem. Soc.*, (1959) 1240, 1247.

(10d) A. Holý, G. T. Rogers, and T. L. V. Ulbricht, *FEBS Lett.*, 7, 335 (1970).

(11) M. P. Schweizer, A. D. Broom, P. O. P. Ts'O, and D. P. Hollis, *J. Amer. Chem. Soc.*, 90, 1042 (1968).

(12) S. S. Danyluk and F. E. Hruska, *Biochemistry*, 7, 1038 (1968).

(12a) S. I. Chan and J. N. Nelson, *J. Amer. Chem. Soc.*, 91, 168 (1969).

(13) T. R. Emerson, R. J. Swan, and T. L. V. Ulbricht, *Biochem. Biophys. Res. Commun.*, 22, 505 (1966).

(14) A. Hampton and A. W. Nichol, *J. Org. Chem.*, 32, 1688 (1967).

(15) M. Ikehara, M. Kaneko, K. Muneyama, and H. Tanaka, *Tetrahedron Lett.*, (1967) 3977; M. Ikehara and M. Keneko, *Chem. Pharm. Bull.* (Tokyo), 15, 2161 (1967).

(15a) G. T. Rogers and T. L. V. Ulbricht, *Biochem. Bio-*

phys. Res. Commun., 39, 419 (1970).
(15b) W. Guschlbauer and Y. Courtois, *FEBS Lett.*, 1, 183 (1968).
(15c) J. F. Chantot and W. Guschlbauer, *FEBS Lett.*, 4, 173 (1969).
(15d) C. E. Bugg and U. Thewalt, *Biochem. Biophys. Res. Commun.*, 37, 623 (1969).
(16) G. T. Rogers and T. L. V. Ulbricht, unpublished results.
(17) W. A. Klee and S. H. Mudd, *Biochemistry*, 6, 988 (1967).
(18) J. T. Yang and T. Samejima, *Progr. Nucleic Acid Res. Mol. Biol.*, 9, 223 (1969).
(19) T. R. Emerson, R. J. Swan, A. M. Michelson, and T. L. V. Ulbricht, *Fed. Eur. Biochem. Soc. Symp. Vienna, Abstr.*, e 59 (1965).
(20) M. M. Warshaw, C. A. Bush, and I. Tinoco, Jr., *Biochem. Biophys. Res. Commun.*, 18, 33 (1965).
(21) M. M. Warshaw and I. Tinoco, Jr., *J. Mol. Biol.*, 20, 29 (1966).
(22) D. W. Miles, R. K. Robins, and H. Eyring, *Proc. Nat. Acad. Sci. U. S.*, 57, 1137 (1967).
(23) G. Snatzke, *Z. Instrumentenk.*, 75, 111 (1967).
(24) A. J. Adler and G. D. Fasman, *Methods Enzymol.*, 12B, 268 (1968).
(25) P. Urnes and P. Doty, *Advan. Protein Chem.*, 16, 401 (1961).
(26) W. Guschlbauer and Y. Courtois, *FEBS Lett.*, 1, 183 (1968).

CHAPTER 6

Infrared Spectroscopy of Nucleic Acid Components

MASAMICHI TSUBOI AND YOSHIMASA KYOGOKU*

FACULTY OF PHARMACEUTICAL SCIENCES,
UNIVERSITY OF TOKYO, TOKYO, JAPAN

I. INTRODUCTION

Infrared absorptions are caused by vibrations of atoms about their equilibrium positions in each molecule of the compound through which the infrared radiation is sent. An infrared spectrum is usually presented as a curve having the wavenumber scale (cm^{-1}) along the abscissa and the percent transmittance along the ordinate. It then consists of a number of minima, whose positions correspond to the characteristic vibrational frequencies of the molecule. The depth (area) of each minimum (or the absorption intensity of each absorption band) is a measure of the amount of the dipole oscillation caused by the vibration in question. The vibrational frequencies of a molecule depend upon the geometrical structure and the

*The authors express their thanks to Dr. Asao Nakamura, Dr. Yuji Iwashita, and Mr. Mitsuru Sakuraba, Central Research Laboratories, Ajinomoto Co., Ltd., Kawasaki, Japan, for their co-operation in preparing the manuscript, and especially for their permission to present Figs. 7–12, which contain their unpublished data.

interatomic force-field of the molecule.[1] The number of characteristic frequencies should be $3n - 6$, where n is the number of atoms in the molecule; thus, in general, the larger the molecule, the more complicated its spectrum. The infrared absorption spectra of nucleic acids are all so complicated that no analyses have yet been made on the basis of the mechanics of the molecules.

II. IDENTIFICATION OF A MOLECULE

The infrared spectrum of a compound may be regarded as a fingerprint of its molecule. Different molecules (except for enantiomorphs) never give identical spectra; their spectral differences are always found with no ambiguity. The identity of the spectra of two identical molecules is also readily established. Therefore, one of the most fruitful uses of infrared spectra is in identifying a sample with one of various known compounds. In this usage, there is no need to interpret any detail of the spectra. Instead, a catalog of infrared spectra is needed; this should contain spectra of as many compounds as possible. In Figs. 1–6, the infrared spectra of some purines, pyrimidines, nucleosides, and nucleotides are given. These were obtained for KBr disks, or Nujol or hexachlorobutadiene mulls.[2] In the curves for mulls, absorption bands caused by the mulling material have been removed. Table I is a list of the purines, pyrimidines, nucleosides, and nucleotides for which infrared absorption curves are given in the literature. In many reports on the organic chemistry of nucleic acids, infrared data are given only as the frequencies observed for absorption bands, instead of by reproduction of the entire absorption curve. These lists of absorption bands are not at all convenient for the purpose of identification; and, therefore, compounds for which only such data are given are not included in Table I.

III. CORRELATION WITH STRUCTURE

In general, most of the vibrations in a molecule are not localized in any bonds or in any atomic groups. Therefore, the correlation of an infrared spectrum with the molecular structure is not simple. At present, it is impossible to assign every absorption band observed for a nucleic acid to a certain vibrational mode in the molecule. It has, however, been found that some of the atomic groups having hydrogen atoms or unsaturated bonds or both, show characteristic absorption bands in certain

TABLE I

Nucleic Acid Components and Related Compounds whose Infrared Absorption Curves have been Recorded in the Literature

Compound	Substituent on Atom Number						References
Pyrimidines	1	2	3	4	5	6	
Pyrimidine							6,14
		NH_2					6
		Cl					12
		OMe					8
		SMe					DMS6216
				OMe			8
					NH_2		6
					OMe		DMS10981
		NH_2		NH_2			6
		NH_2		OMe			DMS10910
		Cl		Cl			6
		OEt		OEt			6
		OPh		OPh			6
		NH_2			NO_2		DMS10904
				NH_2	NH_2		DMS10970
				NH_2	Ph		6
				NH_2		NH_2	6
				NH_2		Cl	6
				Cl		Cl	DMS10983
				Cl		OMe	DMS10988

TABLE I (continued)

Nucleic Acid Components and Related Compounds whose Infrared Absorption Curves have been Recorded in the Literature

Compound	Substituent on Atom Number						References
Pyrimidine	1	2	3	4	5	6	
(continued)				OMe		OMe	DMS10982
	CN	NH_2					DMS12683
		NH_2		Me	NH_2		DMS10971
		NMe_2		Me	COMe		DMS10993
		NPhOCOPh		NH_2	CN		DMS12688
		OEt		OEt	Me		DMS14697
		$OSiMe_3$		OEt	Me		DMS14698
		$OSiMe_3$		$OSiMe_3$	Me		DMS14699
		NH_2		NH_2		NH_2	6
		NH_2		NH_2		Me	DMS10972
		NH_2		NMe_2		Me	DMS10906
		NH_2		NEt_2		Me	DMS10903
		NH_2		NHCOMe		Me	DMS10688
		NH_2		NHNHCHO		Me	DMS10685
		NH_2		NHNHCOMe		Me	DMS10687
		NH_2		Cl		NMe	DMS10914
		NH_2		Cl		NMe_2	DMS10912
		NH_2		Cl		NHPh	DMS10915
		NH_2		Cl		Cl	6
		NH_2		CN		Me	DMS10911

NH_2	Me		Cl		6
NH_2	Me		CN		6
NH_2	Me		$CONH_2$		6
NH_2	Me		Me		6
NH_2	Me		OEt		6
NH_2	Me		piperidino		DMS10900
NH_2	Me		SMe		DMS10901
NH_2	Me		NHNHCOMe	HCl	DMS10682
NH_2	CCl_3		CCl_3		DMS10908
NH_2	OMe		OMe		6
NH_2	OEt		Me		DMS10907
NH_2	$OCH_2CH_2NEt_2$		Me		DMS10909
NH_2	$OCH(Me)_2$		$OCH(Me)_2$		DMS10913
NH_2	$OCH_2CH_2CH_3$		$OCH_2CH_2CH_3$		DMS10902
NMe_2	NH_2		NH_2		DMS10975
NMe_2	Cl		Cl		DMS10990
NMe	Me		Cl		6
NMe	NMe		Cl		DMS10980
NMe_2	Me		NMe_2		DMS10994
Cl	NH_2		Cl		6
Cl	NH_2		Me		6
Cl	NMe_2		Cl		DMS10989
Cl	Cl		Cl		6
Et	NHNHCHO		Me		DMS10686
Me	NH_2		Cl		6
Me	NH_2		Me		6
Me	Cl		Cl		6
Pr	Cl		Me		DMS10987
SMe	Cl		NH_2		DMS10431
SMe	Cl		Me		DMS10991
NH_2		NH_2	NEt_2		DMS10976
	NH_2	NH_2	Me		DMS10973
	NH_2	Br	NH_2		6
	NH_2	Br	OMe		DMS10430

TABLE I (continued)

Nucleic Acid Components and Related Compounds whose Infrared Absorption Curves have been Recorded in the Literature

Compound	Substituent on Atom Number						References
Pyrimidine (continued)	1	2	3	4	5	6	
				Cl	NH_2	Cl	DMS10692
				Cl	Cl	Cl	DMS10984
				Cl	Br	Cl	DMS10985
				Cl	Me	Cl	DMS10986
				OMe	NH_2	OMe	DMS10690
				OMe	NO_2	OMe	DMS10992
		NH_2		Me	NH_2	Me	DMS10974
		NH_2		Me	NHCHO	Me	DMS10684
		NH_2		Me	NO_2	Me	DMS10905
		$NHCH(Me)_2$		Me	NH_2	Me	DMS10979
		Me		NH_2	NH_2	NMe_2	DMS10978
		Me		NH_2	NH_2	NEt_2	DMS10977
		Me		NH_2	NHCHO	NEt_2	DMS10689
		Me		Me	NO_2	Me	DMS10995
		Me		OMe	NH_2	OMe	6
		Me		OMe	NO_2	OMe	6
		OMe		OMe	NH_2	Me	DMS10691
	H	=O					6,8,16
	H	=S					6
	Me	=S					DMS6217
	Me	=O		OMe			16

Cytosine	H	=O		NH_2				3,6,7,12,16,17, Fig. 2
	D	=O		ND_2				16,17, Fig. 2
	H	=O		NH_2			H_2O	4
	H	=O		NH_2			HCl	16,17
	Me	=O		$N(Me)_2$				16
	H	=S		NH_2				6,12
	H	=O		NH_2		NH_2		6
	Me	=O		NH_2				16
	H	=O		NH_2	Me			5
	Me	=O	Me	=NH				16
			H	=O				6,8
			H	=O		Me		6
			H	=S				DMS6215
	Me		H	=S				DMS6218
			H	=O		NH_2		6
			H	=O		OH		see 5–H, 6–=O
		NH_2	H	=O				7
		NH_2	H	=O		Cl		6
		NH_2	H	=O		Me		6
		NH_2	H	=O		Ph	0.5 H_2O	DMS10968
		NH_2	Et	=O		Me		6
		$N(Me)_2$	H	=O		Me		6
		$NHCOCH_3$	H	=O		Ph		DMS10963
		Me	H	=O		NH_2		6
		Me	H	=O		Me		6
		SMe	H	=O		NH_2		6
		SEt	H	=O	Et	Me		6
		NH_2		NH_2	H	=O		6
		NH_2	Et	Me	H	=O		6
	Me	SCH_3		Me	H	=O		6
Uracil	H	=O	H	=O				3,6,12,16,Fig. 1
	D	=O	D	=O				16,Fig. 1
	H	=S	H	=O				6,12

TABLE I (continued)

Nucleic Acid Components and Related Compounds whose Infrared Absorption Curves have been Recorded in the Literature

Compound	Substituent on Atom Number							References
Uracil (continued)	1	2	3	4	5	6		
	H	=S	H	=S				6
	Me	=O	H	=O				Fig. 1
	Me	=O	D	=O				16, Fig. 1
		NH_2	H	=O		OH		see 5-H,6-=O
		Me	H	=O		OH		see 5-H,6-=O
	H	=O	H	=O		Me		Fig. 1
	Me	=O	Me	=O				6,16
	H	=O		$O^{\ominus}$	$N{\equiv}N^{\oplus}$			DMS13507
	H	=O		$O^{\ominus}$	$N{\equiv}N^{\oplus}$		H_2O	DMS13508
Thymine	H	=O	H	=O	Me			3,12,16,Fig. 1
	H	=S	H	=O	Me			12
	Me	=O	H	=O	Me			18
	D	=O	D	=O	Me			18
	H	=O	H	=O	Cl			12
	H	=O	H	=O	NO_2			6
	H	=O	H	=O		NH_2		6
	H	=S	H	=O		NH_2		12
	H	=O	H	=O		COOH		12
	H	=O	H	=O		COOK		12
	H	=S	H	=O		NH_2		6
	H	=O	H	=O		Me		6,12,16,Fig. 1

H	=S	H	=O		Me	6,12
H	=O	H	=O		OH	see 5-H,6-=O
H	=S	H	=O		OH	see 5-H,6-=O
H	=O	H	=O		OEt	12
Me	=O	H	=O		OEt	12
H	=S	H	=O		Pr	12
H	=O	H	=O	H,Me	H	12
H	=O	H	=O	N≡N$^{\oplus}$	O$^{\ominus}$	DMS13506
		H	=O	H	=O	6
	NH_2	H	=O	H	=O	6
	Me	H	=O	Me	=O	6
H	=O	H	=O	H	=O	6,12
H	=S	H	=O	H	=O	12

Purines	References
Adenine	3,12,15,16,DMS6436, Fig. 10
(N-deuterated)	15,16,Fig. 10
hydrochloride	15,16
Adenine-2-d_1	Fig. 10
Adenine-8-d_1	Fig. 10
Adenine-2,8-d_2	Fig. 10
Adenine-6,6,8,9-d_4	Fig. 10
Adenine-2,6,6,8,9-d_5	Fig. 10
2-Amino-6-methylpurine	11
Caffeine	3
8-Chlorocaffeine	DMS11364
6-Chloropurine	Fig. 2
Guanine	3,12,16, DMS6437,Fig. 12
(*N*-deuterated)	16, Fig. 12
Guanine-8-d_1	Fig. 12
Guanine-1,2,2,8,9-d_5	Fig. 12
N-[3-(Hydroxymethyl)-*trans*-2-butenyl]adenine	DMS14897

Class	Compound	Reference
Purines (continued)	7-Hydroxy-8-isopropyltheophylline	DMS13562
	Hypoxanthine	3,12,Fig. 11
	(*N*-deuterated)	Fig. 11
	Hypoxanthine-2-d_1	Fig. 11
	Hypoxanthine-8-d_1	Fig. 11
	Hypoxanthine-2,8-d_2	Fig. 11
	Hypoxanthine-1,2,8,9-d_4	Fig. 11
	8-Methoxycaffeine	DMS11365
	7-Methyladenine	Fig. 3
	9-Methyladenine	18
	(*N*-deuterated)	18
	7-Methylhypoxanthine	Fig. 3
	9-Methyl-9H-purine	16
	Purine	11,16,Fig. 9
	(*N*-deuterated)	Fig. 9
	Purine-6-d_1	Fig. 9
	Purine-8-d_1	Fig. 9
	Purine-2,6-d_2	Fig. 9
	Purine-6,8-d_2	Fig. 9
	Purine-8,9-d_2	Fig. 9
	Purine-2,6,8-d_3	Fig. 9
	Purine-2,6,8,9-d_4	Fig. 9
	1,3,7,9-Tetramethyluric acid	DMS11366
	Theophylline	3
	Theobromine	3
	2-Thioadenine	12
	2,6,8-Trichloropurine	Fig. 2
	Uric acid	12,DMS9090
	Xanthine	3,12,Fig. 2

Nucleosides

Adenosine	15,16,17,Fig. 4
(*N*,*O*-deuterated)	15,16,17
hydrochloride	17
Cytidine	13,16,Fig. 4
(*N*,*O*-deuterated)	16
hydrochloride	16
2'-Deoxyadenosine	15,16,Fig. 3
2'-Deoxycytidine	16
hydrochloride	5,16
2'-Deoxyguanosine	16
[2-Deoxy-D-ribose (2-Deoxy-D-*erythro*-pentose)	Fig. 3]
2'-Deoxyuridine	16
1-(β-D-Glucopyranosyl)-4-ethoxy-2(1*H*)-pyrimidinone	13,DMS2945
1-β-D-Glucopyranosyluracil	13,DMS2944
Guanosine	16,17,DMS6438,Fig. 4
(*N*,*O*-deuterated)	17
hydrochloride	17
(Na salt)	17
Inosine	Fig. 5
2',3'-O-Isopropylideneiso-cytidine	DMS1271
3-Methyluridine	13,DMS2943
D-Ribose	Fig. 5
4-Ethoxy-1-(2,3,4,6-tetra-*O*-acetyl-β-D-glucosyl)-2(1*H*)-pyrimidinone	13,DMS2946
2-Thiouridine	DMS1047
Thymidine	13,16,DMS2948
Uridine	13,16,17,DMS2942,Fig. 4
(*N*,*O*-deuterated)	16,17
(Na salt)	17

Compound	References
Nucleotides	
Adenosine 5'-phosphate	15,16,Fig. 6
Cytidine 2'-phosphate (Cytidylic Acid a)	10,DMS704
Cytidine 3'-phosphate (Cytidylic Acid b)	10,DMS705(alcoholic form),DMS706 (water form)
Cytidine 5'-phosphate	9,10,16
2'-Deoxyadenosine 5'-phosphate	Fig. 6
2'-Deoxycytidine 3'-phosphate	10,16,DMS707
2'-Deoxycytidine 5'-phosphate	10,16
2'-Deoxyguanosine 5'-phosphate	Fig. 6
Guanosine 5'-phosphate	Fig. 6
Inosine 2':3'-cyclic phosphate	Fig. 5
Inosine 5'-phosphate	Fig. 5
Uridine 5'-phosphate	13,DMS2947

[a]DMS = Documentation of Molecular Spectroscopy, Butterworths Scientific Publications, London, and Verlag Chemie GMBH, Weinheim.

ranges of frequency.[20] These will be described for the nucleic acids.

1. Purines and Pyrimidines

a. O-H Stretching Bands. The solid hydroxy derivatives of purines and pyrimidines show no strong absorptions in the normal regions for free OH (3650-3590 cm^{-1}) or hydrogen-bonded OH (3600-3200 cm^{-1}). 2- and 4-Pyrimidinols are considered to exist in the keto form,[6] and 2,4-pyrimidinediol in the diketo form.[6,13] For these solid compounds, therefore, the absence of OH bands is understandable. 4,6-Pyrimidinediol had been considered to have an enolic OH group;[6] however, this compound shows only weak absorptions in the 3500-3300-cm^{-1} region, and it therefore probably exists in the diketo form, too.

b. N-H Stretching Bands. The keto forms of hydroxypurines and hydroxypyrimidines should contain the N^7-H or N^3-H group or both. The stretching vibrations of these NH groups cause a broad absorption band having its center at ~3100 cm^{-1}. The N^9-H stretching vibration of purine causes a broad absorption band in the 3000-2500-cm^{-1} region; this band actually consists of a number of absorption peaks located with nearly equal spacings.

c. NH_2 Stretching Bands. Every amino derivative of purines and pyrimidines shows two or three strong bands in the 3400-3100-cm^{-1} region. Especially, the appearance or absence of a strong band at ~3300 cm^{-1} is a good criterion for judging whether the purine or pyrimidine in question has one or more amino groups or none at all.

d. C-H Stretching Bands. An examination[21] of a few deuterated purines indicates that the C-8-H, C-2-H, and C-6-H stretching frequencies are, respectively, at 3098, 3060, and 3023 cm^{-1} (see Figs. 7, 8, and 9). These correlations are found not only for purine itself but also for adenine, hypoxanthine, and guanine (see Figs. 10, 11, and 12). Most of these C-H bands are sharp, but many of them are very weak. They are usually hidden under the broad, strong absorption bands of the NH groups already mentioned. These NH groups should be removed by converting them into ND, so that the C-H bands are clearly observable. Selective deuteration of the NH groups is readily achieved by use of D_2O, with which every CH group remains undeuterated.

Incidentally, it has been found[21] that the C-8-D, C-2-D, and C-6-D stretching frequencies for purine are at 2303, 2280, and 2264 cm^{-1}, respectively (see Fig. 8).

Corresponding to the four CH groups in pyrimidine, there should be four C-H stretching bands. Only one of them has

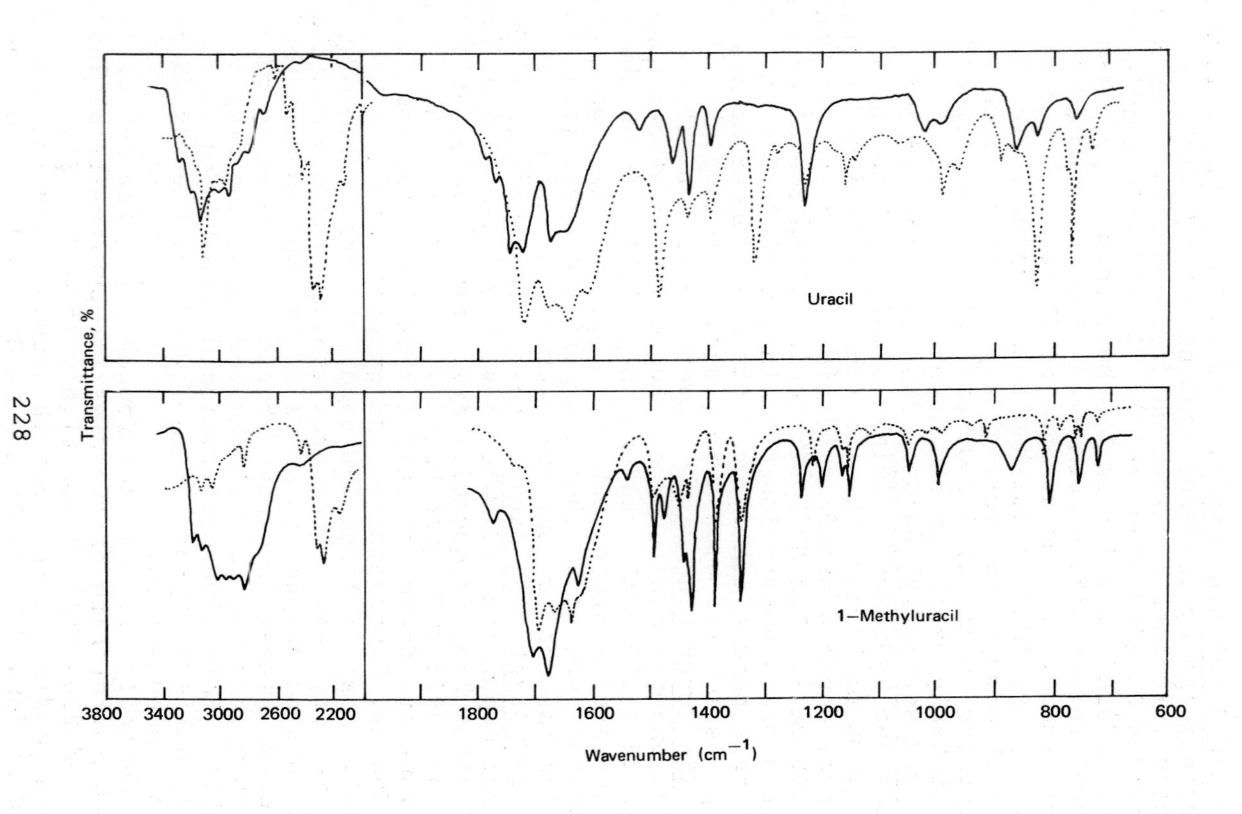
Uracil
1—Methyluracil
Transmittance, %
Wavenumber (cm^{-1})
3800
3400
3000
2600
2200
1800
1600
1400
1200
1000
800
600

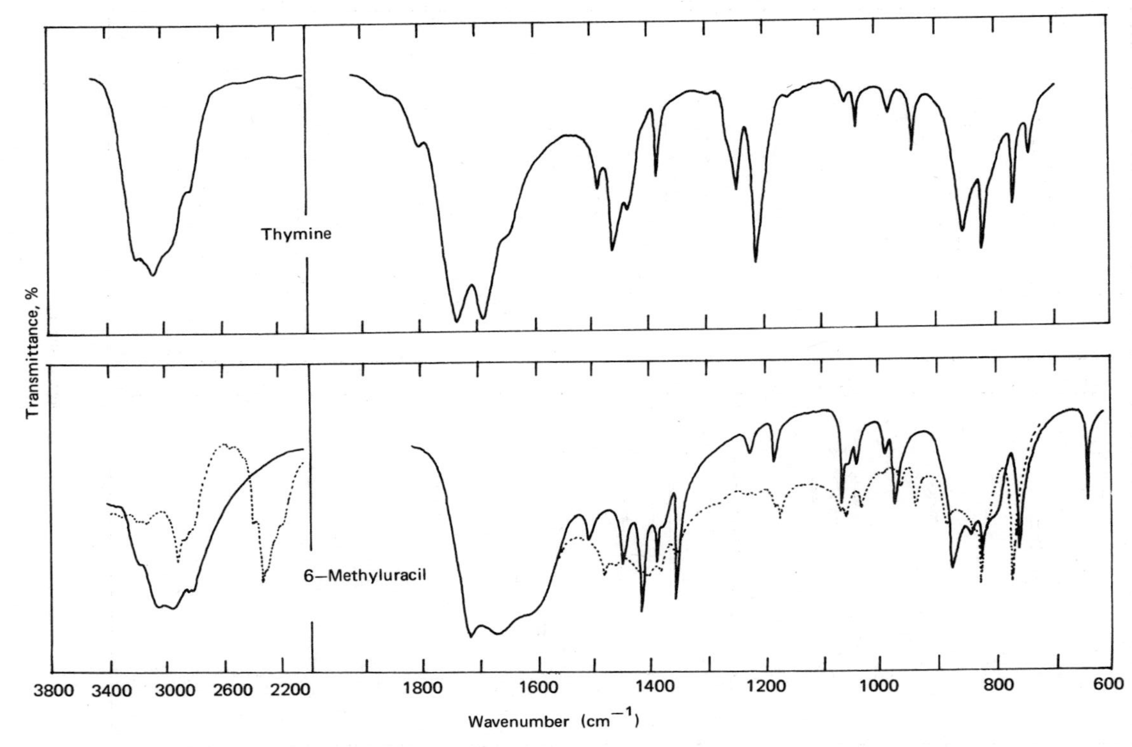

Fig. 1. Infrared Spectra of Uracil, 1-Methyluracil, Thymine, and 6-Methyluracil[Undeuterated (full lines) and N-deuterated products (dotted lines). The ordinate is percent transmittance in every curve in Figs. 1-12.]

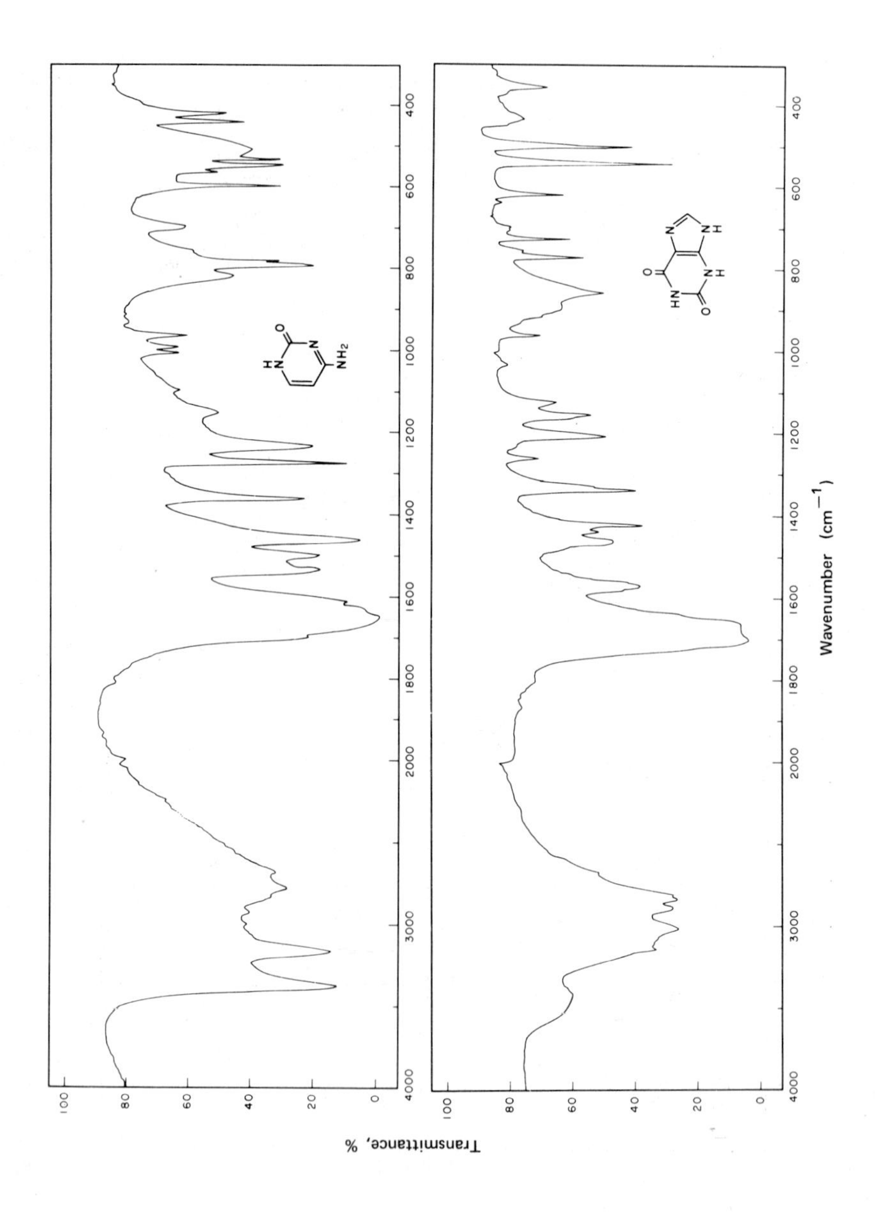

Transmittance, %
Wavenumber (cm^{-1})
100
80
60
40
20
0
4000
3000
2000
1800
1600
1400
1200
1000
800
600
400

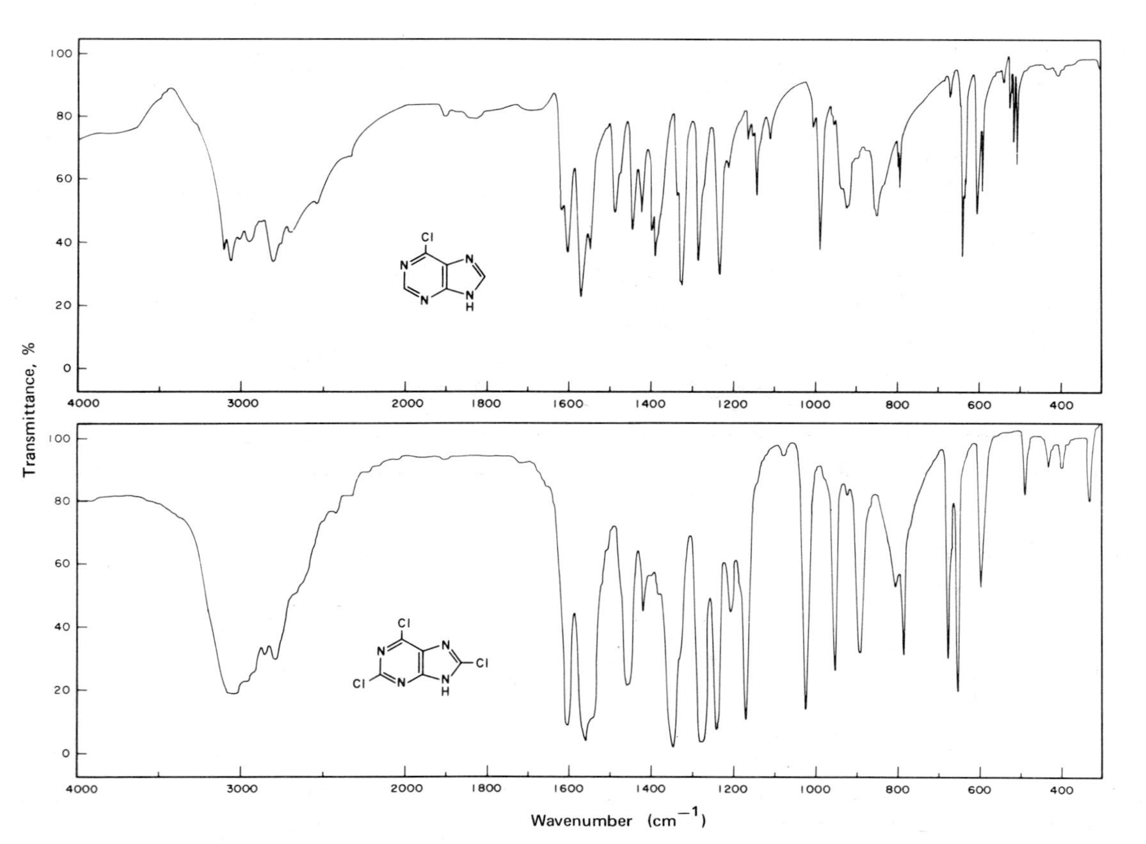

Fig. 2. Infrared Spectra of Cytosine, Xanthine, 6-Chloropurine, and 2,6,8-Trichloropurine.

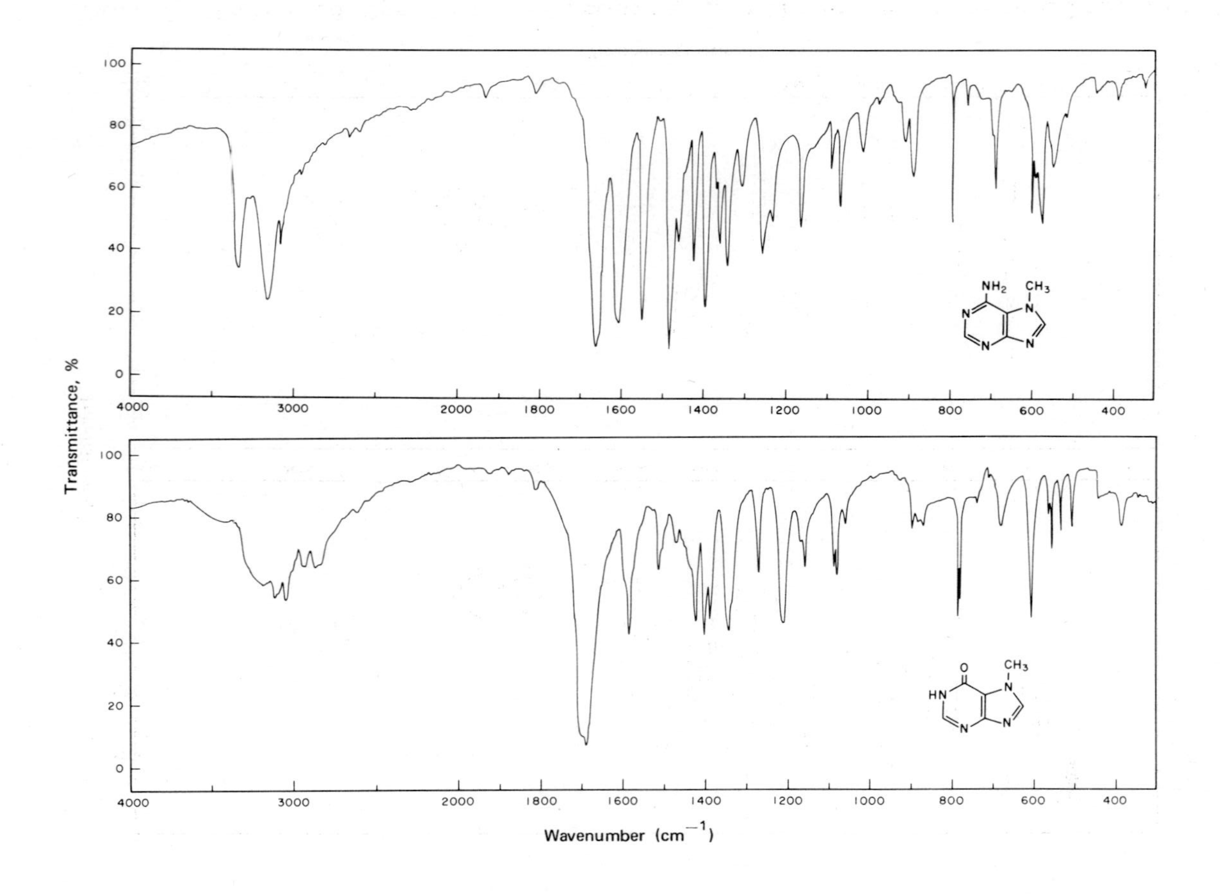
Transmittance, %
Wavenumber (cm^{-1})
100
80
60
40
20
0
4000
3000
2000
1800
1600
1400
1200
1000
800
600
400
NH_2
CH_3
N
O
HN

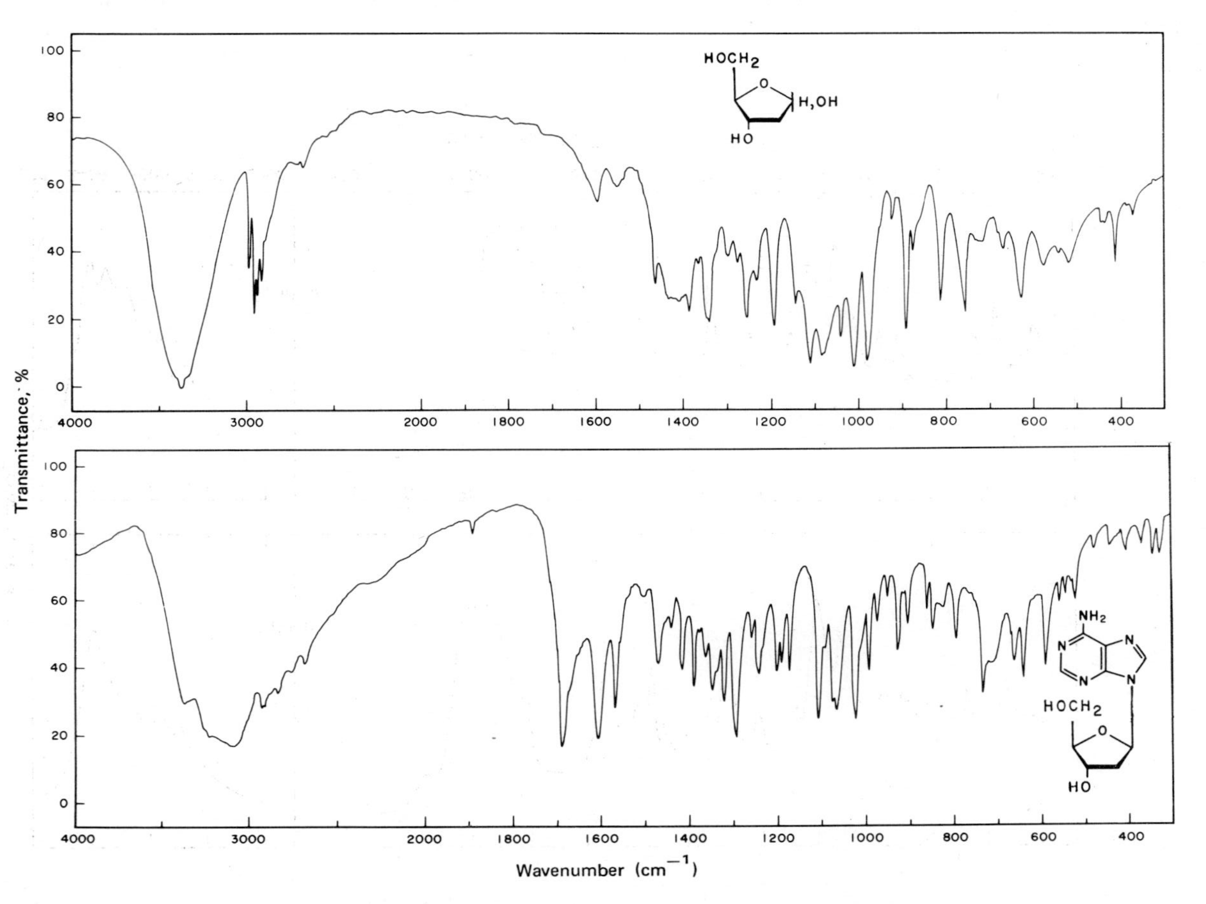

Fig. 3. Infrared Spectra of 7-Methyladenine, 7-Methylhypoxanthine, 2-Deoxy-D-*erythro*-pentose (2-Deoxy-D-ribose), and 2'-Deoxyadenosine.

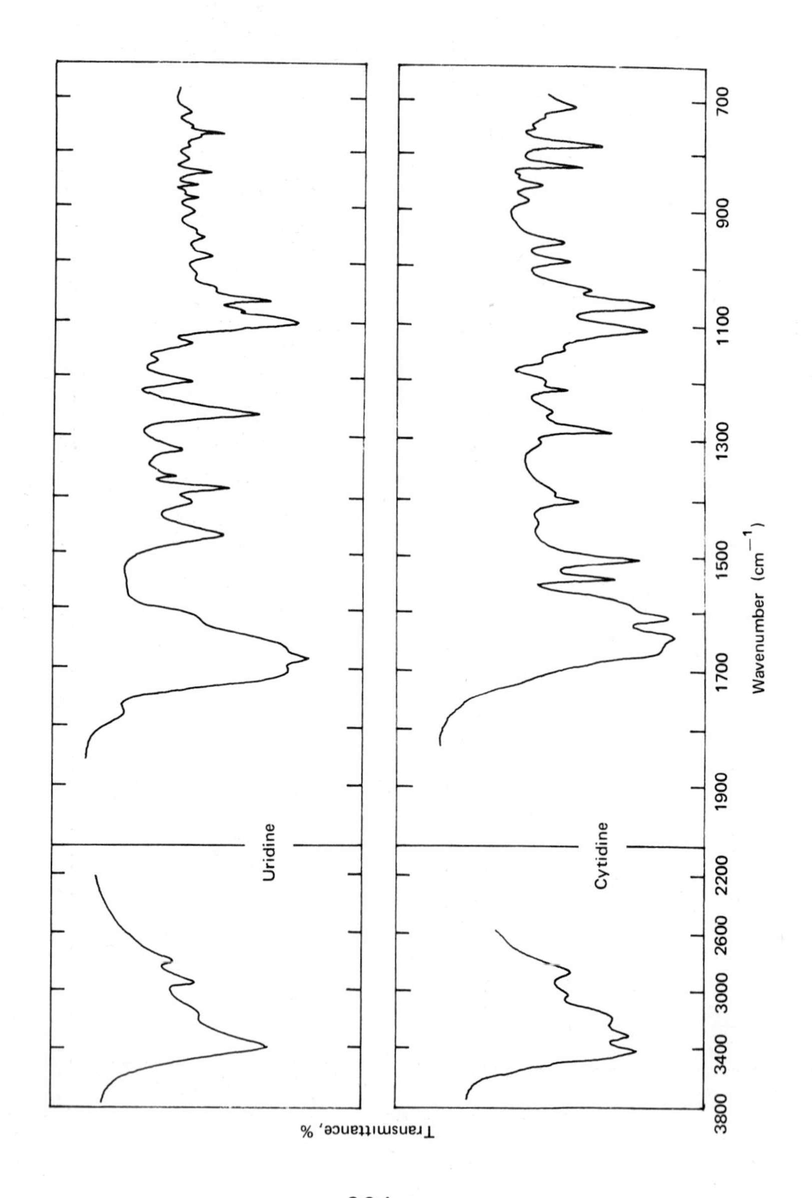
Uridine
Cytidine
Transmittance, %
Wavenumber (cm^{-1})
3800
3400
3000
2600
2200
1900
1700
1500
1300
1100
900
700

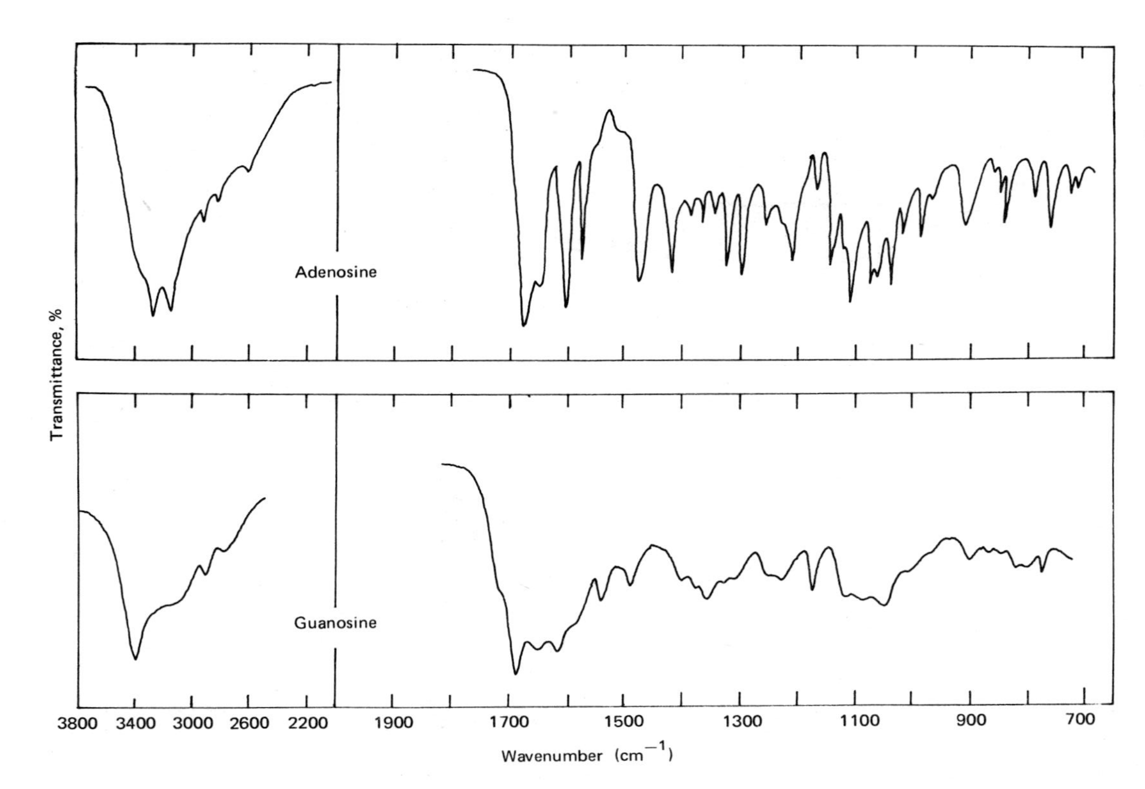

Fig. 4. Infrared Spectra of Uridine, Cytidine, Adenosine, and Guanosine.

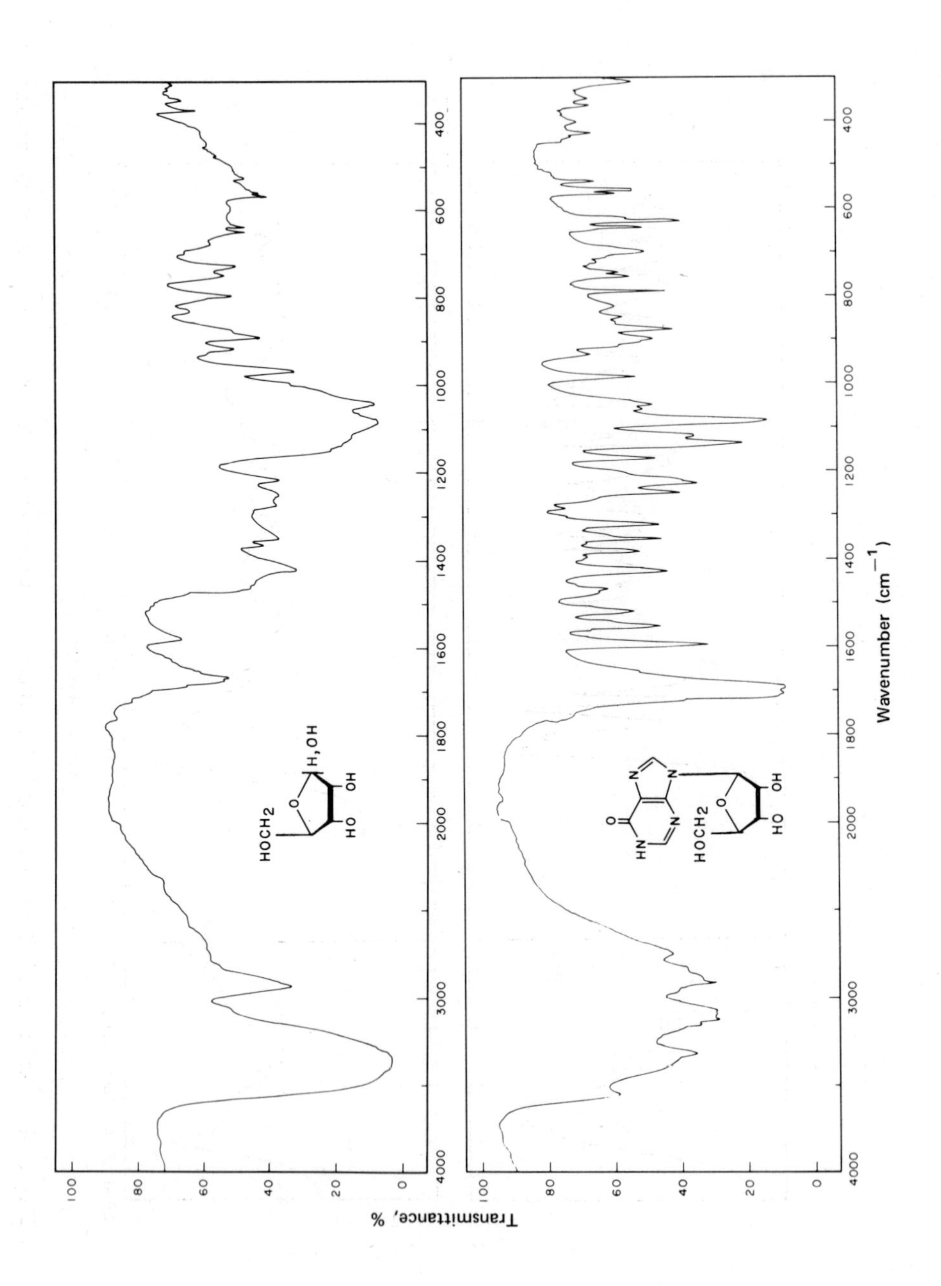

Transmittance, %
Wavenumber (cm^{-1})
100
80
60
40
20
0
4000
3000
2000
1800
1600
1400
1200
1000
800
600
400
HOCH2
H,OH
HO
OH
O
HN
N

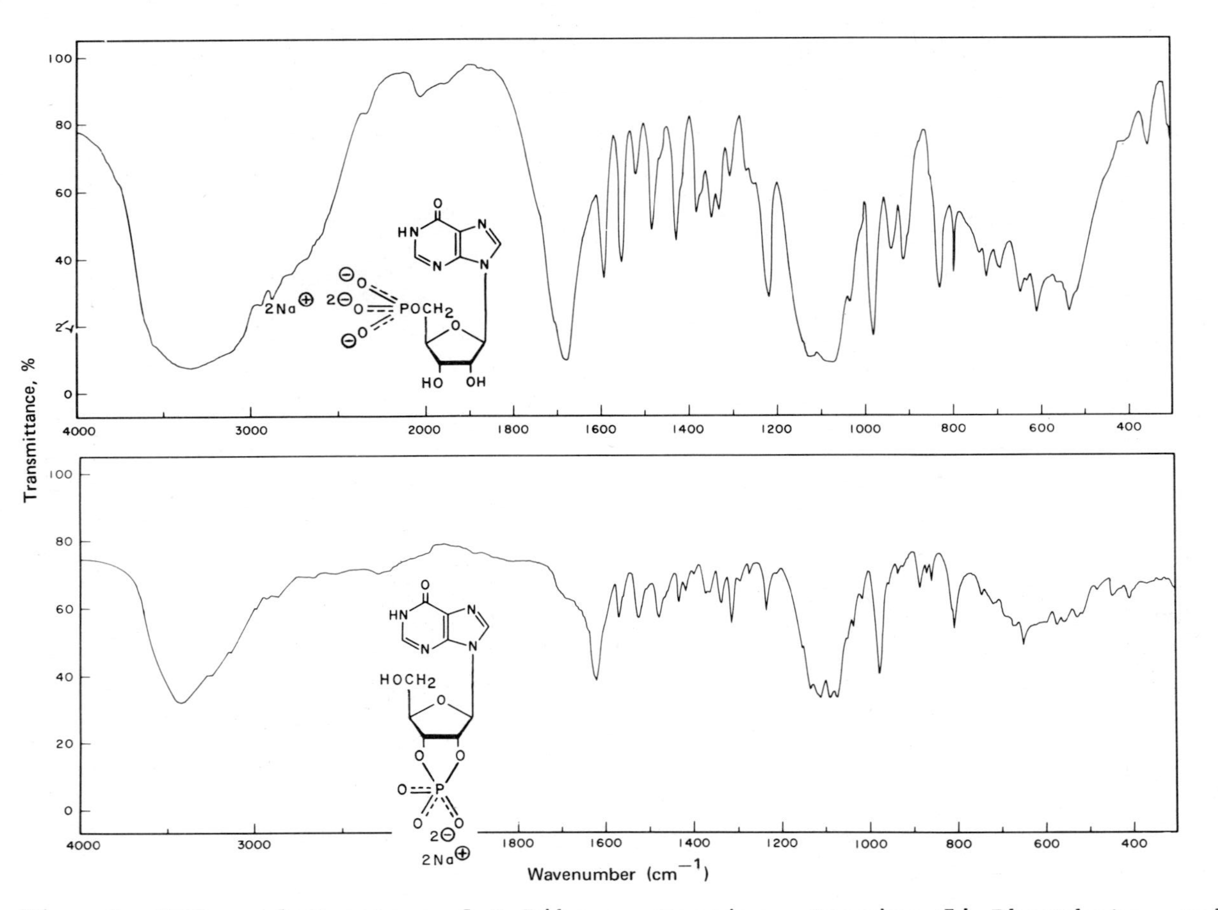

Fig. 5. Infrared Spectra of D-Ribose, Inosine, Inosine 5'-Phosphate, and Inosine 2':3'-Cyclic Phosphate.

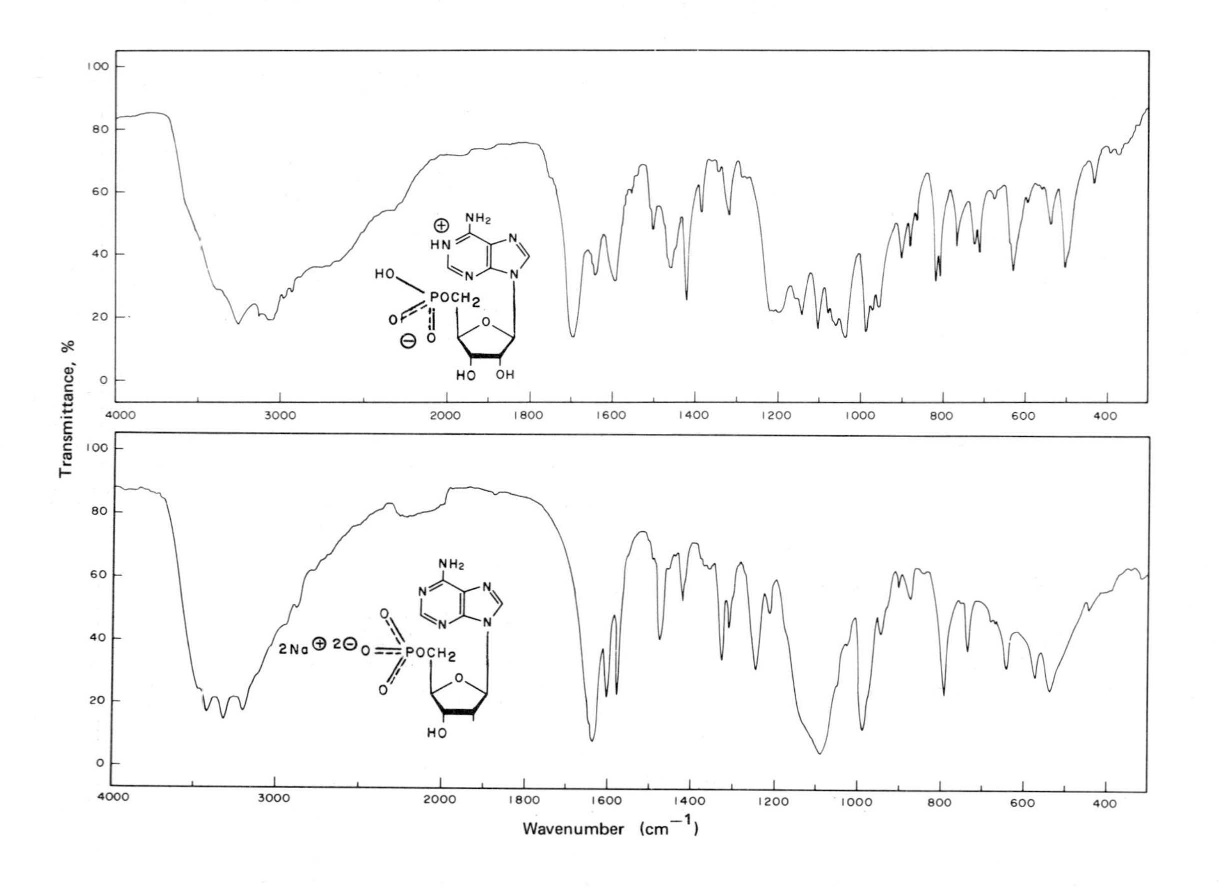
Transmittance, %
Wavenumber (cm^{-1})
4000
3000
2000
1800
1600
1400
1200
1000
800
600
400
100
80
60
40
20
0
NH2
HN
HO
POCH2
HO OH
2Na ⊕ 2⊖
HO

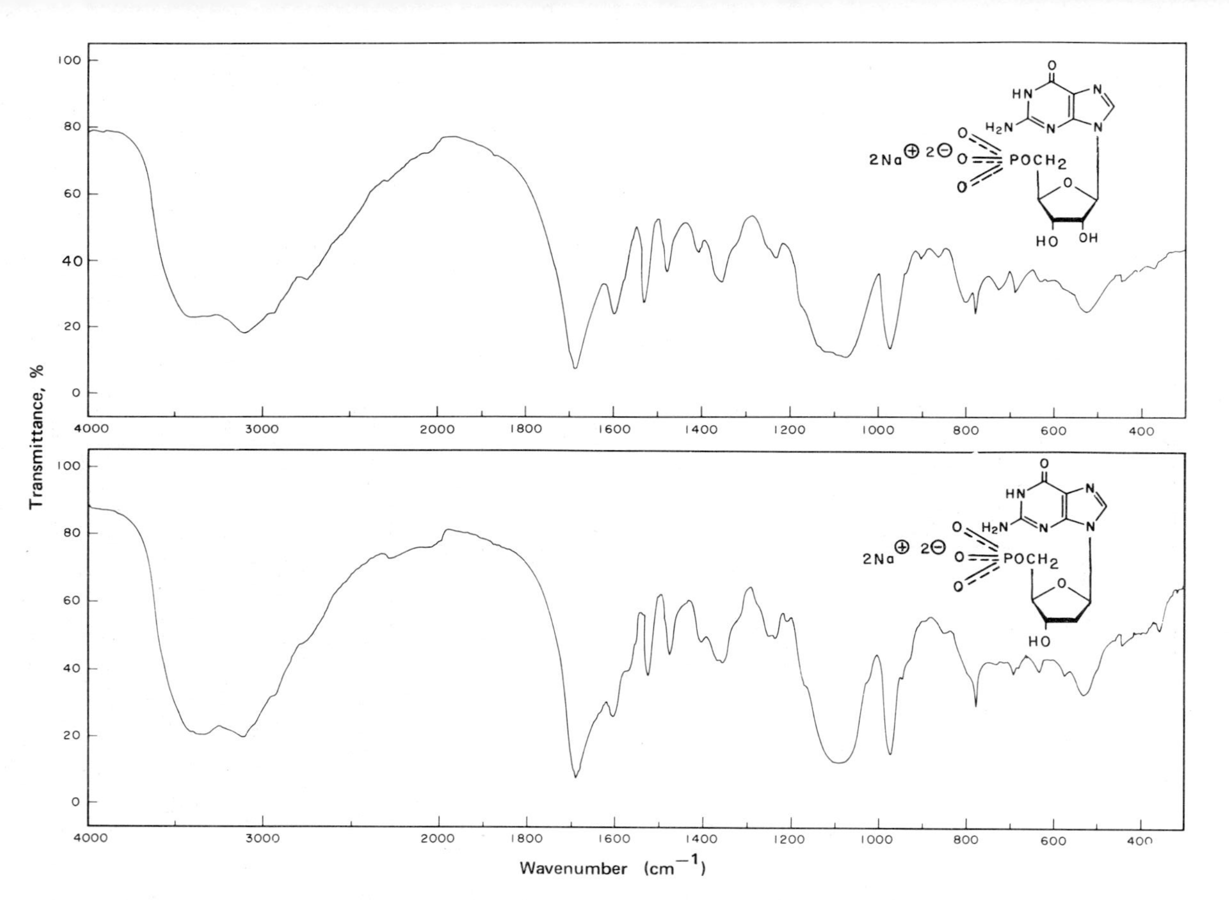

Fig. 6. Infrared Spectra of Adenosine 5'-Phosphate (free), 2'-Deoxyadenosine 5'-Phosphate (Na_2 Salt), Guanosine 5'-Phosphate (Na_2 Salt), and 2'-Deoxyguanosine 5'-Phosphate (Na_2 Salt).

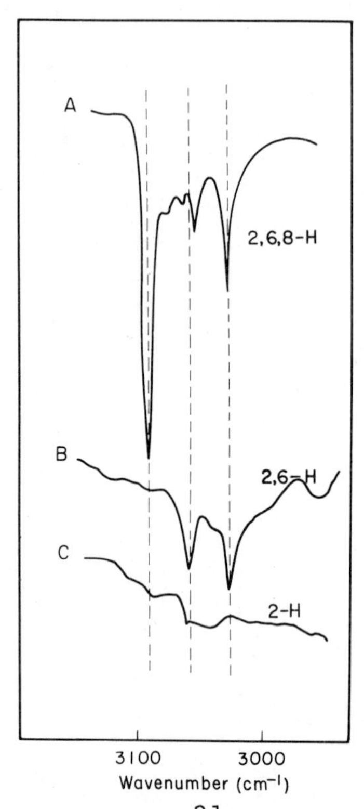

Fig. 7. C—H Stretching Bands[21] of (A) Purine-9-d_1, (B) Purine-8,9-d_2, and (C) Purine-6,8,9-d_3.

been observed, at 3047 cm^{-1}; it is strong and sharp for a solution in carbon tetrachloride. For pyrimidine derivatives, however, the C—H stretching bands are not readily identifiable, because there are a number of overlapping absorptions due to the NH groups and also, probably, due to some combination bands in the 3100—2800-cm^{-1} region.

e. C=O Stretching and Other Bands in the 1550—1750-cm^{-1} Region. The C=O group usually gives a strong absorption band at 1720 cm^{-1}. When it is connected with a nitrogen atom, however, as in the case of uracil, cytosine, hypoxanthine, and guanine, its frequency is lowered to 1700—1620 cm^{-1}. The intrinsic frequencies of the C=N and C=C bonds are considered to be at ~1600 cm^{-1}.

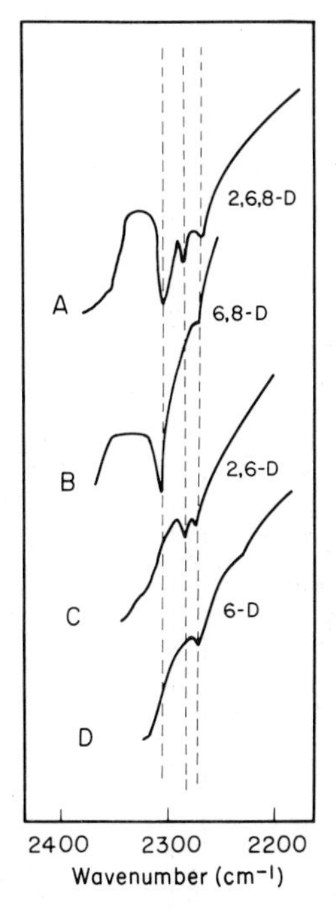

Fig. 8. C–D Stretching Bands[21] of (A) Purine-2,6,8-d_3, (B) Purine-6,8-d_2, (C) Purine-2,6-d_2, and (D) Purine-6-d_1.

Therefore, in the base residues of nucleic acids, the C=O, C=N, and C=C stretching vibrations often couple with one another, resulting in a few, complicated vibrations. Sometimes, however, it is possible to assign, for the sake of convenience, a certain band to a certain "C=O stretching vibration," or to a certain "C=N stretching vibration." In addition to the bands of the stretching vibrations of double bonds, there should appear absorption bands due to the NH in-plane deformation vibrations and the NH_2 scissoring vibrations in the 1750–1550-cm^{-1} region, if the molecule in question has these groups. In any case, the appearance of a few strong absorption bands in the 1750–1550-cm^{-1} region may be regarded as a characteristic of the base residues; the sugar and phosphate parts of a nucleic acid give almost no contribution in this spectral region.

Pyrimidine shows a strong band at 1570 cm^{-1}, whereas purine shows two, medium-intensity bands at 1610 and 1570

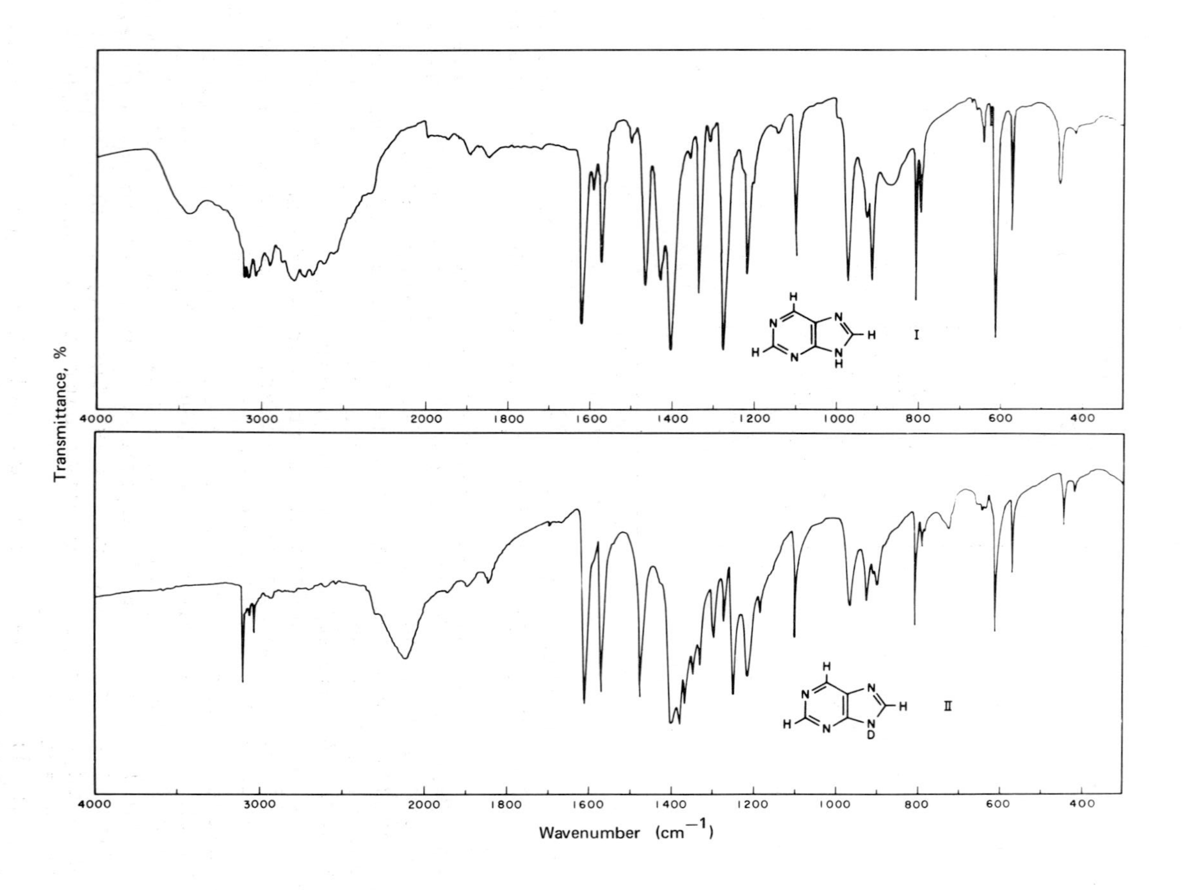

Transmittance, %
I
II
4000
3000
2000
1800
1600
1400
1200
1000
800
600
400
Wavenumber (cm^{-1})

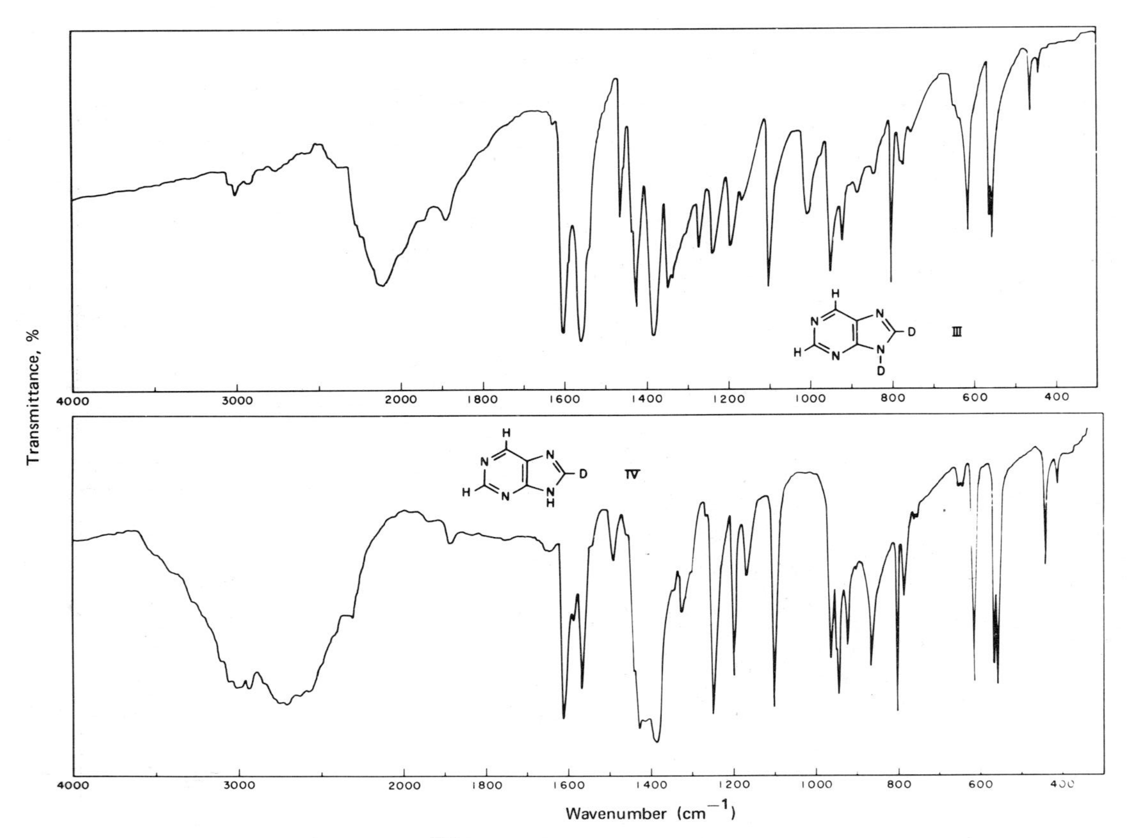

Fig. 9. Infrared Spectra[21] of Purine (I), Purine-9-d_1 (II), Purine-8,9-d_2 (III), Purine-8-d_1 (IV), Purine-6-d_1 (V), Purine-2,6-d_2 (VI), Purine-6,8-d_2 (VII), Purine-2,6,8-d_3 (VIII), and Purine-2,6,8,9-d_4 (IX).

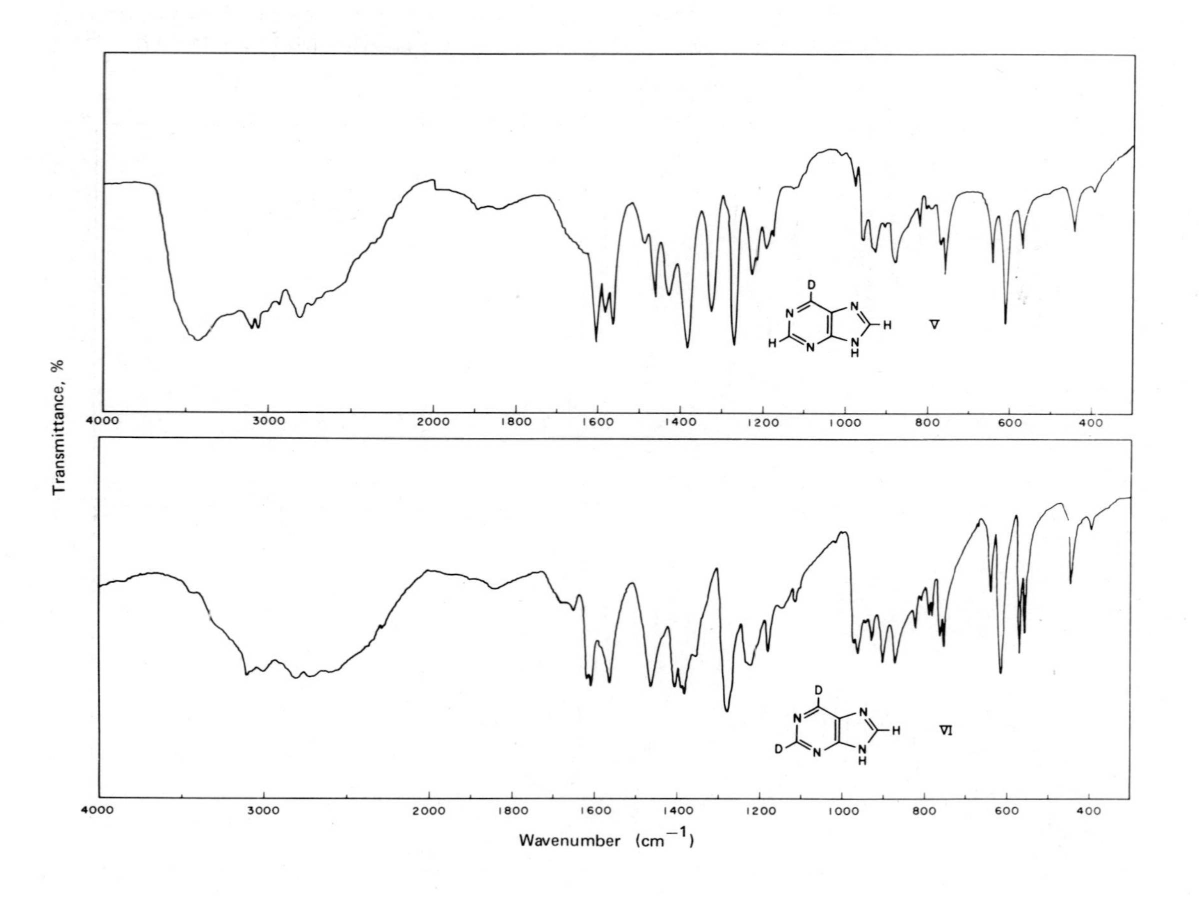

V
VI
Transmittance, %
Wavenumber (cm^{-1})
4000
3000
2000
1800
1600
1400
1200
1000
800
600
400

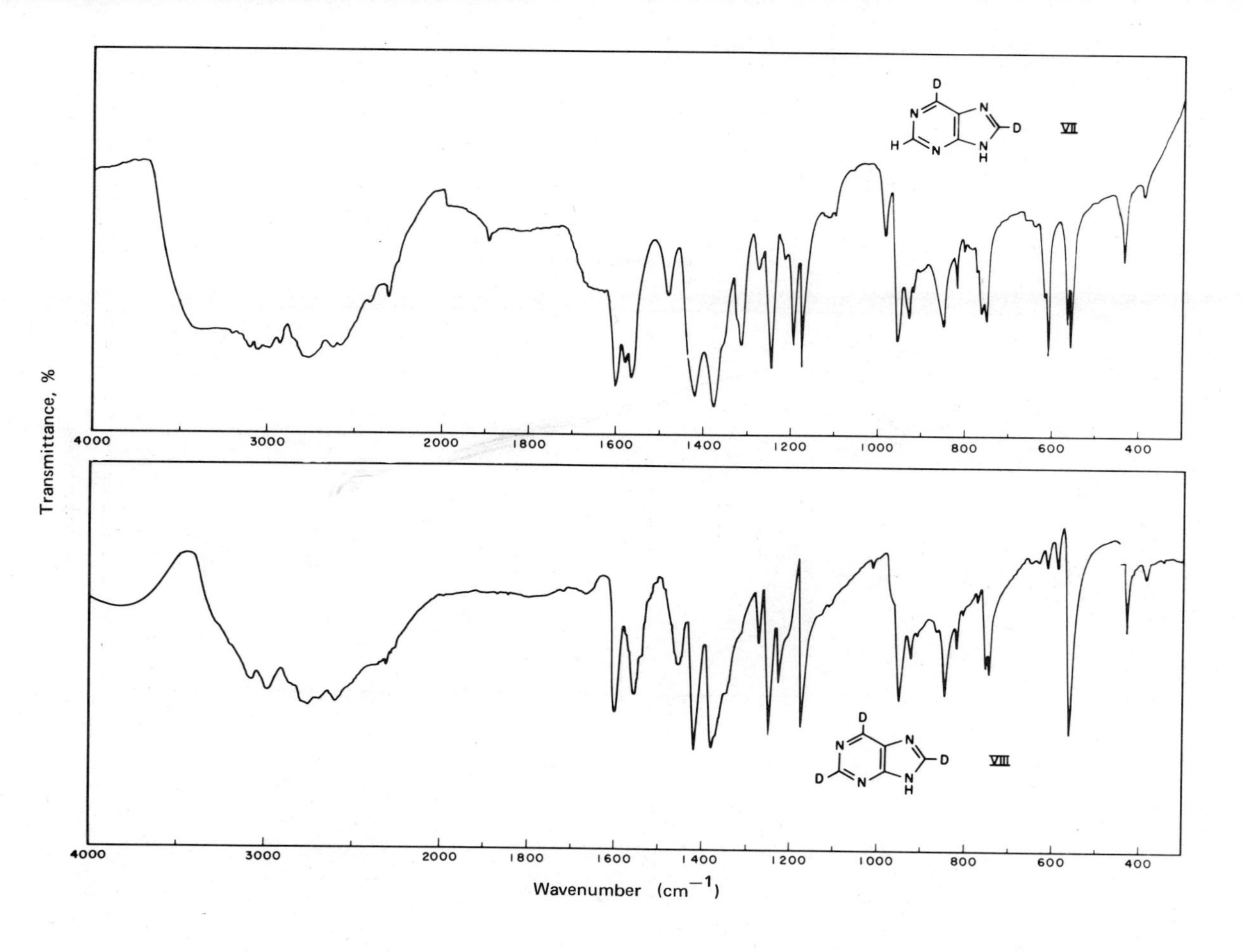
VII
VIII
Transmittance, %
Wavenumber (cm^{-1})
4000
3000
2000
1800
1600
1400
1200
1000
800
600
400

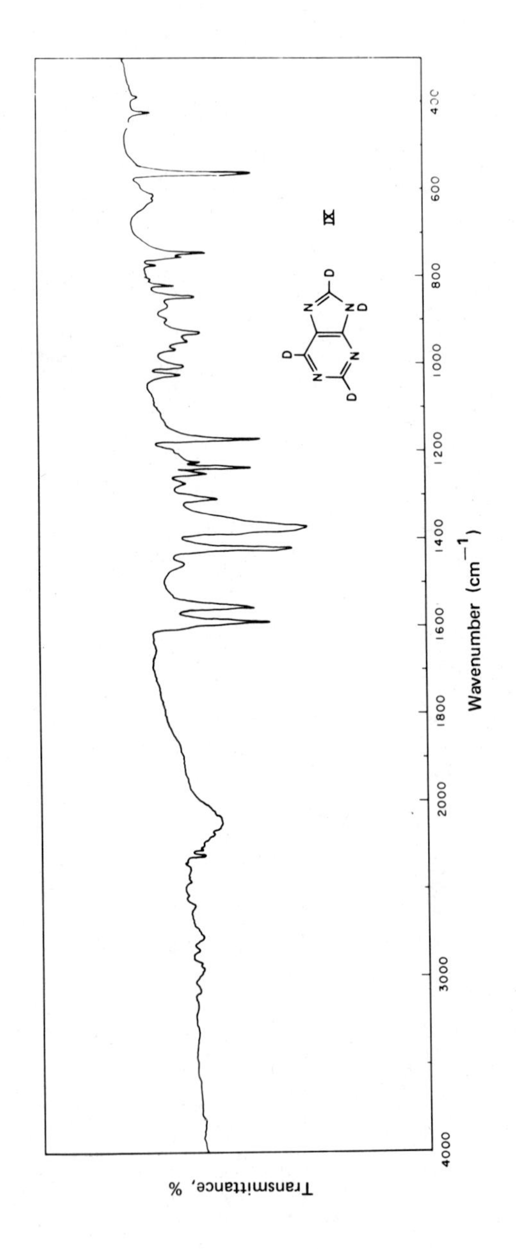

Transmittance, %
Wavenumber (cm^{-1})
4000
3000
2000
1800
1600
1400
1200
1000
800
600
400
IX

cm^{-1}. For purine, a weak, but sharp, band at 1590 cm^{-1} is assignable to the N-9-H in-plane deformation vibration, because this band is removed on *N*-deuteration (see Fig. 9).

Uracil shows two strong absorption bands, at 1741 and 1719 cm^{-1} (see Fig. 1). One of them (probably that at 1741 cm^{-1}) should be assigned to the N^1-H in-plane deformation vibration, because *N*-deuterated uracil (see Fig. 1), 1-methyluracil (see Fig. 1), and uridine (see Fig. 4) each shows only one band in this region (1750–1700 cm^{-1}). The other (probably that at 1719 cm^{-1}) is assigned to the C^2=O stretching vibration. Besides these two bands, uracil shows two strong bands, at 1669 and 1637 cm.$^{-1}$ These bands are ascribed to the C^4=O and C^5=C^6 stretching vibrations.

The spectral features of thymine (see Fig. 1) in this spectral region are similar to those of uracil, as might be expected. In the spectrum of thymine, however, the peaks at 1741 and 1719 cm^{-1} of uracil are brought together. Deprotonation of uracil and thymine residues is considered to take place at position 3, and this causes a considerable lowering of the C=O stretching frequencies (to 1635 cm^{-1} for uridine[17]).

Cytosine shows a strong, broad band in the 1670–1600-cm^{-1} region; it consists of a few absorption peaks (see Fig. 2). On deuteration, the highest frequency peak (at 1670 cm^{-1}) is removed. A shoulder at 1670 cm^{-1} is also observed for cytidine (see Fig. 4), and this is removed on *N*-deuteration. Therefore, the absorption at 1670 cm^{-1} is ascribed mainly to the NH_2 scissoring vibration. It is noticeable that, on protonation of cytosine, cytidine, or 2'-deoxycytidine, two strong and well defined bands appear[16,17] at 1720 and 1680 cm^{-1}. The band at 1720 cm^{-1} is due to a "normal" carbonyl group produced as a result of protonation at N^3. The band at 1680 cm^{-1} may be assigned to the C=N$^{\oplus}$ stretching vibration, in which the π-electrons are more localized than in the neutral cytosine ring. The presence or absence of the two strong bands in the 1750–1680 cm^{-1} region is a criterion for deciding whether the cytosine residue is in a protonated form or not (in a nucleotide, for example; see later).

Adenine shows two strong bands, at 1670 and 1600 cm^{-1}. On *N*-deuteration, the former is removed. No doubt, therefore, the 1670-cm^{-1} band is caused by the NH_2 scissoring vibration, and the 1600-cm^{-1} band by a stretching vibration of double bonds (see Fig. 10). Protonated adenine shows a strong band[15,16] at 1712 cm^{-1} that is assignable to the C=N^1 stretching vibration. Here the π-electrons are more localized than those in neutral adenine, because the protonation takes place at N^1.

Hypoxanthine gives a broad, strong band at 1680 cm^{-1}

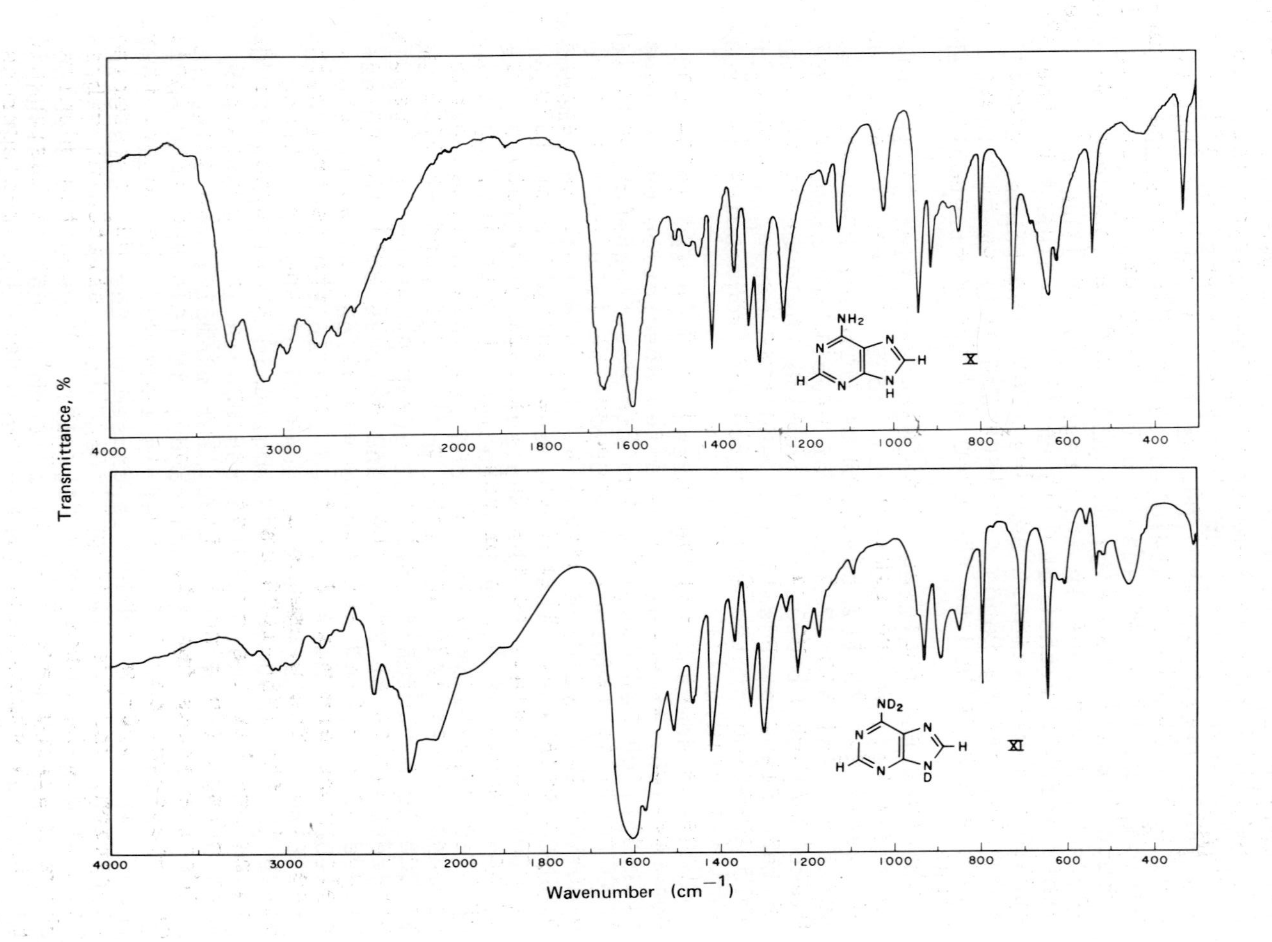
Transmittance, %
Wavenumber (cm^{-1})
X
XI
NH$_2$
ND$_2$
4000
3000
2000
1800
1600
1400
1200
1000
800
600
400

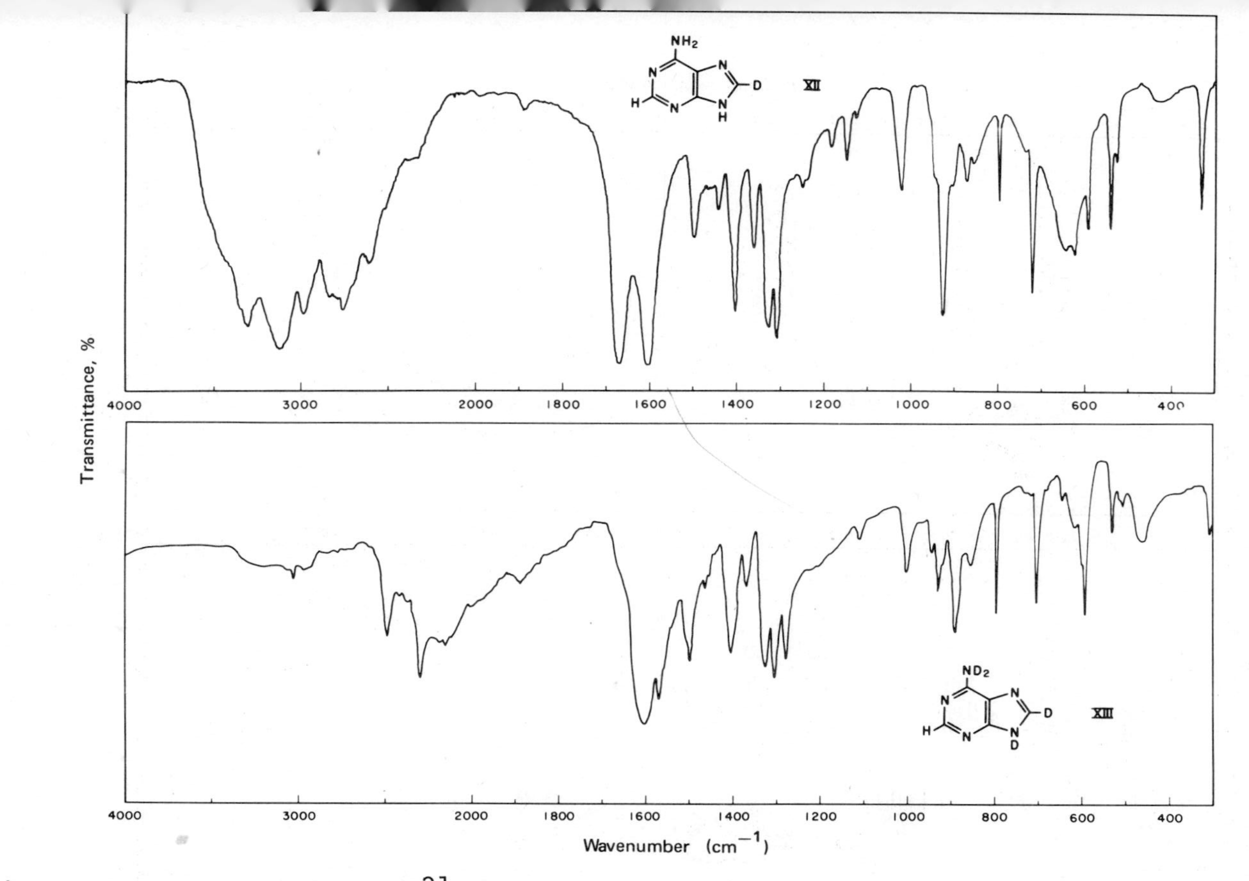

Fig. 10. Infrared Spectra[21] of Adenine (X), Adenine-6,6,9-d_3 (XI), Adenine-8-d_1 (XII), Adenine-6,6,8,9-d_4 (XIII), Adenine-2-d_1 (XIV), Adenine-2,8-d_2 (XV), and Adenine-2,6,6,8,9-d_5 (XVI).

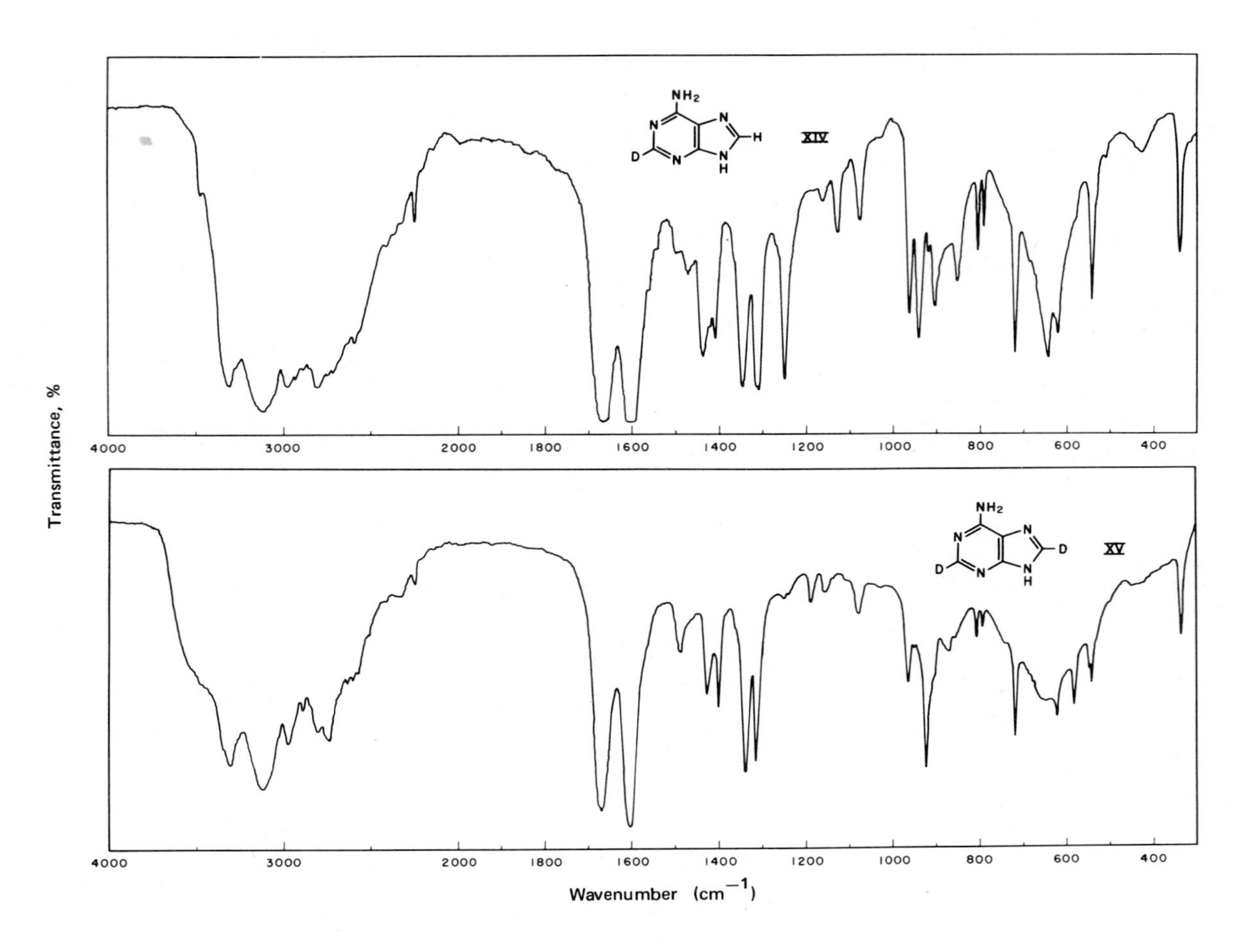
Transmittance, %
XIV
XV
4000
3000
2000
1800
1600
1400
1200
1000
800
600
400
Wavenumber (cm⁻¹)

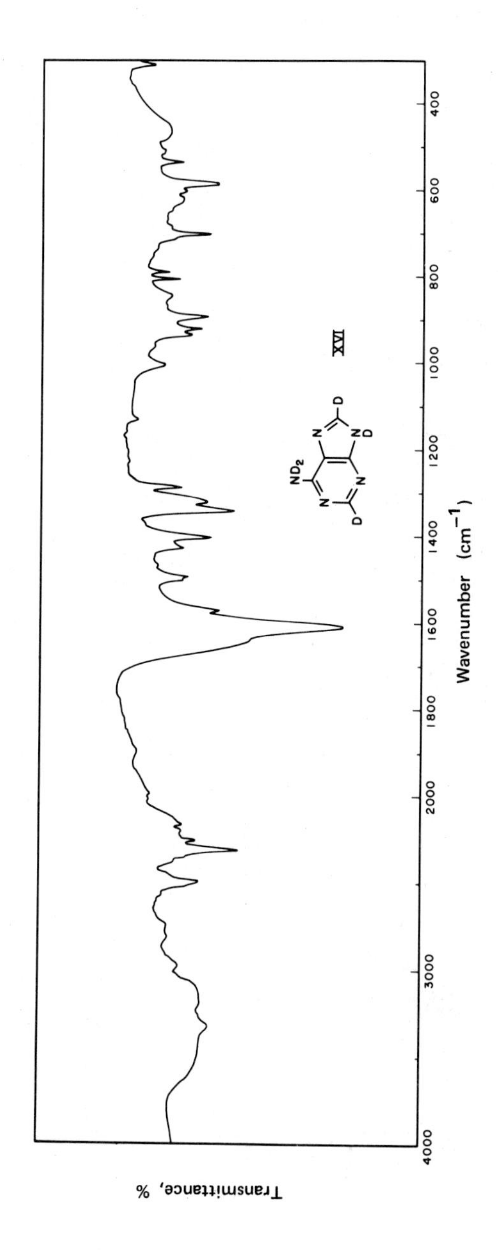
Transmittance, %
Wavenumber (cm^{-1})
4000
3000
2000
1800
1600
1400
1200
1000
800
600
400
XVI
ND$_2$

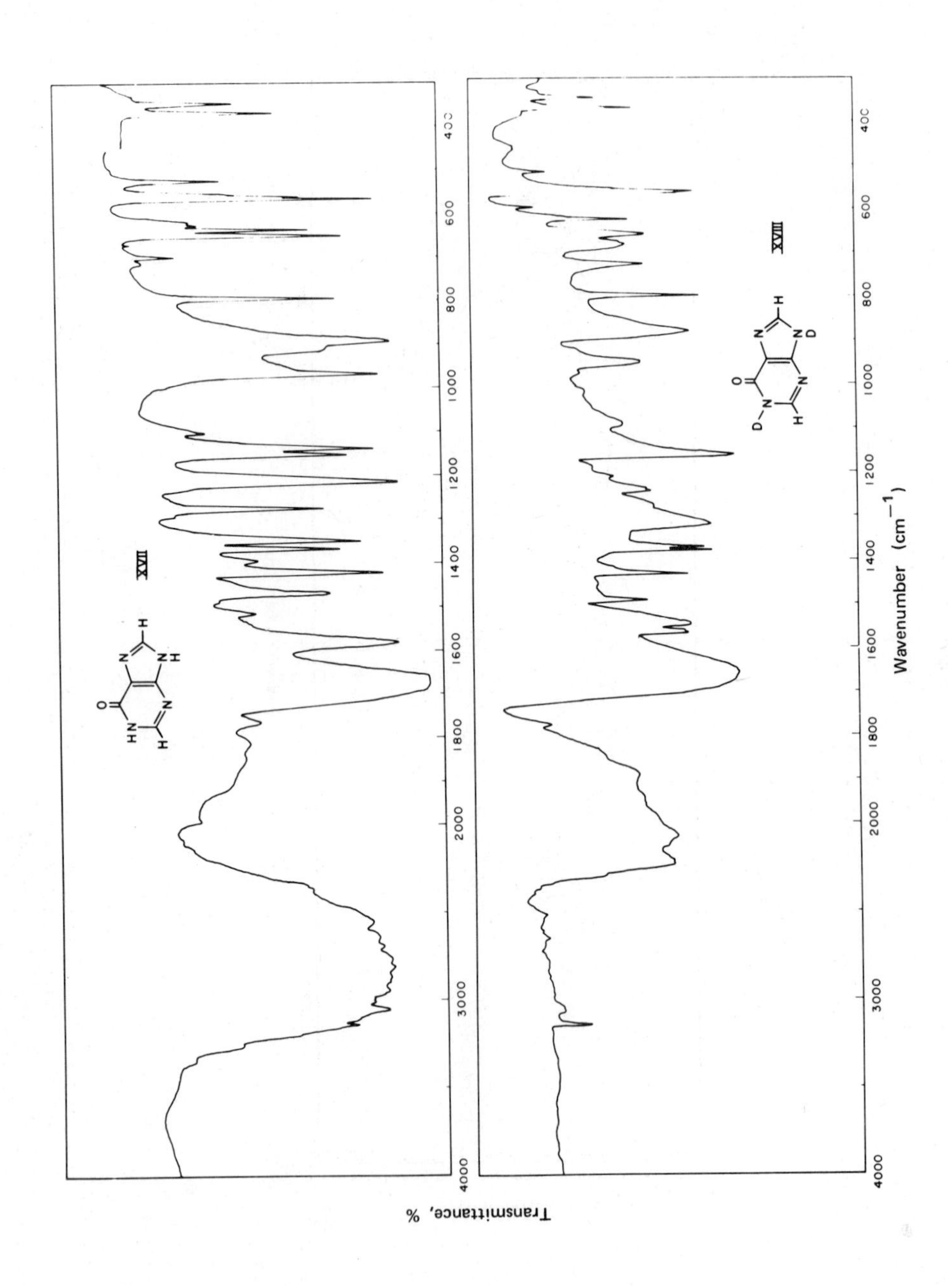
XVII
XVIII
Transmittance, %
Wavenumber (cm^{-1})
4000
3000
2000
1800
1600
1400
1200
1000
800
600
400

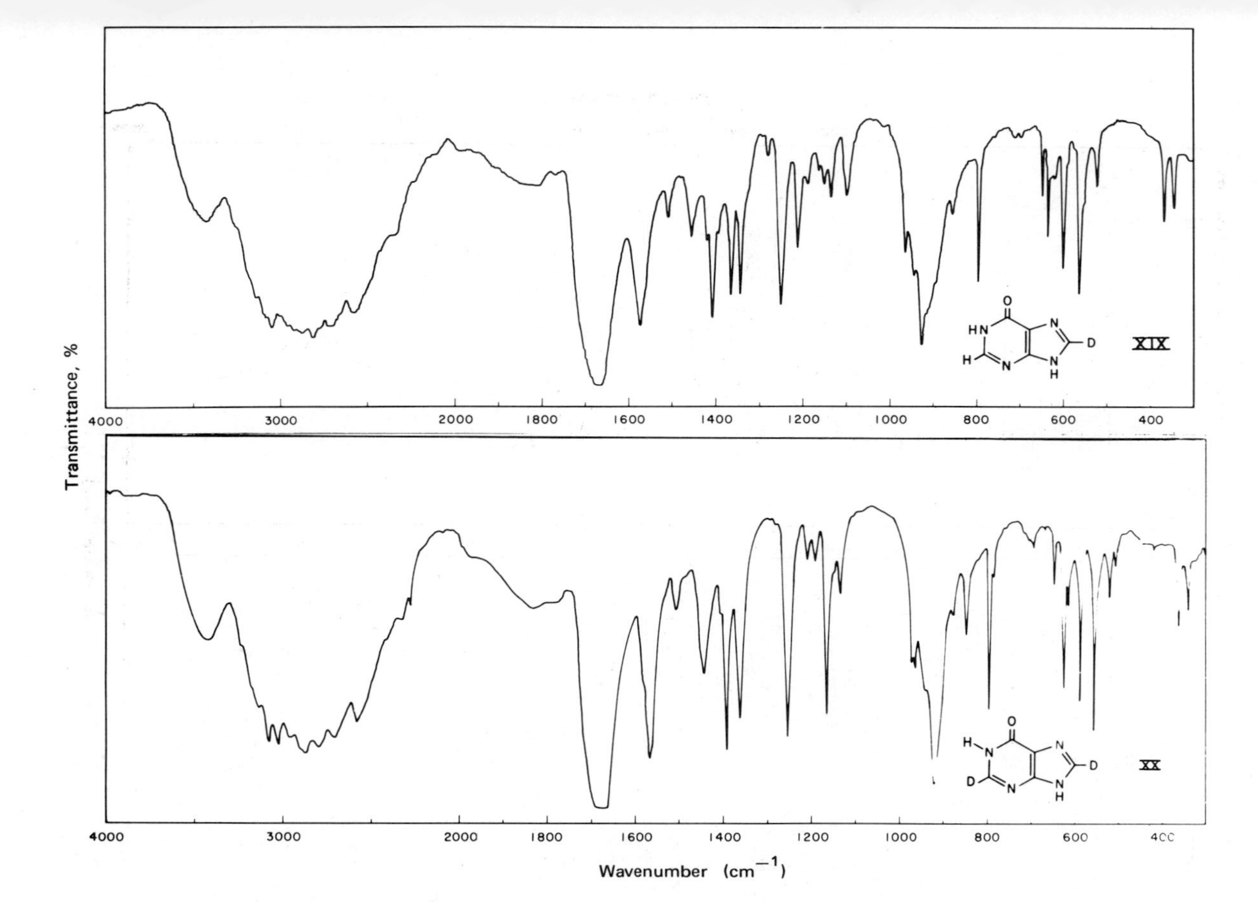

Fig. 11. Infrared Spectra[21] of Hypoxanthine (XVII), Hypoxanthine-1,9-d_2 (XVIII), Hypoxanthine-8-d_1 (XIX), Hypoxanthine-2,8-d_2 (XX), Hypoxanthine-2-d_1 (XXI), and Hypoxanthine-1,2,8,9-d_4 (XXII).

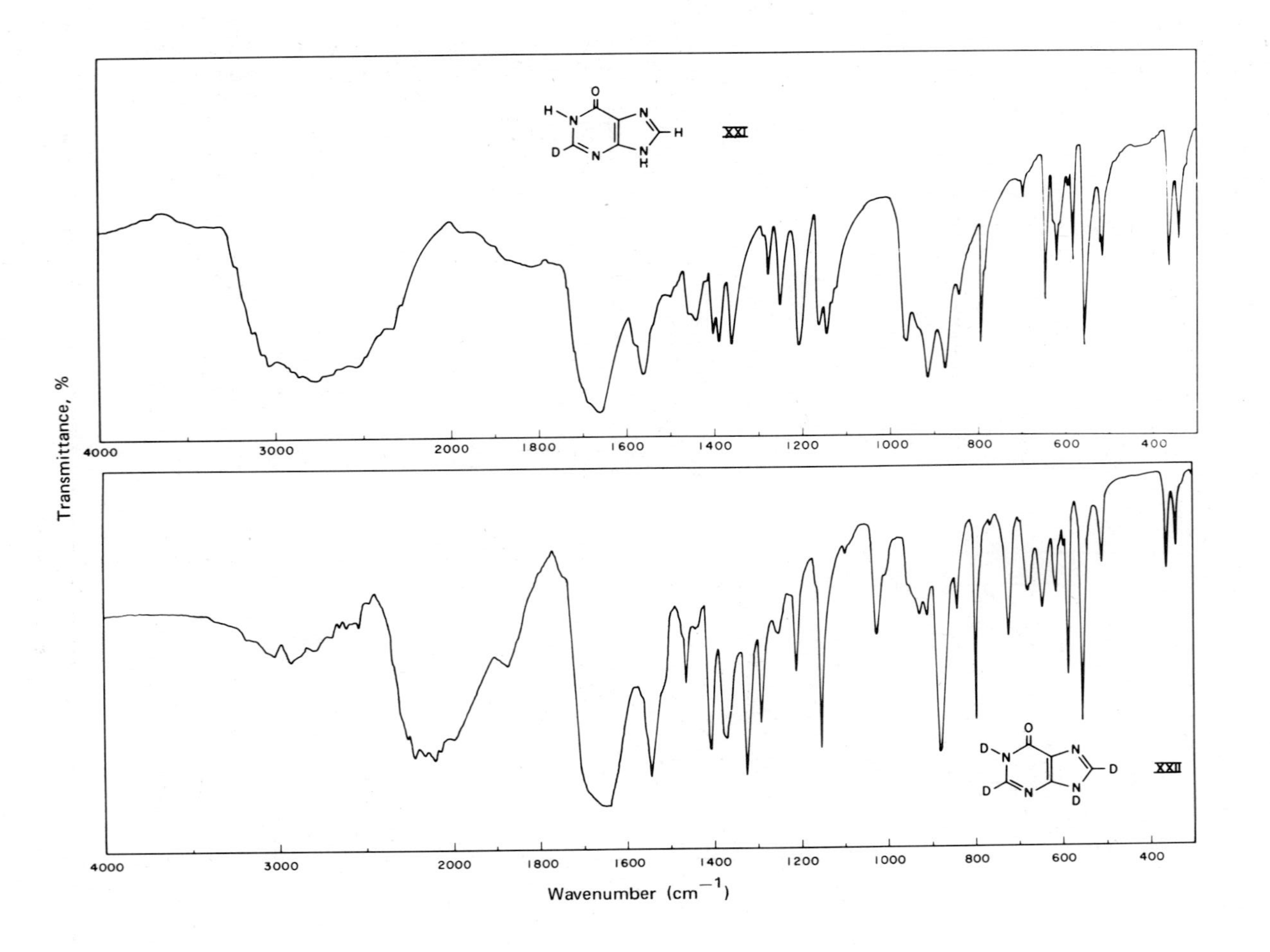
XXI
XXII
Transmittance, %
Wavenumber (cm^{-1})
4000
3000
2000
1800
1600
1400
1200
1000
800
600
400

(see Fig. 11). This band is little affected by *N*-deuteration, and is assigned to the C=O stretching vibration. Besides this band, there is a sharp, medium-intensity band at 1580 cm^{-1}, assignable to a double-bond stretching vibration.

Guanine shows two strong bands, at 1695 and 1670 cm^{-1}, a weak band at 1640 cm^{-1}, and two medium-intensity bands at 1570 and 1560 cm^{-1} (see Fig. 12). On *N*-deuteration, the 1695-cm^{-1} band is removed, and a new, strong band appears at 1600 cm^{-1} (see Fig. 12).

f. NH and NH_2 Out-of-plane Deformation Bands. These bands are mostly broad and of medium intensity. Their frequencies are distributed over the 950-550-cm^{-1} region, depending on the type and strength of the hydrogen bonds with which they are involved. These bands are detected on the basis of examination of the *N*-deuteration effects and the anisotropy of the infrared absorptions for their crystalline state.[18] For example, the N^3-H out-of-plane deformation band of 1-methylthymine is at 882 cm^{-1}, that of 1-methyluracil is at 870 cm^{-1}, the N^9-H out-of-plane deformation band of purine is at 865 cm^{-1} (see Fig. 9), and those of adenine, hypoxanthine, and guanine are all at ~900 cm^{-1} (see Figs. 10, 11, and 12). The NH_2 wagging (out-of-plane deformation) band of adenine is at 650 cm^{-1} (see Fig. 10), that of 9-methyladenine[18] at 690 cm^{-1}, and that of guanine at 580 cm^{-1} (see Fig. 12).

g. CH Deformation Bands. As shown, in Fig. 9, three sharp bands for purine, at 1100, 804, and 790 cm^{-1}, are always observed when the C-6-H group is not deuterated. On the other hand, when this group is deuterated, the three bands are absent, and two bands, at 770 and 760 cm^{-1}, always appear. Hence, the former three are ascribed to vibrations in which the C-6-H deformation motions (in-plane or out-of-plane) are concerned, and the latter two to vibrations in which the C-6-D deformations are concerned. For adenine, hypoxanthine, and guanine, no such absorption bands assignable to certain particular CH groups are found (see Figs. 10, 11, and 12). It is found,[21] however, that the sharp band of adenine at 800 cm^{-1} splits into two components only, and always when the C-2-H group is deuterated to give C-2-D (see Fig. 10). Also, a sharp band at 600 cm^{-1} for hypoxanthine appears only, and always, when the C-8-H group is deuterated to C-8-D (see Fig. 11). Each of these bands is assigned to a certain type of C-2-D or C-8-D deformation vibration.

h. Bands due to Skeletal Vibrations. As may be seen from Figs. 9–12, certain frequencies of purine, adenine, hypoxanthine, and guanine remain almost unaffected by every type of deuteration thus far examined.[21] Such a frequency is assignable to a vibration of the skeleton of

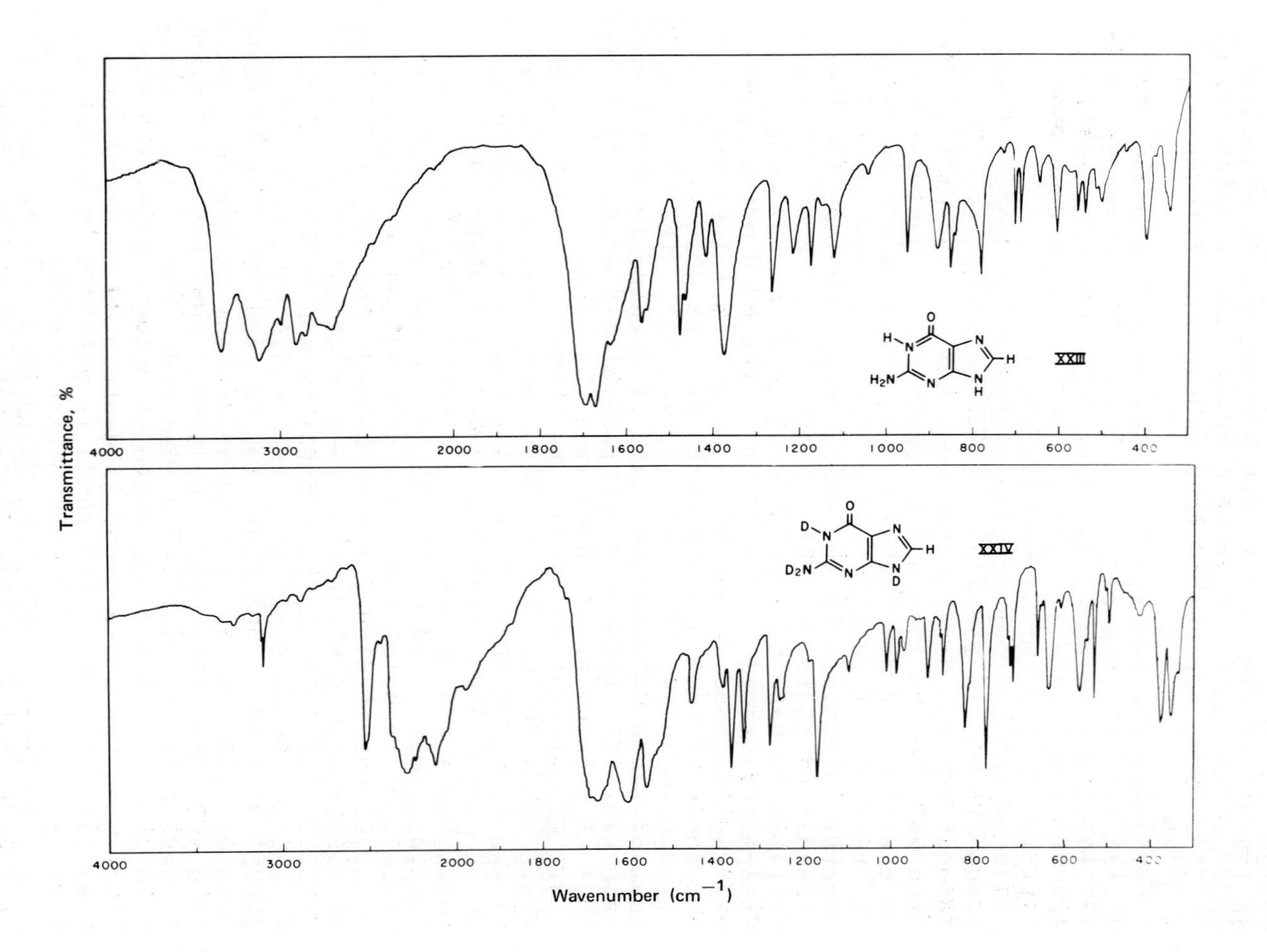
XXIII
XXIV
Transmittance, %
Wavenumber (cm^{-1})
4000
3000
2000
1800
1600
1400
1200
1000
800
600
400

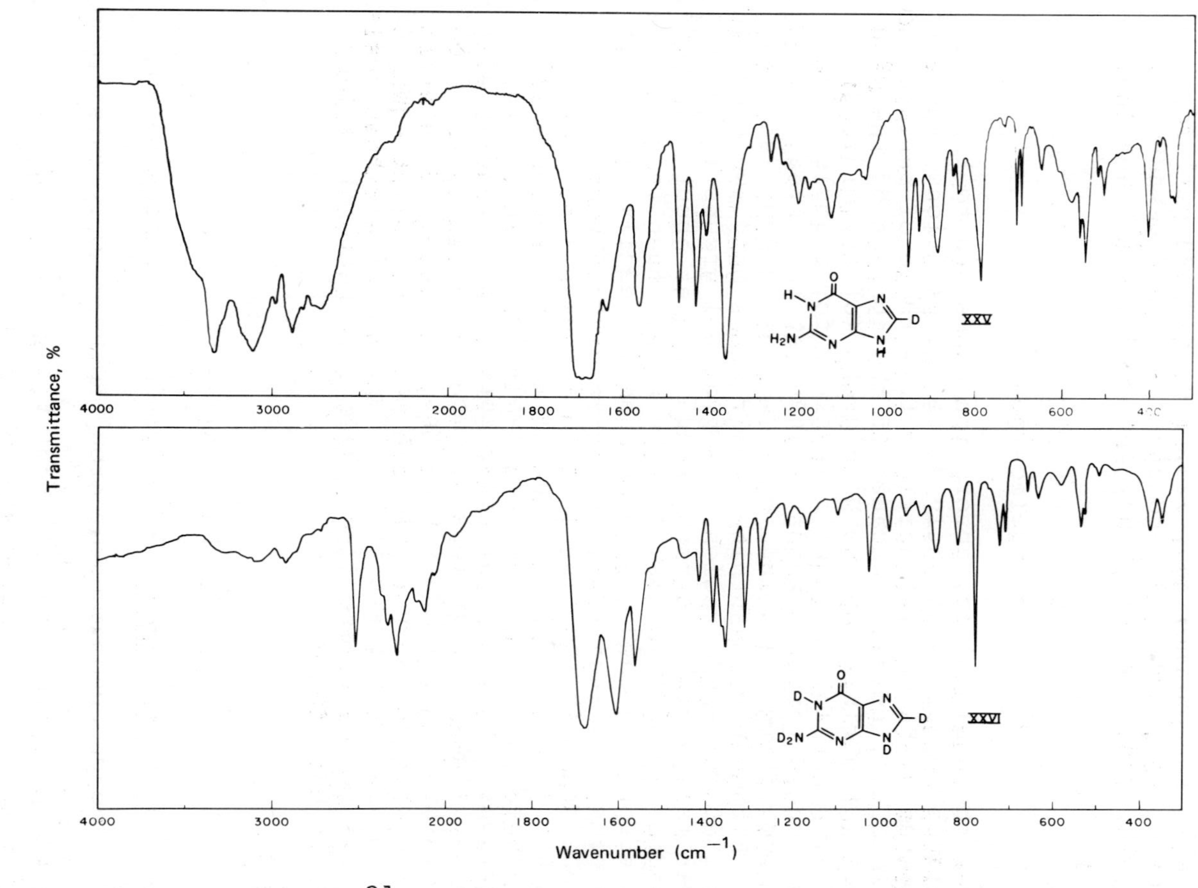

Fig. 12. Infrared Spectra[21] of Guanine (XXIII), Guanine-1,2,2,9-d_4 (XXIV), Guanine-8-d_1 (XXV), and Guanine-1,2,2,8,9-d_5 (XXVI).

the molecule in question (formed by the carbon, nitrogen and oxygen atoms). These are as follow: 570, 450, and 410 cm^{-1} for purine; 800, 720, 540, and 330 cm^{-1} for adenine; 800, 570, 370, and 350 cm^{-1} for hypoxanthine; and 780, 705, 690, 515, 500, and 350 cm^{-1} for guanine.

2. Nucleosides

For the nucleosides, absorption bands due to the base residues appear at slightly different positions, and with somewhat different intensities, from those of purines and pyrimidines, probably because of the absence of the N^9-H group (of purines) or the N^1-H group (of pyrimidines). The hypoxanthine residue in inosine, for example, gives three characteristic bands at 1690 (strong, broad), 1600 (medium, sharp), and 1560 cm^{-1} (medium, sharp) (see Fig. 5). Although hypoxanthine itself shows only two bands in the 1700-1550-cm^{-1} region, the hydroxyl groups in the sugar moiety give a strong, broad band at about 3400 cm^{-1}. Because its position is slightly higher than those of the NH_2 stretching bands, this OH band is readily found. In addition, the sugar residue always gives a number of strong bands in the 1100-1000-cm^{-1} region (see Figs. 3, 4, and 5). The absorption bands of the base residues in this spectral region are relatively weak. Therefore, overlapping of the sugar bands here is also readily detected. These absorptions are mostly caused by the C-O single-bond stretching vibrations, but detailed interpretation of them has not yet been made. No reliable way of distinguishing between the i.r. spectra of a ribose residue and 2-deoxyribose (2-deoxy-*erythro*-pentose) residue has yet been found.

3. Nucleotides

a. Phosphate Bands. A nucleoside monophosphate contains the phosphomonoester group RO-P(=O)(OH)-OH. This group has two p*K* values, one at about 1.5 and the other[22] at about 6.3, and it dissociates as on the next page. Form II shows two medium-intensity bands at about 1200 and 1090 cm^{-1}, whereas form III shows a very strong, broad band at 1100 cm^{-1} and a strong, sharp band at 980 cm^{-1}. The two bands of form II do not always stand out, but the two bands of form III are always conspicuous.[23]

$$RO-P(=O)(OH)-OH \underset{}{\overset{pH\ 1.5}{\rightleftharpoons}} RO-P(\cdots O)(\cdots O^{\ominus})-OH + H^{\oplus}$$

1 2

$$\underset{}{\overset{pH\ 6.3}{\rightleftharpoons}} RO-P(\cdots O)(\cdots O)(\cdots O)^{2\ominus} + 2H^{\oplus}$$

3

b. Zwitterion Structures. In the solid state the cytidylic acids show the phosphate bands at 1200 and 1090 cm^{-1}, and two strong bands[15,16] at 1720 and 1680 cm^{-1}. This fact indicates that each molecule has a zwitterion structure in which the phosphate group is in the $-OP(\cdots O)(\cdots O)(\cdots OH)^{\ominus}$ form, and there is a $(\geqslant N^3-H)^{\oplus}$ group in the cytosine residue. Similarly, in the solid state, the adenylic acids show the phosphate bands at 1200 and 1100 cm^{-1}, and a strong band at about 1700 cm^{-1} (see Fig. 6); this indicates that the adenylic acids are present in the zwitterion form having the $-OP(\cdots O)(\cdots O)(\cdots OH)^{\ominus}$ and $\geqslant N^1H^{\oplus}$ groups.

IV. ANALYSIS OF AQUEOUS SOLUTIONS

Water is not a completely suitable solvent for infrared spectroscopy of the compounds discussed in this Chapter. It shows strong, broad absorption bands at about 3300, 1640, and 700 cm^{-1}. Therefore, measurements of the infrared absorption of nucleic acids in water are not possible in the vicinities of these bands. Measurements are, however, possible in the 1500-900 cm^{-1} region. When deuterium oxide is used as the solvent, infrared absorption measurements can be made in the region of 1800-1400 cm^{-1}. It should be borne in mind, however, that the hydrogen atoms attached to the oxygen or nitrogen atoms have been replaced by deuterium atoms in the molecules dissolved in deuterium oxide. Even from the i.r. spectra in these two limited regions, valuable information on compounds related to the nucleic acids is sometimes obtained. For example, 2'-deoxycytidine 5'-phosphate shows a strong band at 1710 cm^{-1} in solution in deuterium oxide at[24] pD 2.5, indicating that, at this pD, the cytosine residue is deuterated. On raising the pD to 5.1, however, this band becomes weaker, and a neutral cytosine band at 1600 cm^{-1} begins to appear. At pD 7.0 and 11.5, no absorption is found at 1710 cm^{-1}, and only the 1600 cm^{-1} band is observed in the 1750-1600-cm^{-1} region. In the 1300-900-cm^{-1} region, absorptions at 1200 and 1090 cm^{-1}, characteristic of the $RO{-}{-}P{\lessgtr}\begin{smallmatrix}O^{\ominus}\\O\\OH\end{smallmatrix}$ group, are observed at pH 2.5 and 5.1; but, at pH 7.0 and 11.5, absorptions at 1100 and 980 cm^{-1}, characteristic of the $RO{-}P{\lessgtr}\begin{smallmatrix}O2\ominus\\O\\O\end{smallmatrix}$ group, are observed. This nucleotide molecule is thus presumably in a zwitterion form at pH 2.5.

Like their ultraviolet absorption spectra (see p. 47), infrared absorption spectra of compounds related to the nucleic acids in aqueous solution can be used for quantitative analysis. For some nucleosides and nucleotides, curves for the molar extinction coefficient (ε) *vs.* wavenumber have been obtained.[17,25] In the region of 1700-1600 cm^{-1}, ε at each absorption maximum (ε_{max}) lies in the order of magnitude of 1,000. This value is much lower than ε_{max} 15,000-6,000 for the absorption bands in the region of 220-300 nm of the ultraviolet spectra of these compounds.[26] As illustrated in Figs. 13-15, however, overlapping of the absorption bands of different species or different forms is much less in infrared than it is in ultraviolet spectra. This is an advantage of the infrared spectrum as a means of quantitative analysis.

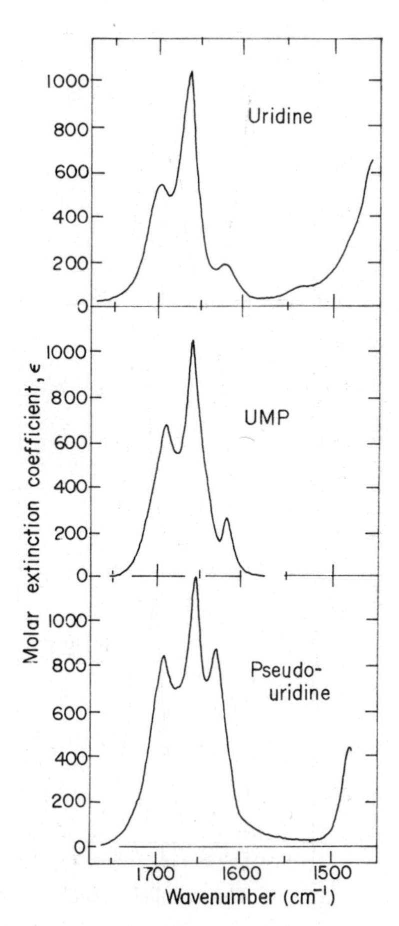

Fig. 13. Infrared Absorption Curves of Uridine,[17] Uridine 5'-Phosphate,[25] and Pseudouridine[27] in Neutral Solutions in D_2O.

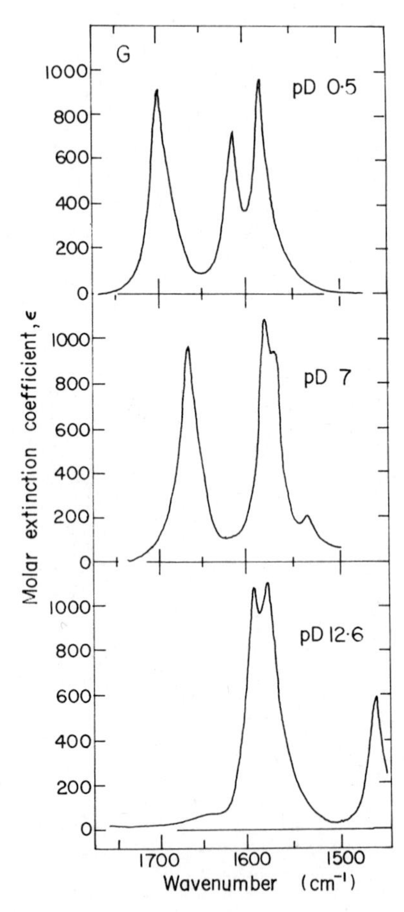

Fig. 14. Infrared Absorption Curves of Guanosine in an Acidic Solution,[17] Guanosine 5'-Phosphate in Neutral Solution, and Guanosine in a Basic Solution[17] (all in D_2O).

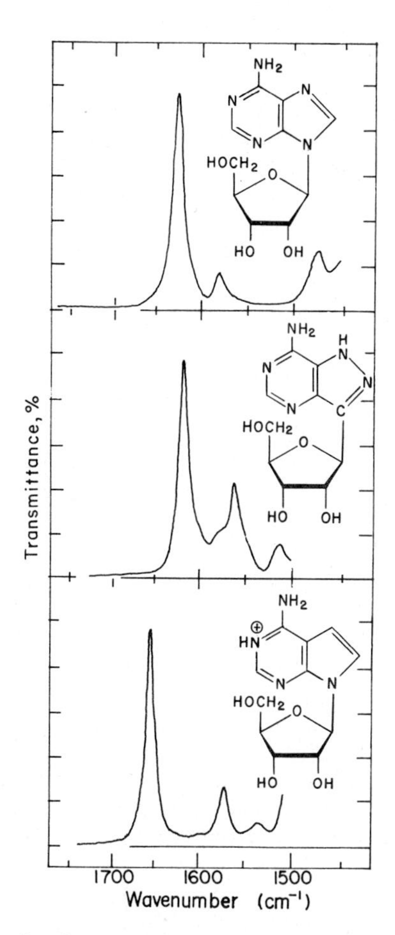

Fig. 15. Infrared Absorption Curves of Adenosine (neutral form),[17] Formycin (neutral form),[27] and Tubercidin (protonated form)[27] in solution in D_2O.

REFERENCES

(1) See M. Davies (Ed.), *Infrared Spectroscopy and Molecular Structure—An Outline of the Principles*, Elsevier Publishing Co., Amsterdam, (1963); A. D. Cross, *An Introduction to Practical Infrared Spectroscopy*, 2nd ed., Butterworth and Co., Ltd., London, England (1963).

(2) For experimental procedures, see W. J. Potts, Jr., *Chemical Infrared Spectroscopy*, Vol. 1, *Techniques*, John Wiley & Sons, Inc., New York, N.Y. (1962); A. E. Martin, *Infrared Instrumentation and Techniques*, Elsevier Publishing Co., Amsterdam (1966).

(3) E. R. Blout and M. Fields, *J. Amer. Chem. Soc.*, 72, 479 (1950).

(4) E. R. Blout and G. R. Bird, *J. Opt. Soc. Amer.*, 41, 547 (1951).

(5) A. Dekker and D. T. Elmore, *J. Chem. Soc.*, (1951) 2864.

(6) L. N. Short and H. M. Thompson, *J. Chem. Soc.*, (1952) 168.

(7) M. M. Stimson and M. J. O'Donnell, *J. Amer. Chem. Soc.*, 74, 1805 (1952).

(8) D. J. Brown and L. N. Short, *J. Chem. Soc.*, (1953) 331.

(9) R. J. C. Harris, S. F. D. Orr, E. M. Roe, and J. F. Thomas, *J. Chem. Soc.*, (1953) 489.

(10) A. M. Michelson and A. R. Todd, *J. Chem. Soc.*, (1954) 34.

(11) C. H. Willits, J. C. Decius, K. L. Dille, and B. E. Christensen, *J. Amer. Chem. Soc.*, 77, 2569 (1955).

(12) J. R. Lacher, J. L. Bitner, D. J. Emery, M. E. Seffl, and J. D. Park, *J. Phys. Chem.*, 59, 615 (1955).

(13) H. T. Miles, *Biochim. Biophys. Acta*, 22, 247 (1956).

(14) R. C. Lord, A. L. Marston, and F. A. Miller, *Spectrochim. Acta*, 9, 113 (1957).

(15) C. L. Angell, *J. Chem. Soc.*, (1961) 504.

(16) C. L. Angell, Ph.D. Thesis, Cambridge University, Cambridge, England, (1955).

(17) M. Tsuboi, Y. Kyogoku, and T. Shimanouchi, *Biochim. Biophys. Acta*, 55, 1 (1962).

(18) Y. Kyogoku, S. Higuchi, and M. Tsuboi, *Spectrochim. Acta*, 23A, 969 (1967).

(19) W. Krohs, *Chem. Ber.*, 88, 866 (1955).

(20) See L. J. Bellamy, *The Infra-red Spectra of Complex Molecules*, John Wiley & Sons, Inc., New York, N.Y. (1958); K. Nakanishi, *Infrared Absorption Spectroscopy—Practical*, Holden-Day, Inc., San Francisco, Calif. (1962).

(21) M. Sakuraba, Y. Iwashita, A. Nakamura, and M. Tsuboi, to be published.
(22) W. D. Kumler and J. J. Eiler, *J. Amer. Chem. Soc.*, 65, 2355 (1943).
(23) M. Tsuboi, *J. Amer. Chem. Soc.*, 79, 1351 (1957).
(24) R. L. Sinsheimer, R. L. Nutter, and G. R. Hopkins, *Biochim. Biophys. Acta*, 18, 13 (1955).
(25) G. J. Thomas, Jr., *Biopolymers*, 7325 (1969).
(26) G. H. Beaven, E. R. Holiday, and E. A. Johnson, in E. Chargaff and J. N. Davidson (Eds.), *The Nucleic Acids*, Vol. I, Academic Press, Inc., New York, N.Y. (1955).
(27) S. Higuchi, K. Morikawa, Y. Kyogoku, and M. Tsuboi, to be published.

CHAPTER 7

Nuclear Magnetic Resonance Spectroscopy in the Study of Nucleic Acid Components and Certain Related Derivatives

LEROY B. TOWNSEND

DEPARTMENT OF BIOPHARMACEUTICAL SCIENCES AND DEPARTMENT OF CHEMISTRY, UNIVERSITY OF UTAH, SALT LAKE CITY, UTAH

I. INTRODUCTION

In the last decade, nuclear magnetic resonance (n.m.r.) spectroscopy has become one of the most useful instrumental techniques available to chemists in many sub-disciplines. The basis for this technique is that certain atomic nuclei, such as ^{1}H, ^{13}C, ^{19}F, and ^{31}P, possess a spin quantum number not equal to zero. Although a number of other nuclei satisfy the requirements for nuclear magnetic resonance absorption, three of those mentioned (^{1}H, ^{13}C, and ^{31}P) are the most useful, especially in the area of nucleic acid chemistry. In fact, most of the work in n.m.r. spectroscopy has been devoted to proton magnetic resonance (p.m.r.) spectroscopy, and it constitutes the major portion of this Chapter. The primary objective of this Chapter is to provide a practical guide for the effective utilization of p.m.r. spectroscopy for investigators in chemistry, biochemistry, and other disciplines. This Chapter has also been designed especially for students or post-doctoral fellows lacking previous experience in the use of p.m.r. spectroscopy as applied to the chemistry of nucleic acids and related derivatives.

Presented first are the techniques involved in obtaining a p.m.r. spectrum suitable for subsequent interpretation. Such aspects as the preparation of a sample, the units used (δ or τ), solvents, and selection of an appropriate solvent are discussed. Also presented is a discussion of an integration curve that provides a method for the determination of the number of protons represented by different absorption bands in the spectrum. A brief discussion of the use of coupling constants and relaxation times completes the basic introduction. This introduction is designed to present, in a simplified way, only a small proportion of the essential theory, with the major emphasis on the practical aspects of p.m.r. as applied to the field of nucleic acids.

The introduction is followed by a discussion involving the application of the basic principles to the actual interpretation of p.m.r. spectra of heterocyclic compounds nucleosides, and nucleotides. Also included are examples of interpreted p.m.r. spectra from several closely related areas. Next presented are brief introductions to proton spin-spin decoupling in the determination of conformation; CAT, 100- and 220-MHz spectra, and other more advanced techniques; and a brief discussion of n.m.r. spectroscopy involving ^{13}C, ^{15}N, ^{19}F, and ^{31}P.

In conclusion, a number of representative structures are included, with p.m.r. spectral data that have been reported in the literature or observed in the author's Laboratory.

II. TECHNIQUES AND PROCEDURES FOR OBTAINING AND INTERPRETING NUCLEAR MAGNETIC RESONANCE SPECTRA

The primary purpose of this Section is to provide a straightforward guide for the application of n.m.r. spectroscopy to problems encountered in the laboratory, as well as in the classroom. The more practical aspects of obtaining an n.m.r. spectrum, along with its interpretation, will be presented. The principles and theory of n.m.r. spectroscopy have been covered in a number of excellent texts,[1-7] and, therefore, only a small amount of basic theory is included in the present discussion.

The first problem encountered in practical n.m.r. spectroscopy involves the preparation of a sample. Every effort should be made, by checking melting point, boiling point, appropriate chromatography (thin-layer, paper, gas, or column), or other techniques, to obtain a pure sample before recording a spectrum, as the presence of impurities will usually afford a number of extra, often unexplainable, resonance peaks. If these extra peaks are not recognized as caused by an impurity, a misinterpretation of the spectrum can result, leading to assignment of the wrong structure to the compound studied.

It is important to decide the scale that should be chosen when submitting a sample for recording of its n.m.r. spectrum, in order not to exclude any peaks. Peaks will occur in various regions of the spectrum and, with experience, the majority of these can be predicted with some degree of accuracy and they may require the use of different scales. Such peaks are usually recorded in units relative to a reference or standard, and are designated as chemical shifts relative to the reference. These chemical shifts (or resonance peaks) can be expressed as occurring at certain cycles per second (c.p.s.) but are now usually given as Hertz (Hz). Currently, however, two additional designations are used extensively in the literature;

these are delta (δ) and tau (τ) values. Both δ and τ may be values relative to the peak observed for tetramethylsilane (TMS) used as a reference point. The δ scale uses the TMS signal as zero, whereas the τ scale assigns to the same signal (TMS) the number ten. A δ value can be obtained by reading the value directly from the bottom scale; however, it may be calculated much more accurately by using the following equation.

$$\delta = \frac{\text{(Chemical shift in Hz)} \times 10^6}{\text{Frequency of the spectrometer}}$$

Therefore, should a peak centered at 249 Hz be observed in a 60-MHz spectrum, then, by using the above equation, the value calculated for δ is 4.15. In p.m.r. spectroscopy, all peaks that are observed at a field lower than the standard are assigned a positive sign, and peaks occurring at a higher field are assigned a negative sign. A chemical shift (δ) larger than the standard (*e.g.*, TMS) indicates that the proton being observed is more deshielded and, hence, resonates at a lower field. The effect of an increased deshielding of a proton can be equated to a shift to lower field (larger δ value) for the resonance peak.

Tau (τ) values may be converted into delta (δ) values by simply subtracting the τ value from ten ($\delta = 10 - \tau$), and the reverse situation is applicable for converting δ values into τ values ($\tau = 10 - \delta$). In this Chapter, the precalibrated spectral charts purchased from Varian Associates, used for the spectra reported, have been labeled with the τ scale at the top and the δ scale at the botton, and, by first-order treatment, may be read directly from the chart when a 500-Hz field-sweep is used [see Fig. 1(A)]. There are a number of items of information recorded on the right-hand side of the spectrum; however, those of major practical importance are the sweep-width, solvent, and sweep-offset data. On visual examination of Fig. 1(A), it is evident that the peak observed at δ4.5 on the bottom scale could also be recorded as τ5.5. When a possibility exists that a certain proton may exhibit a peak that would not be observed on the spectrum at a 500-Hz sweep-width, a spectrum should be obtained at a 1,000-Hz sweep-width. To record these peaks from a 1000-Hz sweep-width spectrum as δ, the observed value is merely multiplied by two; for example, the peak observed at δ6.5 [see Fig. 1(B)] was observed on a 1000-Hz sweep-width spectrum and is actually recorded as δ13.

As already mentioned, τ and δ values are recorded relative to a common standard, which is generally tetra-

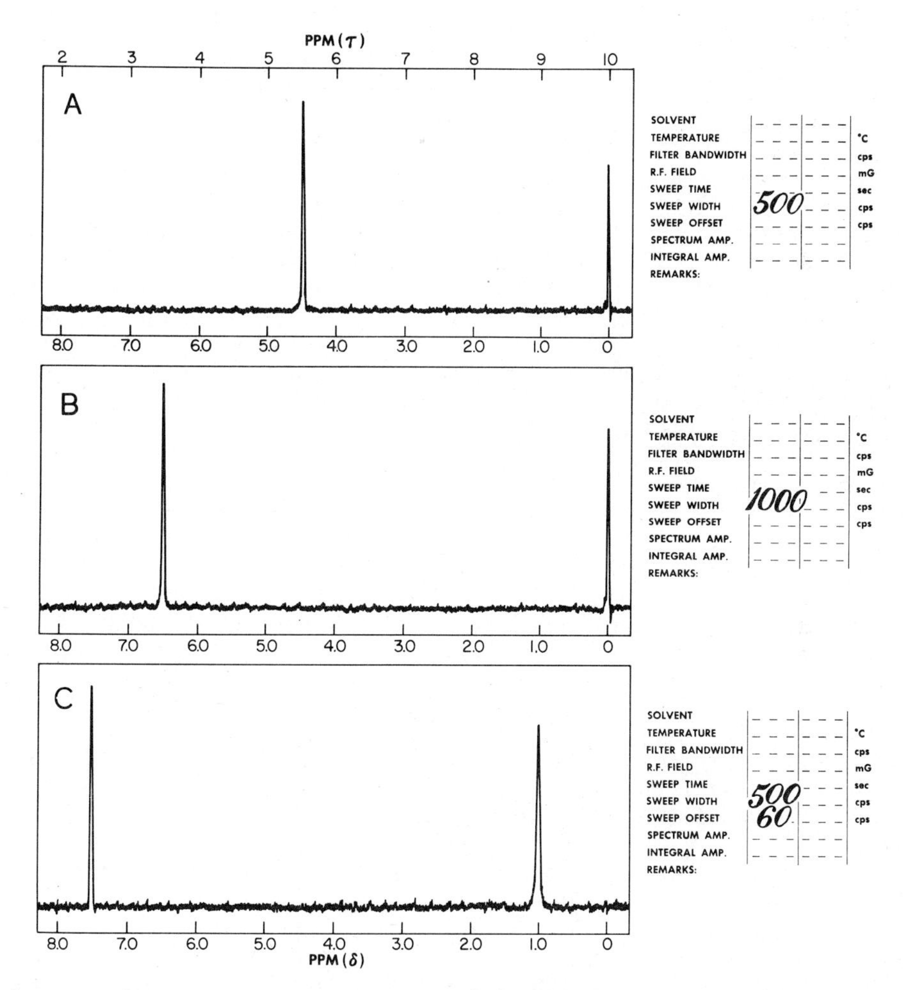

Fig. 1. A p.m.r. spectrum on (A) 500-Hz sweep-width; (B) 1000-Hz sweep-width; (C) 500-Hz sweep-width, with a 60-Hz offset.

methylsilane (Me_4Si, TMS), and thus peak is adjusted to zero Hz on the chart paper. If, on visual examination, it is seen that the TMS signal had not been set at zero, all of the peaks in the spectrum must be adjusted accordingly. For example, if the TMS signal occurs at 5 Hz, then 5 Hz must be subtracted from the value for every peak in the spectrum observed.

The sweep-offset technique may be used to definite advantage when a compound displays a peak at high field (for example, at $\delta 2$) and yet exhibits another peak at a lower field (for example, at $\delta 8.5$). Without use of the sweep-offset technique, this problem would preclude the use of a 500-Hz sweep-width spectrum. The p.m.r. spectrum in Fig. 1(C) has a peak at $\delta 1.0$ and another at $\delta 7.5$. However, on the right-hand side of the chart paper in the sweep-offset line, a notation had been made that there had been a 60-Hz offset; this indicates that the entire spectrum had been shifted 60 Hz to the right, and that the peaks shown at $\delta 1$ and $\delta 7.5$ actually occur at $\delta 2$ and $\delta 8.5$. Had the sweep-offset line shown an offset of 120 Hz, it would indicate a shift of 120 Hz to the right, on a 60-MHz instrument, and the peaks at $\delta 1$ and $\delta 7.5$ would actually have been recorded as $\delta 3$ and $\delta 9.5$. <u>All spectra that are depicted in this Chapter were obtained with a Varian A 60-MHz or Varian 56/60 MHz spectrophotometer, and the values given are δ values unless otherwise specified.</u>

Tetramethylsilane may be used either as an internal or an external standard. An internal standard is a reference compound, such as TMS, that is added to a liquid sample or to a solution of a solid sample in a solvent. An external standard is a reference compound, such as TMS, that is enclosed in a small, sealed, glass capillary tube which is placed in the p.m.r. tube containing the sample. The external standard is generally used when the standard is immiscible with the solvent chosen. A recent innovation is the introduction of commercial deuterated solvents that already contain 1% by volume of TMS. Of course, the same effect may be accomplished in the laboratory by adding the calculated quantity of TMS to a suitable volume of the solvent selected.

Another very good standard is sodium 3-(trimethylsilyl)-propane-1-sulfonate (DSS); it is water-soluble, and gives

$$Me_3Si(CH_2)_3SO_3^{\ominus}\ Na^{\oplus}$$

a strong signal for the methyl protons. However, at a very high gain, multiplets may be observed that are caused by the methylene protons.

The use of other organic compounds as standards has also been described, especially in the early reports on p.m.r. spectral studies. An example of such a compound is <u>p</u>-dioxane, a small proportion of which is added to the sample; the peak at $\delta 3.68$ (which may be assigned to <u>p</u>-dioxane) is then the standard and is, therefore, assigned the chemical shift of 0 Hz. Consequently, all peaks observed at a lower field would be assigned a positive δ value, and all peaks occurring at a higher field would be assigned a negative δ value. Other organic compounds, including acetone, acetonitrile, benzene, and cyclohexane, have also been used as standards.

The samples used for obtaining n.m.r. spectra are usually either liquid or solid. Gaseous samples, not encountered in the field of nucleic acid chemistry, present a number of difficulties, but these will not be discussed here.

A p.m.r. spectrum may be obtained for an undiluted (neat) liquid if a sufficient quantity of sample is available. Normally, use of 0.4 ml of neat liquid or solution is essential, in order that the region of the sample tube in the vicinity of the sample coil be kept filled; the liquid level must be well above the receiver coil, so as to prevent the vortex (created by spinning the sample) from descending into the vicinity of the coil. A p.m.r. spectrum may also be obtained for a small amount of a liquid sample by dissolving it in a suitable solvent, to give 0.4 ml of a 10% solution. With a solid sample, the p.m.r. spectrum is recorded for a solution of the solid in an appropriate solvent (usual concentration ~10%). Because only a small fraction of the total volume of liquid in a sample tube is actually observed, several types of microcell can be used for obtaining a p.m.r. spectrum on a small sample. One of these (Varian Associates) consists of two plastic (Teflon or nylon) plugs having hemispherical cavities, in opposition to one another, which contain the liquid; the cell may be placed inside a standard-size p.m.r. tube. Two other types of microcell are available (Kontes Glass Co., Wilmad Glass Co., and NMR Specialties) that are made entirely of glass. A more recent innovation[8] involves sealing the solution of the sample (4-8 mg in 40 μl of solvent) in a capillary tube (1.5-2.0 x ~90 mm), placing the sealed ampoule in a standard p.m.r. tube, and covering it with solvent. The optimal working conditions (acquired by adjusting the amplitude and filter band-width) have been discussed,[8a] and some p.m.r. spectra obtained by this technique have been presented.[8b] There are, however, several disadvantages in the technique, one of which is the confusion brought about by having spinning side-bands, and the other, the problem of having to adjust the spin-rate of the p.m.r. tube in order to maintain capillary symmetry. The latter problem is a result of the fact that the surrounding solvent is the only support for the ampoule. A technique used with considerable success in the author's laboratory[9] is as follows:

Two flat, cylindrical, Teflon plugs that fitted very

snugly inside a standard n.m.r. tube were obtained. A hole was drilled in the exact center of each plug, and two very small pinholes were then made on each side of the large hole to allow for temperature equilibration; the plugs were then placed as shown in Fig. 2. A com-

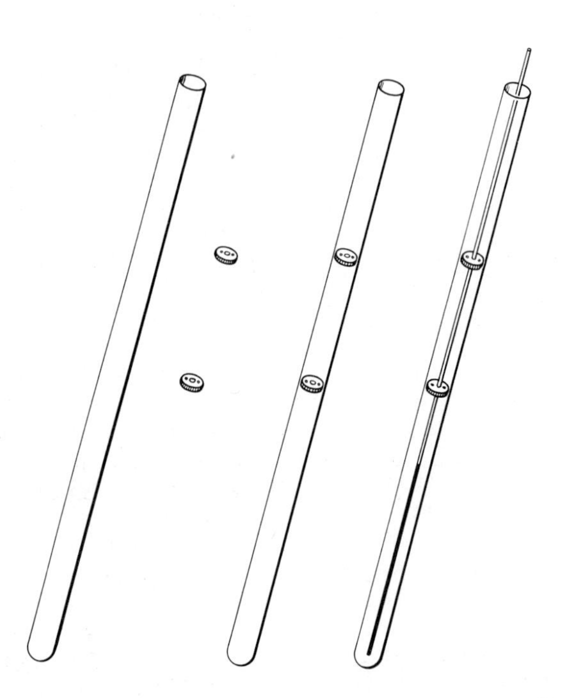

Fig. 2. Schematic drawing of apparatus used in the author's laboratory for obtaining p.m.r. spectra on a micro scale.

mercially available, capillary tube (~20 cm long), sealed at one end, was inserted through the holes in the Teflon plugs, the small amount of liquid needed to raise the level of the liquid to above the receiver coil was introduced with a syringe, and the tube was centrifuged. This procedure has the distinct advantage of diminishing the time required for the exact positioning of the cell in the probe and for adjusting the spin-rate. There is also an increase in efficiency and an economy of time, because the capillary tube may be removed from the n.m.r. tube and replaced with a new capillary tube within a few minutes.

An essential step in obtaining a p.m.r. spectrum of

either a liquid or a solid in solution is the selection of an appropriate solvent, that is, one in which the compound is soluble to the extent of 10%; this is often a difficult problem; furthermore, there should be no overlap between the resonance peaks observed for the solvent and for the compound studied. A suitable solvent may be selected from the list included in Table I, provided that the approximate chemical-shift of the peaks given by the compound can be predicted. However, should there be no basis for predicting where these peaks will appear in the spectrum, a deuterated solvent should be selected, because there will be little interference from the deuterium atoms. The difference between the spectra obtained for solutions in pyridine and perdeuterated pyridine (pyridine-d_5, C_5D_5N) may be seen by inspection of Figs. 3(A) and 3(B), which illustrate the marked difference (and, therefore, advantage) in the use of the deuterated as compared with the nondeuterated solvent. The spectra of other deuterated solvents are shown in Fig. 3(C) (deuterated acetic acid), Fig. 4(A) (deuterated benzene), Fig. 4(B) (deuterated chloroform), Fig. 4(C) (deuterated methanol), Fig. 5(A) (deuterated methyl sulfoxide), Fig. 5(B) (deuterated acetonitrile), and Fig. 5(C) (deuterated acetone). These spectra should prove of considerable utility in the interpretation of p.m.r. spectra, because each not only provides the chemical shift for the peak or peaks of the solvent studied, but also gives a general idea of the shape, general appearance, and splitting patterns. The peaks observed for a deuterated solvent usually occur at a field slightly lower (δ~0.02–0.05) than that for the peaks observed for the nondeuterated compound. However, this change is so small that, for practical purposes, it may be disregarded. All interference from solvent peaks may be eliminated by also recording a spectrum for the same sample in a different solvent, and then using both spectra, which complement each other, in the interpretation. An excellent solvent for general use is methyl sulfoxide-d_6, in which the majority of nucleic acid components dissolve; and yet it does not interfere in the interpretation, because it has only a small absorption band, at δ2.5 (due to the presence of a small proportion of methyl sulfoxide-d_5). An effective technique is to record a p.m.r. spectrum of a compound in methyl sulfoxide-d_6, and then to add one or two drops of D_2O to the sample in the p.m.r. tube; the solution is either kept for a few minutes or gently heated, and a new spectrum is recorded. The second spectrum will generally lack all those peaks attributable to readily replaceable protons (from amino, sulfhydryl, and hydroxyl groups); however, an additional peak, for HDO

TABLE I. Characteristic Resonance Bands of Solvents for Use in N.m.r. Spectroscopy

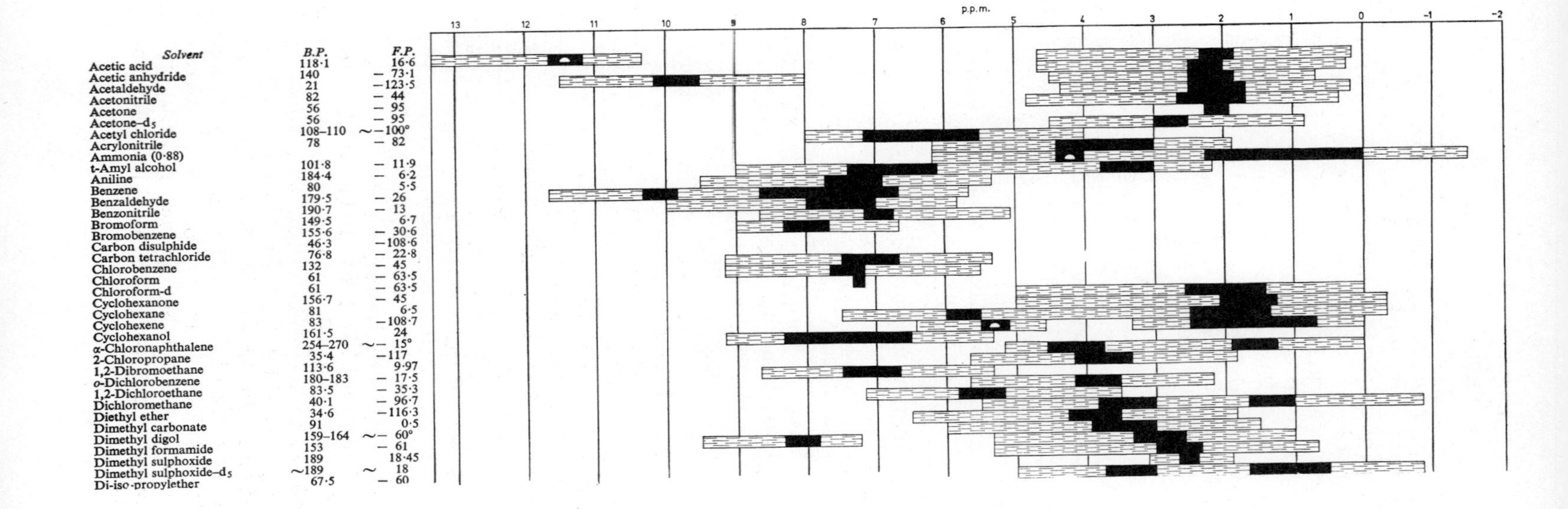

Solvent	B.P.	F.P.
Acetic acid	118·1	16·6
Acetic anhydride	140	− 73·1
Acetaldehyde	21	−123·5
Acetonitrile	82	− 44
Acetone	56	− 95
Acetone-d_5	56	− 95
Acetyl chloride	108–110	~−100°
Acrylonitrile	78	− 82
Ammonia (0·88)		
t-Amyl alcohol	101·8	− 11·9
Aniline	184·4	− 6·2
Benzene	80	5·5
Benzaldehyde	179·5	− 26
Benzonitrile	190·7	− 13
Bromoform	149·5	6·7
Bromobenzene	155·6	− 30·6
Carbon disulphide	46·3	−108·6
Carbon tetrachloride	76·8	− 22·8
Chlorobenzene	132	− 45
Chloroform	61	− 63·5
Chloroform-d	61	− 63·5
Cyclohexanone	156·7	− 45
Cyclohexane	81	6·5
Cyclohexene	83	−108·7
Cyclohexanol	161·5	24
α-Chloronaphthalene	254–270	~− 15°
2-Chloropropane	35·4	−117
1,2-Dibromoethane	113·6	9·97
o-Dichlorobenzene	180–183	− 17·5
1,2-Dichloroethane	83·5	− 35·3
Dichloromethane	40·1	− 96·7
Diethyl ether	34·6	−116·3
Dimethyl carbonate	91	0·5
Dimethyl digol	159–164	~− 60°
Dimethyl formamide	153	− 61
Dimethyl sulphoxide	189	18·45
Dimethyl sulphoxide-d_5	~189	~ 18
Di-iso-propylether	67·5	− 60

The original Table was furnished by, and is reproduced with the permission of, Dr. D. N. Henty and the General Secretary of the Society of Chemical Industry, 14 Belgrave Square, London, England. Published originally by D. N. Henty and S. Vary, *Chem. Ind.* (London), 1782 (1967).

TABLE I (Cont). Characteristic Resonance Bands of Solvents for Use in N.m.r. Spectroscopy

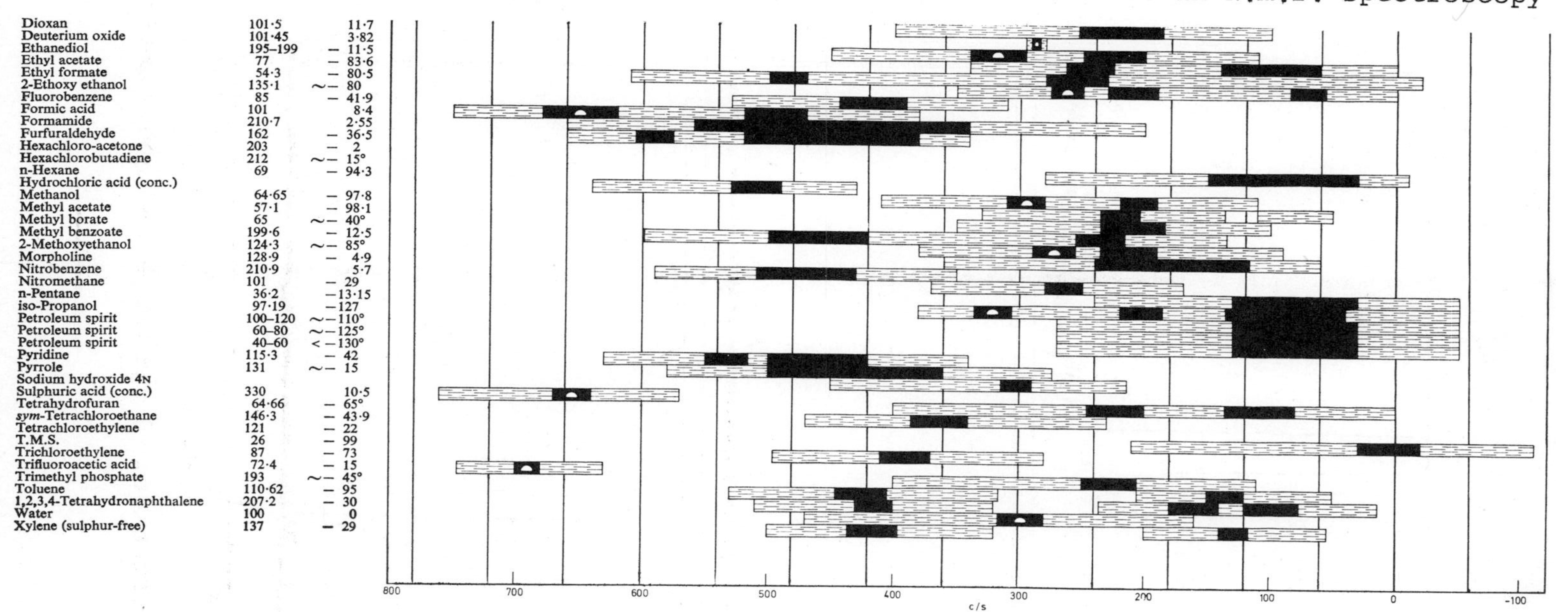

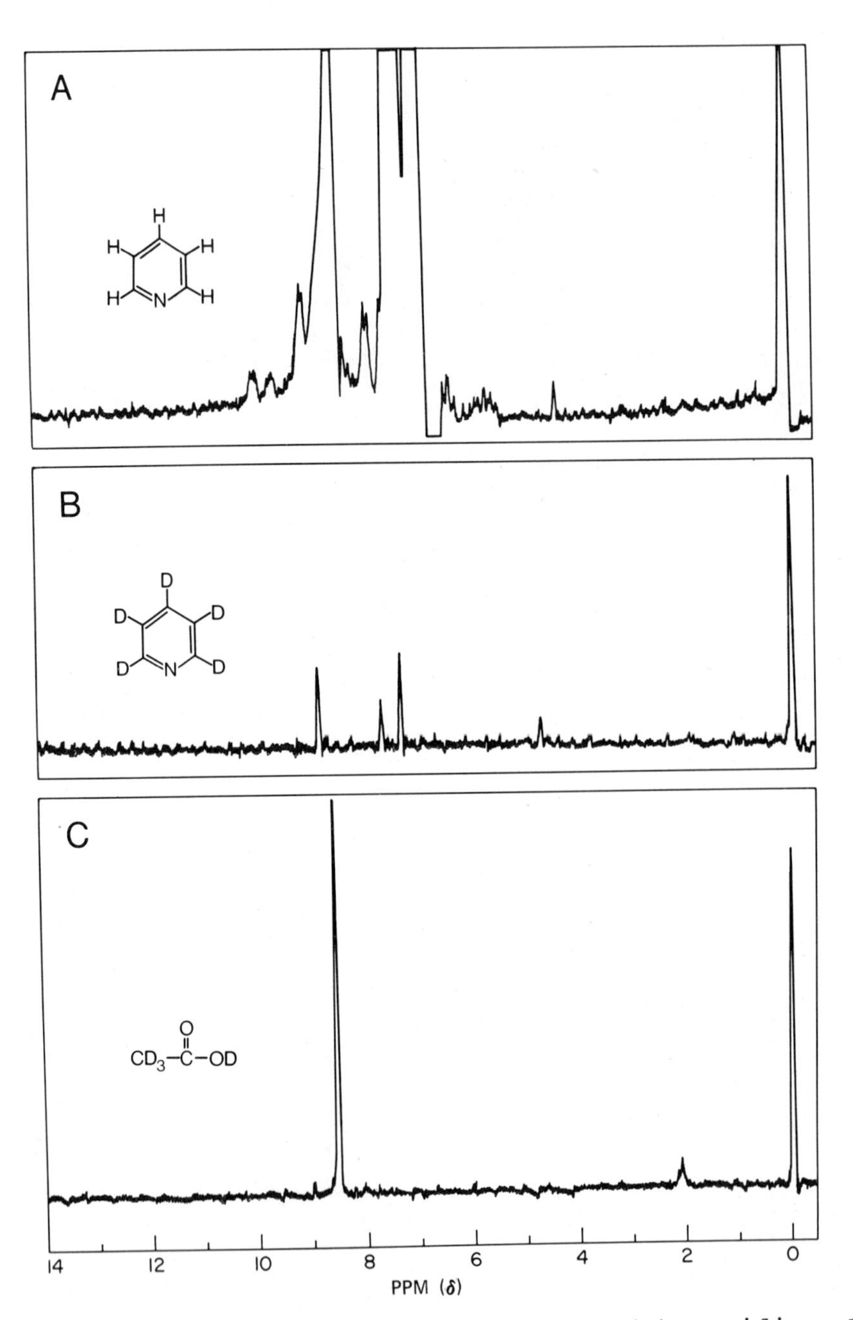

Fig. 3. P.m.r. spectra of (A) pyridine; (B) pyridine-d_5; (C) acetic acid-d_4.

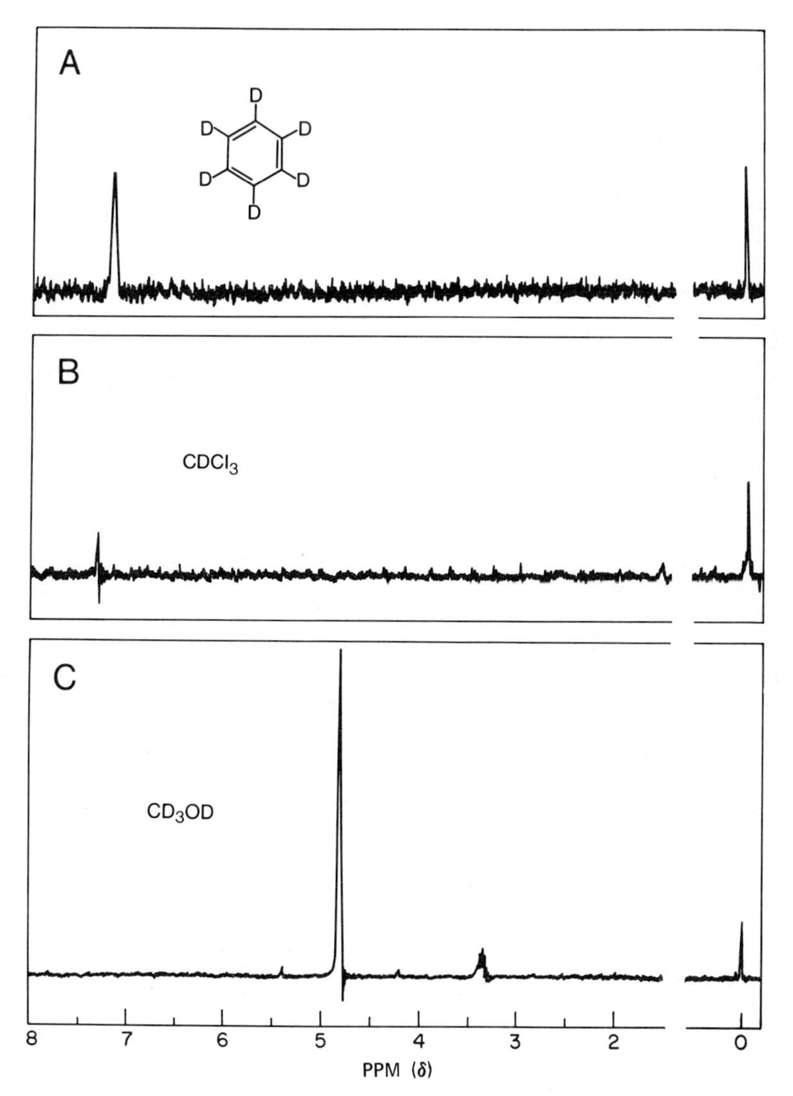

Fig. 4. P.m.r. spectra of (A) benzene-d_6; (B) chloroform-d; (C) methanol-d_4.

(ranging from δ3.0 to δ4.5), is then observed in the spectrum, depending on the concentration of HDO present [see Fig. 6(A, B, and C)]. The chemical shift of this peak is also dependent on pH and temperature, and will be discussed subsequently. Another excellent solvent is

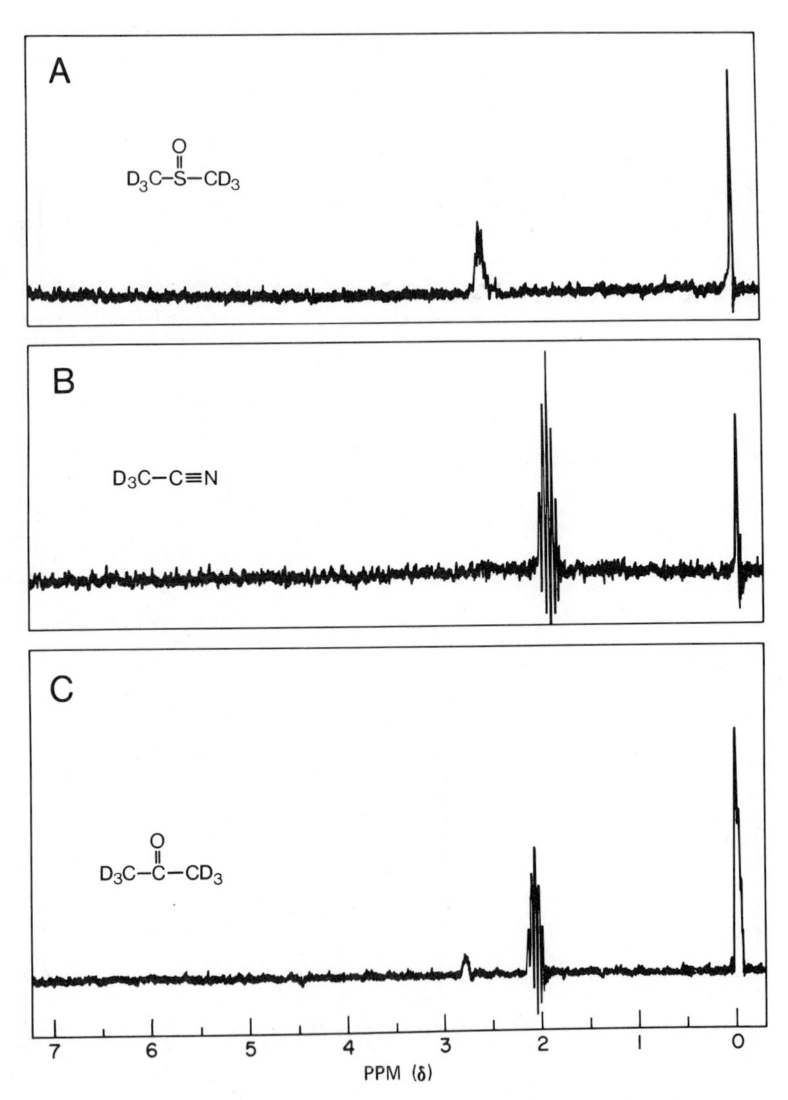

Fig. 5. P.m.r. spectra of (A) methyl sulfoxide-d_6; (B) acetonitrile-d_3; (C) acetone-d_6.

D_2O, its major disadvantages being that (1) the replaceab protons cannot be accounted for, and (2) the peak for HDO usually overlaps the peaks observed for the carbohydrate moiety of nucleosides and nucleotides. A character istic that can be utilized is the mobility of the HDO peak as a function of temperature [see Fig. 7 (A, B, and

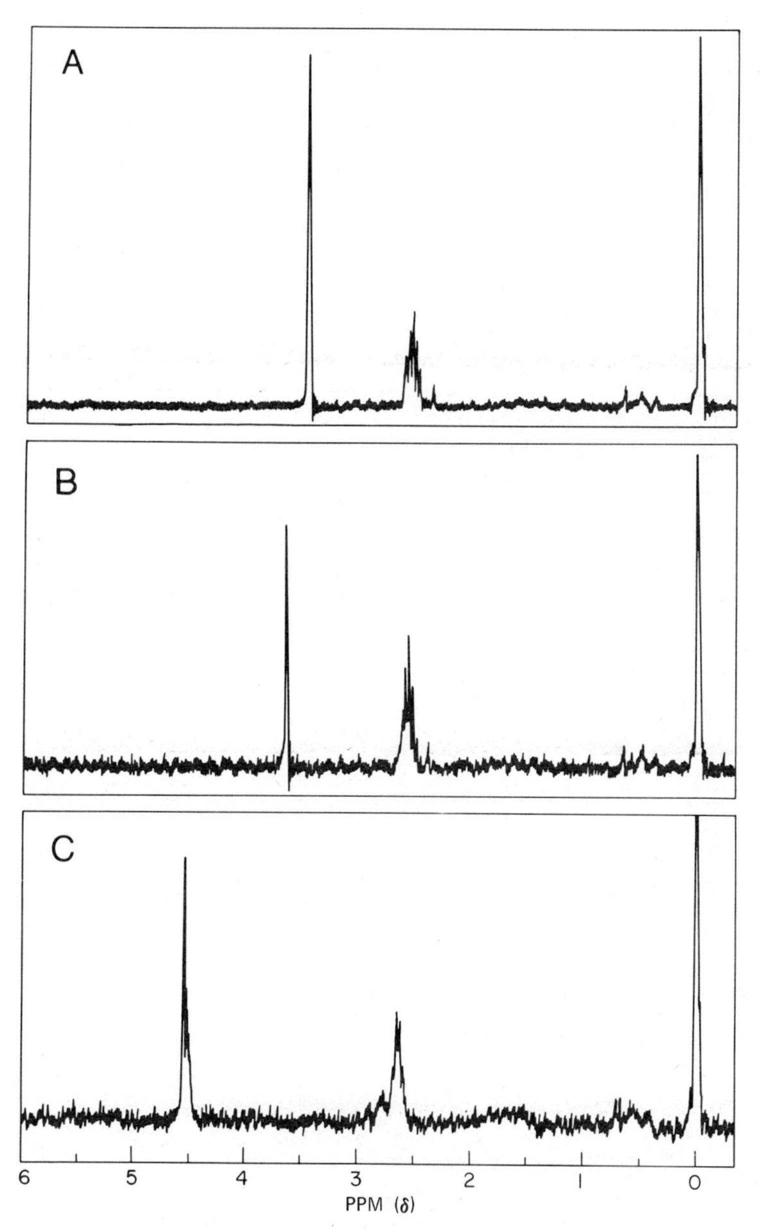

Fig. 6. P.m.r. spectrum of methyl sulfoxide-d_6 with (A) 1 drop of D_2O and (B) 2 drops of D_2O; (C) 1:1 (v/v) methyl sulfoxide-d_6-D_2O.

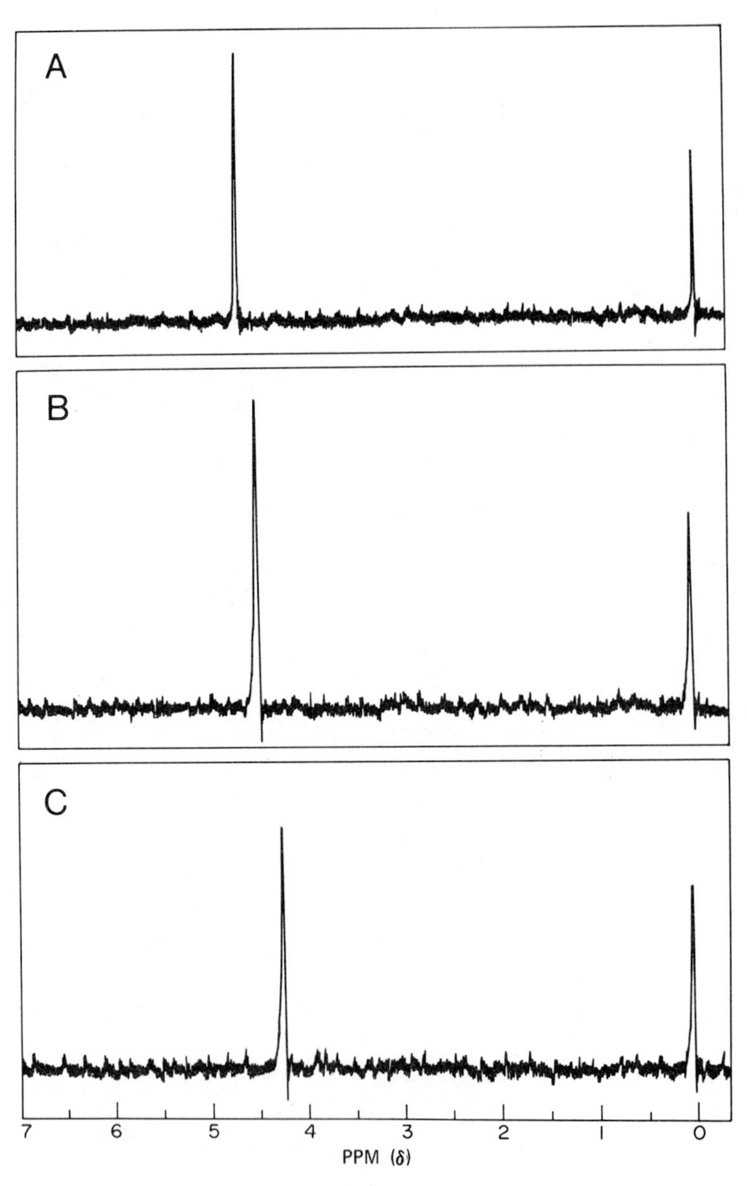

Fig. 7. P.m.r. spectrum of deuterium oxide at (A) 25-30°; (B) 50°; (C) 70°.

C)], measurable if a temperature probe is available. An additional chemical shift can also be effected by increasing the temperature to 90-95°. The temperature of the sample can be adjusted between -100 and +200° in most n.m.r. spectrophotometers, merely by passing either preheated or precooled nitrogen past the spinning sample-tube in the probe. Another technique employed, for compounds that are soluble and stable in aqueous acid, is a shifting of the HDO peak by the addition of gaseous hydrogen chloride or, preferably, deuterium chloride. In addition to methyl sulfoxide-d_6 and D_2O, acetic acid-d_4 is an excellent solvent for nucleosides, providing a range of absorbance from δ2.0 to δ9.0 that is free from any solvent peaks. A spectrum of a base-soluble compound can be obtained for the anionic species by using D_2O containing a small proportion of sodium peroxide.

With regard to the information that may be obtained from an integration curve, the first step to be taken is the assignment of a peak to a specific proton, protons, or group. This assignment serves as a reference, and permits assigning of the relative number of protons to each of the different peaks in the spectrum, because the area under each peak is proportional to the number of protons that give rise to the peak. For example, as shown in Fig. 8,

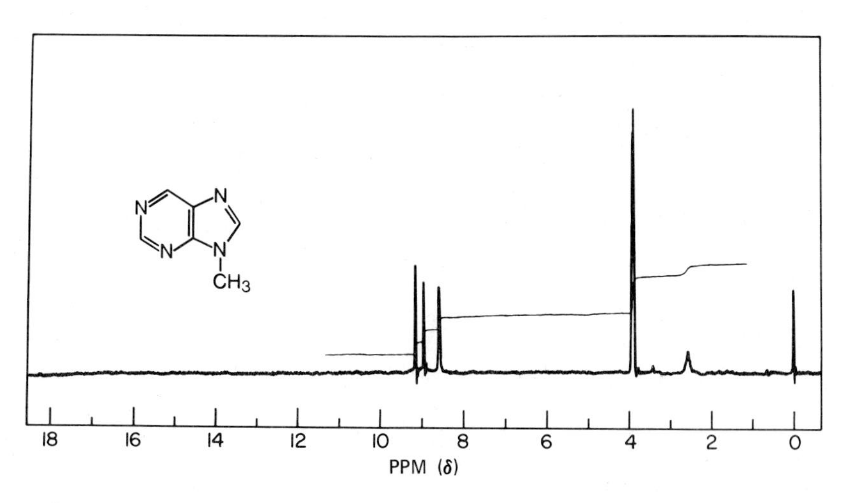

Fig. 8. P.m.r. spectrum of 9-methylpurine, in methyl sulfoxide-d_6.

the peak at $\delta 3.9$ may be assigned to the methyl group at the 9-position; therefore, the number of protons attributable to this peak is three. The peaks at $\delta 8.6$, $\delta 9.0$, and $\delta 9.2$ (Fig. 8) are seen to have only one third the height of the integration curve for the methyl peak at $\delta 3.9$; consequently, each gives rise to only one third of the area under the curve, and represents one third of the number of protons of the methyl group on N-9. Each of these peaks is, obviously, caused by a single proton, and may be assigned to positions 2, 6, and 8, although not necessarily in that order. The height of each elevation in the integration curve may be measured in any units desired (for example, mm, cm, or squares on the paper), so long as the same units are used throughout for a given curve. A point that requires clarification is that the height of the integration curve may vary, but the *relative* heights for the observed peaks always remain constant.

Another way in which the integration curve may be used to advantage depends upon a predetermination of the number of protons assigned to the sample. In this method, the height of the entire curve is measured, and the increase in height attributable to the solvent, and to known or suspected impurities, is subtracted. The numerical value for the remaining height of the curve is divided by the total number of protons involved, into arbitrary units as discussed in the foregoing, and provides a constant for a determination of the number of protons that can be assigned to each peak. If the height of the entire integration curve in Fig. 8 was 70 mm, subtraction of 10 mm for the solvent (methyl sulfoxide-d_5 at $\delta 2.5$) leaves 60 mm. Because the empirical formula for the compound studied contains six protons, 60 mm is divided by six, giving 10 mm; this is the height of the integration curve that is equivalent to one proton. Hence, the bands at $\delta 8.6$, $\delta 9.0$ and $\delta 9.2$ could be assigned to one proton each, and the peak at $\delta 3.9$ (30 mm) must be the result of three equivalent protons, the peak at $\delta 2.5$ (10 mm) being assigned to methyl sulfoxide-d_5, and the peak at $\delta 0$ to the standard (DSS). With nucleosides, a proton attached to a carbon atom of the heterocyclic moiety, the anomeric proton, or the absorption band for the acetyl groups of acetylated nucleosides, is generally used as the reference peak for the evaluation of the entire integration curve by one of the foregoing methods. Because there is such a large increase in height of the integration curve, namely, $\delta 2.0$–2.5 for acetylated nucleosides, $\delta 7.0$–7.5 for benzylated or benzoylated nucleosides, and $\delta 4.5$–5.5 when D_2O is used as the solvent, it may become necessary to reset the integrator to the base line. This adjustment does not, in any way, invalidate the foregoing methods for the interpretation of the integration curve, because the total height of the curve may be obtained merely by adding the heights of the two separate curves. Interpretation of the integration curve is only as reliable as the initial assignment of a certain absorption peak to a specific proton, protons, or functional group, or as the determination of the total number of protons attributable to the sample.

A complication is introduced when an integration curve indicates that a certain peak is caused by 0.5 proton, with the requirement that the area under two such peaks (a doublet) should account for one proton. The splitting of a peak attributed to one proton may afford a doublet, a triplet, or a quartet, but, for illustrative purposes, only the case of a doublet will be discussed. The splitting of a peak into a doublet is attributed to a phenomenon known as spin—spin coupling, which is a result of the interaction of the spinning proton that is being measured with an adjacent, spinning nucleus. Spin causes the nucleus to behave as a small bar-magnet having two possible orientations, resulting in the appearance of two absorption peaks, the midpoint between which is the position where the absorption peak expected would appear were there no coupling. The distance between these two peaks is referred to as the coupling constant, and is represented by the letter J. Coupling constants are given in Hertz (Hz), and may have either a negative or a positive sign. The coupling constant is independent of the frequency and is, therefore, the same regardless of whether the spectrum is recorded at 40, 60, 100, or 220 MHz.

Spin—spin coupling occurs between two protons on the same atom or on contiguous atoms. In general, spin—spin coupling is not observed between protons on nonadjacent atoms, unless this arrangement of atoms contains unsaturation, as, for example, in an allylic or aromatic system. Instances have been encountered in which it has been firmly established that some coupling constants (or splitting patterns) are the result of long-range, spin—spin interactions or nonbonded interactions, in addition to vicinal interaction, but this situation will not be discussed in this Chapter.

The coupling constant is an extremely valuable tool in conformational studies, and is particularly applicable to the carbohydrate moiety of nucleosides and nucleotides, because the coupling constant may be directly related to the dihedral angle between two protons situated on adjacent atoms. An illustration is the dihedral angle between the protons, H_A and H_B, in the Böeseken ("Newman") projection shown in Fig. 9. Here, the dihedral angle between H_A and H_B is 60°. Were this an isolated system lacking any other

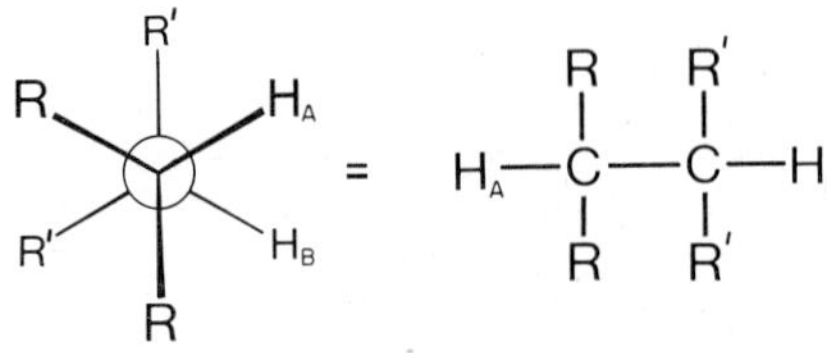

Fig. 9. Böeseken ("Newman") projection illustrating the dihedral angle (between vicinal protons) used in the Karplus equation.

coupling, and were R and R' the same, a doublet would be observed, because each proton would have the same chemical shift. Therefore, the coupling constant would be equal to the distance (in Hz) between the two peaks of the doublet, and would be expressed as $J_{A,B}$ = X Hz. However, were R and R' different groups, the doublet for each proton would possess a different chemical shift, and two separate doublets would be observed (AB system), having the *same* coupling constant. If the coupling constant is arrived at experimentally, the dihedral angle between the protons may be ascertained by use of the modified Karplus equation[10], as follows:

$$J_{H,H'} = A\cos^2\emptyset - 0.28, \text{ where } A = 8.5 \text{ for } 0° \leq \emptyset \leq 90°, \text{ or}$$

$$J_{H,H'} = A\cos^2\emptyset - 0.28, \text{ where } A = 9.5 \text{ for } 90° \leq \emptyset \leq 180°,$$

where Ø is the angle between the protons H and H'. From this equation, a curve (Fig. 10) has been constructed, by use of which the dihedral angle may be directly related to the observed coupling-constant of protons on adjacent carbon atoms. However, it must be emphasized that the coupling constant may also be affected by a number of other factors,[11] such as variation in temperature, or the presence of electronegative substituents. Additionally, there are a number of factors that must be considered when the Karplus equation is used; some of these lessen the value of the results obtained from the equation. Coupling constants, together with use of the spin–spin decoupling technique, have been employed in many investigations of the conformation of carbohydrate derivatives. It is re-emphasized that the Karplus equation is generally used only as a *guide* in conformational analysis, and that it should be used with caution when an attempt is made to obtain a *precise* relationship between dihedral angles and coupling constants.

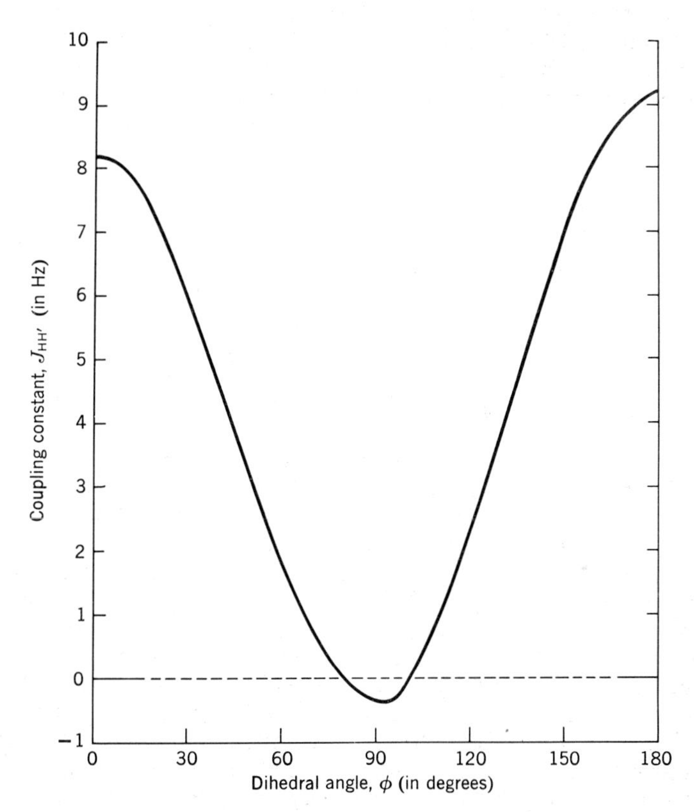

Fig. 10. Karplus curve, relating coupling constant (J) and dihedral angle, as calculated by the modified equation given in Ref. 10.

III. THE APPLICATION OF PROTON MAGNETIC RESONANCE SPECTROSCOPY IN THE FIELD OF NUCLEIC ACID COMPONENTS AND CERTAIN RELATED DERIVATIVES

This Section consists of a discussion of applications of p.m.r. spectroscopy, and of the interpretation of selected p.m.r. spectra of certain imidazoles, pyrimidines, purines, and corresponding nucleosides. Also included is a discussion on carbohydrates, as well as of certain relatives of the afore-mentioned heterocyclic compounds and their nucleosides, with a brief subsection on nitrogen-heterocycle antibiotics that should be of interest to investigators in the field of nucleic acid chemistry.

1. Imidazoles and Imidazole Nucleosides

The p.m.r. spectrum of imidazole (1) in methyl sulfoxide-d_6 [See Fig. 11(A)] has a low, broad peak at δ10.7 due to the proton on the ring-nitrogen atom. The peak at δ7.77, representing one proton, may be assigned to the proton at the 2-position, the remaining peak (singlet, 2 protons), at δ7.12, being assigned to the protons at the 4- and 5-positions. The number of protons attributable to each peak in the p.m.r. spectrum may be determined in the following way. Inspection of the integration curve for imidazole [Fig. 11(A)] reveals a total height of 7.5 cm (the units are arbitrarily chosen), and subtraction of 0.7 cm for the peak at δ2.5 (given by methyl sulfoxide-d_6) gives 6.8 cm, which, when divided by four (the number of hydrogen atoms in imidazole) affords a value of $\sim$1.7 cm height for each proton. On this basis, one proton can be assigned to each of the peaks at δ10.7 (NH) and δ7.77 (C-2-H), with 2 protons (C-4-H, C-5-H) assigned to the singlet at δ7.12. Another procedure involves assigning the singlet at δ7.77 to the proton on C-2 and then, on a comparative basis, the low, broad peak at δ10.7 is assigned to the proton of NH, and the singlet at δ7.12 to the protons on C-4 and C-5, with the exclusion of the solvent peak (δ2.5). The appearance of a low, broad peak for the proton on the ring-nitrogen atom is a result of bonding to an atom (nitrogen) whose nucleus has a nuclear spin (I) of unity. Nitrogen has a nonspherical charge-distribution and, therefore, an electric quadrupole moment (Q), which leads to a rapidly fluctuating electric field about the nucleus, and effects a rapid relaxation of the electric field about the nucleus. This situation results in a very broad peak, in the p.m.r. spectrum, for a proton bonded to a nitrogen atom. Peaks for NH protons are also broadened slightly if a small proportion of water is present in the solvent.

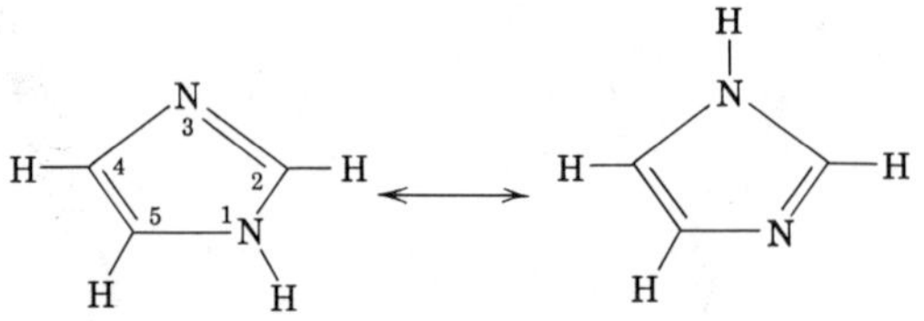

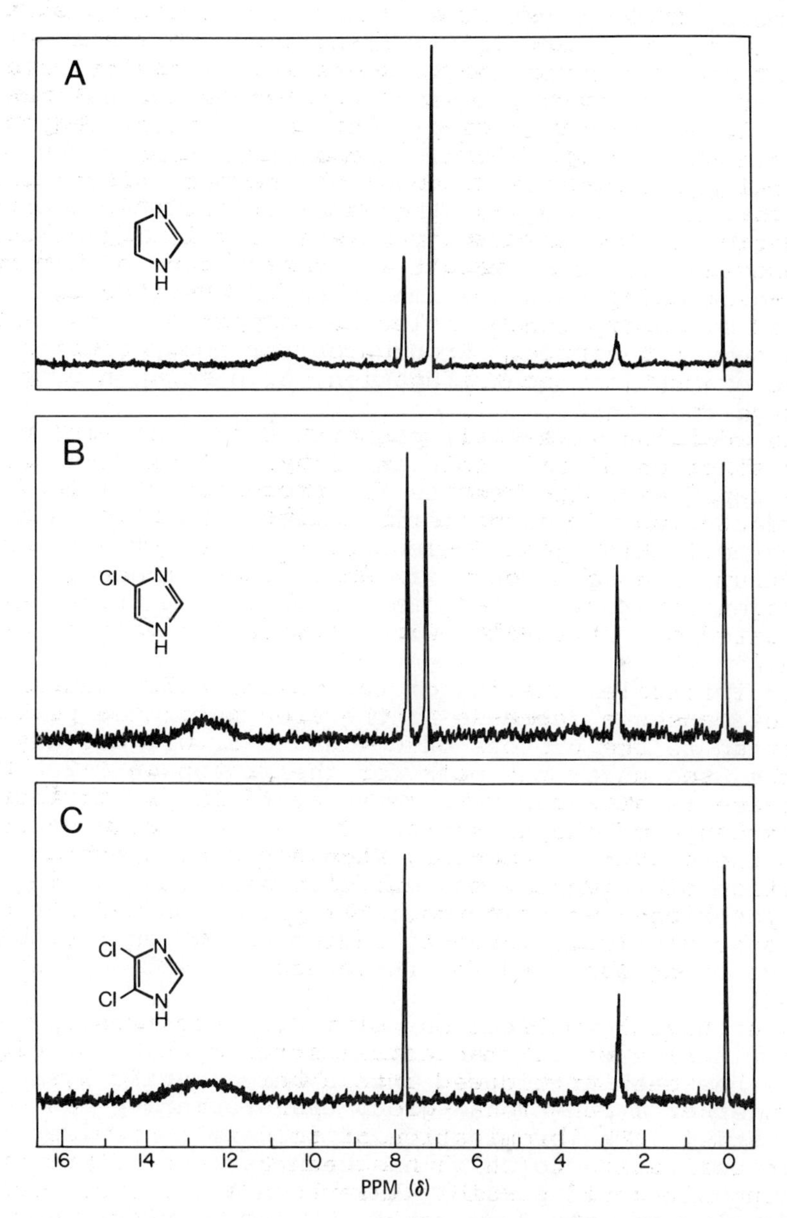

Fig. 11. P.m.r. spectra of (A) imidazole; (B) 4-chloroimidazole, and (C) 4,5-dichloroimidazole, in methyl sulfoxide-d_6.

Line-broadened spectra in which most or all of the peaks (including those caused by protons on carbon) are affected may be produced by effects other than nuclear quadrupole relaxation of nuclei with I=1; the most common cause is the presence of paramagnetic materials (usually, such ions as ferric and cupric, having unpaired electrons) that cause line broadening because of the decrease they cause in spin–spin or spin-lattice relaxation-time. There are several procedures for removal of the particles; micropore filtration also removes other small, insoluble (nonparamagnetic) particles which may cause local, magnetic-field inhomogeneities that result in a broadening of the peaks. Peak-broadening may sometimes also be observed that is caused by absorption of molecular oxygen from the atmosphere by the sample. Procedures for deoxygenation include evacuation, centrifugation, or passing nitrogen through the liquid.

The addition of a small proportion of deuterium oxide to a solution of imidazole in methyl sulfoxide-d_6 results in a rapid exchange between the proton of NH (δ10.7) and the deuterium of the deuterium oxide, and affords a p.m.r. spectrum lacking peaks below δ8. However, the p.m.r. spectrum of a solution of imidazole, or, indeed, of any nitrogen-heterocycle compound in deuterium oxide must be evaluated or interpreted with a certain degree of caution.[12]

For imidazole, heating of the solution (in deuterium oxide) causes a decrease in the area under the peak attributed to the protons on C-4 and C-5, without any change in the area under the peak for the proton on C-2. This decrease in area under the peak at δ7.12 is attributable to exchange of the protons at C-4 and C-5 with deuterium from the solvent. However, when a p.m.r. spectrum of a solution of imidazole in deuterium oxide containing sodium hydroxide-d was recorded, the proton on C-2 was observed to be slowly replaced by deuterium, no appreciable change being observed for the other singlet (C-4—H and C-5—H).

Other useful applications of p.m.r. spectroscopy are determination of (a) the actual site at which an electrophile has been introduced into a heterocyclic system, and (b) whether a functional-group transformation has been successful. The bromination of imidazole provides a tribromo derivative to which has been assigned structure 2, although it could possibly have been a N-bromodisubstitut imidazole, such as 3, 4, or 5. A p.m.r. spectrum of the imidazole in methyl sulfoxide-d_6 would have firmly established the correct structure as 2, because of the complete absence of any absorption peaks between δ7.0 and 9.0 for "aromatic" protons (on C-2, C-4, and C-5), and

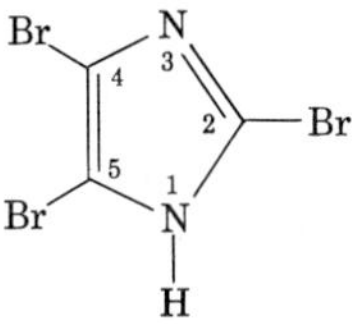

2 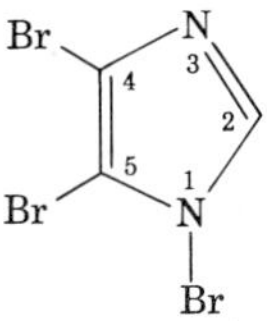 3

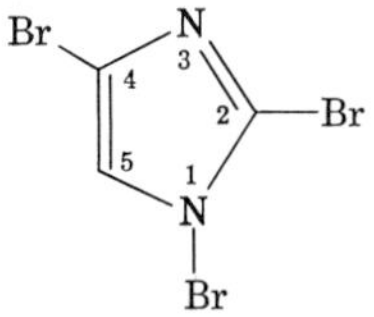

4 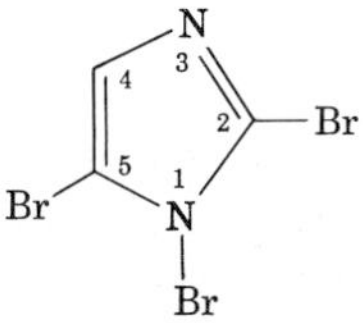 5

the presence of a low, broad absorption peak at ~δ12.5 for the NH proton. Had dihalogenation occurred (instead of trisubstitution), one peak, assignable to a proton on either C-2 or C-4, would be observed, in addition to the NH proton. This problem could have been resolved by merely recording a p.m.r. spectrum, because 2,4-dibromoimidazole exhibits a sharp singlet at δ7.38 for the proton at C-5, whereas 4,5-dibromoimidazole has an absorption peak at δ7.81 caused by the proton at C-2. The same pattern may be observed for 4,5-dichloroimidazole (6) [see Fig. 11 (B)], where only a low, broad peak for the NH proton and a sharp singlet at δ7.73 for the proton

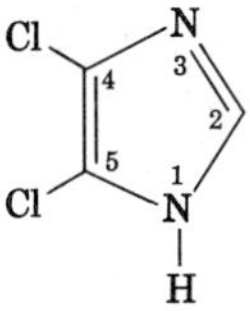

6

on C-2 are observed. These chemical shifts are in very close correlation with those for the protons of imidazole, since it is observed that the proton on C-2 (between the two nitrogen atoms) is more deshielded than the protons on C-4 and C-5. Consequently, the peak for the proton on C-2 will occur at a lower field than the peaks observed for the protons on C-4 and C-5. A good example of this situation is the difference in the p.m.r. spectra of 2-chloroimidazole (7) and 4-chloroimidazole (8). The p.m.r. spectrum of 2-chloroimida-

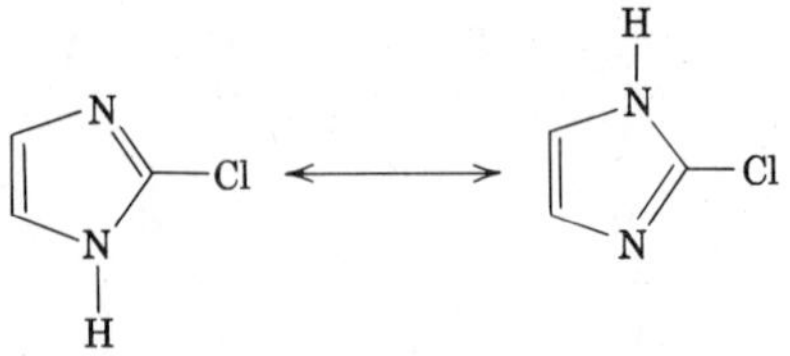

7 7

zole[13] has only one peak (for 2 protons, at C-4 and C-5), at δ7.07, which may be attributed to the symmetry of the compound. However, when the chlorine atom is on C-4, there is a definite change in the appearance of the p.m.r. spectrum. An examination of the p.m.r. spectrum of 4-chloroimidazole (8) [see Fig. 11(C)] reveals two peaks, at δ7.72 and 7.25, for the protons on C-2 and C-4, respectively. The peak at the lower field (δ7.72) may be as-

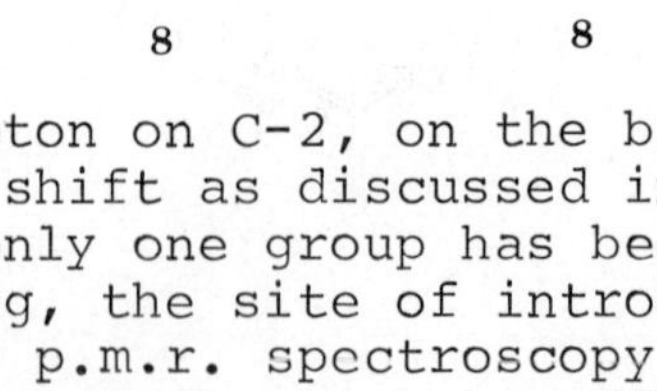

8 8

signed to the proton on C-2, on the basis of the difference in chemical shift as discussed in the foregoing. Therefore, when only one group has been introduced onto the imidazole ring, the site of introduction may readily be ascertained by p.m.r. spectroscopy.

P.m.r. spectroscopy may also be effectively utilized in the structural assignment of imidazoles substituted at C-2 and C-4, or at C-4 and C-5. This technique may be illustrated by a discussion of a re-investigation of the site of monoiodination of histidine (9). It had been reported[14] that monoiodination affords 2-iodohistidine (10); however, it was subsequently established[15] that the iodine atom entered the 4-position, by p.m.r. studies of the chemical shift observed for the "aromatic" proton. Data for the chemical shift of the "aromatic" protons of histidine are δ7.66 (C-2—H) and δ7.0 [C-4(5)—H]. The peak for the "aromatic" proton of the monoiodinated imidazole lies at δ7.7, and is evidently caused by the proton on C-2. Hence, the actual structure is that of 4-iodohistidine (11). The same investigation also established, on the basis of chemical shifts, that the 2-iodoimidazole (12) previously reported[16] is, in fact, 4-iodoimidazole (13), with peaks of equal intensity at δ7.1 and 7.7, instead of one peak at ~δ7.0 for two, equivalent protons. These assignments may be made with a reasonable degree of certainty, because it has been found that, in general, there is a more pronounced chemical shift to lower field for the proton on C-2 than is observed for the protons on C-4 or C-5. Another interesting example of the utilization of p.m.r. spectroscopy in the area of biologically important imidazoles is a re-investigation[17] of the structure of "2-*iso*-histamine" [2-(2-aminoethyl)imidazole, 19]. This derivative was supposedly prepared[18] from 1-

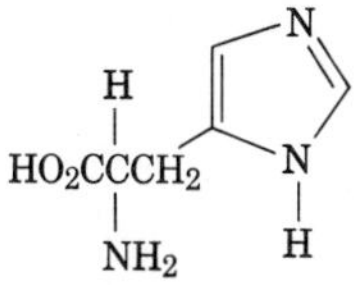

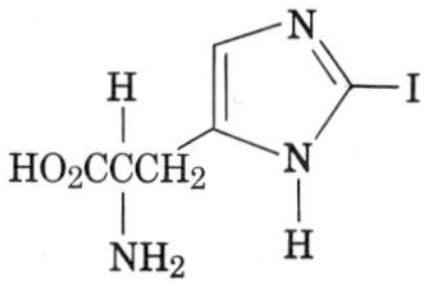

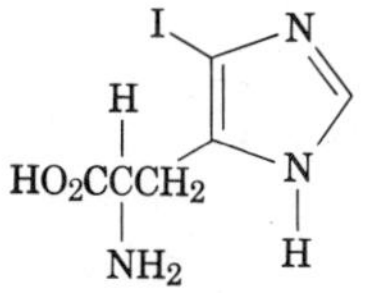

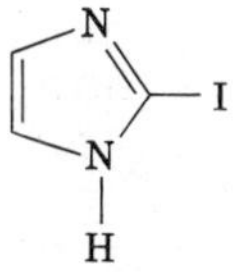

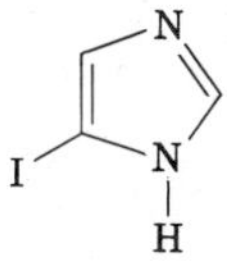

benzyl-2-(chloromethyl)imidazole (14) by the sequence shown in Scheme 1. However, the p.m.r. spectrum of the

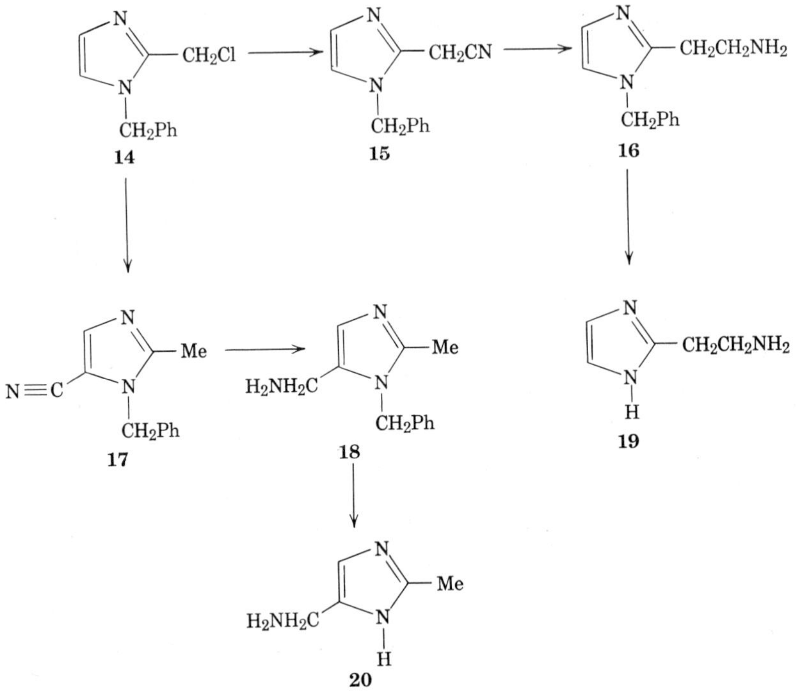

Scheme 1

product had one peak in the δ7.8 region, as expected, but an integration revealed that it had a relative intensity for only one proton (instead of for two protons, at C-4 and C-5). The spectrum also had a singlet (3 protons) at δ2.69, indicative of the presence of a methyl group. It was found that treatment of 14 with cyanide ion in ethanol results in an interesting rearrangement that affords 4-(aminomethyl)-2-methylimidazole (20) by the route 14→ 17→ 18→ 20 in Scheme 1. The unequivocal synthesis of 19 was then accomplished[17] by a modification of the original reaction conditions. The compound had a p.m.r. spectrum having an absorption peak (singlet, 2 protons) at δ7.42, as originally expected for "*iso*-histamine" (19).

A p.m.r. spectrum [see Fig. 12(A)] of the antibiotic azomycin (2-nitroimidazole, 21) is typical of that of a 2-substituted imidazole, the protons on C-4 and C-5 appearing as a single peak in the δ7–8 region (singlet, 2 protons), with a low, broad peak at lower field (~δ12-13)

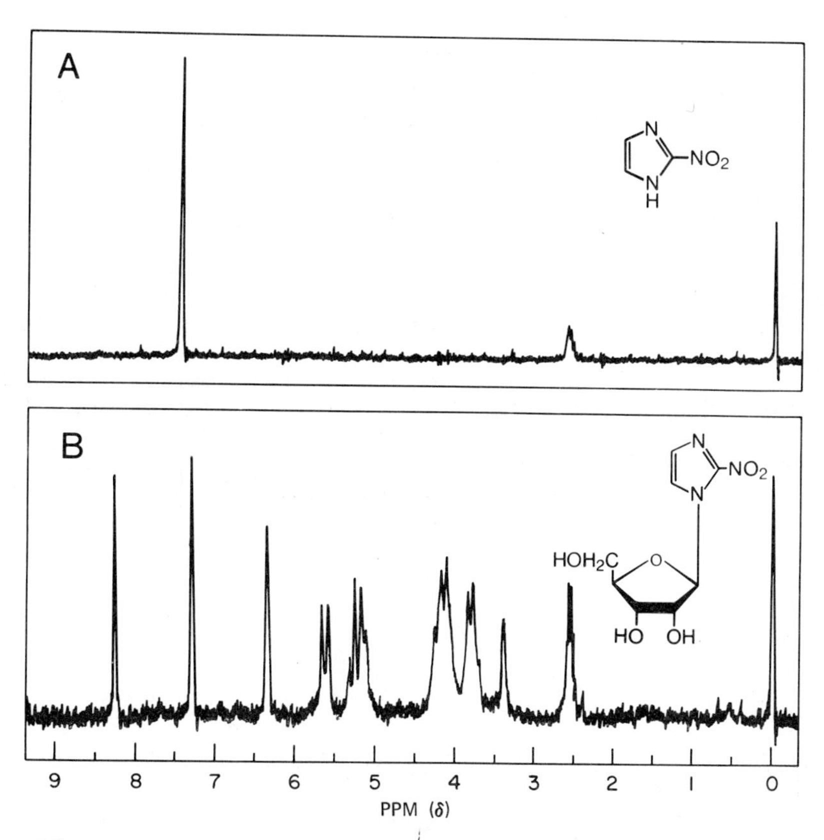

Fig. 12. P.m.r. spectra of (A) 2-nitroimidazole (azomycin), (B) 2-nitro-1-β-D-ribofuranosylimidazole, in methyl sulfoxide-d_6.

for the NH proton not shown in Fig. 12(A). However, a substituent on a ring-nitrogen atom exerts a definite change in the p.m.r. spectrum. The major change generally observed on *N*-substitution is the emergence of two absorption peaks in the δ7–8 region, instead of a singlet, and the disappearance of a broad peak in the δ12–13 region. This change is caused by the nonequivalence of the protons on C-4 and C-5, created by the introduction of a substituent on a ring-nitrogen atom replacing the NH proton, and is illustrated by the p.m.r. spectrum [see Fig. 12(B)] of 2-nitro-1-β-D-ribofuranosylimidazole (22), which exhibits, essentially, a singlet at δ8.25 (C-5-H) and a

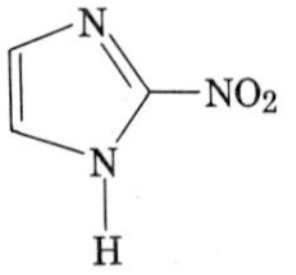

21

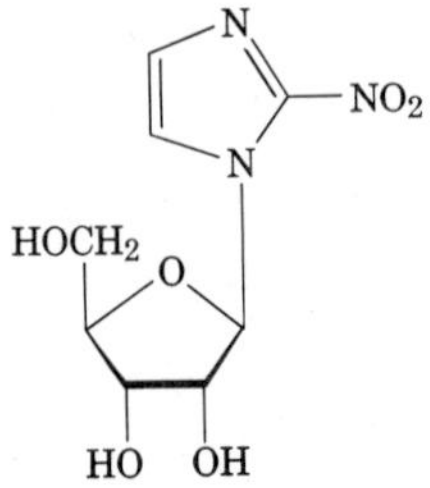

22

singlet at δ7.3 (C-4—H). In general, two doublets would have been observed; however, in this instance, the coupling constants between the protons on C-4 and C-5 are essentially zero, and so two singlets are observed.

At this juncture, the terminology used for nucleosides[19] and the p.m.r. spectrum of the D-ribosyl or 2-deoxy-D-ribosyl (2-deoxy-D-*erythro*-pentosyl) moiety will be briefly discussed, a more detailed discussion being presented in the Section on carbohydrates. Numerals will be used for the positions on the aglycon (heterocyclic moiety), and primed numerals for the positions on the carbohydrate

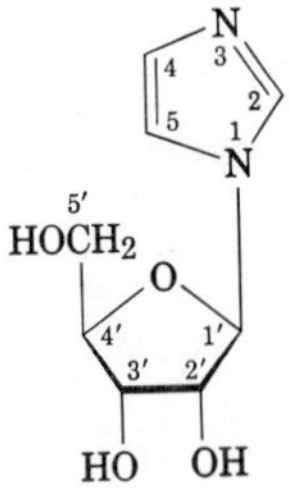

23

moiety. The numbering of the carbohydrate moiety starts at the anomeric carbon atom (C-1'), which is involved in the glycosyl bond (C-1'—N-1), and, for the numbering of the various heterocyclic moieties, the reader is referred to several sources.[20-22] The singlet at δ6.36 may be attributed to the anomeric proton on C-1' and, in fact, it was the basis for the original assignment[23] of anomeric configuration for "azomycin riboside" (22). The peaks at δ5.0-5.3 and the doublet centered at δ5.6 are attributed to the protons of the hydroxyl groups of the carbohydrate moiety. This assignment was established because the addition of D_2O to the sample caused complete disappearance of these peaks, with retention of those (singlets) at $\delta 8.2$ (C-5 proton) and $\delta 6.38$ (C-1' proton), and the protons on the carbon atoms of the carbohydrate moiety.

5-Amino-1-β-D-ribofuranosylimidazole-4-carboxamide ("AICA riboside") is an especially interesting imidazole nucleoside, because it is an integral intermediate in the *de novo* biosynthetic pathway of purines. The p.m.r. spectrum [See Fig. 13(A)] of this nucleoside exhibits a sharp singlet at $\delta 7.41$ (C-2 proton), a broad singlet at $\delta 6.83$ (protons of the NH_2 of the carboxamido group), a broad singlet at $\delta 6.00$ (of protons of NH_2 at C-5), and a doublet centered at $\delta 5.56$ (anomeric proton at C-1', $J_{1,2}$ 6.0 Hz). The addition of D_2O to the sample caused a rapid exchange of the protons of OH and NH with deuterium of the solvent, resulting in the disappearance of all of the peaks in the $\delta 4.5$–8.5 region, except peaks for the proton at C-2 (singlet at $\delta 7.46$), and the proton at C-1' ($\delta 5.58$ (d), $J_{1,2}$ 6.0 Hz). The rest of the C-bound protons of the carbohydrate moiety were observed in the $\delta 3.5$–4.5 (multiplet) region. A more-detailed discussion of the chemical shift and splitting patterns for specific protons will be given in the Section on carbohydrates (see p. 323), and in connection with the p.m.r. spectra of adenosine obtained on 60-,100-, and 220-MHz spectrometers.

The p.m.r. spectrum of the isomer, 4-amino-1-β-D-ribofuranosylimidazole-5-carboxamide ("*iso*-AICA riboside"), has revealed some interesting facts about the effect of a D-ribofuranosyl group on the chemical shift of certain groups of the aglycon. The p.m.r. spectrum of "*iso*-AICA riboside" [see Fig. 13(B)] has a sharp singlet at $\delta 7.85$ (C-2 proton), a broad singlet at $\delta 6.96$ (protons of the NH_2 of the carboxamido group), a broad singlet at $\delta 5.48$ (protons of NH_2 at C-4), and a doublet centered at $\delta 5.94$ (anomeric proton at C-1', $J_{1,2}$ 5.0 Hz). The addition of D_2O to the sample caused the disappearance of peaks in the $\delta 4.5$–8.5 region, except for the proton at C-2 (singlet at $\delta 7.86$) and the proton at C-1' [$\delta 5.94$ (d), $J_{1,2}$ 5.0 Hz). The rest of the peaks in the $\delta 3.5$–4.5 region may be assigned to the carbohydrate moiety, with the exception of the solvent peak (methyl sulfoxide-d_5 at $\delta 2.6$). There was a pronounced effect on the chemical shift for the exocyclic amino groups attached directly to the aromatic ring; this may presumably be attributed to the increased deshielding effect that the D-ribofuranosyl group has on an adjacent substituent (amino group in "AICA riboside," 22) in comparison to the exocyclic amino group in "*iso*-AICA riboside." The same relative effect (proximal deshielding) was observed for the carboxamido groups, although it was not

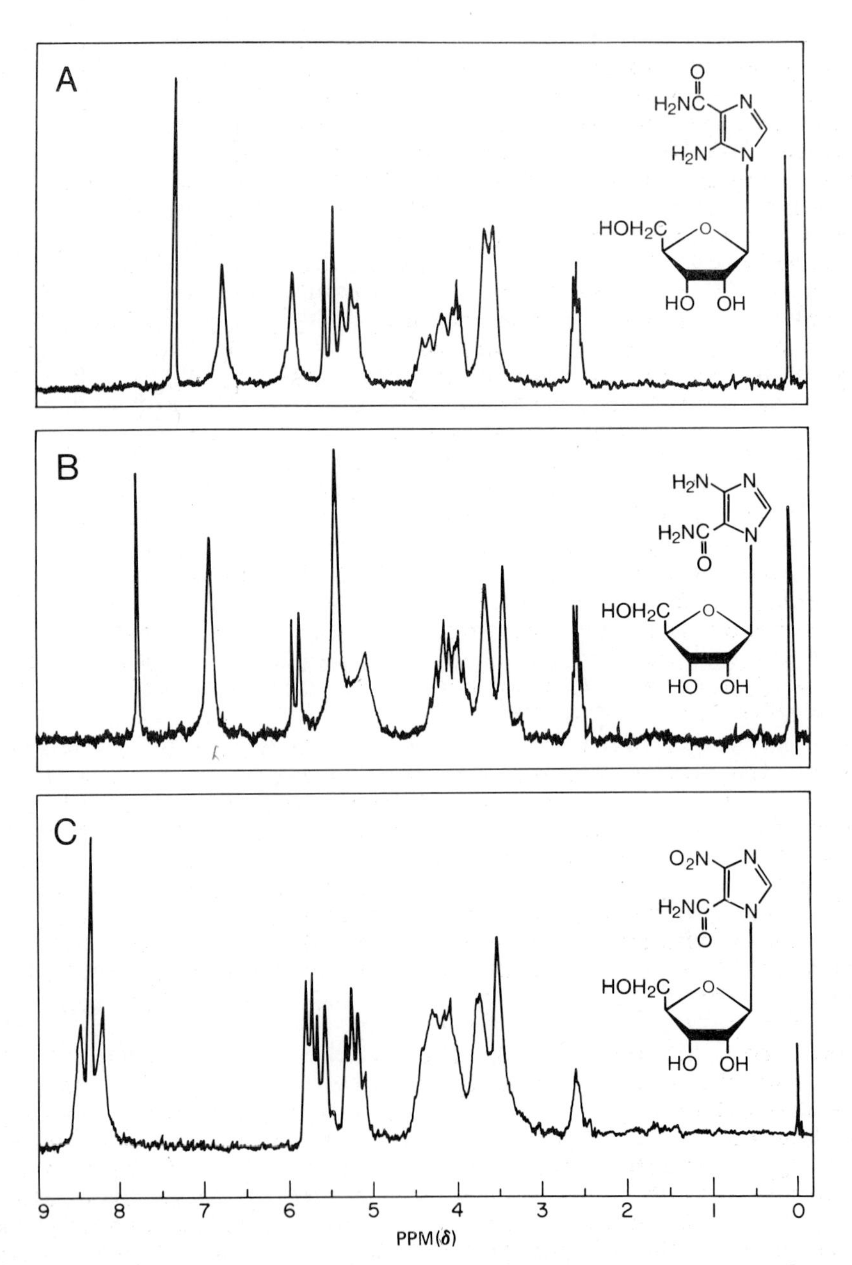

Fig. 13. P.m.r. spectra of (A) 5-amino-1-β-D-ribofuranosylimidazole-4-carboxamide, (B) 4-amino-1-β-D-ribofuranosylimidazole-5-carboxamide, and (C) 4-nitro-1-β-D-ribofuranosylimidazole-5-carboxamide, in methyl sulfoxide-d_6.

so pronounced as for the amino groups. The p.m.r. spectrum of 4-nitro-1-β-D-ribofuranosylimidazole-5-carboxamide[See Fig. 13(C)] is very similar to that of "*iso*-AICA riboside," with two major exceptions. The first is evidently the absence of a broad singlet for the 4-amino group which has been replaced by a nitro group; the second is the occurrence of the peak for the protons of the NH_2 of the carboxamido group as a doublet at δ7.3. The sharp singlet observed at δ7.4 may be assigned to the proton at C-2. However, the point of interest in this spectrum is the doublet for the carboxamido protons, which indicates that the protons are nonequivalent. This nonequivalence could be the result of restricted rotation around the carbon-nitrogen bond, because of a number of factors, such as the inductive effect of the nitro group or the D-ribosyl group or both; intramolecular bonding with the nitro or the D-ribosyl group or both; or intermolecular bonding with the solvent or solute. The cause has not yet been determined, although it has now been established[24] that the D-ribosyl group is not involved. However, by changing the nitro group to an amino, cyano, or carboxamido group, disappearance of the doublet for the carboxamido group and the appearance of a broad singlet were observed. Therefore, it appears that the nitro group is directly involved, possibly via intramolecular bonding, in a restricted rotation of the carbon-nitrogen bond of the carboxamido group. This effect has also been observed in the author's laboratory[24] for several other, related, heterocyclic systems.

The same general type of spectrum that is exhibited by "*iso*-AICA riboside" was also observed for the 2'-deoxy derivative, with two major exceptions. The first is the appearance of a broad multiplet centered at δ2.33, which is assigned to the protons at C-2' on the basis of previous p.m.r. investigations on 2-deoxy-D-*erythro*-pentofuranosyl nucleosides.[25-28] The second is the difference in the splitting patterns observed for the anomeric proton. In fact, the assignment of anomeric configuration to the anomers of 4-amino-1-(2-deoxy-D-*erythro*-pentofuranosyl)-imidazole-5-carboxamide (*iso*-AICA 2'-deoxy-riboside) was based on p.m.r. spectra.

The following is a brief introduction to the use of p.m.r. spectroscopy for the assignment of anomeric configuration for 2-deoxy-D-*erythro*-pentofuranosyl nucleosides, a more-detailed discussion being given in the Section on carbohydrates (see p.323). The preparation of "*iso*-AICA 2'-deoxy-riboside" is accomplished *via* the fusion procedure, which gives an anomeric mixture (α,β) resolvable into the pure anomers. The anomeric assignments were established by p.m.r. spectroscopy, as one anomer [see Fig. 14(A)] displayed a "*pseudo*-triplet" centered at δ6.37, having a peak-width of 13.2 Hz and an apparent splitting-constant of 6.6 Hz. This absorption was assigned to the anomeric proton, and it was found to correlate closely with

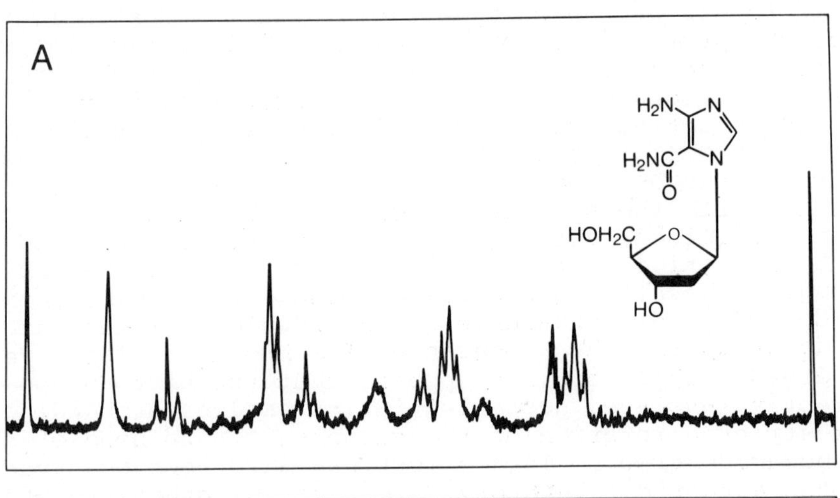

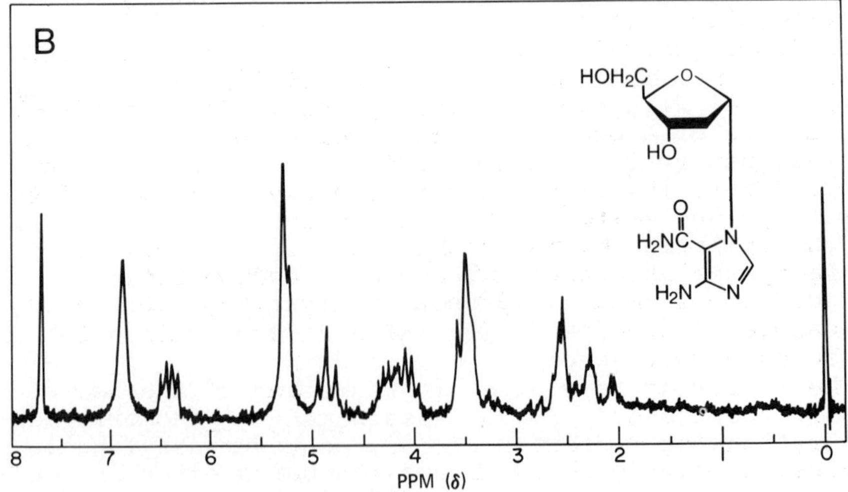

Fig. 14. P.m.r. spectra of (A) 4-amino-1-(2-deoxy-β-D-*erythro*-pentofuranosyl)imidazole-5-carboxamide, and (B) α-anomeric form, in methyl sulfoxide-d_6.

the data reported for other 2-deoxy-β-D-*erythro*-pentofuranosyl derivatives, but to differ from those for 2-deoxy-α-D-*erythro*-pentofuranosyl derivatives. Consequently, on the basis of the splitting pattern observed [see Fig. 14(A)] for the proton at C-1 of the nucleoside, the β-D anomeric configuration was assigned, and the compound is, therefore, 4-amino-1-(2-deoxy-β-D-*erythro*-pentofuranosyl)imidazole-5-carboxamide. The p.m.r. spectrum [Fig. 14B] of the other anomer has a quartet, centered at δ6.41, with coupling constants of 3.5 Hz and 7.0 Hz, and a peak width of 10.5 Hz, and this was assigned to the anomeric proton. Also, this pattern is in close agreement with the data previously published[25] for 2-deoxy-α-D-*erythro*-pentofuranosyl nucleosides, and, on this basis, the assignment of the structure for the α-D anomer is made. Although such splitting patterns have been employed in a number of investigations for the assignment of anomeric configuration, they must be used with caution, because there are factors that may affect the patterns; the exceptions that have been reported will be discussed in detail in the Section on carbohydrates (see p. 323).

The foregoing principles and p.m.r. spectra provide a foundation for preliminary interpretations of the p.m.r. spectra of other substituted imidazoles and imidazole nucleosides. The p.m.r. spectra or p.m.r. spectral data for some related compounds have been reported: halogenated imidazoles,[29] imidazole carboxaldehydes,[30] protonated nitroimidazoles,[31] methylimidazoles,[31a] and certain imidazole nucleosides.[32,33] This list is not comprehensive, but it affords some supplemental references.

2. Pyrimidines and Pyrimidine Nucleosides

For pyrimidine, the parent ring-system may be numbered either as in 23 or 24, because, prior to substitution, the ring system is symmetrical. On this basis, it might be supposed that the p.m.r. spectrum of pyrimidine would show that the protons at C-4 and C-6 are magnetically equiv-

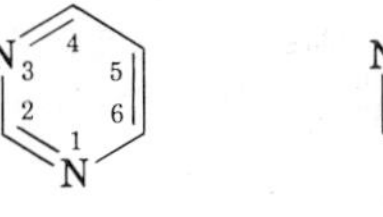

23

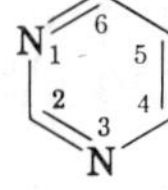

24

alent. This equivalence has been established,[34] and the spectrum of pyrimidine does, in fact, have a doublet centered at δ8.78 for the protons at C-4 and C-6 (coupled with the proton at C-5), a broad (but not too well defined) singlet centered at δ9.26 for the C-2 proton, and a sextet centered at δ7.36 for the proton at C-5. The downfield shift of the protons at C-2, C-4, and C-6, relative to the proton at C-5, is probably due not only to the electronic effects of the ring system, but also to the anisotropic effect from the adjacent, ring-nitrogen atoms. These assignments were corroborated by the results of p.m.r. studies on 2-methylpyrimidine (25), 5-methylpyrimidine (26) and 4-methylpyrimidine (27); as might be expected, because 25 and 26 are symmetrical, the protons at C-4 and C-6 are magnetically equivalent and have the same chemical shift. However, for 27 a chemical shift is observed between the peaks assigned to the protons on C-2, C-5, and C-6. For the protons on the parent ring-system, the chemical shift will, in general, be to a large extent dependent on the type of group or groups intro-

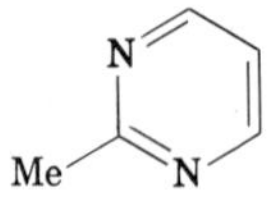

25 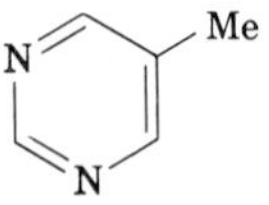 **26** 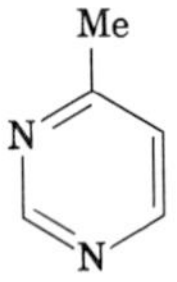 **27**

duced onto the ring. The site of introduction will also have a pronounced effect on the p.m.r. spectrum. A group at the 2-position should produce the same relative effect on the protons on C-4 and C-6, but a slightly different effect on the proton on C-5. If the group introduced at the 2-position is electron-donating, a chemical shift to higher field (smaller δ) will usually be observed for the protons on C-4, C-5, and C-6. If the group at the 2-position is electron-withdrawing, in general, a chemical shift to lower field (larger δ value) will be observed owing to a deshielding of the protons on C-4, C-5, and C-6.

The p.m.r. spectrum [see Fig. 15(A)] of 2-aminopyrimidine (28) in methyl sulfoxide shows a doublet centered at δ8.33 assigned to the protons on C-4 and C-6. However, the peak for the proton on C-5 occurs at the same chemical shift as that for the exocyclic amino group at the 2-position. To determine the splitting pattern for the

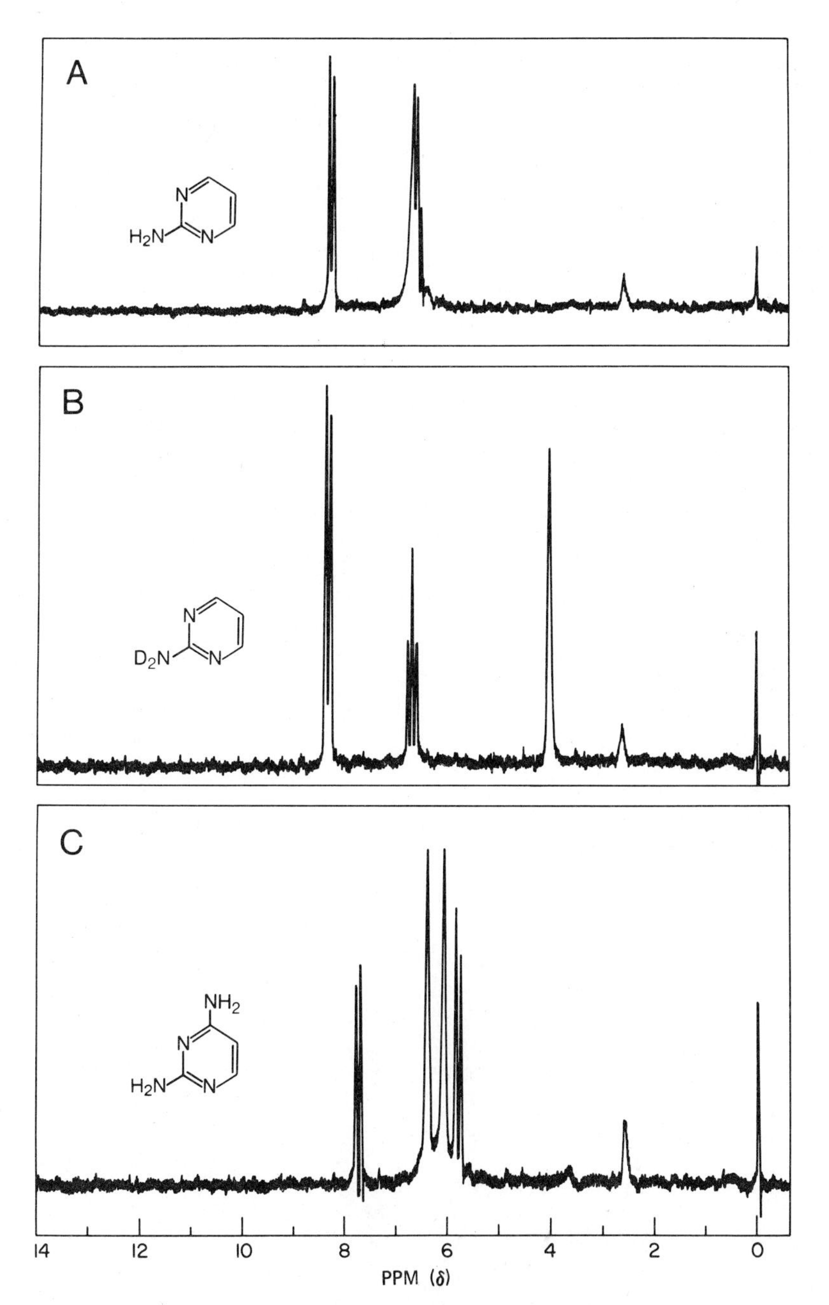

Fig. 15. P.m.r. spectra of (A) 2-aminopyrimidine in methyl sulfoxide-d_6, (B) 2-aminopyrimidine, in methyl sulfoxide-d_6-D_2O, and (C) 2,4-diaminopyrimidine, in methyl sulfoxide-d_6.

proton on C-5, a few drops of D_2O were added to the

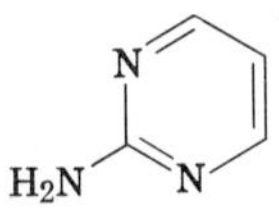

28 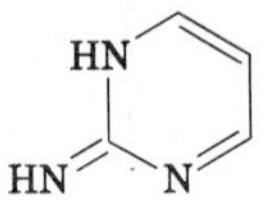 **29**

sample, and a new spectrum was obtained [see Fig. 15(B)]. A rapid exchange between the solvent and the protons of the exocyclic amino group occurred, and disappearance of a broad singlet at δ6.7 due to the protons of the amino group and the appearance of another singlet at δ4.03, attributable to HDO, were observed. This revealed the splitting pattern for the proton at C-5 as a triplet, centered at δ6.7. Also observed was a slight shift (to lower field) due to a solvent effect caused by the addition of D_2O. A re-examination of the p.m.r. spectrum of <u>28</u> in methyl sulfoxide-d_6, revealed that the proton at C-5 could be assigned to the triplet centered at δ6.63. These results lend support to the foregoing statement that an electron-donating group in the 2-position shifts the other protons in the ring system to higher field, because a triplet at δ6.63 (assigned to the proton on C-5) and a doublet at δ8.33 (assigned to the protons on C-4 and C-6) are observed.

The tautomeric form (<u>29</u>) of 2-aminopyrimidine was excluded, because a singlet (2 protons) is observed at δ6.7, assigned to the protons of the amino group. This assignment was made on the basis of a loss of two protons (by integration) for the peaks at δ6.4-6.8, on adding D_2O to the solution in methyl sulfoxide-d_6. Had the imino form (<u>29</u>) been the preponderant species, two separate peaks for the NH protons, having different chemical shifts, would have been observed, instead of one broad singlet.

The introduction of another amino group at C-4 results in an additional chemical-shift upfield (smaller δ value) for the peaks for the protons at C-5 and C-6, in comparison with those observed for <u>28</u>. The proton at C-6 (H-6) is observed [see Fig. 15(C)] at δ7.71 (doublet, $J_{5,6}$ 5 Hz), the H-5 signal occurring at δ5.77 (doublet, $J_{5,6}$ 5 Hz). Introduction of the amino group at the 4-position simplifies the splitting patterns, by eliminating the peaks attributable to the coupling of H-6 and H-5 with H-4. Because the amino group is electron-donating, an upfield shift for the peak assigned to the 2-amino group is also observed.

If the amino group is on C-5, instead of C-2, a completely different p.m.r. spectrum is found. Not only is there an upfield chemical-shift for the adjacent proton on C-6, but a significant upfield shift is found for the amino group (in comparison with the chemical shift when the amino group is on C-2). The p.m.r. spectrum of 4,5-diaminopyrimidine [see Fig. 16(C)] in methyl sulfoxide-d_6 has a singlet at δ7.83 (H-2), a singlet at δ7.58 (H-6), a broad singlet at δ6.35 (4-NH_2), and a broad singlet at δ4.73 (5-NH_2). This upfield shift for the amino group on C-5 may be attributed to the fact that C-5 of the pyrimidine ring is the center having the highest electron density.

The foregoing discussion has primarily been concerned with the effect of an electron-donating group on the chemical shift of ring protons and protons of exocyclic groups. The introduction of an electron-withdrawing group generally produces the opposite effect. *A priori*, the introduction of a chlorine atom at C-2 of 4,5-diaminopyrimidine should result in a downfield chemical-shift for both of the exocyclic amino groups and for the "aromatic" proton. Inspection of the p.m.r. spectrum of 4,5-diamino-2-chloropyrimidine in methyl sulfoxide-d_6 [see Fig. 16(B)] reveals that the peak for the 4-amino group (singlet) occurs at δ6.87 (2 protons), and that for the 5-amino group, at δ4.87 (singlet, 2 protons), the most significant chemical shift being observed for the 4-amino group. When the chlorine atom is introduced at C-6 in 4,5-diamino-6-chloropyrimidine, the p.m.r. spectrum [see Fig. 16(A)] shows that there is only a very small effect on the 5-amino group (singlet at δ4.92) and the 4-amino group (singlet at δ6.70). However, the most pronounced effect (greatest deshielding) is, in this instance, that for the 5-amino group, presumably because of the difference in spatial proximity between the 5-amino and 4-amino groups relative to the 6-chlorine atom. Although the electron-withdrawing effect of the chlorine atom may be the primary cause of the downfield shift observed for the absorption peaks discussed, other factors, including anisotropy and steric considerations, may be important.

Most nitrogen heterocyclic compounds have exocyclic groups capable of existing in various tautomeric forms. Uracil (30), the nitrogenous base of the nucleoside uridine, is theoretically capable of existing in several tautomeric forms, for example 30, 30a, 30b, 30c, but the dioxo form 30 is the main tautomer, as shown by X-ray crystallographic,

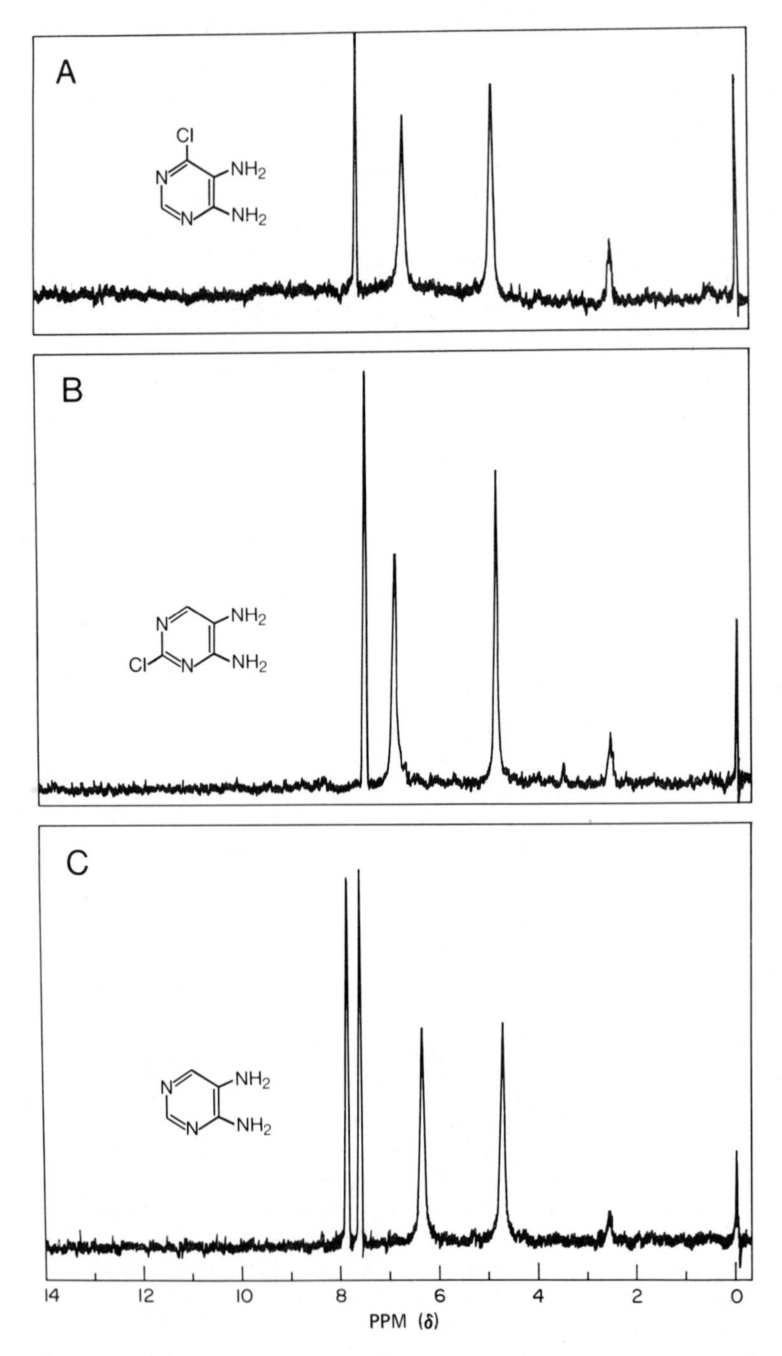

Fig. 16. P.m.r. spectra of (A) 4,5-diamino-6-chloropyrimidine, (B) 4,5-diamino-2-chloropyrimidine, and (C) 4,5-diaminopyrimidine, in methyl sulfoxide-d_6.

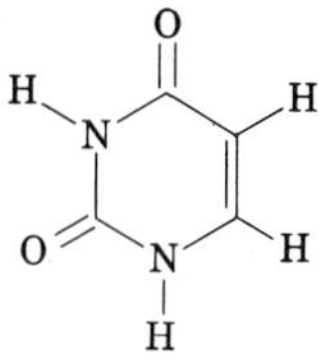

30 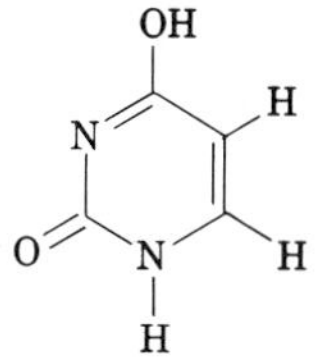 30a

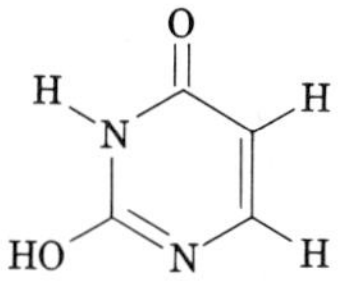

30b 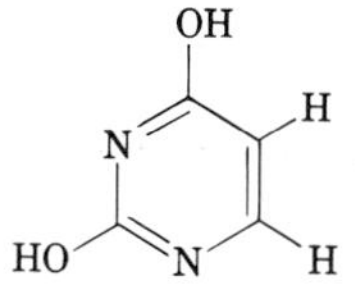 30c

ultraviolet absorption, and infrared studies, and this assignment has been corroborated[35] by p.m.r. spectroscopy. The p.m.r. spectrum [see Fig. 17(A)] of uracil (30) in methyl sulfoxide-d_6 has a doublet ($J_{1,2}$ 8.0 Hz) centered at δ5.5 for the proton at C-5, a similar doublet, centered at δ7.4, for the proton at C-6 (with a coupling constant the same as that observed for the doublet of the C-5 proton), and a singlet at δ10.8, assigned to the protons on N-1 and N-3. On addition of two drops of D_2O to the methyl sulfoxide-d_6 solution, a rapid exchange between the solvent (D_2O) and the protons on NH of uracil (30) is observed. This technique gives a spectrum [see Fig. 17 (B)] in which the peak at δ10.8 has disappeared, and the replaced protons of NH are at δ3.9, and may be assigned to the absorption peak for HDO. The doublet centered at δ5.6 can be assigned to the proton at C-5, and the doublet centered at δ7.4 to the proton at C-6.

It is of interest that the isomeric base, 4,6-pyrimidinediol (31), has been shown by p.m.r. spectroscopy to exist preponderantly in the oxo-hydroxy form 31b in methyl sulfoxide-D_2O. P.m.r. studies were made on *N*- and *O*-methyl derivatives of 31. These findings were supported by ultraviolet and infrared spectral comparisons.[36,37]

Another pyrimidine that has been the subject of several investigations is cytosine (32), which may exist as one of several possible tautomeric forms. The main forms to be considered are 32, 32a, 32b, and 32c, although other

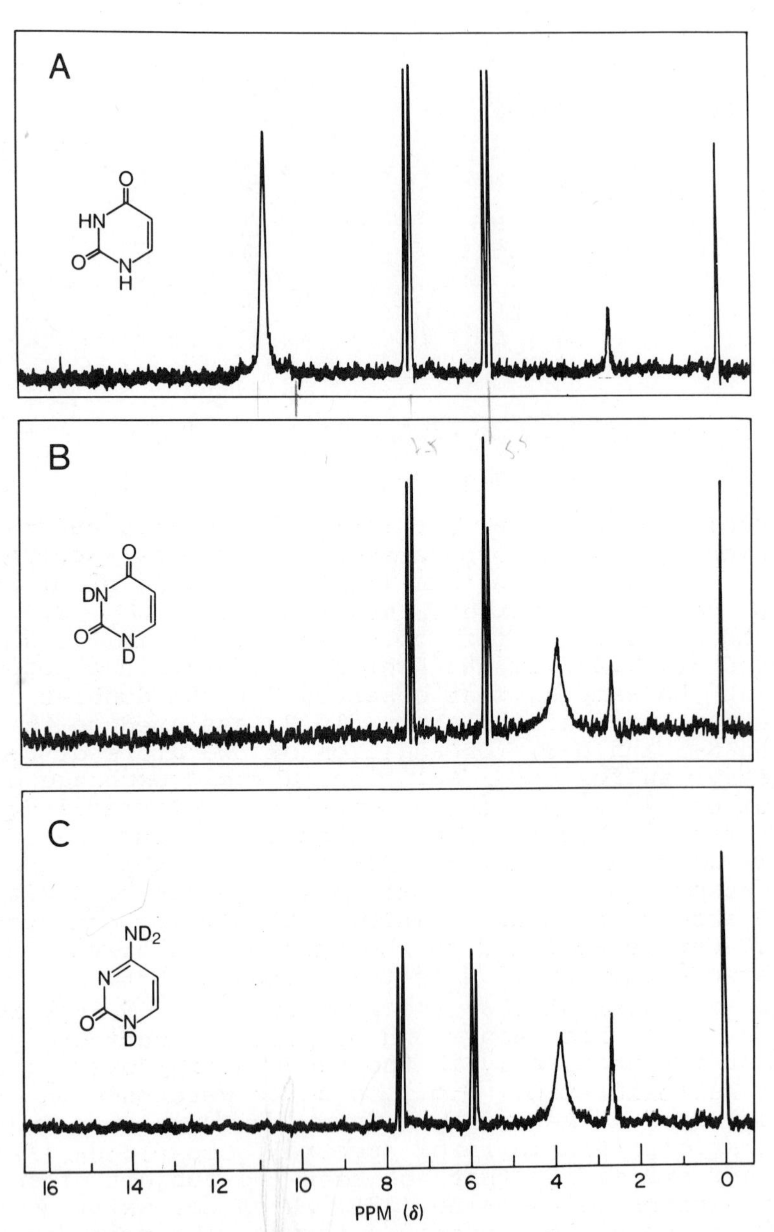

Fig. 17. P.m.r. spectra of (A) uracil in methyl sulfoxide-d_6; (B) uracil in methyl sulfoxide-d_6/D_2O; (C) cytosine, in methyl sulfoxide-d_6/D_2O.

minor structural variations (charged types) could be proposed. The p.m.r. spectrum [see Fig. 17(C)] of cytosine

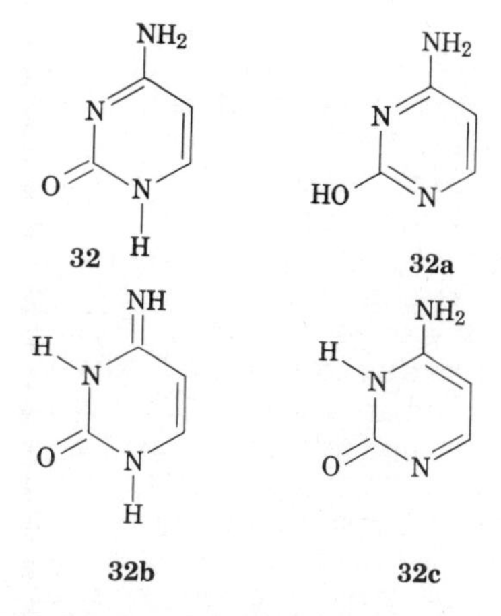

(32) in methyl sulfoxide-d_6-D_2O has a sharp doublet ($J_{5,6}$ 7.0 Hz) centered at δ5.93, which may be assigned to the proton at C-5, and a similar doublet, having the same J value, centered at δ7.72, for the proton at C-6. A broad absorption peak at $\sim\delta$3.9 is assigned to the HDO peak. The p.m.r. spectrum of cytosine hydrochloride in methyl sulfoxide has been recorded,[38] and, on the basis of the splitting patterns observed for the hydrochloride, it was suggested that cytosine (as the base, 32) exists preponderantly as a zwitterion in methyl sulfoxide. The

protonation of cytosine by hydrochloric acid was considere to occur at the exocyclic amino group; however, a later investigation[39] established that, in trifluoroacetic acid,

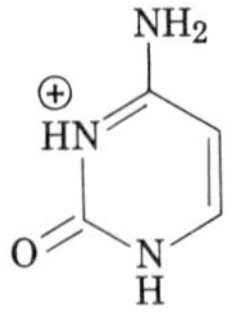

33

cytosine exists as the structure depicted in 33. This finding is also in agreement with the infrared absorption spectra reported[40] for the cationic species of cytosine. On the basis of ultraviolet and infrared spectral studies,[41,42] it had been proposed that cytosine (as the base exists as structure 32. It has been reported that cytosine exists mainly as 32, not 32c, although there appears to be an equilibrium between the two structures.

The D-ribosyl and "2-deoxy-D-ribosyl" derivatives of cytosine, namely, cytidine and 2′-deoxycytidine, have been the subject of several p.m.r. investigations.[43–46] The p.m.r. spectrum of cytidine [see Fig. 18(A)] has a patter of peaks that would be expected for this pyrimidine D-ribonucleoside. The peaks attributable to the D-ribofuranosyl group display the usual pattern and chemical shifts; the doublet centered at δ8.17 may be assigned to the proton at C-6, and the doublet for the proton at C-5 at δ6.03 overlaps the peak for the anomeric proton. The peak at δ7.5 (2 protons) may be assigned to the exocyclic amino group; it disappears when D_2O is added to the solution. These results established that, in methyl sulfoxide-d_6, cytidine exists in the amino, not the iminc form. It had been reported[45] that, although cytidine exists in the amino form, 2′-deoxycytidine favors the imino form. This claim resulted in a re-investigation which established that the report was incorrect because the spectrum of the protonated species of 2′-deoxycytidir had been recorded. The p.m.r. spectrum of 2′-deoxycytid (for those protons assigned to the heterocyclic moiety) is essentially the same as that of cytidine; hence, 2′-deoxycytidine exists in the amino form. However, the p.m.r. spectrum of 2′-deoxycytidine hydrochloride, in methyl sulfoxide-d_6, has the same p.m.r. spectrum as had been reported for 2′-deoxycytidine, two peaks occurri

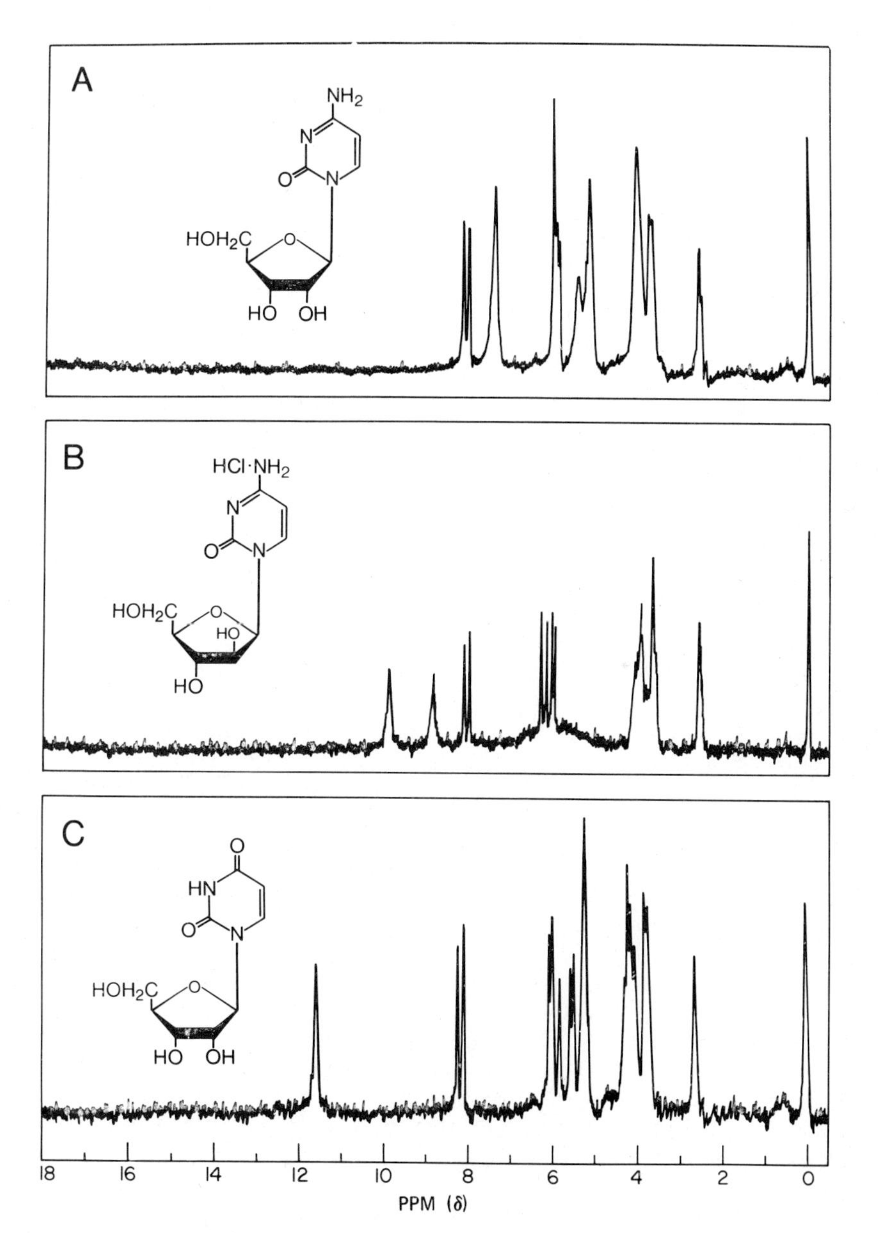

Fig. 18. P.m.r. spectra of (A) cytidine, (B) 1-β-D-arabinofuranosylcytosine hydrochloride, and (C) uridine, in methyl sulfoxide-d_6.

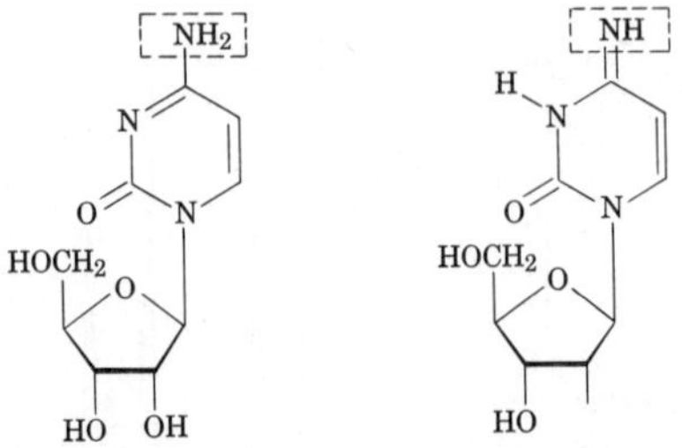

NH$_2$ N O N HOCH$_2$ O HO OH

Cytidine
(amino form)

NH H N O N HOCH$_2$ O HO

2′-Deoxycytidine
(imino form)

downfield from the doublet observed for the proton at C-6, with a very low, broad absorption peak in the NH region, and this situation has been found to be the direct result of protonation that occurs at N-3 (not glycosylated). The two protons on the amino group at C-4 are then deshielded to different extents, and occur as two equal, but separate peaks.

The p.m.r. spectrum of 1-β-D-arabinofuranosylcytosine hydrochloride [see Fig. 18(B)] has a similar distribution of peaks, and the rather broad singlets at δ8.93 and 9.92 are attributed to the NH protons, the doublets centered at δ8.07 and 6.23 being assigned to the protons at C-6 and C-5, respectively. No overlap between the proton at C-5 and the anomeric proton is observed, the doublet ($J_{1,2}$ 4 Hz) centered at δ6.03 being assigned to the anomeric proton. The p.m.r. spectrum of 1-β-D-arabinofuranosylcytosine, in methyl sulfoxide-d_6, lacks the two separate peaks at δ8.93 and 9.92, but has a rather broad peak (2 protons) at δ7.2 which may be assigned to the exocyclic amino group. Of especial significance is the observed upfield shifts from δ6.73 and 8.07 to δ5.8 and 7.68, for the doublets attributed to the protons at C-5 and C-6, respectively, no significant chemical shift being observed for the anomeric proton. This observation lends considerable support to the hypothesis that protonation does, indeed, occur on the heterocyclic moiety, as evidenced by the differences observed in the chemical shifts produced by the protons thereof.

The p.m.r. spectrum of uridine in methyl sulfoxide-d_6 [see Fig. 18(C)] has the expected pattern, the proton of the ring NH occurring at δ11.63 as a rather broad singlet. The rest of the spectrum is similar to that of cytidine under similar conditions, except that it lacks the peak for the amino group at C-4.

Obviously, a change in the heterocyclic moiety, or in

the glycosyl group, will change the p.m.r. spectrum, but, in general, the foregoing examples may be used as a basis for further extrapolation. Other selected examples of p.m.r. spectra, or p.m.r. spectral data, for pyrimidines and pyrimidine nucleosides are as follows: studies of the replacement of the proton at C-6 of derivatives of 5-fluorouracil by deuterium,[47] various substituted uracils,[48] the anomeric configuration of amicetin,[49] certain nitropyrimidine derivatives,[50] 2-substituted pyrimidines and their derivatives,[51] effects of solvent on the spectra of a number of halogenopyrimidines and certain derivatives thereof,[52] a deuterium–hydrogen exchange at C-5 of uridines,[53] certain (halogenomethyl)pyrimidines,[54] 4-benzyloxy-1-glycosyl-6(1*H*)-pyrimidinones,[55] 2',3'-anhydro-D-lyxosyl nucleosides,[56] tautomeric forms of derivatives of (hydroxylamino)pyrimidines,[57] 5-(carboxymethyl)uracil and 6-(carboxymethyl)uracil,[58] data for a number of halogenated, methylated, and alkoxylated pyrimidines,[59] anhydronucleosides,[60] 6-methyl-1(*H*)-pyrimidine D-ribonucleosides,[61] 1-(2-deoxy-D-*ribo*-hexofuranosyl)thymine,[62] 4-substituted 6(1*H*)-pyrimidinone D-ribonucleosides,[63] 5-(deoxy-D-hexosyl)pyrimidines,[64] 5,6-dihydro-1(*H*)-pyrimidine D-ribonucleosides,[65] precursors of gougerotin,[66] and 2-(butenylamino)-3-methylpyrimidines.[67]

3. Purines and Purine Nucleosides

Assignment of the chemical shifts for the protons at C-2, C-6, and C-8 of purine has been accomplished[66,68–70] with specifically deuterated derivatives. These investigations have established that the relative positions of the peaks in the p.m.r. spectrum occur in the order 6, 2, and 8. The proton at C-6 is more deshielded than the protons at C-2 and C-8. Hence, in the p.m.r. spectrum of purine [see Fig. 19(A)], the peak at δ9.23 may be assigned to the proton at C-6, that at δ9.03 to the proton at C-2, and that at δ8.73 to the proton at C-8.

A procedure that has been used for the assignment of protons, not only for purine, but also for other heterocyclic systems, involves systematic replacement of each "aromatic" proton by a methyl group, and recording of the p.m.r. spectrum of each methylated derivative. The introduction of a *C*-methyl group has only a small effect on the chemical shift observed for the remaining "aromatic" protons;[70] hence, the introduction of a methyl group at a specific position may be used to make an assignment for at least one peak in the original spectrum. This process may then be repeated, until an assignment has been made for every peak in the p.m.r. spectrum of the parent ring compound or of the original compound.

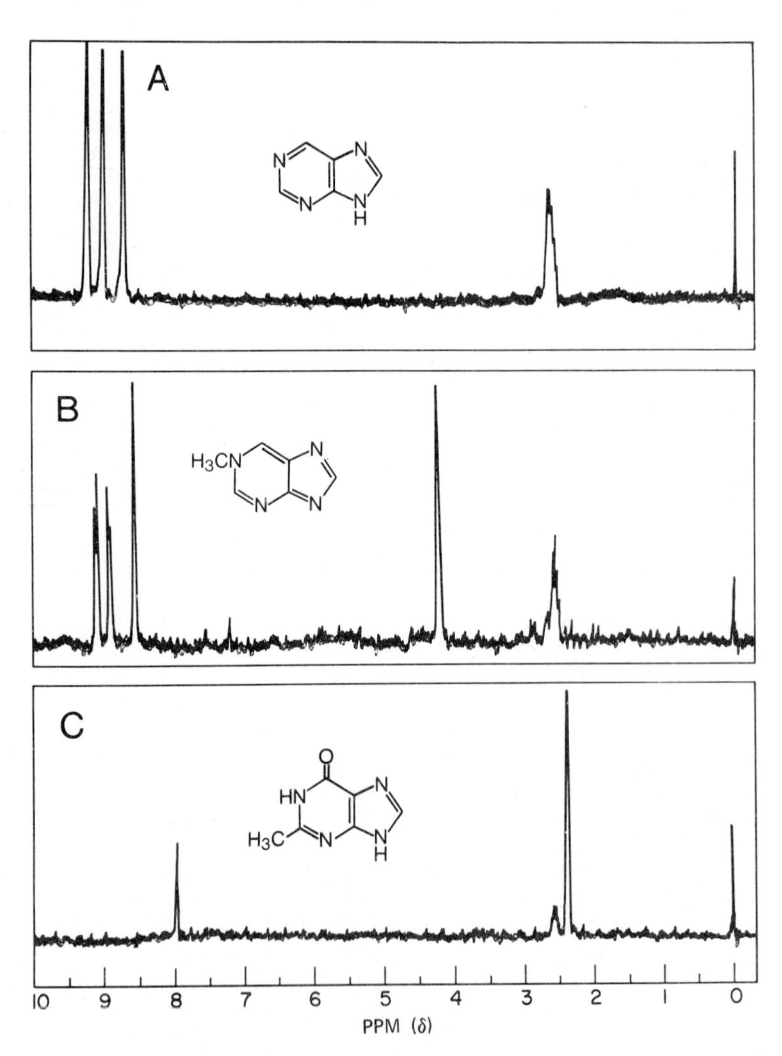

Fig. 19. P.m.r. spectra of (A) purine, (B) 1-methylpurine, and (C) 2-methylhypoxanthine, in methyl sulfoxide-

Another procedure that has been used with considerable success consists in effecting a ring closure of the appr priate starting material with a reagent containing deute ium instead of hydrogen, thus affording a product having deuterium at a specific site. A variation of this procedure involves the use of a nondeuterated reagent to effect annulation of a compound already containing deuterium at a position other than the site of ring closure. Specifically deuterated compounds often provide valuable information for studies on the chemical shift.

The p.m.r. spectrum of purine in an acidic solvent has revealed that spin–spin coupling between the protons on C-2 and C-6 does occur, and this has been suggested to be the result of protonation, presumably at N-1, which inhibits (or decreases) the nuclear quadrupole effect and permits spin–spin coupling of the protons on C-2 and C-6. The effect was found to occur at pH <2.4, whereas, for the neutral species (pH >2.4), no spin–spin coupling between the protons on C-2 and C-6 was observed. The same effect (spin–spin coupling) may presumably be effected by the introduction of a group at the position of protonation (N-1). The p.m.r. spectrum of 1-methylpurine (neutral species) in methyl sulfoxide-d_6 [see Fig. 19(B)] showed that this was, indeed, correct.[71] The protons on C-2 and C-6 produce definite doublets, at δ9.07 and 9.20, with a coupling constant of ~2 Hz, and the proton at C-8 and the methyl group at N-1 are assigned to the sharp singlets at δ8.67 and 4.28, respectively. This situation is in direct contrast to that for the other *N*-methylpurines, namely, 3-, 7-, and 9-methylpurine. The p.m.r. spectra (recorded in methyl sulfoxide-d_6) of these compounds revealed a peak for the methyl group, and three other distinct, separate singlets for the protons at C-2, C-6, and C-8 (in the δ8.0–9.5 region), with no evidence of spin–spin coupling.

Another generalization that can be made is concerned with the use of chemical-shift data in conjunction with other instrumental methods, for aid in locating the position of a methyl group in a methylated, nitrogen heterocycle. In the p.m.r. spectrum of 1-methylpurine, the peak assigned to the methyl group on the ring-nitrogen atom (N-1) is at δ4.28. However, the chemical shift for a methyl group of a *C*-methylpurine (for example, 2-methylhypoxanthine) appears at a much higher field (smaller δ value).

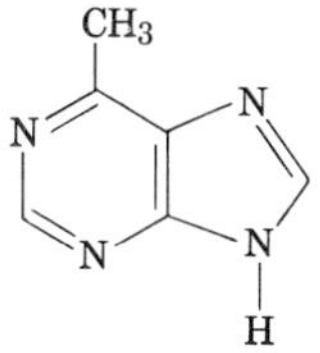

CH_3 N N N N H

34

The p.m.r. spectrum of 2-methylhypoxanthine in methyl sulfoxide-d_6 [see Fig. 19 (C)] has a peak at δ2.38 which may be assigned to the methyl group on C-2. A comparison between the value for 1-methylpurine and that for 6-methyl-purine (34), for which the value reported[70] for the 6-meth group is δ2.26 (in D_2O), is more valid, because they are isomers, and yet they have different p.m.r. spectra. There is also a significant chemical shift for the methyl group when a hetero atom is inserted between the methyl group an the heterocyclic ring.

The p.m.r. spectrum of 6-(methylthio)purine in methyl sulfoxide-d_6 [see Fig. 20(A)] has two singlets, at δ8.5 and 8.75, which may be assigned to the protons at C-2 and C-8, respectively, the singlet (3 protons) at δ2.7 being assigned to the exocyclic methyl group. Hence, insertion of a sulfur atom between the methyl group and the ring effects a downfield shift (larger δ value). An additional downfield shift is caused by replacing the sulfur atom by a nitrogen atom. The p.m.r. spectrum of 6-(methylamino)-purine in methyl sulfoxide-d_6 [see Fig. 20(B)] has two sharp singlets, at δ8.2 and 8.33, that can be assigned to the protons at C-2 and C-8, respectively. In this instance, the peak centered at δ3.13 (for the methyl group) is a doublet, owing to spin–spin coupling with the proton on the exocyclic nitrogen-atom at C-6. The low, broad doublet centered at δ7.67 may be assigned to the proton of NH. The peak for the NH proton is generally observed as a low, broad singlet, not a doublet, owing to the nuclear quadrupole relaxation of the nitrogen atom, o to presence of a trace of water in the sample or in the solvent. The addition of a few drops of D_2O causes complete disappearance of the peaks for the proton on N-9 an (at δ7.67) for the NH proton at C-6. The peaks for th protons on C-2 and C-8 are still evident as singlets in the δ8–9 region, whereas that for the methyl group is now a singlet. An additional, downfield shift for the methyl group is effected by replacing the nitrogen atom by an oxygen atom. The peak for the protons of the methyl grou of 6-methoxypurine is observed at δ4.17 in the p.m.r. spe trum [see Fig. 20(C)]. Although other factors are involved, the major influence appears to be the electronega tivity of the atom to which the methyl group is bonded.

The unequivocal assignment of the peaks at δ8.47 and 8.57 to the protons at C-8 and C-2, respectively, of 6-methoxypurine has been accomplished by study of the p.m.r spectrum of 6-methoxypurine-*2-d* (35). The p.m.r. spectrum of 35 has only one singlet for an "aromatic" proton bonded to a carbon atom, and this was unequivocally assig ed to the proton on C-8. In fact, this assignment is included in a study[72,73] on the chemical shift observed for

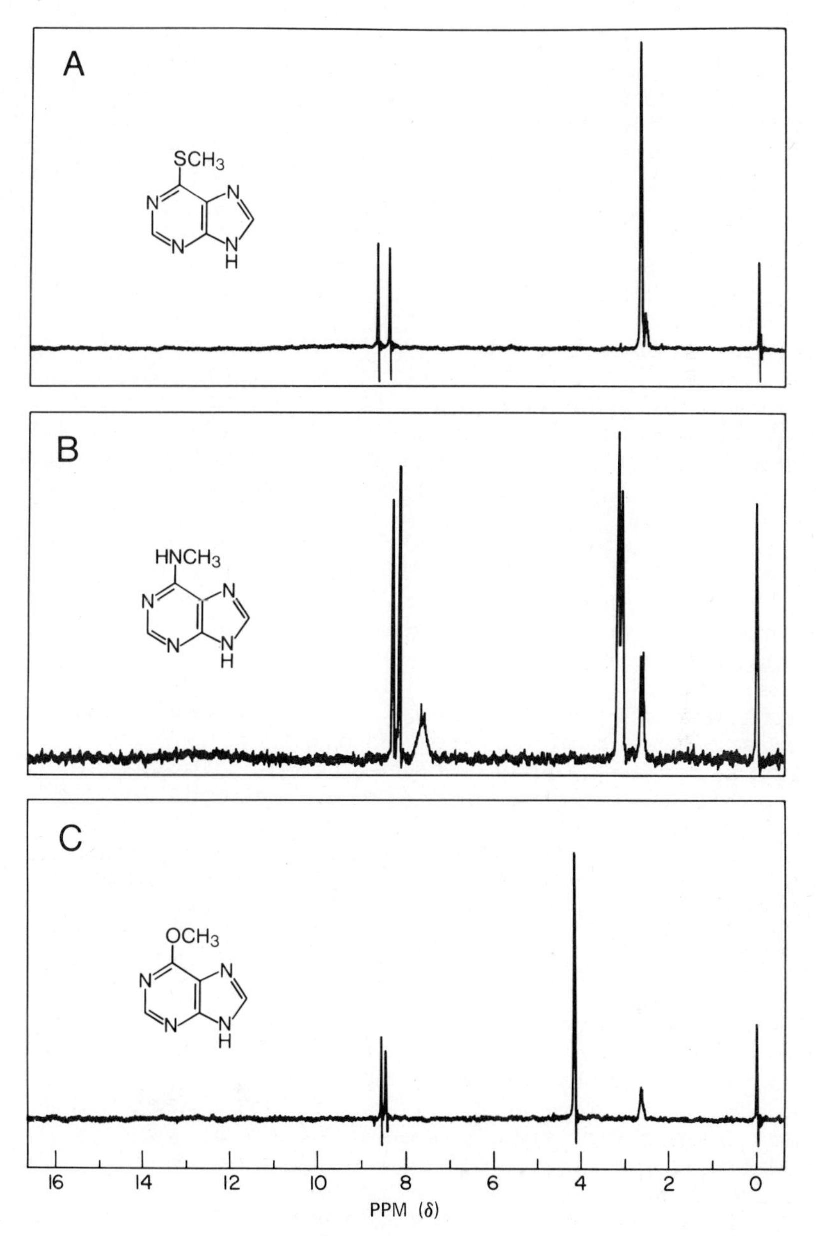

Fig. 20. P.m.r. spectra of (A) 6-(methylthio)purine, (B) 6-(methylamino)purine, and (C) 6-methoxypurine, in methyl sulfoxide-d_6.

the proton at C-8 of various purine derivatives. It has been established that the chemical shift for the proton at C-8 is directly affected by various substituents at C-2 and C-6. If the groups are electron-donating [for example, $-N(CH_3)_2$], a shielding effect (upfield shift to a smaller δ value) is observed and, conversely, if the groups are electron-withdrawing (for example, $-CF_3$) a deshielding effect (downfield shift to a larger δ value) results. Care must, however, be taken in employing this generalization, because a number of other factors, such as resonance, could negate or completely outweigh the proposed inductive effect. An equation has been derived, incorporating an electrophilic-substituent constant[74] in a multiple-regression equation, for use in predicting, with remarkable success, the chemical shift for the proton at C-8 of additional purines. This work involved the compiling of data for values of the chemical shift observed for the protons at C-2 and C-8 of various 6-substituted and 2,6-disubstituted purines in methyl sulfoxide-d_6 and in trifluoroacetic acid. In essence, the study has shown that the chemical shifts for the protons bound to carbon atoms of a derivative of purine are definitely a function of the other groups attached to the ring.

In this respect, it appears that, in general, the D-ribofuranosyl group of a nucleoside exerts a deshielding effect on ring positions (of the base) that are directly adjacent to the site of D-ribosylation. The p.m.r. spectrum of adenine exhibits two sharp singlets at δ8.11 and 8.14, which are assigned[73] to the protons at C-2 and C-8. The p.m.r. spectrum[75] of adenosine (9-β-D-ribofuranosyladenine) exhibits two singlets for the protons at C-2 and C-8, which are observed at δ8.27 and 8.45, and so, on D-ribosylation, a downfield shift occurs for both protons. The most significant shift is, presumably, that observed for the proton at C-8, primarily because of the proximity to the site of D-ribosylation, although these chemical shifts are also dependent on solute—solute and solute—solvent interactions.

P.m.r. spectroscopy has been used as an aid in the determination of the actual site of alkylation or D-ribosylation of compounds isolated from the direct alkylation or D-ribosylation of adenine and certain related derivatives.[76] It has been shown that *N*-substitution of adenine in the "imidazole" portion [as in 9-methyladenine (36) and 7-methyladenine (37)] results in a small $\Delta\delta$ (difference, in Hz, between the δ values observed for the peaks from the protons at C-2 and C-8) in the p.m.r. spectrum, whereas, under similar conditions, 1-methyladenine (38) and 3-methyladenine (39) exhibit a much larger $\Delta\delta$.

Chemical evidence has indicated that the introduction o

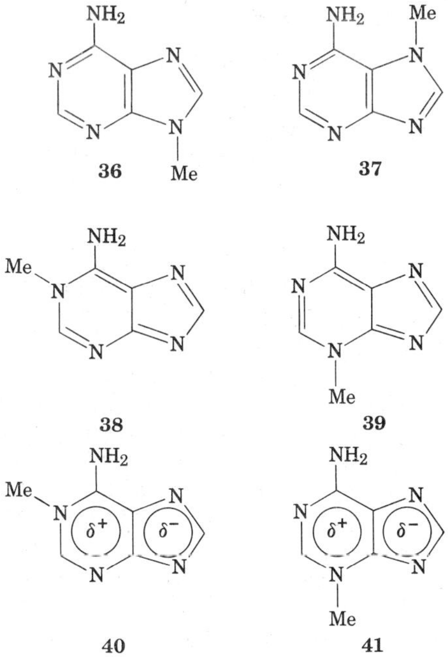

Scheme 2

a substituent on a ring-nitrogen atom in the "pyrimidine" portion of purine results in a pyrimidine ring that appears to be a π-deficient and an apparently π-excessive imidazole ring (40 and 41). This situation would result in a deshielding effect for the proton at C-2 and would produce a shielding effect for the proton at C-8, as well as a larger $\Delta\delta$ than for a compound in which the effect is essentially nonexistent (substitution in the imidazole moiety). The foregoing empirical rule has been used in several investigations, but it should be employed with caution, and only for *preliminary* structural assignments.[77] It has been also established that the methylated adenines (36-39) exist in the amino, not the imino, form, in methyl sulfoxide-d_6. The same situation has been demonstrated also for the ring-*N*-methyl isomers of 6-(methylamino)purine (42), where, in each instance, the peak for the methyl group attached to the exocyclic nitrogen atom is a doublet, which indicates coupling with the proton on the exocyclic nitrogen atom on C-6. This fact demonstrates that, in solution, such purine derivatives also exist in the amino form. The same general trend was observed for the

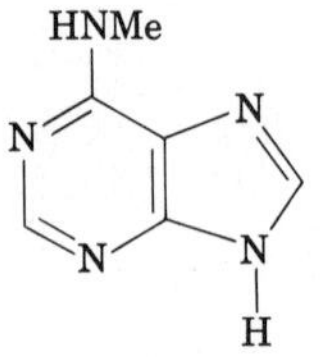

42

$\Delta\delta$ values for 1-, 3-, 7-, and 9-methyl-6-(methylamino)-purine, and 1-, 3-, 7-, or 9-methyl derivatives of 6-(dimethylamino)purine, as described in the foregoing for the corresponding adenine derivatives.

Another interesting point concerning ring *N*-methylpurines (derived from adenine or purine) is that, in general, the peaks observed for the methyl group on a ring-nitrogen atom of the pyrimidine moiety occur downfield from that observed for a methyl group on a ring-nitrogen atom of the imidazole moiety. The chemical shift produced could result from a number of factors, the two most probable ones being (*1*) the difference between the methyl group on a ring-nitrogen atom in a π-deficient ring-system (pyrimidine) and a π-excessive ring-system (imidazole), and (*2*) the difference between such partial zwitterionic structure as 40 and 41, and 36 and 37. 7-Methylguanosine (43) is an example of a ring-methyl group that causes an additional,

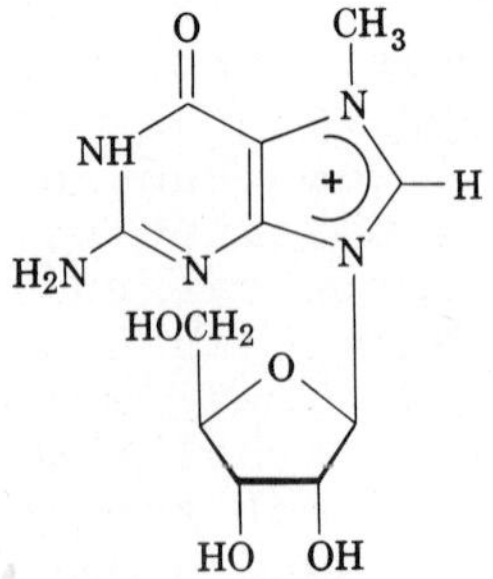

43

downfield chemical-shift by quaternization, which results in a considerable deshielding of the proton at C-8, shifting the peak downfield to δ9.6. This displacement constitutes a downfield shift of δ1.3, relative to the peak observed (δ8.3) under similar conditions for the proton at C-8 of guanosine. The peak for the proton at C-8 of 43 occurs as a sharp singlet at δ9.6 in the p.m.r. spectrum, recorded in methyl sulfoxide-d_6, but, on addition of D_2O to the solution (or with D_2O as the sole solvent), a rapid exchange of the proton at C-8 with a deuterium atom from the solvent occurs. This result is corroborated by the disappearance of the peak assigned to the proton at C-8. The deshielding effect was equated with an electron deficiency in the imidazole ring, caused by quaternization of the ring-nitrogen atom (N-7), when it was found[80] that very dilute base effects facile ring-opening at C-8.

Additionally, p.m.r. spectroscopy has been a valuable tool in the analysis of certain mixtures. The direct glycosylation of *N*-benzoyladenine with 2,3,5-tri-O-benzyl-D-arabinofuranosyl chloride gives,[78] after removal of the benzoyl group at N-6, 9-(2,3,5-tri-O-benzyl-β-D-arabinofuranosyl)adenine. Another nucleoside may be isolated from the filtrate (as a minor component), and this was at first assumed to be the α anomer. In a separate investigation[79] that used p.m.r. and u.v. spectroscopy, it was established that the mixture is, in fact, composed of isomers, not anomers. The p.m.r. spectrum of the mixture [see Fig. 20(A)] has a complex pattern of peaks in the δ6.7–7.7 region that may be assigned to the protons of the benzyl groups, the exocyclic amino group, and the anomeric proton. The peaks in the δ3.8–5.0 region are attributable to the protons of the carbohydrate moiety and the methylene protons of the benzyl groups. Of particular interest in the spectrum are the peaks in the δ8.1–8.7 region, assigned to the "aromatic" protons on C-2 and C-8. The peaks (singlets) at δ8.47 and 8.43 (Δδ 2 Hz) are assigned to the protons at C-2 and C-8 of 9-(2,3,5-tri-O-benzyl-β-D-arabinofuranosyl)adenine, and the smaller peaks at δ8.1 and 8.67 (Δδ 34 Hz) are assigned to the protons at C-2 and C-8 of the isomeric nucleoside. Separation of the nucleoside mixture was followed by debenzylation to afford the known 9-β-D-arabinofuranosyladenine; the isomeric nucleoside was subsequently shown to be 3-β-D-arabinofuranosyladenine. The p.m.r. spectrum of 9-β-D-arabinofuranosyladenine [see Fig. 21(B)] has the absorption peaks expected for the carbohydrate moiety (δ3.8–6.5), a singlet at δ7.37 for the exocyclic amino group, and two singlets at δ8.2 and 8.3 (Δδ 6 Hz). The small Δδ value indicates that the site of glycosylation is on the imidazole portion, suggestive of a 9-substituted adenine, a conclusion corroborated by u.v. spectral data (for discerning between N-7 and N-9). The p.m.r. spectrum of 3-β-D-arabinofuranosyladenine [see Fig. 21(C)] shows some significant differences from Fig. 21(B). Although the carbohydrate region (δ3.8–δ6.6) is typical, the absorption for the exocyclic amino group is observed as a broad peak at δ8.2, which

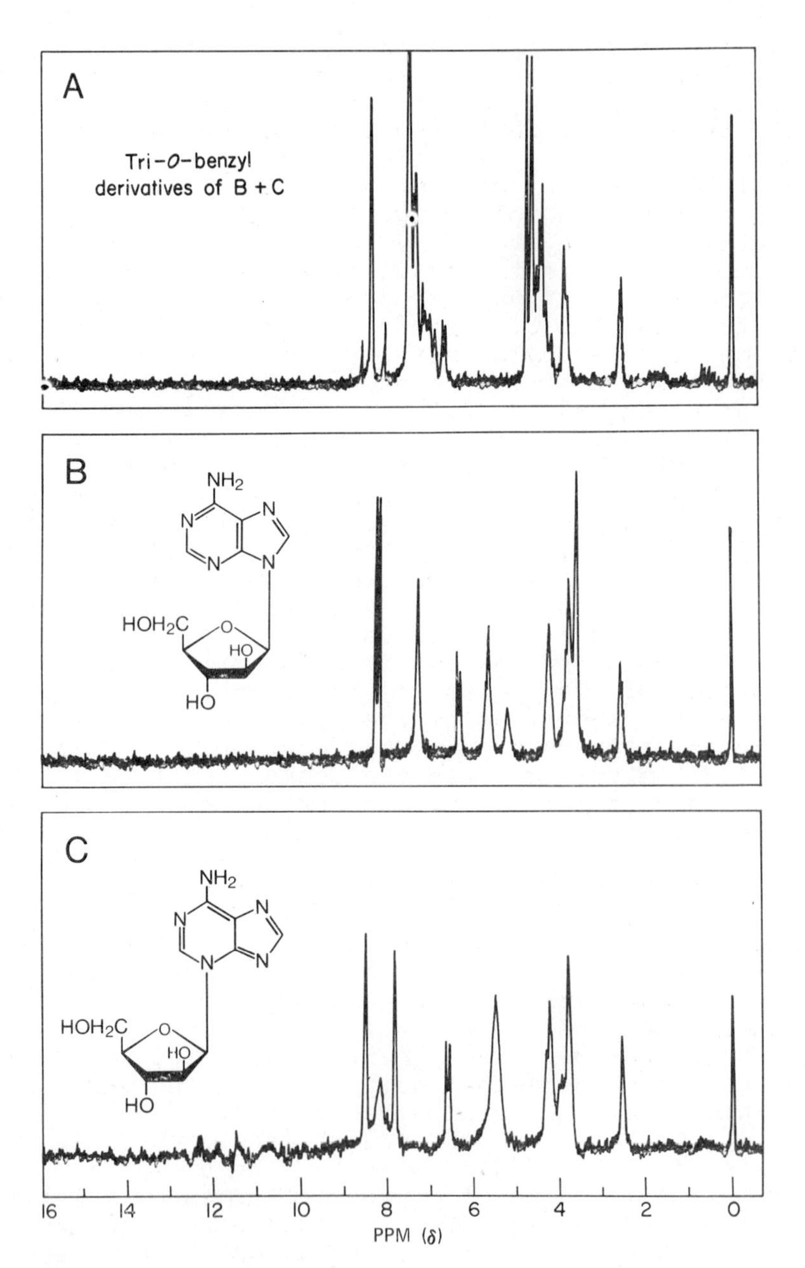

Fig. 21. P.m.r. spectra of (A) a mixture of the 2',3',5'-tri-O-benzyl derivatives of B and C, (B) 9-β-D-arabinofuranosyladenine, and (C) 3-β-D-arabinofuranosyladenine, in methyl sulfoxide-d_6.

constitutes a shift downfield of δ0.8. The protons on C-2 and C-8 are observed at δ7.93 and 8.60 (Δδ 4.0 Hz), indicating that the D-arabinofuranosyl group is on N-3 of the pyrimidine portion, and, on the basis of u.v. spectral data, this assignment has been confirmed. The foregoing study illustrates the use of p.m.r. spectroscopy in the detection of isomeric mixtures, a common problem encountered when direct glycosylation is used for the synthesis of nucleosides.

The preceding presentation, supported by the discussions on functional groups presented in the sections on imidazoles (see p. 288) and pyrimidines (see p. 301) provides a starting point for the interpretation of the p.m.r. spectra of purines and purine nucleosides, especially with regard to changes associated with the heterocyclic moiety. (The carbohydrate moiety is discussed in more detail in the next Section.) Other selected examples of p.m.r. spectra or p.m.r. spectral data for certain purines and purine nucleosides are as follows: methyl-, (methylthio)-, and (methylsulfonyl)-purines,[81] 2-chloro-9-[3-*C*-(hydroxymethyl)-β-D-*glycero*-tetrofuranosyl]adenine,[82] the effect of pH on the p.m.r. spectra of several nucleosides,[83] the effect of alkyl substituents on the stacking of purine derivatives in solution,[84] 2-methyladenosine,[85] dimethylalkyl derivatives of purine α-D-ribonucleosides,[86] adenine and adenosine, [87] 3-(3-deoxy-β-D-ribofuranosyl)-adenine,[87,88] 7*H*- and 9*H*-purine nucleosides,[89] purine 2',8-anhydronucleosides,[90] a study of the hydrogen bonding between inosine and other nucleosides,[91] purine nucleosides derived from 5-deoxy-D-*ribo*-hexose,[92] the adduct of guanosine with 2,3-butanedione,[93] *N*-("homoribose") (3-methyl-2-butenyl)adenosine,[94] 2'-*C*-methyladenosine,[95] various acetylnucleosides,[96] methoxymethylidene derivatives of nucleosides,[97] theophylline 3'-amino nucleosides,[98] 2-fluoro-6-methylpurine nucleosides,[99] 5',5'-di-*C*-methyladenosine and derivatives,[100] an analog of puromycin,[101] various purine nucleosides,[102] bis(D-glucopyranosyloxy)-purines,[103] mono-, di-, and tri-deoxyadenosines and certain unsaturated derivatives,[104] tri-substituted purine nucleosides,[105] 7-D-ribofuranosyl-7*H*-purines,[106-109] 9-(5-deoxy-β-D-allofuranosyl)adenine,[110] 2'- and 3'-(aminodeoxy)-D-ribofuranosylpurines,[111] 9-(4-thio-D-xylofuranosyl)adenine and 9-(4-thio-D-arabinofuranosyl)adenine,[112] studies on the conformation and interaction of certain nucleosides,[113] and several alkyl-carbonyl derivatives of mercaptopurines.[114]

4. Carbohydrates and the Carbohydrate Moiety of Nucleosides

In several of the preceding Sections, p.m.r. spectra of

the carbohydrate moiety of nucleosides have been touched on only briefly. This Section attempts to complete a fundamental introduction for the use of p.m.r. spectroscopy as it relates to the carbohydrate moiety of nucleosides and to carbohydrate derivatives used in the synthesis of nucleosides.

The most general type of synthesis of nucleosides involves the condensation of a heterocyclic compound with the appropriate carbohydrate derivative, and p.m.r. spectroscopy is a useful tool in the structural assignment of the products (heterocyclic, carbohydrate, or nucleosidic) isolated from this type of reaction. The p.m.r. spectra of various heterocyclic compounds, and their nucleosides have been discussed in previous Sections, and these types may readily be differentiated by the chemical shifts observed in the p.m.r. spectrum for the "aromatic" protons of the heterocyclic compounds, as compared with those for the "aromatic" protons of the heterocyclic moiety (*combined* with the pattern of peaks observed for the carbohydrate moiety, for the nucleosides). However, in addition to the foregoing two possibilities, some of the unreacted carbohydrate derivative used in the synthesis of the nucleoside may also be isolated from the reaction mixture. The p.m.r. spectrum of a few of the more common carbohydrates used in the synthesis of nucleosides will therefore be discussed; this should aid in their identification.

The p.m.r. spectrum of 1,2,3,5-tetra-O-acetyl-β-D-ribose [see Fig. 22(A)], in chloroform-*d*, has a strong peak at δ≅ 2.1, which may be assigned to the methyl protons of the four acetyl groups. The singlet at δ6.13 may be assigned to the anomeric proton, the remaining protons of the carbohydrate moiety being assigned to the peaks at δ4.0-5.5. The acetylation of adenosine can afford a derivative that has three O-acetyl groups (determined on the basis of acetic ester analysis), and these are assigned to the hydroxyl groups on C-2', C-3', and C-5'. Its p.m.r. spectrum [see Fig. 22 (B)], in methyl sulfoxide-d_6, corroborates this assignment. The peaks (9 protons) at δ2.0-2.2 may be assigned to the methyl protons of the three acetyl groups by means of the integration curve [by comparison of the peaks with those observed for the protons on C-2 and C-8 (δ8.7 and 8.53, respectively)]. That the exocyclic amino group on C-6 had not been acetylated is established conclusively by the presence of a broad singlet in the spectrum at δ7.6, which corresponds to two protons, assigned to the amino group. The remaining peaks are assignable to the protons of the carbohydrate moiety, with the exception of the small singlet at δ3.47, assigned to a trace of water

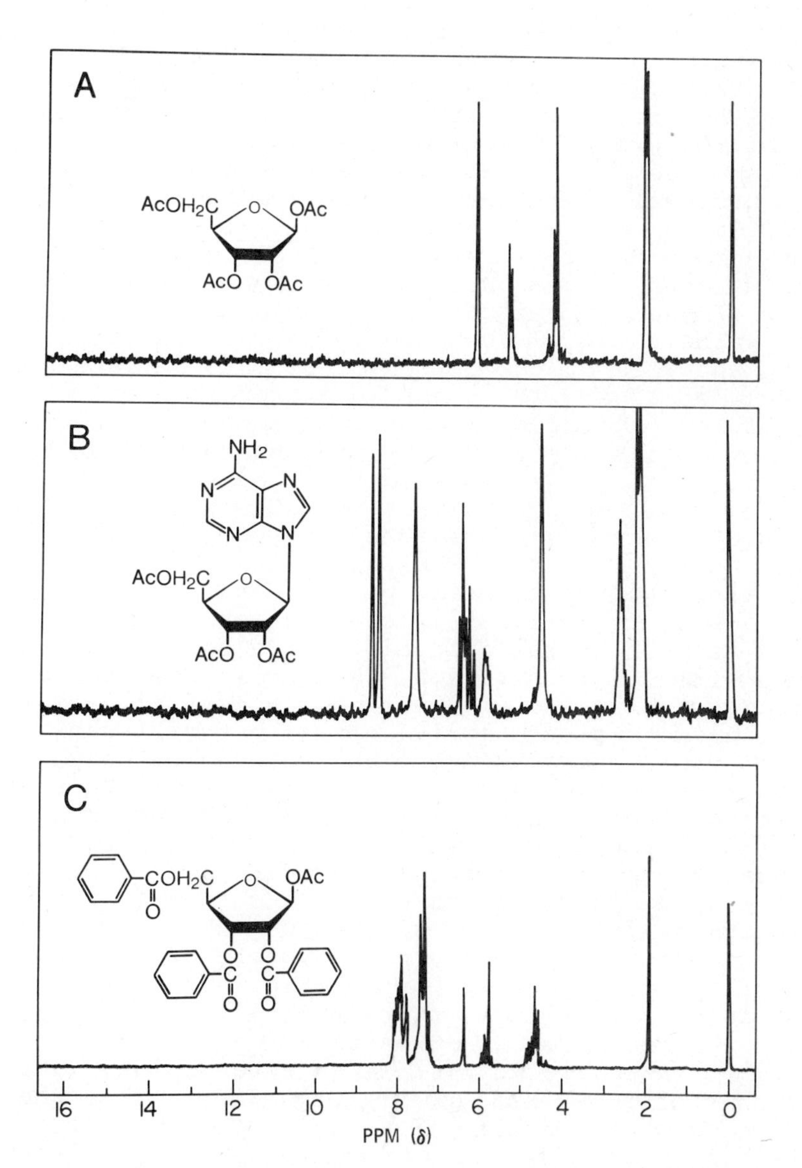

Fig. 22. P.m.r. spectra of (A) 1,2,3,5-tetra-O-acetyl-β-D-ribose in chloroform-*d*; 2',3',5'-tri-O-acetyl-adenosine, in methyl sulfoxide-d_6; (C) 1-O-acetyl-2,3,5-tri-O-benzoyl-D-ribose in chloroform-*d*.

associated either with the nucleoside or with the solvent (methyl sulfoxide-d_6).

The p.m.r. spectrum of the carbohydrate moiety of 2′,3′,5′-tri-O-acetyladenosine is typical of that obtained for the carbohydrate moiety of nucleosides prepared by the condensation of a heterocyclic compound with 1,2,3,5-tetra-O-acetyl-β-D-ribose or with 2,3,5-tri-O-acetyl-D-ribosyl chloride (or bromide). The results of an examination of the p.m.r. spectrum of 1,2,3,5-tetra-O-acetyl-β-D-ribose [see Fig. 22(A)] generally exclude the structure for a nucleoside, because peaks are not observed in the δ7-10 region for "aromatic" protons. However, several carbohydrate derivatives used in nucleoside syntheses contain aromatic groups for protection of the hydroxyl groups of the carbohydrate component, and will, therefore give peaks in the δ7-10 region. In this instance, the p.m.r. spectrum of the carbohydrate intermediate should be recorded as an aid in the interpretation of the spectrum of the product or mixture of products. The p.m.r. spectrum of 1-O-acetyl-2,3,5-tri-O-benzoyl-β-D-ribose [see Fig. 22 (C)] has an absorption peak at δ2.1 for the acetyl group, and two complex multiplets (15 protons), at δ7.2-8.2, which may be assigned to the "aromatic" protons of the benzoyl groups. The peak (singlet) at δ6.4 may be assigned to the anomeric proton, and the other peaks, at δ4.4-6.0, to the rest of the protons attached to carbon atoms. If the foregoing acylated D-ribofuranose is used in a D-ribosylation, the p.m.r. spectrum of the resulting nucleoside will not show a peak at δ2.1 for the methyl protons of the acetyl group, but will show the rest of the peaks just described for the carbohydrate derivative, in addition to the peaks assignable to the heterocyclic aglycon. For the protected nucleosides, only two significant changes might occur in the pattern observed for the carbohydrate moiety; these are a slight chemical shift for the anomeric proton and a change from a singlet to a doublet for the anomeric proton caused by a change in the conformation of the carbohydrate part. Another type of protecting group encountered is the benzyl group, which is useful in the synthesis of 1'2'-*cis* β-D-nucleosides. For example, tri-O-benzyl-1-O-*p*-(nitrobenzoyl)-β-D-arabinose (44) is an intermediate in the synthesis of 9-β-D-arabinofuranosyladenine (see the Section on purine nucleosides, p. 313), which gives a pattern of peaks for the carbohydrate moiety [see Fig. 21(A)] similar to that observed for 44 except that peaks for the protons of a *p*-nitrobenzoyl group are not present Instead, peaks are observed that are assignable to the adenine moiety, appearing as two singlets at

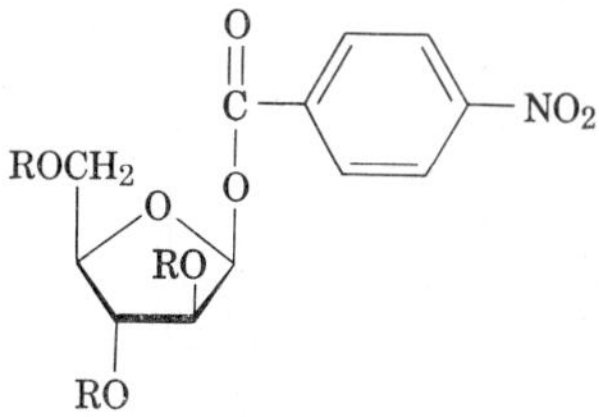

44

where R = $PhCH_2$

δ8.47 and 8.43 for the protons at C-2 and C-8, respectively. The peak for the exocyclic amino group is overlapped by the peaks for the "aromatic" protons of the benzyl groups (δ6.7-7.7 region).

Another common protecting group is the isopropylidene group. The p.m.r. spectrum of 2',3'-O-isopropylideneadenosine in methyl sulfoxide-d_6 [see Fig. 22(A)] is similar to that of adenosine. The significant difference is the presence of two singlets (3 protons each) at δ1.43 and 1.66 for the methyl groups of the isopropylidene group, and the absence of two protons (determined by integration) from the δ4-6 region. The different chemical shifts observed for the methyl groups have been rationalized on the basis of steric environment. The individual methyl groups have been designated α, β, or γ, depending on their environment; therefore, if R = R' = H, the methyl group on the same side of the ring is α. If R = H and R' = alkyl, the methyl group is β, and if R = R' = alkyl, the methyl is γ. An α-methyl group is found to be more shielded (smaller δ value) than a β-methyl group, which, in turn is more shielded than a γ-methyl group. The methyl group of 2',3'-O-isopropylideneadenosine that is on the same side of the 1,3-dioxolane ring as the protons on C-2' is designated an α-methyl group, and may be assigned to the peak at δ1.43. The other methyl group is designated γ, because it has a

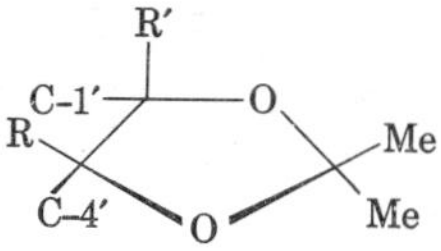

44a

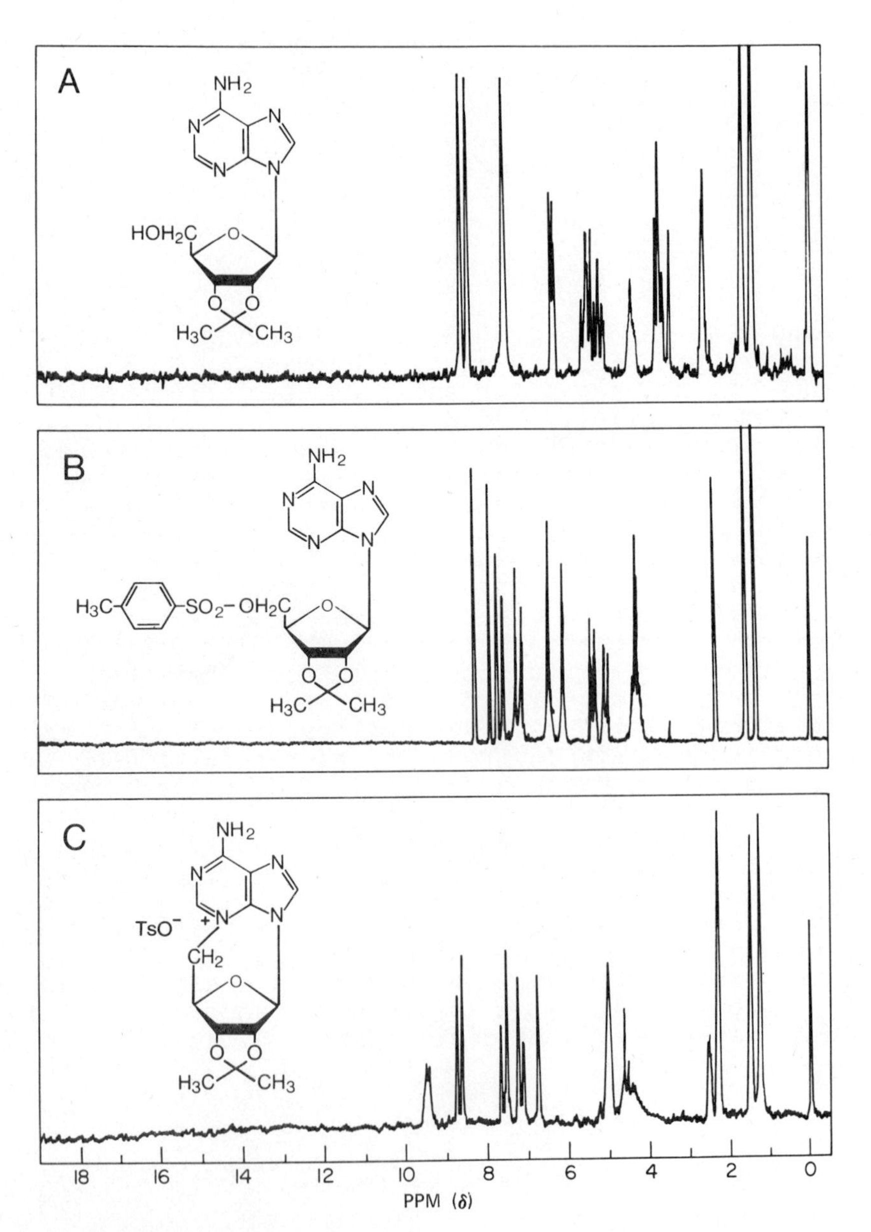

Fig. 23. P.m.r. spectra of (A) 2',3'-O-isopropylidene-adenosine in methyl sulfoxide-d_6; (B) 2',3'-O-isopropylidene-5'-O-*p*-tolylsulfonyladenosine, in chloroform-*d*; (C) 2',3'-O-isopropylideneadenosine *p*-toluenesulfonate in methyl sulfoxide-d_6.

cis relationship to C-1' and C-4', and may, therefore, be assigned to the downfield peak at δ1.66. An extensive discussion of this rule, as applied to a number of other isopropylidene acetals, has been reported.[115] *p*-Toluenesulfonylation of the D-ribofuranosyl group of 2',3'-O-isopropylideneadenosine obviously occurs at O-5'. The p.m.r. spectrum of 2',3'-O-isopropylidene-5'-O-*p*-tolylsulfonyladenosine, in chloroform-*d* [see Fig. 23(B)], is very similar to that of 2',3'-O-isopropylideneadenosine, except for the presence of a singlet (3 protons), at δ2.37, for the methyl protons of the *p*-tolylsulfonyl group, and a pair of doublets centered at δ7.18 and 7.65 for the "aromatic" protons, also of the *p*-tolylsulfonyl group, with the absence of a peak (determined by integration) for the proton of the 5'-hydroxyl group. The peaks for the methyl protons of the isopropylidene group are essentially unchanged; the two singlets, at δ7.9 and 8.25, may be assigned to the protons on C-2 and C-8, the broad singlet (2 protons) at δ6.47 to the exocyclic amino group (on C-6), and the doublet ($J_{1',2'}$ 2 Hz) centered at δ6.1, to the anomeric proton. The small peak at δ7.32, which appears to be caused by an impurity, may be assigned to the proton from a trace of chloroform present in the solvent. However, a p.m.r. spectrum of the nucleoside in methyl sulfoxide-d_6 [see Fig. 23(C)], shows some very significant changes which indicate that anhydronucleoside formation has occurred, owing to the increased polarity of the solvent. The peaks for the methyl protons (δ1.27, 1.5, and 2.33) of the isopropylidene and the *p*-tolylsulfonyl groups, and the two doublets centered at δ7.2 and 7.6, assigned to the "aromatic" protons of the *p*-tolylsulfonyl group, remain essentially unchanged. However, the peaks assigned to the protons on C-2 and C-8 are now observed at δ8.66 and 8.78, a significant chemical shift (downfield). Also observed is a broad doublet centered at δ9.48, which is assigned to the exocyclic amino group on C-6, and which represents

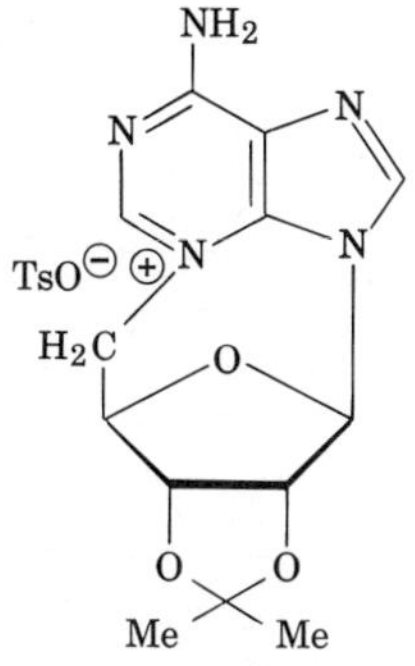

45 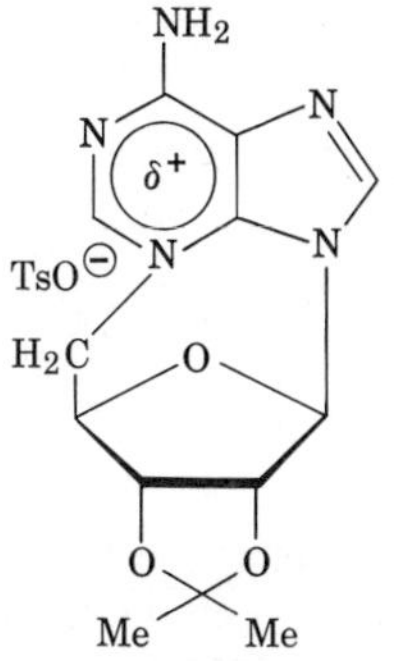 46 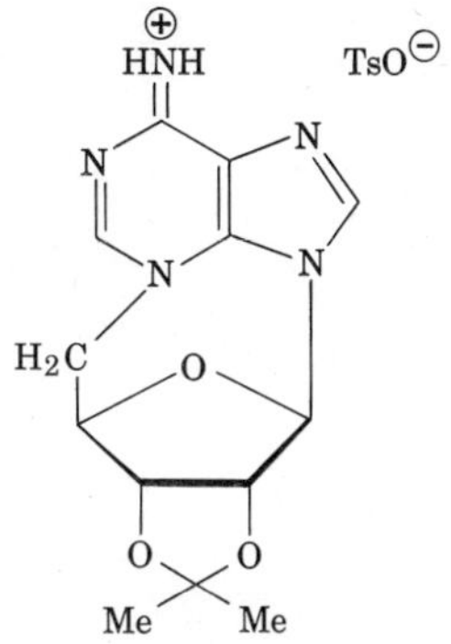 47

a downfield chemical-shift of ≅120 Hz. This may be explained by anhydronucleoside formation, a new bond being formed between C-5' and N-3. As previously mentioned, a downfield shift in the peaks assigned to the protons on C-2 and C-8 (δ8.66 and 8.78) is also observed. The structure of the anhydronucleoside *p*-tolueneusulfonate may be visualized as existing in several different forms (45, 46, 47); yet, on the basis of its p.m.r. spectrum, in methyl sulfoxide-d_6, it is logical to suppose that 47 is the preponderant species, as it is the only structure shown that would explain a larger deshielding effect (downfield chemical-shift) observed for the exocyclic amino group than for the proton on C-2. The structure also accounts for the broad doublet observed at δ9.48 for the exocyclic amino group, presumably because of restricted rotation about the C—N bond.

P.m.r. spectroscopy is an invaluable tool in the determination of anomeric configuration (the spatial relationship of the groups attached to C-1 of the sugar residue, designated either α-(D or L) or β-(D or L). The determination of anomeric configuration is a problem of considerable importance in the field of nucleoside synthesis, because, in the direct glycosylation of a heterocyclic compound, the simultaneous formation of both anomers is often a possibility. It has been determined[116,117] that, for furanoid derivatives (five-membered ring forms) of sugars, the dihedral angle between neighboring, or vicinal, *cis*-hydrogen atoms (as with the α-D-ribosyl derivative 48) and neighboring *trans*-hydrogen atoms (as with the β-D-ribosyl derivative 49), may vary from 0-45° and 75—165°, respectively. The Karplus equation,[118] or a modification thereof, makes possible the prediction that the observed coupling-constants ($J_{1',2'}$) will be in the approximate region of 3.5—8.0 Hz for the α-D, and 0.0—8.0 Hz for the β-D, configuration. Consequently, an assignment of anomeric configuration on the basis of the foregoing data, excluding certain conforma-

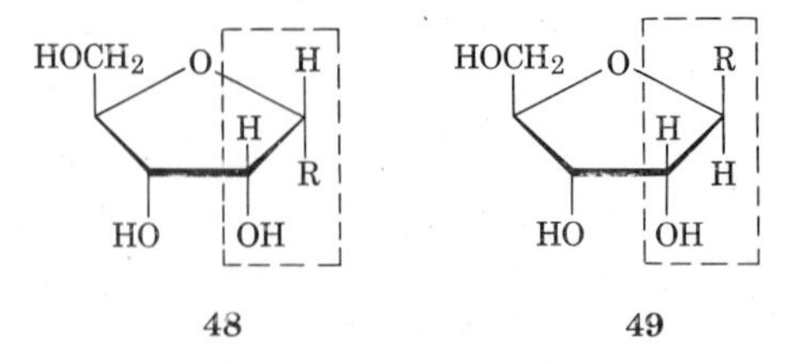

tional changes, may be made only for the β-D anomer (the *trans* arrangement for the protons on C-1' and C-2'), and then only if the coupling constant is found to be less than 3.5 Hz. However, subsequent studies[119,120] established that this assignment should be applied for neighboring, *trans* hydrogen atoms, only when the coupling constant is ≤1.0 Hz. A good example of how the coupling constant for the anomeric proton ($J_{1,2}$) may be modified to conform to the foregoing requirements has previously been given in this Section. The coupling constant ($J_{1,2}$) observed for adenosine is ~6 Hz, and formation of its 2',3'-O-isopropylidene derivative diminishes this coupling constant to ~2.5 Hz, which is less than 3.5 Hz and indicative, therefore, of the β-D configuration. However, because the value is greater than 1.0 Hz, the assignment of the configuration as β-D is not unequivocal. *p*-Toluenesulfonylation of the isopropylidene acetal [see Fig. 23(B)] has little or no effect on the coupling constant observed for the anomeric proton. However, formation of the N-3,C-5'-anhydronucleoside *p*-toluenesulfonate has a significant effect on the coupling constant, because a singlet (δ6.8) [see Fig. 23(C)] for the anomeric proton is observed. Anhydronucleoside formation (there are various routes for anhydronucleoside formation with other ring systems) has been used extensively for the assignment of anomeric configuration to certain nucleosides, owing to the fact that it lowers the coupling constant to within acceptable limits, as well as establishing certain spatial relationships.

It has also been demonstrated that the coupling constant observed for the anomeric proton ($J_{1,2}$) may be solvent-dependent. A decrease in the coupling constant observed may, in certain instances, be effected by the addition of another solvent (1 or 2 drops) to the sample, followed by the recording of a new spectrum. Obviously, a number of different solvents may be used, either alone or in a mixture, in an effort to lower the coupling constant to within acceptable limits.

P.m.r. spectroscopy is likewise invaluable in the assignment of the anomeric configuration to nucleosides having the carbohydrate moiety in its pyranoid form.[121–123] The pyranoid derivatives might exist in one of several conformations, two of which (*C1* and *1C*) are depicted in Scheme 3. It has been established[121] that β-D-xylofuranosyl nucleosides exist mainly in the *C1*(D) conformation, on the basis of the large coupling-constant observed for the anomeric proton. A visual examination of a molecular model of the *C1*(D) conformation of a β-D-xylofuranosyl nucleoside reveals a *trans*-diaxial arrangement for the protons on C-1' and C-2' (dihedral angle, 180°), and, from the Karplus equation, a large coupling-constant for the anomeric proton would be predicted. In the *1C*(D) conformation of the same compound, the protons at C-1' and

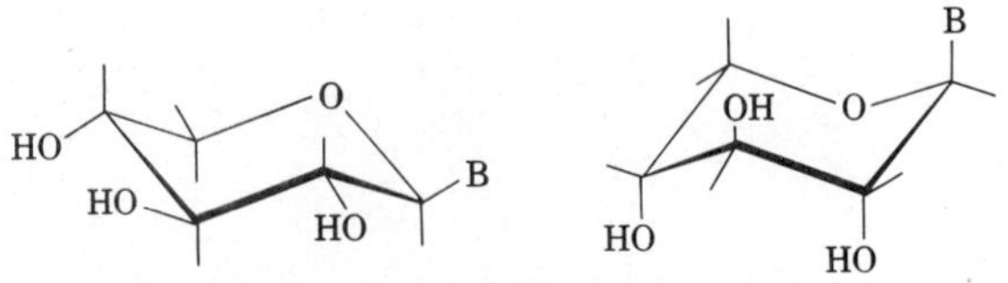

C1(D) *1C*(D)

β-D-Xylopyranosyl

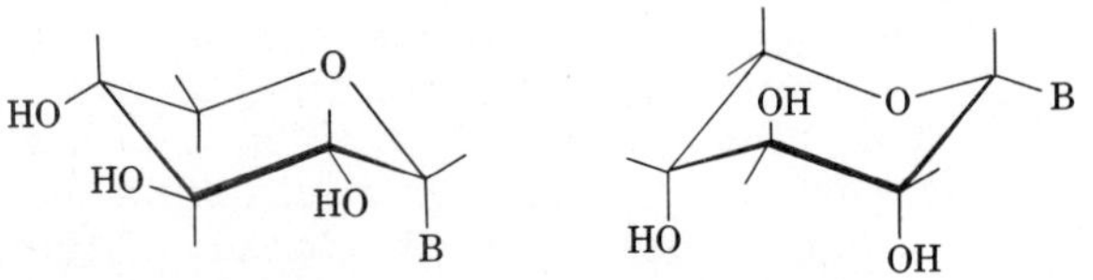

C1(D) *1C*(D)

α-D-Xylopyranosyl

Scheme 3

C-2' have the equatorial–equatorial (angle, 60° dihedral) arrangement, and so should exhibit a smaller coupling-constant. It is also evident, as shown in Scheme 3, that the *1C*(D) and *C1*(D) conformations of the α-D-xylopyranosyl nucleosides have the protons on C-1' and C-2' in the equatorial–axial (dihedral angle, 60°) arrangement, which also precludes a large coupling-constant for the anomeric proton. However, some inherent difficulties are associated with the assignment of anomeric configuration to pyranosyl compounds by p.m.r. spectroscopy, because, sometimes, a conformational equilibrium may exist between the two ideal chair forms, *C1*(D) and *1C*(D), and this would result in a spectrum that would reflect neither form. This complication could result in a wrong conclusion, not only as regards the conformational assignment, but also as regards the assignment of the anomeric configuration; this is especially true of deoxypyranosyl derivatives, for which a number of the stability factors[124] that tend to result in an equilibrium containing mainly one chair conformer are absent. A study of the conformation of both anomers of certain 2-deoxy-D-*erythro*-pentopyranosyl-9*H*-purines has established that both exist in a flexible conformation, not a chair form. However, it appears that the purine moiety is sufficiently bulky to restrict any large conformational changes at the anomeric and adjacent carbon atoms, and so it assumes the equatorial position. The conformations of the pyranosypurines were eventually established by double-resonance studies.

Another point of interest in the same study[125] involves the use of p.m.r. spectroscopy to estimate the proportions of the anomers of a nucleoside in a mixture of them. This procedure is based on the pattern of peaks observed for the anomeric proton, because previous studies had shown that the anomers of certain nucleosides exhibit significant differences in the chemical shift for the anomeric proton. Although a difference in chemical shift for the anomeric proton of certain 2'-deoxynucleosides has been observed (*e.g.*, in the foregoing example), this difference is not so definitive as that for anomers of a nucleoside having a hydroxyl group on C-2'. The technique is, therefore, not recommended for making a definitive assignment of anomeric configuration of a 2'-deoxynucleoside.

As indicated previously, it has been found[79,126–130] that the peak assigned to the anomeric proton of a C-1'–C-2'-*trans*-nucleoside (that is, a β-D-ribofuranosyl nucleoside) appears at higher field (usually at δ~0.5) than the peak observed for the anomeric proton of the corresponding α-D anomer (C-1'–C-2'-*cis*-nucleoside). This analysis usually requires a comparison between the p.m.r. spectra of the anomers before a definite anomeric assignment may be made, because the chemical shift of a proton may be affected by a number of factors. Obviously, a close visual inspection of the p.m.r. spectrum of the nucleoside under consideration is also necessary, because the orientation of the hydroxyl group at C-2' determines whether the α or β anomer is the *cis*- or *trans*-nucleoside; for example, the β anomer of 9-D-ribofuranosyladenine is the *trans*-nucleoside, whereas the α anomer is the *cis*-nucleoside. Conversely, for 9-D-arabinofuranosyladenine, the β anomer is the *cis*-nucleoside and the α anomer is the *trans*-nucleoside. This difference in the chemical shift observed for the anomeric proton of *cis*- and *trans*-nucleosides may be clearly seen by inspection of the p.m.r. spectra of the four β-D-aldopentofuranosyladenines; namely, 9-β-D-ribofuranosyladenine (Fig. 27, p.356), 9-β-D-arabinofuranosyladenine [Fig. 24(A)], 9-β-D-xylofuranosyladenine [Fig. 24(B)], and 9-β-D-lyxofuranosyladenine [Fig. 24(C)]. It is evident that the peaks assigned to the anomeric proton of the *cis*-nucleosides (9-β-D-arabinofuranosyladenine and 9-β-D-lyxofuranosyladenine) occur downfield (at δ~0.5) from the peaks assigned to the anomeric proton of the *trans*-nucleosides, namely, 9-β-D-ribofuranosyladenine and 9-β-D-xylofuranosyladenine. The α-D anomers of these four nucleosides exhibit the opposite effect, owing to the change in juxtaposition between the heterocyclic

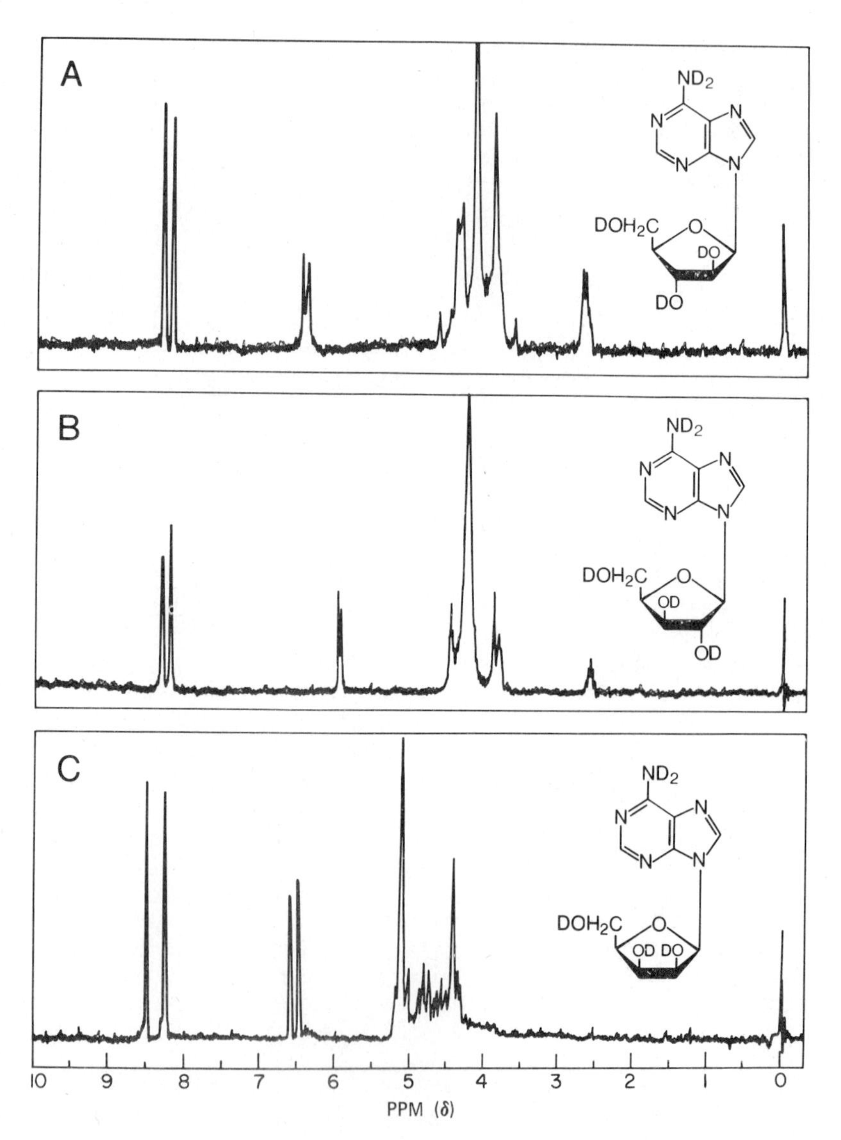

Fig. 24. P.m.r. spectra of (A) 9-β-D-arabinofuranosyladenine and (B) 9-β-D-xylofuranosyladenine, in methyl sulfoxide-d_6-D_2O; (C) 9-β-D-lyxofuranosyladenine, in D_2O. (The original spectra were furnished by Dr. Leon Goodman, Department of Chemistry, University of Rhode Island, Kingston, R.I.)

aglycon and the hydroxyl group on C-2'.

Another method employed for the determination of anomeric configuration (for certain acetylated pyrimidine nucleosides) involves hydrogenation of the double bond between C-5 and C-6 of the pyrimidine moiety, followed by observation of the chemical shift for the acetoxyl group on C-2'. This method has shown some promise, especially with pyranosyl nucleosides. It had been established, and later

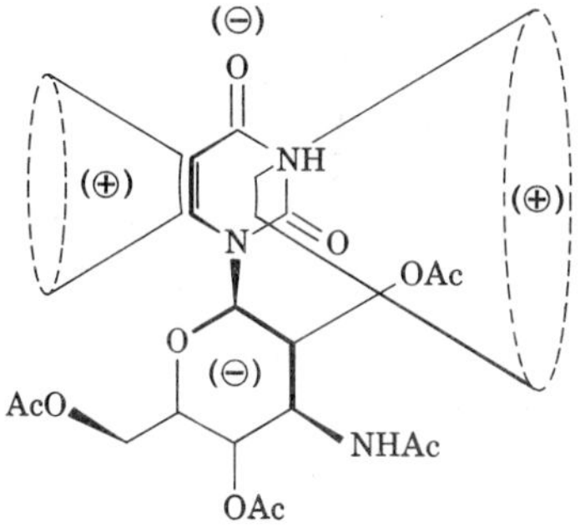

50 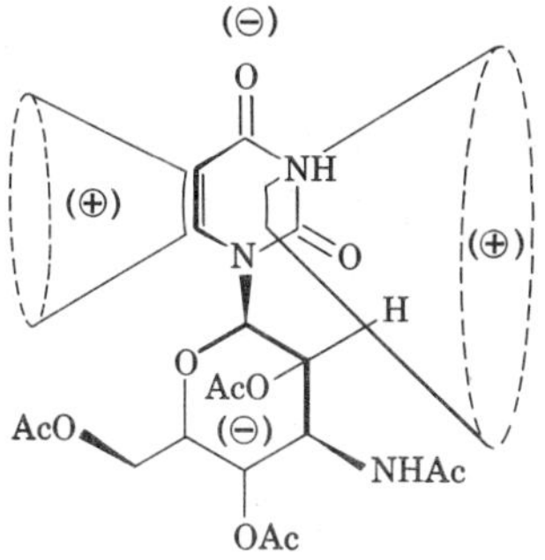 51

confirmed,[131-133] that, for a six-membered ring, such as that of a pyranoid sugar, the absorption peak for an axial acetoxyl group at C-2' occurs at a lower field (larger δ value) than when the same group at C-2' is equatorially attached. Several exceptions[134] to this empirical rule have been found, and it is, therefore, of little use, *per se*, in the determination of the anomeric configuration; however, the chemical shifts observed for acetoxyl groups have been used[135] for the determination of the anomeric configuration of several acetylated aldopyranosyl pyrimidine nucleosides. Visual examination of the p.m.r. spectra of these nucleosides reveals that, for nucleosides having a *trans* relationship at C-1' and C-2', such as D-glucopyranosyl nucleosides, the acetoxyl group at C-2' is in an area of positive shielding (50) whereas, if it has the *cis* relationship (51), the acetoxyl group is no longer shielded. Hydrogenation of the double bond between C-5 and C-6 of the uracil moiety removes this shielding effect ("cone"), and results in a downfield shift for the absorption peak for the acetoxyl group on C-2' in the *trans* relationship. If the acetoxyl group at C-2' has the *cis* relationship, the disappearance of the anisotropic effect (hydrogenation of the double bond at C-5 and C-6) results in a general upfield shift for the group. The same method has been investigated for the assignment of the anomeric configuration of some aldopentofuranosyl pyrimidines, and it appears to have some merit. It is found, for example, that, with acetylated aldopentofuranosyl pyrimidine nucleosides in which the acetoxyl group at C-2' has the *cis* arrangement, hydrogenation of the double bond at C-5 and C-6 results in a small downfield shift, whereas, for the alternative situation, in which the acetoxyl group has the *trans* arrangement, either no shift, or a very small upfield shift, is observed.

The same procedure is useful for the assignment of the anomeric configuration to several nucleosides of indole and indoline, which may be considered as being "1,3,7-trideazapurine" nucleosides. On this basis, they were used as model compounds for an investigation[136] on the possible use of the foregoing method for the assignment of the anomeric configuration of purine nucleosides. From the p.m.r. spectra of 52 and 53, it was established that the rules

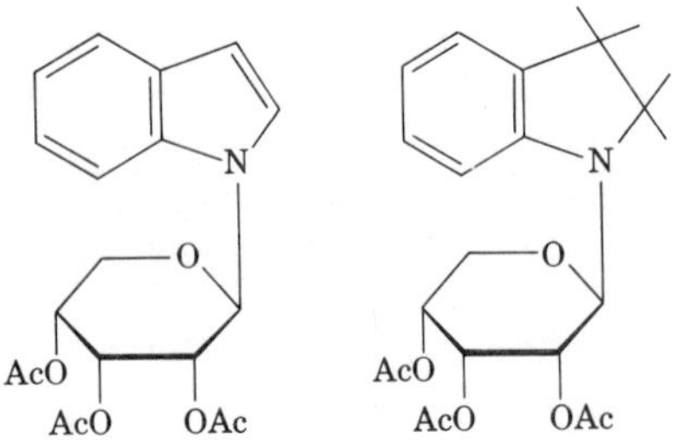

AcO AcO OAc AcO AcO OAc

52 53

previously formulated for the pyrimidine nucleosides were also applicable to a determination of the anomeric configuration of acetylated aldopyranosyl- and acetylated aldofuranosyl-indoles. However, in order to apply the procedure in determination of the anomeric configuration of purine nucleosides, a facile method must be available for the preparation of the 7,8-dihydropurine nucleosides needed, either by total synthesis, or, preferably, by a general procedure for the reduction of a preformed purine nucleoside.

As discussed briefly in the Section on imidazole nucleosides, p.m.r. spectroscopy has been extensively used in the assignment of the anomeric configuration of 2'-deoxyaldofuranosyl nucleosides. The procedure used is based not only on the difference in the pattern of peaks observed for the anomeric proton but also on the peak-width (width of the pattern, in Hz). It is observed that the α-D anomer displays [see Fig. 14(B), p. 300] a multiplet of four (J 3.1 $\pm$0.4 and 7.2 $\pm$0.3 Hz) having a peak-width of 10.4 $\pm$0.4 Hz for the anomeric proton, whereas the β-D anomer displays [see Fig. 14(A)] a pseudotriplet (J 6.8 $\pm$0.3 Hz), having a peak-width of 13.7 $\pm$0.5 Hz. This difference was observed[26,28,137] for several anomeric pairs of purine and pyrimidine nucleosides containing the 2-deoxy-D-*erythro*-pentofuranosyl group. This method has been shown to have utility for the assignment of the anomeric configuration to several mixtures of anomers of 2-deoxy-D-*erythro*-pentofuranosyl nucleosides secured by the direct glycosylation of a preformed heterocyclic compound, and it appears, therefore, to be generally applicable.

A subsequent investigation[138] revealed that the pattern of peaks for the anomeric proton may be altered by a change in the substituents on the heterocyclic moiety. The p.m.r. spectrum of 6,8-diamino-9-(2-deoxy-β-D-*erythro*-pentofuranosyl)purine (54) has a quartet (J 6.6 and 8.6 Hz),

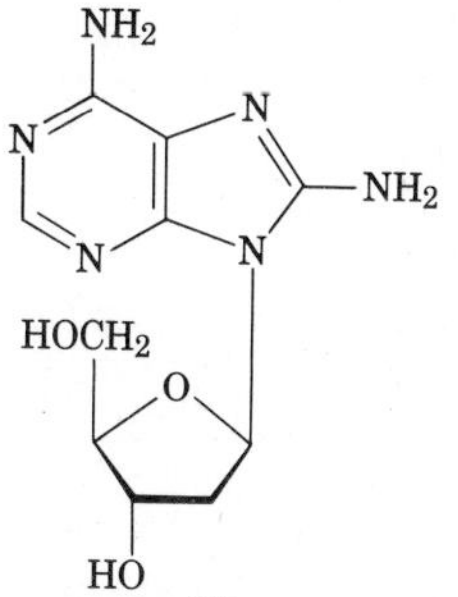

NH_2 N N NH_2 N N $HOCH_2$ O HO

54

instead of the pseudotriplet expected for the anomeric proton. This result prompted the synthesis of several other 8-substituted derivatives of 2'-deoxyadenosine, namely, the 8-(methylamino)-, 8-(ethylamino), and 8-(dimethylamino) derivatives. For each compound a quartet having a peak width of 15.15 $\pm$0.15 Hz for the anomeric proton was observed. On the basis of the p.m.r. spectrum obtained for the dimethylamino derivative, hydrogen bonding between a proton on the exocyclic nitrogen atom on C-8 and the carbohydrate moiety was eliminated as a possible cause for a conformational change in the carbohydrate ring. Hydrogen bonding between the anomeric proton and the amino group at C-8 was tentatively eliminated, owing to the fact that no significant chemical shift for the anomeric proton, relative to the peaks observed for the anomeric proton of 2'-deoxyadenosine, was observed.

The occurrence of a quartet instead of a pseudotriplet for the anomeric proton of certain 2′-deoxy-D-*erythro*-pentofuranosyl pyrimidines has also been reported.[139] For the purine series, it has been suggested[138] that this change in splitting pattern and peak-width may be caused by steric hindrance imposed by the introduction of a large group onto C-8 of the purine ring. This hindrance could, possibly, change the conformation of the nucleoside (from the presumed *anti* to *syn*, or, possibly, an intermediate conformer). However, because the changes noted in the foregoing are very small, it appears that the anomeric configuration of 2′-deoxy-D-*erythro*-pentofuranosyl nucleosides may still be assigned on the basis of a close examination of the splitting pattern for the anomeric proton, because a significant difference in the quartet observed for the α-D anomer and that observed for the β-D anomer is manifest; restrictions on the conformation of the furanoid ring must also be taken into account.

The peak-width of the splitting pattern for the anomeric protons appears to remain within the experimental limits, and may, presumably, still be used to ascertain the anomeric configuration, although this procedure may prove to be invalid, once the α-D anomers of the aforementioned 8-substituted derivatives of 2'-deoxyadenosine are prepared, as suggested,[138] for additional p.m.r. studies. In fact, there is some preliminary evidence[140] that the empirical rule concerning the peak-width displayed by an anomeric proton may be invalid in a related series of 2′-deoxy-D-*erythro*-pentofuranosyl nucleosides. Nevertheless, the foregoing empirical rule may be used judiciously for the assignment of anomeric configuration, particularly if both anomers are available for comparison and their p.m.r. spectra exhibit the classical quartet and pseudotriplet for the anomeric proton.

It has been found that the aglycon, also, may have a significant effect, not only on the pattern of peaks and peak-width of the peaks for the anomeric proton, but also on the observed chemical shift. In a study[141] of the effect that groups (other than amino or alkylamino) on C-8 of 2'-deoxyadenosine would have on the splitting pattern and peak-width for the anomeric proton, a significant downfield shift was observed for the peaks (quartet, centered at δ7.0) assigned to the anomeric proton of 6-amino-9-(2-deoxy-β-D-*erythro*-pentofuranosyl)purine-8-thione (2'-deoxy-8-mercaptoadenosine) (55). Several additional 8-substituted derivatives of 2'-deoxyadenosine have been prepared by nucleophilic displacement of the bromine atom from 8-bromo-2'-deoxyadenosine, and the chemical-shift data obtained for the anomeric proton of all of the nucleosides compared favorably with the chemical shift for the

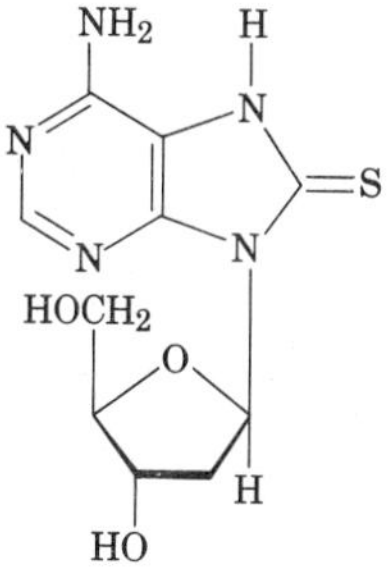

55

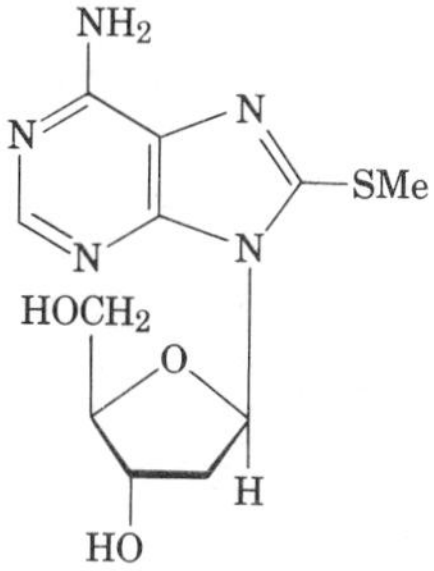

56

anomeric proton of 2'-deoxyadenosine. Methylation of 55 afforded 2'-deoxy-8-(methylthio)adenosine (56), which also gave a quartet centered at δ6.25 (the region for the anomeric proton of 2'-deoxyadenosine). This effect (downfield chemical-shift of δ0.5-0.7) for the anomeric proton has also been observed for several other nucleosides containing the D-ribofuranosyl-N-C=S arrangement, and it appears to be caused by the proximal effect of the C=S group, because other groups (including C=O) in the same juxtaposition have shown essentially no effect on the chemical shift observed for the anomeric proton. On the basis of preliminary results, it appears likely that this effect may prove to be a powerful diagnostic tool for studies in structural elucidation.

The chemical-shift data observed for the peak assigned to the anomeric proton have been used[141] in determining whether the 2'- or 3'-hydroxyl group of certain D-ribofuranosyl nucleosides is acylated. On the basis of p.m.r. spectral data obtained for several 2'- and 3'-O-acylnucleosides, the configurations of which had been previously determined by chemical methods, two rules for distinguishing between the isomeric 2'-O-acyl-D-ribonucleosides (57) and 3'-O-acyl-D-ribonucleosides (58) were

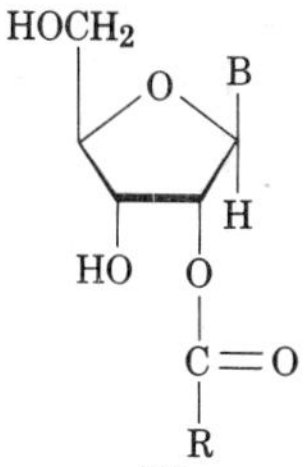

57

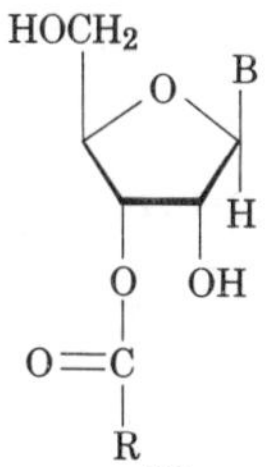

58

formulated:[142] *(1)* for a pair of 2'- and 3' -isomers, the peak assigned to the anomeric proton will be observed at lower field for the 2'-isomer (57) than for the 3'-isomer (58); and (2) the coupling constant ($J_{1,2}$) is larger for the 3'-isomer (58) than for the 2'-isomer (57). Subsequent investigations have verified the validity of these rules, and it has now been shown[143] that the peaks assigned to the protons at C-4' are more deshielded in the spectrum of the 3'-isomer (58) than in that of the 2'-isomer (57), an expected consequence of the proximal effect of the 3'-O-acyl group.

A number of interesting areas of carbohydrate and nucleoside chemistry in which p.m.r. spectroscopy has been applied with success are not discussed in this Section. Some examples of such p.m.r. spectra, p.m.r. spectral data, or p.m.r. spectral studies are as follows: an extensive study of certain derivatives of D-arabinose, D-lyxose, D-ribose, and D-xylose;[144] 3-*C*-methyl-D-ribofuranose and derivatives, including 3'-*C*-methyladenosine;[145] 2-*C*-methyl-D-ribofuranose and derivatives, including 2'-*C*-methyladenosine;[146] 2-deoxy-2-*C*-(hydroxymethyl)-D-erythrofuranose, and adenosine analogs prepared therefrom;[147] 9-[3-deoxy-3-*C*-(hydroxymethyl)-α(and β)-D-allofuranosyl- and -D-ribofuranosyl] adenine ;[148] a p.m.r. study of aqueous solutions;[149] 2',3'-O-(dimethoxymethylidene)-D-ribonucleoside derivatives;[150] 2,5-di-O-benzoyl-3-deoxy-β-D-*erythro*-pentose;[151] some acetylated monosaccharides;[152] conformational equilibrium of 1,2,3,5-tetra-O-acetyl-β-D-ribose;[153] the conformation of 2-deoxy-D-*erythro*-pentose;[154] the conformation of D-ribose;[155] 2,3-anhydro-D-glycopyranosides;[156] partially acylated β-D-glucopyranosides;[157] a nucleoside derived from 4-O-acetyl-2,6-dideoxy-3-*C*-methyl-L-*xylo*-hexose;[158] 2,3:5,6-di-O-isopropylidene-α-D-mannofuranosyl nucleosides and derivatives;[159] p.m.r. spectra of 5-deoxy-1,2-O-isopropylidene-3-O-methyl-α-D-xylo-*hex*-4-enofuranose;[1 acetylated 3-deoxy-3-*C*-(hydroxymethyl)-α-D-glucose and derivatives;[161] several derivatives of 3-deoxy-D-*erythro*-pentofuranose;[162] the addition product of α-toluenethiol and 1,5-anhydro-2-deoxy-D-*arabino*-hex-1-enitol (D-glucal);[163] β-D-ribofuranosyl β-D-ribofuranoside;[164] nucleosides containing 2-deoxy-D-*arabino*-hexopyranuronic acid;[165] 2-acetamido-2-deoxy-β-D-glucopyranosides;[166] acetyl- and formyl-substituted nucleosides;[167] 4-O-(2-acetamido-2-deoxy-β-D-galactopyranosyl)-D-galactopyranose;[168] p.m.r. studies on pyranosides;[169] the conformation of methyl 5-thio-α- and β-D-xylopyranosides;[170] a study of the proton of the hydroxyl group in relation to several factors;[171] a study of the chemical shifts observed for specific protons of certain carbohydrates;[172] and a study on the favored conformation of 1,2: 3,4-di-O- isopropylidene-β-D-arabinopyranose.[173]

5. Nucleoside Antibiotics and Related Derivatives

A number of minor nucleosides have been isolated from *s*-RNA; they have been found[174-178] to be very similar in structure to the major nucleosides occurring in nucleic acids. P.m.r. spectroscopy has been most useful in the structural elucidation of these minor nucleosides, and of various nucleoside antibiotics and other antibiotics structurally related to the major nucleosides and their aglycon moieties. The isolation of *N*-isopentenyladenosine from *s*-RNA was followed by its structural elucidation, in which p.m.r. spectroscopy played an important role. A number of methylated nucleosides have also been isolated, and p.m.r. spectroscopy was applied to certain of them to help elucidate the actual site of methylation; this use of p.m.r. spectroscopy has been discussed in previous Sections (see pp. **318** and **321**).

The isolation of pseudouridine (59) from various nucleic acids was followed by structural elucidation without the use of p.m.r. spectroscopy. P.m.r. spectroscopy would have been a valuable tool in these studies, because examination of the p.m.r. spectra obtained in subsequent studies[179-184] of 59 has revealed the absence of a peak for the proton on C-5 and the appearance of a peak downfield, assigned to the two N-H protons. As the empirical formula of 59 was determined to be the same as that of uridine, these findings would have immediately suggested that the D-ribosyl group is attached to C-5. Another significant feature of this p.m.r. spectrum was the upfield chemical-shift observed for the peak assigned to the anomeric proton; this peak has now been observed for several other *C*-glycosyl nucleosides (formycin, formycin B, showdomycin, and pyrazomycin).

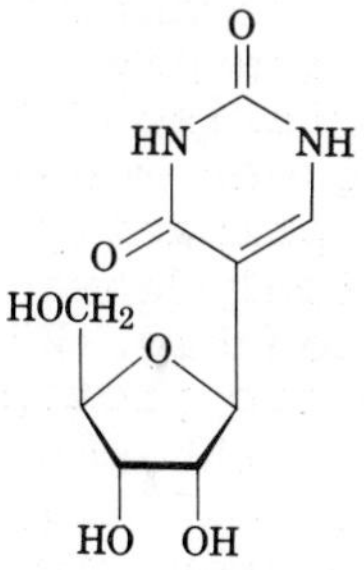

A p.m.r. study[185] of the conformation of pseudouridine (59) by use of a 100-MHz spectrophotometer and double irradiation has furnished a complete assignment of rotame: population about the C-4'—C-5' bond, the conformation of the D-ribosyl group, and the torsion angle between it and the heterocyclic aglycon.

Isolation of the nucleoside antibiotics formycin and formycin B was followed by their structural elucidation a *C*-glycosyl nucleosides, namely, 7-amino-3-β-D-ribofurano-sylpyrazolo[4,3-*d*]pyrimidine and 3-β-D-ribofuranosylpyra-zolo[4,3-*d*]-7-pyrimidinone,[186-190] respectively. P.m.r. spectroscopy was used extensively in the structural as-signments of these nucleoside antibiotics, and it has bee established[191] that mass spectroscopy would also have bee a valuable tool in the original structural-elucidation studies. The p.m.r. spectrum [see Fig. 25(A)] of formyci (which has the same empirical formula as adenosine) has a peak for only one CH aromatic proton, at δ8.17, and an upfield chemical-shift for the anomeric proton (δ5.0). These p.m.r. data, together with the increase in acid stability observed for the glycosyl bond, established tha formycin is a *C*-glycosyl nucleoside. The broad peak at δ7.45 can be assigned to the exocyclic amino group at C-7, and the low, broad peak at δ12—13, to the NH proton in the pyrazole moiety. The remaining peaks in the p.m.r spectrum (δ3.2—5.0) may be assigned to the other protons of the carbohydrate moiety.

The p.m.r. spectrum [see Fig. 25(B)] of formycin B (whi has the same empirical formula as inosine) exhibits esse tially the same pattern as that observed for formycin, except for the disappearance of the peak at δ7.45 for the exocyclic amino group and the appearance of another peak δ12—13, assigned to the proton at N-6. The peak for the anomeric proton is again observed upfield at δ4.95 (as a doublet).

A *C*-glycosyl nucleoside antibiotic has been isolated[19 from *Streptomyces showdoensis*, and named showdomycin. Showdomycin was characterized[193,194] as 3-β-D-ribofurano sylmaleimide (60), and was considered to be a uridine or pseudouridine analog, primarily on the basis of p.m.r. spectroscopy and chemical degradation. The p.m.r. spec-trum of showdomycin [see Fig. 25(C)] has a doublet at δ6 which was assigned to the proton at C-4. The doub-let can be explained by the coupling of this proton with the anomeric proton at δ4.75 (q). Once again, the upfie chemical-shift for the anomeric proton of a *C*-glycosyl nucleoside was evident, and here, it was observed as a quartet (instead of a doublet) owing to coupling, not on with the proton on C-2' but also with that on C-4. The remaining peaks (δ3.5 to δ4.2) were assigned to the

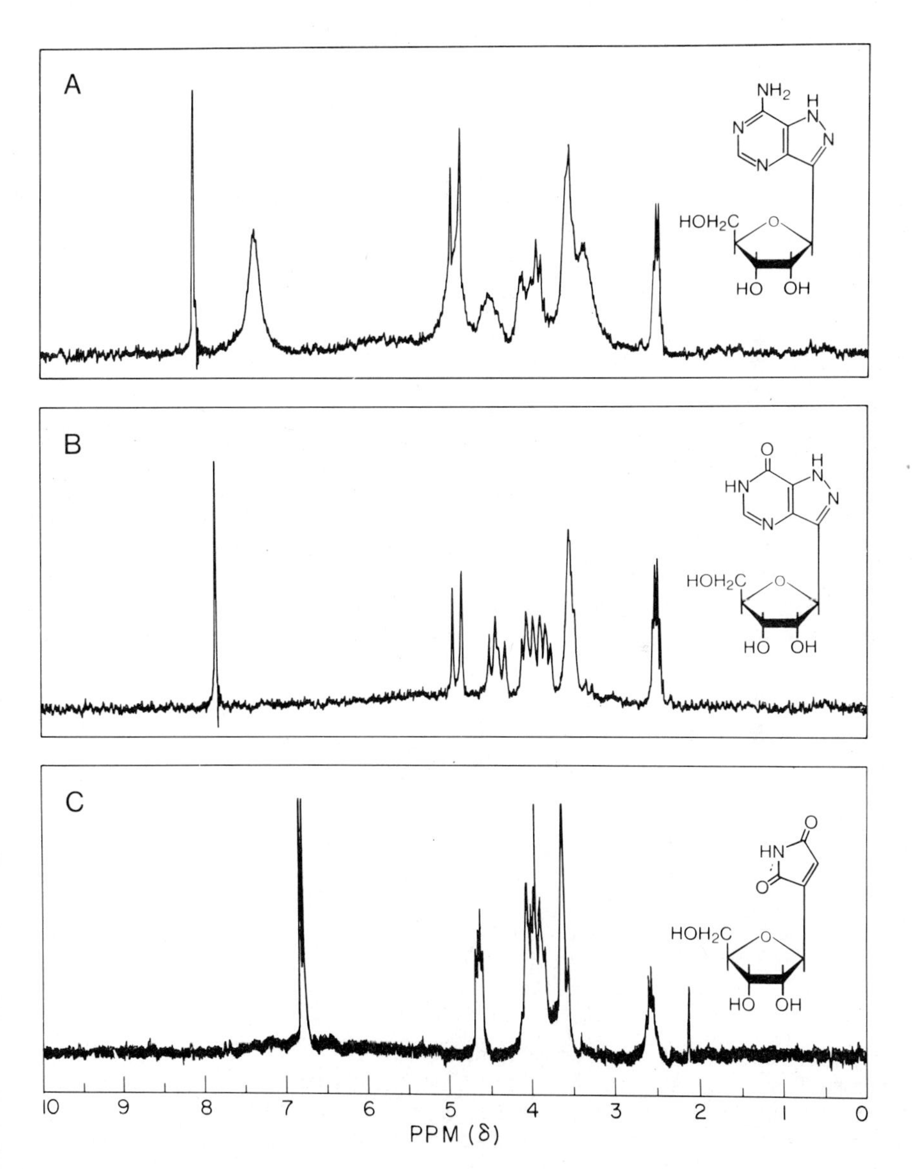

Fig. 25. (A) 7-Amino-3-β-D-ribofuranosylpyrazolo[4,3-*d*]-pyrimidine (Formycin), in methyl sulfoxide-d_6; (B) 3-β-D-ribofuranosylpyrazolo[4,3-*d*]-7-pyrimidinone (Formycin B) in methyl sulfoxide-d_6; and (C) 3-β-D-ribofuranosylmaleimide (Showdomycin) in methyl sulfoxide-d_6–acetic acid-d_4.

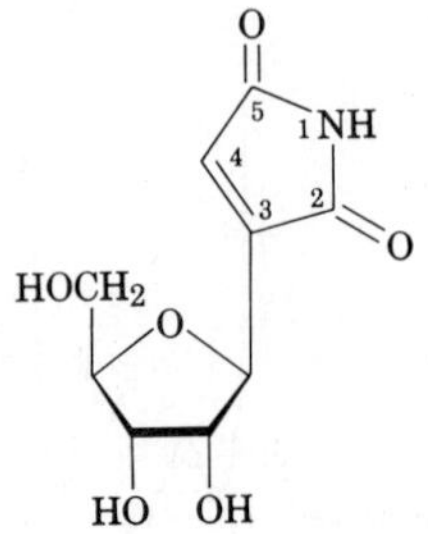

60

carbohydrate moiety. Although the peak for the NH proton is not observed in the p.m.r. spectrum in Fig. 25 (C), because of exchange with the solvent (deuterated acetic acid), the proton has been observed as a broad peak (singlet at δ10.9) in a p.m.r. spectrum of showdomycin in methyl sulfoxide-d_6.

Another antibiotic substance that has been isolated and then characterized[195] as a *C*-glycosyl nucleoside is pyrazomycin (61).

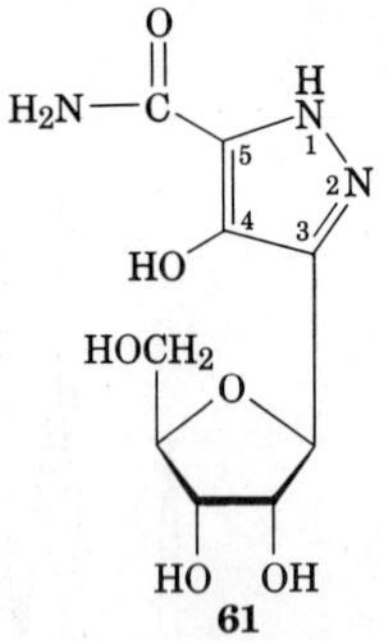

61

The p.m.r. spectrum[195] of 61 has a definit upfield chemical-shift for the anomeric proton (δ4.7) that is typical of *C*-glycosyl nucleosides (having a C—C glycosyl bond), in comparison to nucleosides (having a C—N glycosyl bond).

Another class of nucleoside antibiotics that have a close structural similarity to purine nucleosides compris the pyrrolo[2,3-*d*]pyrimidine nucleoside antibiotics tuber cidin (62a), sangivamycin (62b), and toyocamycin (62c).[19 In fact, these antibiotics are often referred to as 7-deazapurine nucleosides. The only structural difference

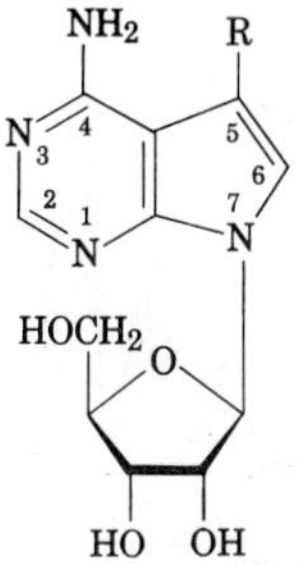

62a, R = H

62b, R = C(=O)—NH_2

62c, R = CN

between tubercidin (62a) and adenosine is that, in 62a, N-7 of adenosine is replaced by a carbon atom; this results in a definite change in the p.m.r. spectrum of tubercidin [see Fig. 26(A)]. The peaks at δ8.25 and 7.03 may be assigned to the proton on C-2 and to the exocyclic amino group, respectively. That portion (δ3.3-5.3) of the p.m.r. spectrum attributable to the D-ribofuranosyl group is very similar to that observed for adenosine. The major difference between the spectra of adenosine and tubercidin is that, in place of the singlet at δ8.53 observed for the proton on C-8 of adenosine, there is a pair of doublets (δ7.35 and δ6.6, *J* 4 Hz) for the protons on C-6 and C-5, respectively. This upfield chemical-shift (in comparison to the proton on C-8 of adenosine) can be attributed primarily to the increase in electron density of the pyrrole ring in comparison to that of the imidazole ring. The relative positions of the protons on C-5 and C-6 was determined by the use of p.m.r. spectra of appropriately substituted pyrrolo[2,3-*d*]pyrimidines.[198-199]

Toyocamycin (62c) has a structure very similar to that of tubercidin (62a), except that a cyano group replaces the proton on C-5 of tubercidin; the p.m.r. spectrum [see Fig. 26(B)] is almost identical to that of adenosine. The pair of doublets observed for the protons on C-5 and C-6 of tubercidin are replaced by a singlet at δ8.25 for the proton on C-6 (which corresponds very closely to the proton on C-8 of adenosine).

The p.m.r. spectrum of sangivamycin (62b) is essentially the same as that for toyocamycin, except that, in methyl sulfoxide-d_6, there is an additional, broad, absorption peak at δ8.3 that may be assigned to the NH protons of the 5-carboxamide group.

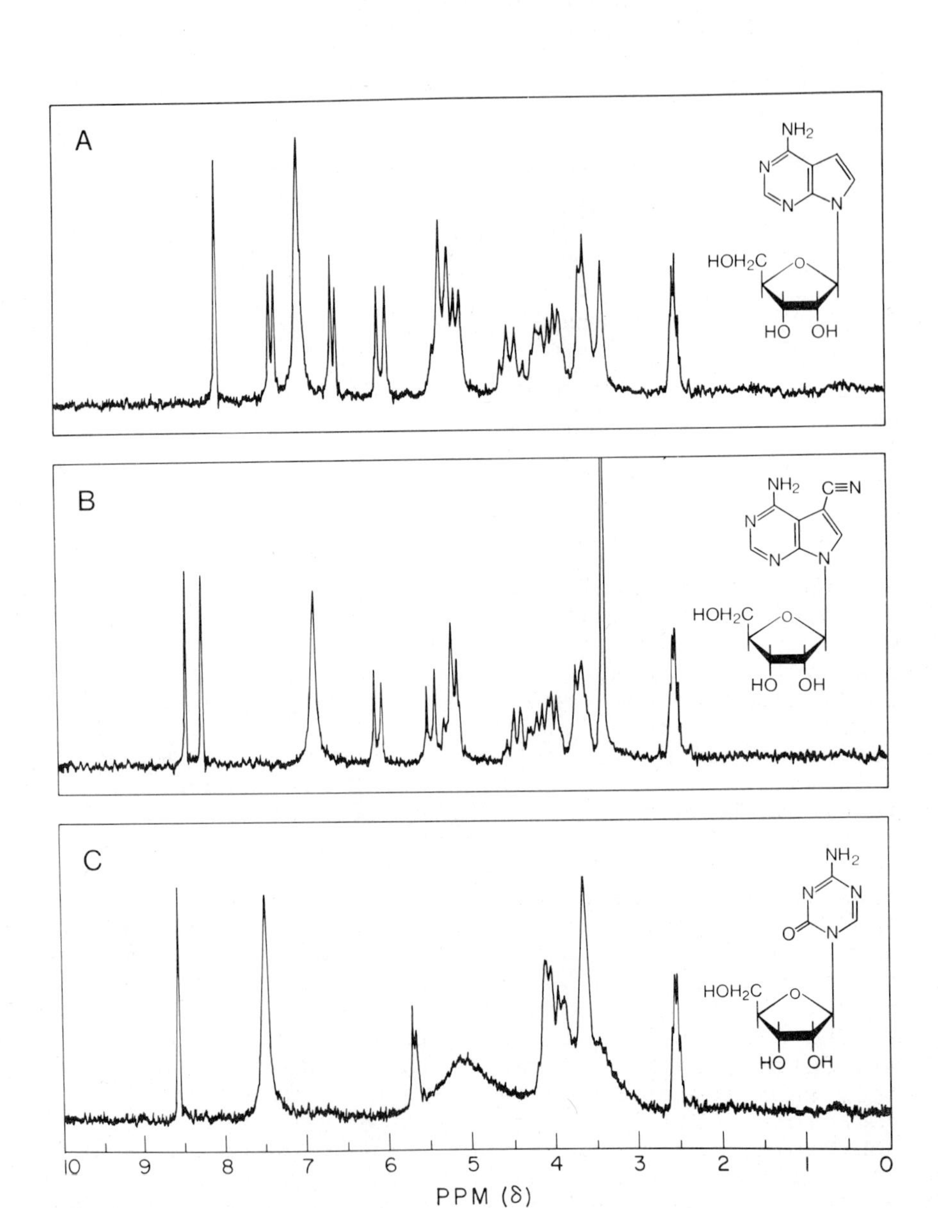

Fig. 26. (A) 4-Amino-7-β-D-ribofuranosylpyrrolo[2,3 -*d*]-pyrimidine (Tubercidin), (B) 4-amino-5-cyano-7-β-D-ribofuranosylpyrrolo[2,3-*d*]pyrimidine (Toyocamycin), and (C) 4-amino-1-β-D-ribofuranosyl-*s*-triazin-2-one (5-Azacytidine all in methyl sulfoxide-d_6.

These nucleoside antibiotics, which are deaza analogs of adenosine, manifest, in their p.m.r. spectra, upfield chemical-shifts, especially for the protons in the deaza moiety (the pyrrole ring). The converse is observed in the p.m.r. spectrum [see Fig. 26(C)] of the antibiotic nucleoside 5-azacytidine.[200,201] Here, instead of a nitrogen atom of the ring of the naturally occurring nucleoside cytidine being replaced by a carbon atom, C-5 is replaced by a nitrogen atom. The major difference observed in the p.m.r. spectrum may be attributed to the absence of a proton on position 5; this results in the occurrence of only a singlet, at δ8.6, for the proton on C-6, instead of the two doublets observable for the protons on C-5 and C-6 of cytidine. Also observed is a downfield chemical-shift for the proton on C-6 (δ8.6) of 5-azacytidine, in comparison with that for the proton on C-6 (δ8.17) of cytidine; this may be attributed to the decrease in electron density of the entire ring-system (*s*-triazine).

A number of other antibiotics, nucleoside antibiotics, and minor nucleosides are structurally related to the nucleoside components of the nucleic acids. However, the structural changes are, in general, very similar to those already discussed, and they will, therefore, not be discussed. P.m.r. spectra, or p.m.r. spectral data and interpretations, have been reported for: pyrazomycin,[202,203] polyoxin C,[204] polyoxin A,[205] nucleocidin-(^{19}F),[206] gougerotin,[207] decoyinine,[208] cordycepin,[209] 3'-acetamido-3'-deoxyadenosine,[210] the C-substance from gougerotin,[211] sparsomycin,[212] and several other nucleoside antibiotics.[213]

IV. NUCLEAR MAGNETIC RESONANCE SPECTROSCOPY OF OTHER NUCLEI, AND ADDITIONAL TECHNIQUES AND PROCEDURES USED IN PROTON MAGNETIC RESONANCE SPECTROSCOPY

Proton magnetic resonance spectroscopy has been used extensively in the area of nucleic acids and related derivatives. Other nuclei (*e.g.*, ^{19}F, ^{13}C, ^{15}N, and ^{31}P) have now received attention, and will be discussed very briefly.

1. Fluorine-19 Nuclear Magnetic Resonance Spectroscopy

Fluorine has a spin of 1/2, and it can provide an n.m.r. spectrum. A fluorine n.m.r. spectrum shows a pattern of peaks for the fluorine atoms present in the compound. However, at the present time, the effect that ^{19}F produces in the p.m.r. spectrum of heterocyclic compounds, including nucleosides, is used much more, and it has provided some interesting results. An excellent example is a p.m.r. spectral study[214] of various 5-fluorouracil derivatives.

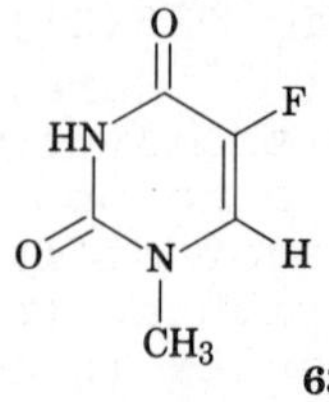

63

The p.m.r. spectrum of 5-fluoro-1-methyluracil (63) revealed a pattern of peaks very similar to that expected for a 5-substituted 1-methyluracil derivative, except that the pattern for the proton on C-6, which would normally be observed as a singlet, was now observed as a doublet (J_{H6-F} 6.0 Hz) due to spin–spin coupling with the fluorine atom on C-5. The same pattern was observed for the heterocyclic moiety of several 5-fluoropyrimidine nucleosides [*e.g.*, 5-fluorouridine (64) and 2'-deoxy-5-fluorouridine (65)]. However, in addition to the doublet for *H*-6 long-

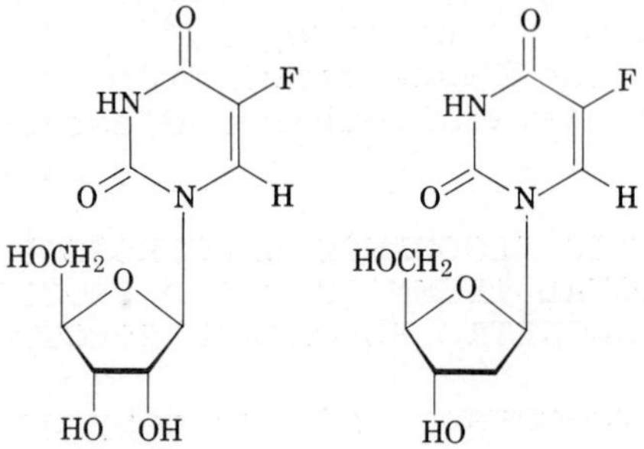

64 65

range, spin–spin coupling between the fluorine atom on C-5 and the anomeric proton (H-1') was also observed. A doublet was expected for H-1' of 64, but a pair of doublet was actually observed. This additional splitting was attributed to the coupling effect of the fluorine atom on C-5 with the anomeric proton. Six peaks (three doublets) were observed for the anomeric proton of 65, instead of the triplet expected. It was suggested[214] that this phenomenon was caused by a long-range coupling between fluorine and the anomeric proton through the bonds instead of through space.

The p.m.r. spectra of 2'-deoxy-2'-fluorouridine (66) and 1-(3-deoxy-3-fluoro-β-D-arabinofuranosyl)uracil (67) have revealed some interesting changes in the conformation of the furanosyl group, as determined[215] by the use of coupling constants. The large coupling-constants associated with proton–fluorine were used for the spectral assignments. The pattern of peaks for the anomeric proton of 66 was observed as a pair of distinct doublets, instead of the pseudotriplet observed for the anomeric proton of 2'-deoxyuridine. For 67, the pattern of peaks for the anomeric proton was a quartet, due to long-range coupling with the fluorine atom on C-3'. There was also a very pronounced, downfield chemical-shift for a proton on a carbon atom bearing a fluorine atom; this would appear to correlate the chemical shift rather closely with the electron density of the carbon atom.

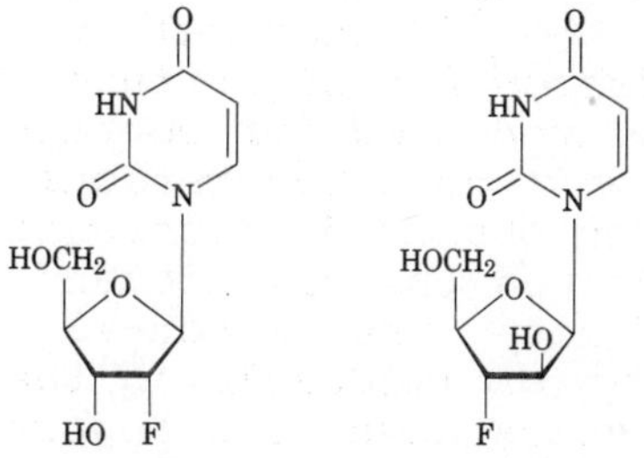

A few additional studies include those on carbohydrates containing fluorine,[216–218] heterocyclic fluorine compounds,[219] the ^{19}F n.m.r. spectral data for certain fluoroheterocyclic compounds,[220,221] the use of ^{19}F n.m.r. spectra for structural elucidation of the nucleoside antibiotic nucleocidin,[206] fluoro sugar analogs of arabinosyl- and xylosyl-cytosines,[222] and some pentofuranosyl fluorides.[223]

2. Carbon-13 Nuclear Magnetic Resonance Spectroscopy

The use of ^{13}C n.m.r. spectroscopy was hampered for a number of years by the paucity of ^{13}C, which, in natural abundance, is only 1.1% of the total carbon content. This situation presented a number of difficulties related directly to the experimental techniques involved in obtaining a ^{13}C spectrum. Many of these difficulties have now been resolved by such techniques as use of large sample-tubes, rapid passage to avoid saturation, internuclear double-resonance, signal enhancement by use of the Overhauser effect, and the use of CAT (Computer of Average Transients). Such techniques improved the resolution of ^{13}C n.m.r. (c.m.r.) spectra, thus increasing the use of ^{13}C n.m.r. spectroscopy in several fields (including study of heterocyclic compounds and heterocyclic nucleosides and nucleotides).[224]

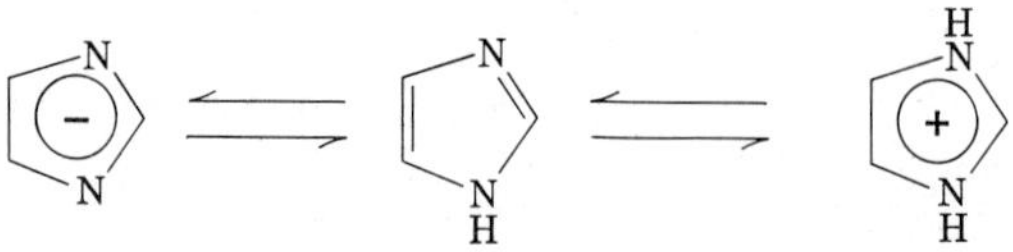

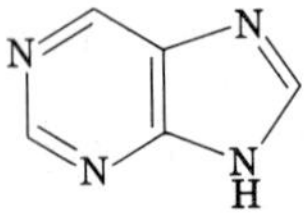

The ^{13}C chemical shifts are usually reported relative to the peak observed for carbon disulfide (other standards are benzene, *p*-dioxane, and methyl sulfoxide). Several studies have been reported[225,226] on substituted pyridines that correlated the ^{13}C chemical shifts with those observed for the corresponding substituted benzene. The ^{13}C chemical shifts predicted for certain pyrimidine derivatives were found[227] to agree reasonably well with the experimental values observed. These studies were later extended[228] to include the anionic and cationic species of several pyrimidines. The carbon-13 chemical shifts of C-2, C-4, and C-5 of imidazole in the neutral, anionic, and cationic species have been reported,[229] together with those for the same species for pyrrole and pyrazole. A c.m.r. study[230] on purine and certain deuterated derivatives permitted definite assignment of peaks in the spectrum to specific carbon atoms by use of the decoupling technique. The π-electron charge-density of the specific carbon atoms was correlated with the chemical shifts for the peaks observed in the c.m.r. spectrum. It was found that the relative positions of the peaks for the carbon atoms of purine are in the order C-4>C-2>C-8>C-6>C-5, with C-4 being deshielded to a greater extent than any of the other carbon atoms. The relative order observed for the carbon atoms (2>8>6) is of considerable interest, as the p.m.r. spectrum has revealed a different order (6>2>8) for the corresponding hydrogen atoms, and this suggests that chemical shifts in n.m.r. spectra should be used with caution for predicting charge densities. These ^{13}C n.m.r. studies of pyrimidines and purines have been extended to include pyrimidine and purine nucleosides and nucleotides. The natural-abundance ^{13}C n.m.r. spectra of the naturally occurring purine and pyrimidine nucleosides have revealed[231] a clean separation between the peaks observed for the carbon atoms of the glycosyl group and those of the aglycon. The peaks for the glycosyl group were found to occur in the range of +37 to +89 p.p.m, and those for the aglycon in the range of -36 to +33 p.p.m. with benzene as

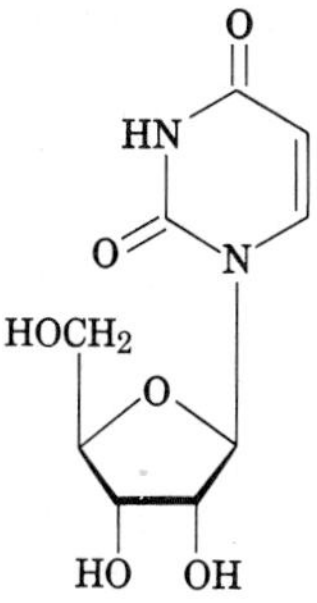

68 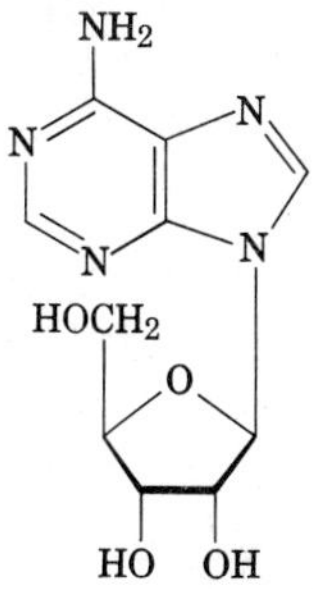 69

the reference standard, *e.g.*, uridine (68) (aglycon -36 to +25, D-ribosyl +37 to +65); adenosine (69) (aglycon -28.5 to to +8.1, D-ribosyl +37 to +65). Atom C-2' of the 2'-deoxynucleosides was observed upfield at approximately +90 p.p.m. Atoms C-1' and C-4' gave rise to peaks having very similar chemical shifts, and were assigned with the help of a report[232] on the corresponding 5'-nucleotides, for which the peak for C-4' could be assigned unequivocally, owing to the observed phosphorus coupling with C-4'. In fact, this study[232] revealed a very close analogy between the nucleosides and their corresponding 5'-phosphates. Also observed was a definite effect on the chemical shifts of several carbon atoms on changing the pH.

It has been suggested[232] that, as advances in instrumentation and techniques permit enhanced sensitivity in the detection of natural-abundance ^{13}C resonance signals, significant insight into the inter- and intra-molecular interactions of nucleosides, nucleotides, and oligonucleotides will be obtained.

A few references involving ^{13}C n.m.r. spectroscopy include: long-range, spin-spin couplings,[233] a study of carbon-13 chemical shifts in various esters,[234] a natural-abundance c.m.r. study of a mixture of pinenes, sucrose, squalene, and cholesterol,[235] a c.m.r. study of amino acids and peptides with ^{13}C enrichment,[236] and ^{13}C indole derivatives.[237]

3. NITROGEN-15 NUCLEAR MAGNETIC RESONANCE SPECTROSCOPY

The major isotope of nitrogen is ^{14}N (almost 100% natural abundance); it has a very low magnetic moment, with a quadrupole moment, and a spin of unity, and it is, therefore, essentially insensitive to n.m.r. detection. The nitrogen-15 isotope has a spin of 1/2, with no quadrupole moment, and it would appear to be a prime candidate for n.m.r. studies, except for the fact that ^{15}N occurs in such a very low natural abundance (0.37%). The use of compounds enriched in ^{15}N, advanced measuring techniques, and devices for enhancement of sensitivity has increased the number of studies involving nitrogen magnetic resonance.

A study[238] on the site of protonation of certain pyrimidines prompted the synthesis of several ^{15}N-enriched pyrimidines related to the naturally occurring pyrimidines cytosine and uracil. Of special interest in this group of compounds were the ^{15}N-enriched uracil (70) and 1-methylcytosine (71). The protons bonded to the nitrogen-15 atoms at positions 1 and 3 of 70 give rise to two doublets centered at δ10.78 and δ10.96 in the p.m.r. spectrum. The values are those expected for a proton directly bonded to a sp^2-hybridized nitrogen atom, and they furnish additional support for the diketo structure. The high-field doublet was further split into a pair of doublets, and this effect can be ascribed to coupling with the proton at position six. The p.m.r. spectra of several other pyrimidines were reported in this investigation, as well as the use of ^{15}N n.m.r. spectroscopy for the determination of the site of protonation in several ^{15}N-enriched pyrimidines. Another study[239] established that

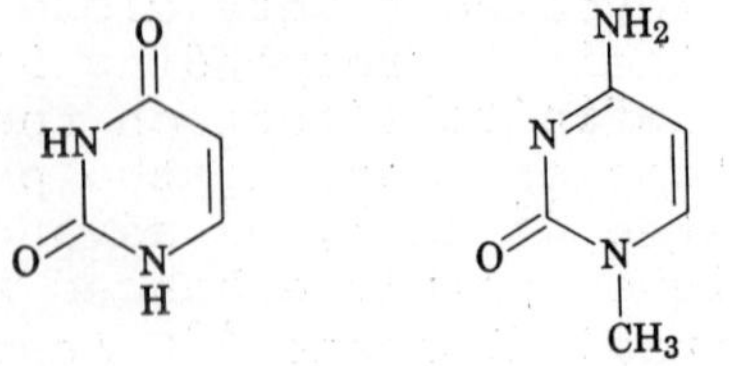

70 71

^{15}N-enriched 1-methylcytosine (71) exists in the amino form, and that protonation definitely occurs at N-3. The dependence of the chemical shifts and coupling constants on such factors as the solvent and the temperature was also discussed.

A ^{15}N n.m.r. study[240] of the binding of Mg^{2+} with adenosine 5'-triphosphate (72) has been accomplished by using 70%-enriched ^{15}N for the nitrogen atoms of 72. This

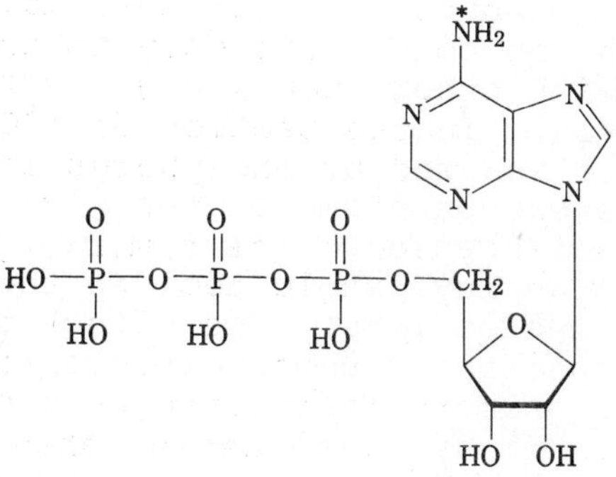

72

study was initiated in an effort to determine whether Mg^{2+} actually forms a complex with any of the nitrogen atoms of 72. The n.m.r. spectrum for the ^{15}N-enriched 72 had five well-defined peaks, assigned as follows. The exocyclic, 6-amino group was assigned to the peak at highest field, and was observed as a very broad peak, owing to incomplete collapse of the nitrogen-proton, spin-spin coupling-interaction; N-9 was observed as a singlet; N-7 was observed as a doublet, owing to coupling with the proton on C-8 and was assigned to the peak at lowest field; N-1 and N-3 were observed as doublets, owing to coupling with the proton on C-2, and were assigned to the remaining two peaks, N-1 being assigned to the peak at highest field. The addition of Mg^{2+} to a solution of 72 caused no change in the chemical shift of the peaks assigned to the nitrogen atoms, and this indicates that there is no complexing between the nitrogen atoms and Mg^{2+}. However, the addition of Zn^{2+} resulted in a definite interaction, and, on the basis of the n.m.r. spectrum, it was suggested that complex formation had occurred at N-7 and at the exocyclic amino group at position six. There has been an increase in the number of reports concerning ^{15}N n.m.r. spectroscopy; examples include examination of the effect of the nitrogen lone-pai orientation on geminal ^{15}N-H spin-spin coupling[241] and a study of some ^{15}N-H coupling constants.[242]

4. PHOSPHORUS-31 NUCLEAR MAGNETIC RESONANCE SPECTROSCOPY

The use of ^{31}p n.m.r. spectroscopy in the area of nucleic acids has, for several reasons, been of limited value. Nevertheless, an increasing number of reports on the use of ^{31}p n.m.r. spectroscopy have been published, presumably because of recent advances in techniques and instrumentation. Another factor that may be involved is that phosphorus can couple not only to another phosphorus atom but also with other atoms, *e.g.* hydrogen, fluorine-19, and carbon-13. In the section on ^{13}C n.m.r. (see p.349), the role assumed by phosphorus in the assignment of peaks to the carbon atoms of the glycosyl group of nucleotides was discussed. ^{31}P n.m.r. spectroscopy will probably become an invaluable tool in the structure elucidation of nucleotides. The effect that ^{31}P exerts on the p.m.r. spectra of phosphorus-containing organic compounds is of primary importance in the field of nuclei acid chemistry. It has been established[243] that phosphorylation of a hydroxyl group usually results in a

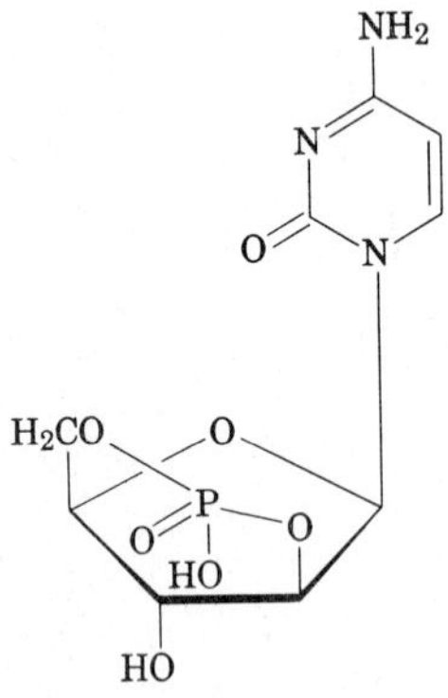

73

change in the chemical shift observed for the protons on C-5'. It has been found[244] that the phosphate group exer an anisotropic effect on the C-5' protons to which it is directly bound. Of considerable interest is the finding that the phosphate group is also responsible for a very significant, downfield chemical-shift for H-3' of D-arabinofuranosylcytosine 2',5'-cyclic phosphate (73). The structure of 73 was definitely established not only b decoupling studies made on the p.m.r. spectrum and by chemical studies but also by examination of its ^{31}P n.m.r spectrum; the deshielding mentioned must be due to anisotropy. The deshielding effect observed for the protons on the carbon atom to which a phosphate group is directly attached is most probably caused by the electron-

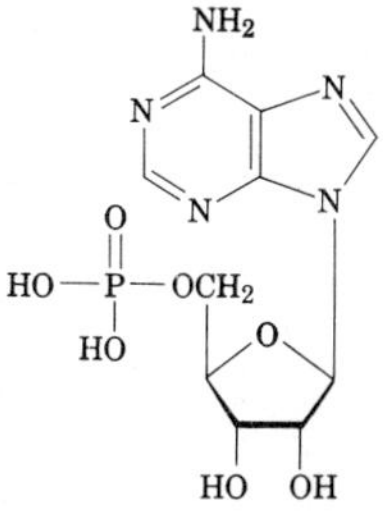

74

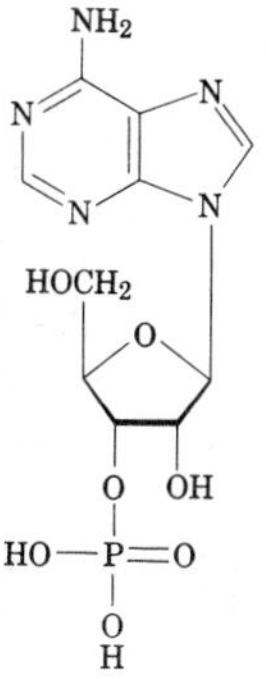

75

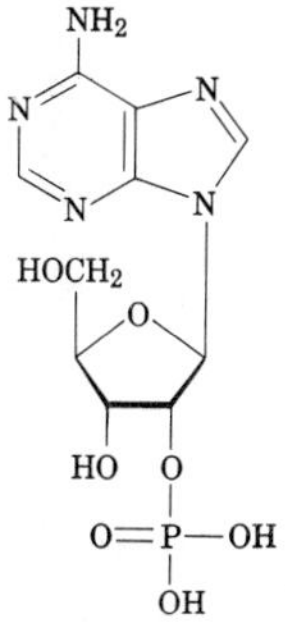

76

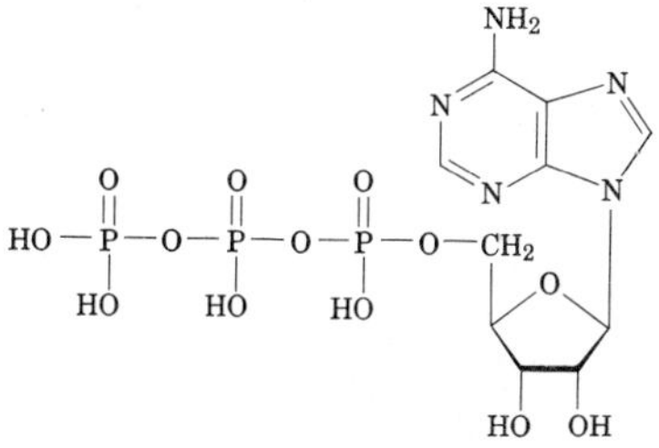

72

withdrawing phosphate group. A study[245] of the p.m.r. spectra of adenosine 5'-phosphate (74), adenosine 3'-phosphate (75), and adenosine 2'-phosphate (76) has unequivocally established that this is so, by making a definite assignment for the peaks assigned to the various protons of the D-ribofuranosyl residue by spin-spin decoupling; *e.g.*, the p.m.r. spectra of 74, 75, and 76 revealed deshielding for H-5', 3', and 2', respectively, in comparison to the non-phosphorylated compound (adenosine). In addition, this study established that there is also a very small, downfield shift for the vicinal protons, *e.g.*, for H-1' and H-3' of 76 and H-2' and H-4' of 75. Other factors discussed included structure and stacking in relation to concentration and pH. Another point of interest is that the p.m.r. spectrum of adenosine 5'-triphosphate (72) is very similar to that of 74, indicating that the triphosphate moiety has very little effect on the rest of the molecule. ^{31}P n.m.r. spectroscopy appears to be assuming an increasingly important

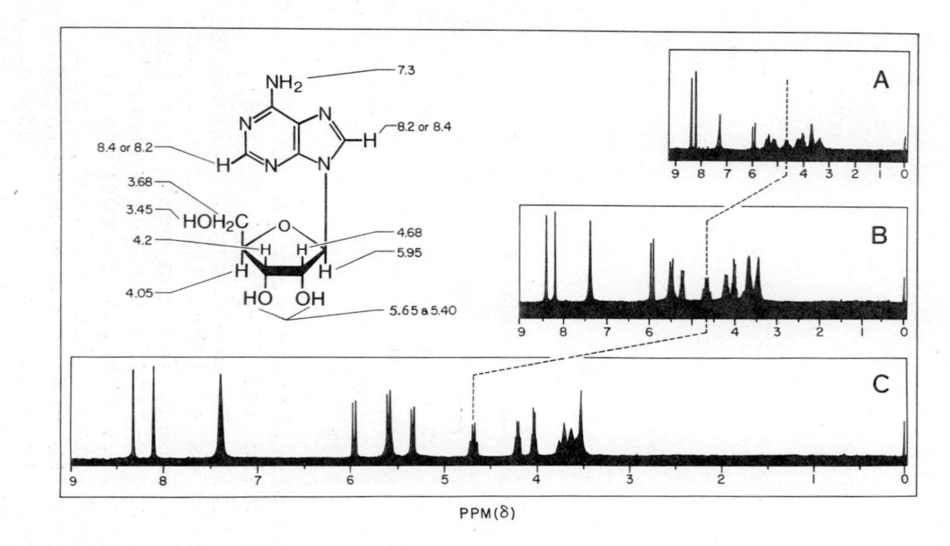

Fig. 27. 6-Amino-9-β-D-ribofuranosylpurine (adenosine), in methyl sulfoxide-d_6; (A) at 60 MHz; (B) at 100 MHz; and (C) at 220 MHz. (The spectra and assignments were furnished through the courtesy of Dr. James N. Shoolery, Varian Associates, 611 Hansen Way, Palo Alto, California.)

role in conformational and structural studies in the area of nucleic acid chemistry, especially as applied to the cyclic phosphates.

V. NUCLEAR MAGNETIC RESONANCE AT HIGHER FIELD STRENGTHS

Throughout this Chapter, 60-MHz spectra have been discussed, primarily because 60 MHz is the field strength still mainly used and still encountered most frequently in the literature. However, the number of reports involving the use of 100- and 220-MHz spectra is increasing. The 220-MHz instrument has resolved problems not amenable to solution by use of the 60- and 100-MHz instruments. Considerable expense is involved in the purchase, operation, and maintenance of a 220-MHz n.m.r. spectrophotometer, but the resolution is greatly enhanced. The p.m.r. spectra of adenosine (see Fig. 27) at 60, 100, and 220 MHz show the advantages provided by higher field-strength. Thus, the unresolved multiplet at $\delta 4$-4.5 in the 60-MHz spectrum has been assigned to H-3' and H-4', although there is a considerable overlap that prevents the assignment of peaks to any specific pattern. At 100 MHz, the patterns for these individual protons have been separated, and they could probably be used with considerable confidence for assignment of the first-order patterns for H-3' and H-4'. At 220 MHz, the separation of these protons can be considered to be absolute, and the patterns observed can readily be assigned on a first-order basis. Another example of the tremendous advantage of 220-MHz spectroscopy in comparison to 60-MHz spectroscopy is provided by a visual examination of Fig. 27, where the overlapped resonances at $\sim\delta 5.5$ in the 60-MHz spectrum is observed as two separate, distinct doublets in the 220-MHz spectrum. The use of 220-MHz n.m.r. spectroscopy in studies involving conformation, structure, and composition of various macromolecules has also been reported. The changes involved in the p.m.r. spectrum of yeast *t*-RNA with changes in pH, and of unfractionated yeast *t*-RNA on varying the temperature have been reported.[246,247] The use of 220-MHz spectroscopy has furnished[248] a measure of the occurrence of specific bases in relation to thymidine for samples of 2'-deoxyribonucleic acid from different organisms; more-advanced techniques or instrumentation will be needed[249,250] in order to permit assignments of individual peaks to the heterocyclic bases of alanine *t*-RNA, although several gross correlations were made. It is clear that, on the basis of these preliminary studies, 220-MHz n.m.r. spectroscopy will provide useful information concerning a large number of complex organic compounds, information that cannot be obtained by use of other n.m.r. instruments or other instrumental methods.

VI. NUCLEAR MAGNETIC DOUBLE-RESONANCE

The double-resonance technique can be described in terms of spin–spin decoupling; this is accomplished by irradiating the sample with a second radio-frequency field at exactly the resonance position of the proton to be decoupled. In essence, this procedure affords the same effect as replacement, with a deuterium atom, of the proton studied. Therefore, if the doublet at δ7.4 assigned to H-6 of uracil (see Section III,2, Fig. 17, p. 308) is irradiated, the doublet usually observed at δ5.5 for H-5 is observed as a singlet, due to the spin-spin, decoupling effect, with complete disappearance of the peaks for H-6. Thus, instead of the pair of doublets usually observed for H-5 and H-6, there is now observed only a singlet in the δ5.5 region (for H-5).

This technique has been used extensively in structural and conformational studies of various carbohydrate derivatives, *e.g.*, the anomeric proton of a nucleoside usually occurs downfield from the remaining carbohydrate protons, and irradiation of the anomeric proton results in simplification of the pattern observed for H-2'. The coupling constant can then be used in the Karplus equation, in order to calculate the dihedral angle between H-2' and H-3'. Further decoupling studies can then establish the conformation of the entire carbohydrate moiety. This technique is now so widely used that it is considered almost a necessity for the interpretation of complex p.m.r. spectra involving structural elucidation or conformational studies. The proton–proton decoupling technique can be classified as *homonuclear* decoupling.

The decoupling technique has taken on a new dimension with the increased study of nuclei other than the proton. The ^{13}C n.m.r. spectrum of a compound can be simplified by decoupling a specific proton or protons on a particular carbon atom, *e.g.*, decoupling H-6 of uridine from C-6 results in simplification of the splitting pattern for one of the carbon atoms in the ^{13}C n.m.r. spectrum. As decoupling H-6 caused this simplification, this effect obviously affords an unequivocal assignment of at least one carbon atom in the ^{13}C n.m.r. spectrum. This type of decoupling between different types of nuclei can be classified as *heteronuclear* decoupling. The difference between homonuclear and heteronuclear decoupling is obvious, and will not be discussed further.

However, the potential of this technique is evident from the reports of n.m.r. studies involving use of ^{31}P, ^{77}Se, ^{19}F, ^{13}C, and ^{29}Si, all of which should be amenable to decoupling studies for the determination of their chemical shifts.

The reader should consult the recent literature for modifications and new innovations concerning the aforementioned techniques.

VII. OTHER TECHNIQUES AND APPLICATIONS IN THE AREA OF NUCLEAR MAGNETIC RESONANCE SPECTROSCOPY

The number of reports on new techniques and their application to the field of n.m.r. will undoubtedly expand at an increasing rate. Even a brief description of the various techniques at present available is outside the scope of this Chapter; it would require at least as much space as has been allotted to this entire Volume. The reader should consult some of the general references and recent literature for further references pertaining to such subjects as enhancement of signals and resolution, anisotropy of various groups and ring systems, n.m.r. spectra in liquid-crystal solvents, solvent effects, chemical-shift non-equivalence, conformational studies of carbohydrates and hexahydro heterocyclic compounds, equilibrium studies, hydrogen bonding, long-range and vicinal coupling, "chemically induced dynamic nuclear polarization" (CIDNP), nuclear Overhauser effect, effects of stacking on chemical shifts, the effect of *syn* or *anti* conformation on the n.m.r. spectra of nucleosides and nucleotides, and the use of the n.m.r. spectrometer as an analytical instrument, as well as those that will undoubtedly be reported during the time that has elapsed between the completion and the publication of this manuscript.

VIII. TABLE OF REPRESENTATIVE STRUCTURES, WITH CHEMICAL-SHIFT VALUES

TABLE II. REPRESENTATIVE STRUCTURES, WITH δ VALUES FOR CHEMICAL SHIFTS

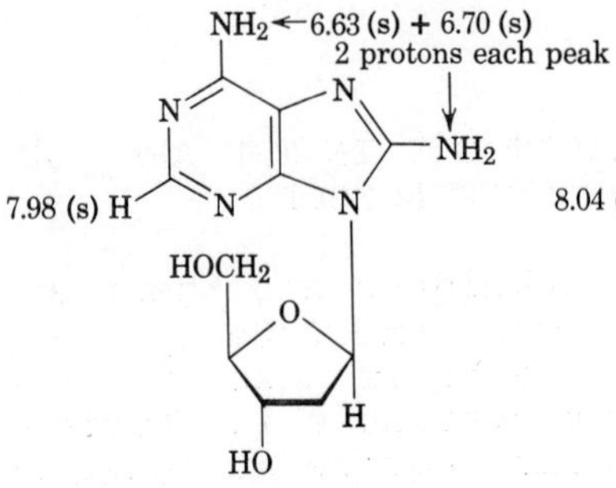

NH_2 ← 6.63 (s) + 6.70 (s)
2 protons each peak
N N NH_2
7.98 (s) H N N
$HOCH_2$
O
H
HO

in methyl sulfoxide-d_6

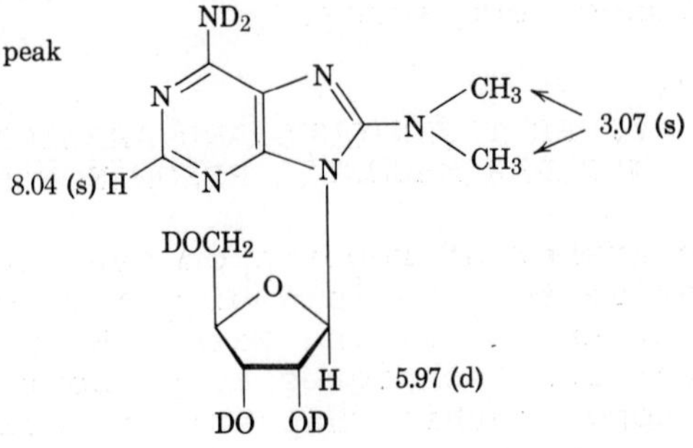

ND_2
N N CH_3 3.07 (s)
N
8.04 (s) H N N CH_3
$DOCH_2$
O
H 5.97 (d)
DO OD

in D_2O

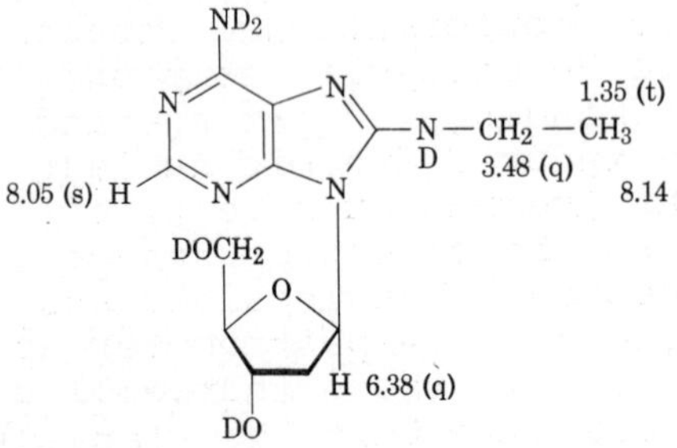

ND_2
N N 1.35 (t)
N — CH_2 — CH_3
D 3.48 (q)
8.05 (s) H N N
$DOCH_2$
O
H 6.38 (q)
DO

in D_2O

O
DN N CH_3
N 2.95 (s)
8.14 (s) H N N CH_3
$DOCH_2$
O
H 5.87 (d)
DO OD

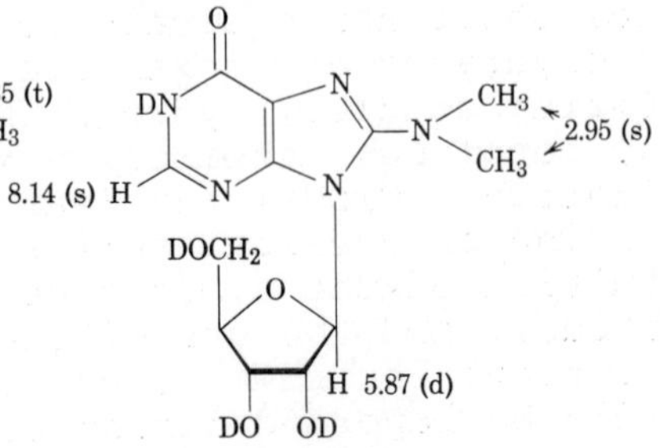

in methyl sulfoxide-d_6–D_2O

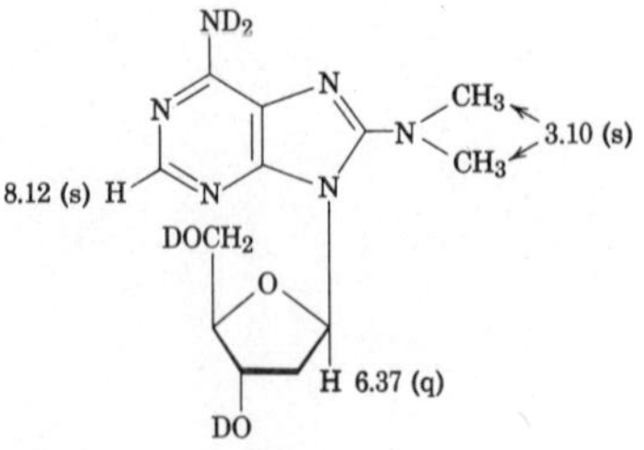

ND_2
N N CH_3
N 3.10 (s)
8.12 (s) H N N CH_3
$DOCH_2$
O
H 6.37 (q)
DO

in D_2O

O
HN N
NH_2 6.65 (s)
7.98 (s) H N N
$HOCH_2$
O
H 5.99 (d)
HO OH

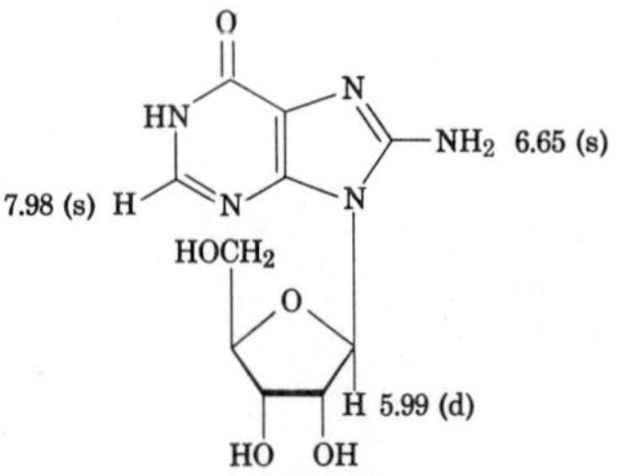

in methyl sulfoxide-d_6

(Ref. 251)

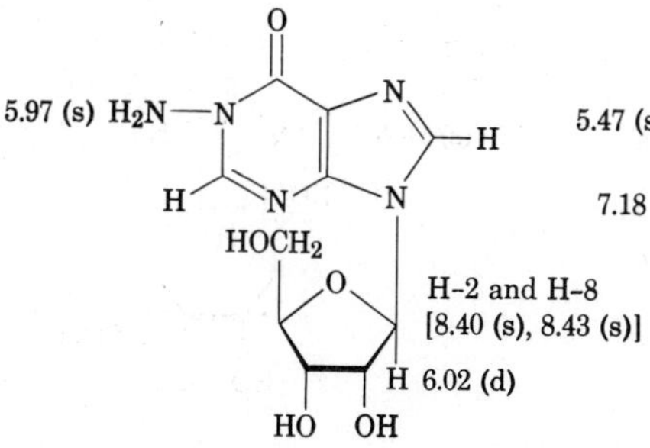
O
5.97 (s) H2N—N
N
H
H
N
N
HOCH2
O
H-2 and H-8
[8.40 (s), 8.43 (s)]
H 6.02 (d)
HO OH

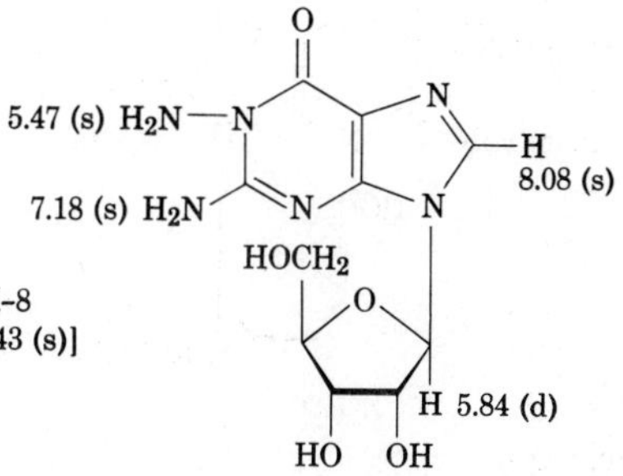
O
5.47 (s) H2N—N
N
H
8.08 (s)
7.18 (s) H2N
N
N
HOCH2
O
H 5.84 (d)
HO OH

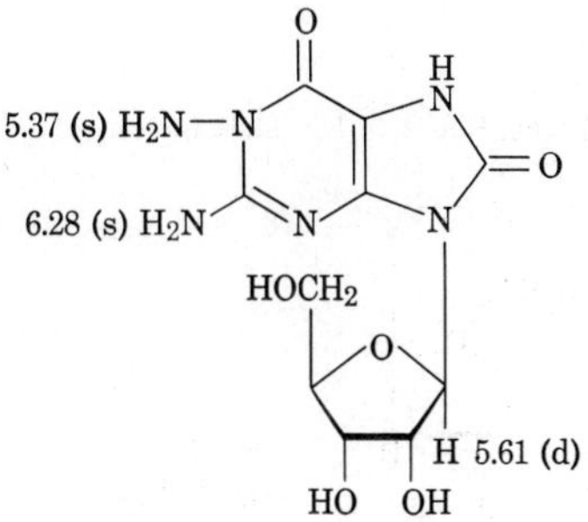
O
H
N
5.37 (s) H2N—N
O
6.28 (s) H2N
N
N
HOCH2
O
H 5.61 (d)
HO OH

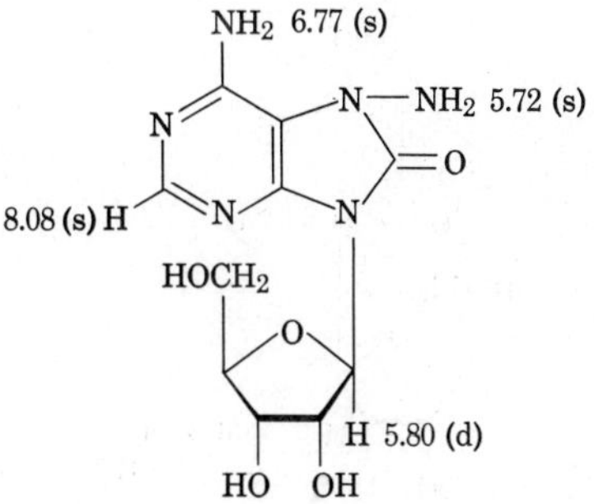
NH2 6.77 (s)
N—NH2 5.72 (s)
N
O
8.08 (s) H
N
N
HOCH2
O
H 5.80 (d)
HO OH

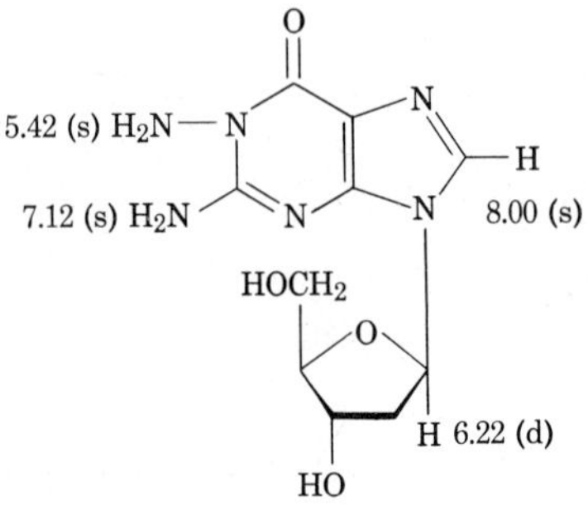

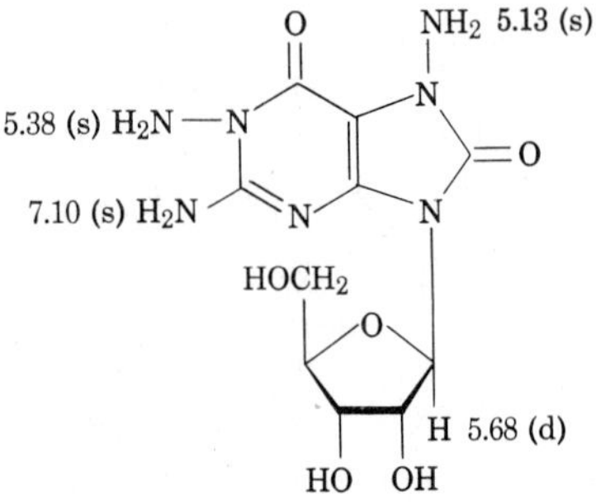

5.42 (s) H_2N—N; 7.12 (s) H_2N; 8.00 (s); $HOCH_2$; H 6.22 (d); HO

(Ref. 252)

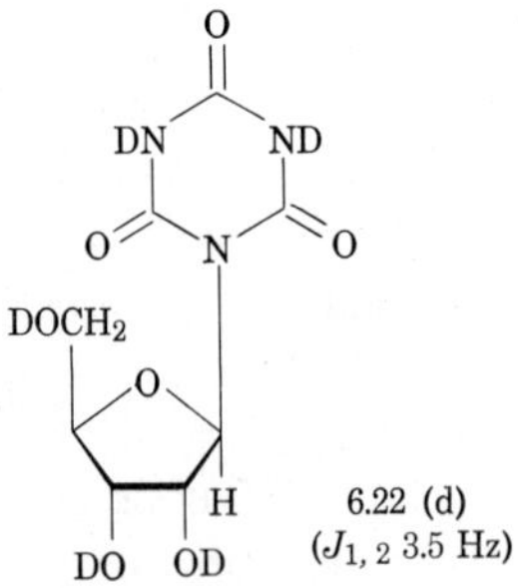

NH_2 5.13 (s); 5.38 (s) H_2N—N; 7.10 (s) H_2N; $HOCH_2$; H 5.68 (d); HO OH

DN ND; $DOCH_2$; H; DO OD; 6.22 (d) ($J_{1,2}$ 3.5 Hz)

in D_2O

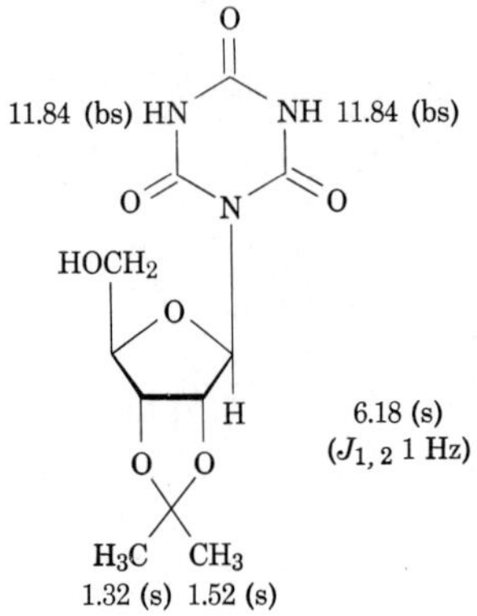

11.84 (bs) HN NH 11.84 (bs); $HOCH_2$; H; 6.18 (s) ($J_{1,2}$ 1 Hz); H_3C CH_3 1.32 (s) 1.52 (s)

in methyl sulfoxide-d_6

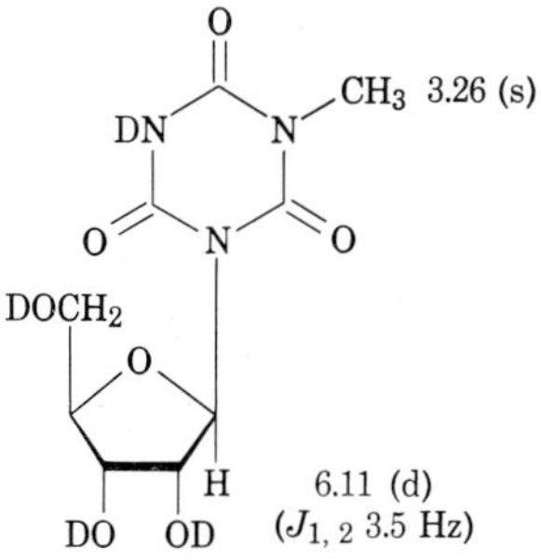

O
DN N CH3 3.26 (s)
O N O
DOCH2
O
H 6.11 (d)
($J_{1,2}$ 3.5 Hz)
DO OD

in D_2O

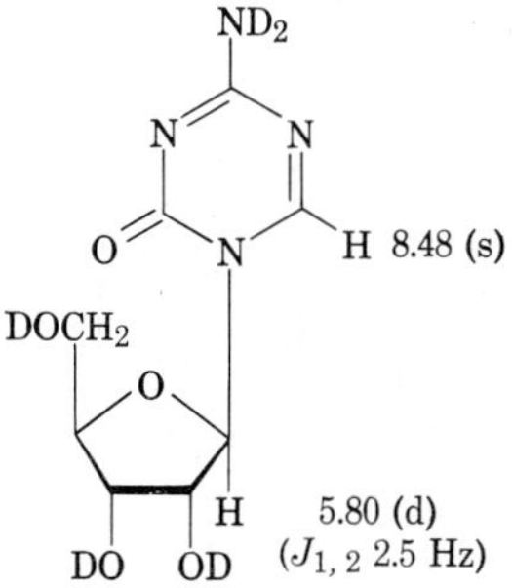

ND2
N N
O N H 8.48 (s)
DOCH2
O
H 5.80 (d)
($J_{1,2}$ 2.5 Hz)
DO OD

in D_2O

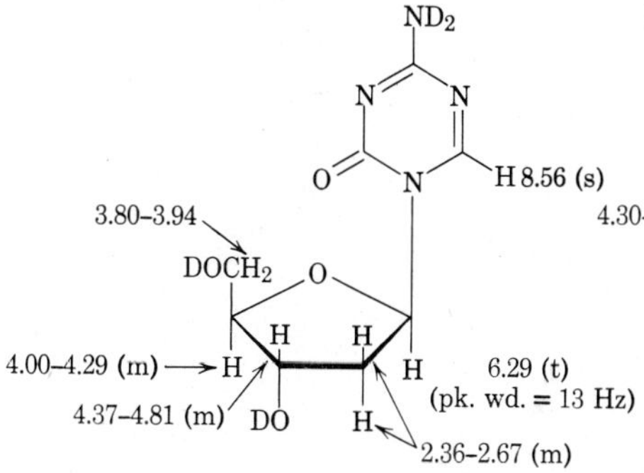

ND2
N N
O N H 8.56 (s)
3.80–3.94
DOCH2 O
H H
4.00–4.29 (m) → H H 6.29 (t)
(pk. wd. = 13 Hz)
4.37–4.81 (m) DO H
2.36–2.67 (m)

in D_2O

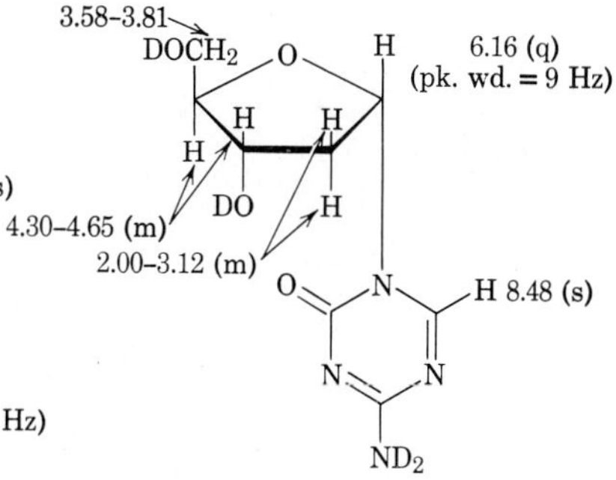

3.58–3.81
DOCH2 O H 6.16 (q)
(pk. wd. = 9 Hz)
H H
H
4.30–4.65 (m) DO H
2.00–3.12 (m)
O N H 8.48 (s)
N N
ND2

in D_2O

(Ref. 253)

3.43 (s) H_3C
10.67 (bs)
6.93 (s) H_2N
$HOCH_2$
H 5.63 (d)
HO OH
3.31 + 3.40 (s)
CH_3
H_3C
7.06 (s) H_2N
$HOCH_2$
H 5.65 (d)
HO OH
CH_3
11.61 (s) HN
6.65 (s) H_2N
$HOCH_2$
H 5.70 (d)
HO OH
CH_3 3.38 (s)
5.36 (s) H_2N
7.08 (s) H_2N
$HOCH_2$
H 5.63 (d)
HO OH
NH_2 6.73(s)
N—CH_3 3.55(s)
8.08(s) H
$HOCH_2$
H 5.80 (d)
HO OH

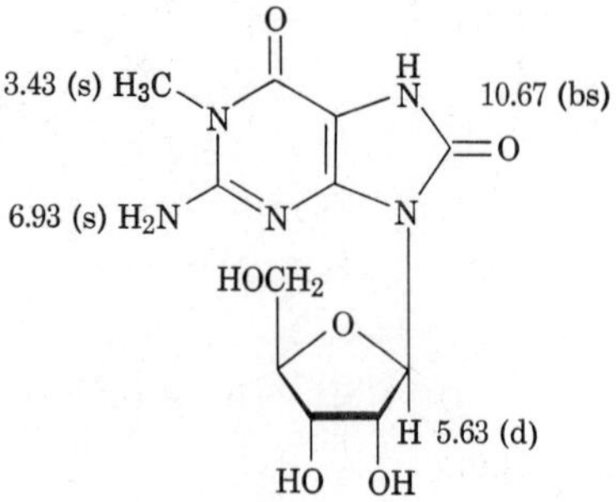

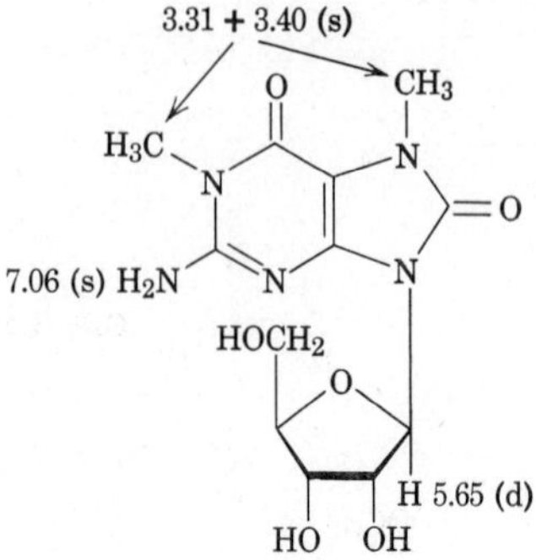

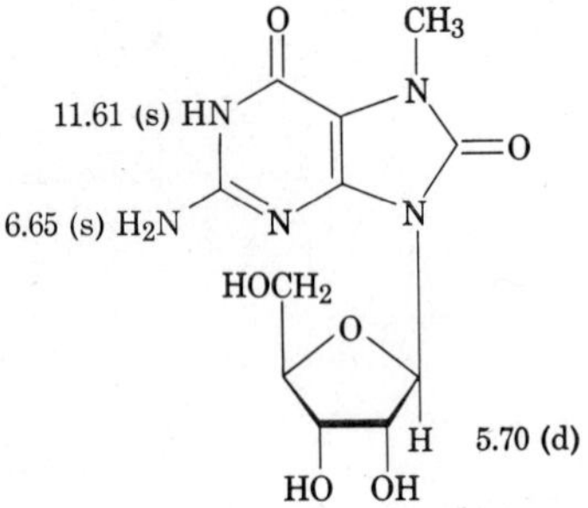

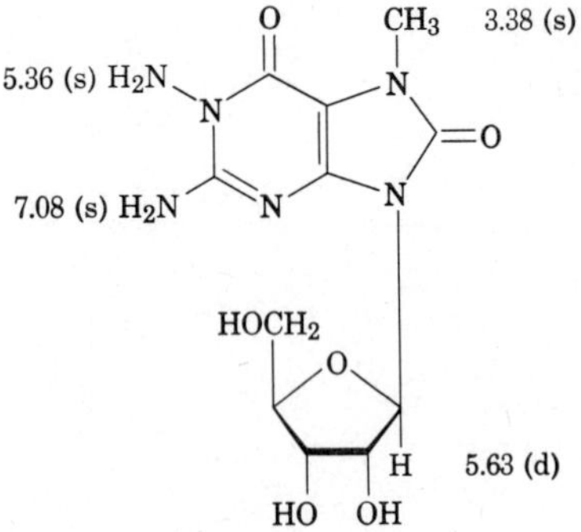

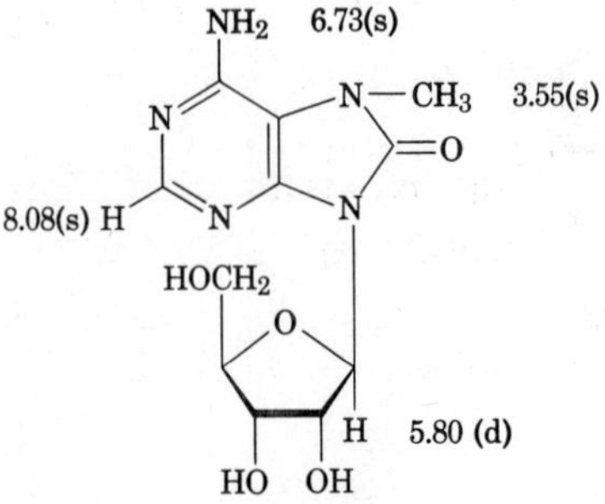

(Ref. 254)

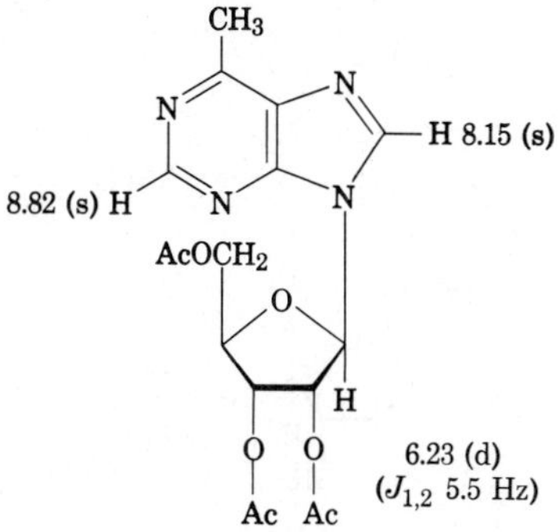

CH_3
N N
8.82 (s) H N N — H 8.15 (s)
$AcOCH_2$
O
H
O O 6.23 (d)
Ac Ac ($J_{1,2}$ 5.5 Hz)
in $CDCl_3$

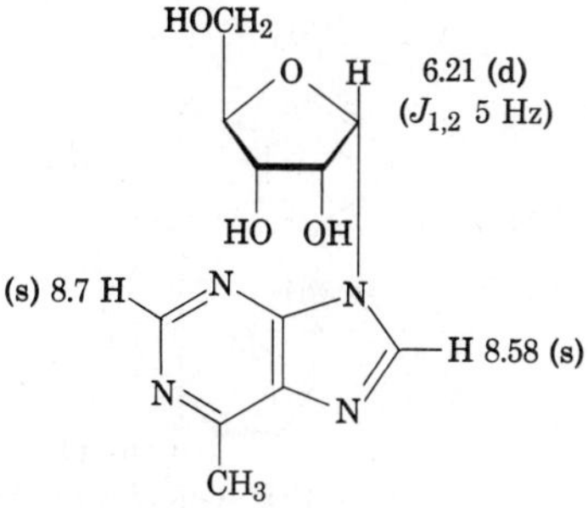

$HOCH_2$
O H 6.21 (d)
($J_{1,2}$ 5 Hz)
HO OH
(s) 8.7 H N N
N N — H 8.58 (s)
CH_3
in methyl sulfoxide-d_6

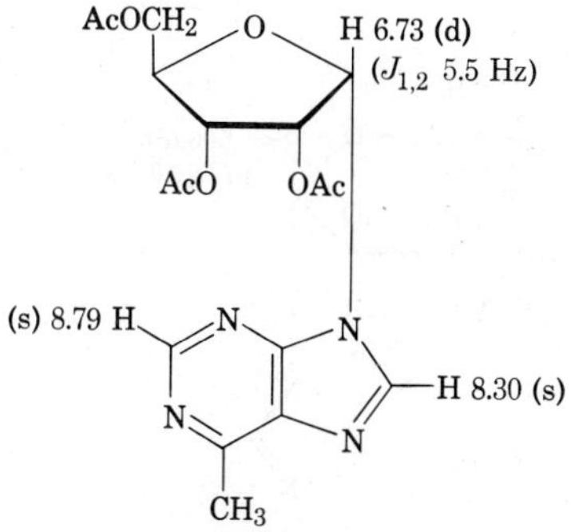

$AcOCH_2$ O H 6.73 (d)
($J_{1,2}$ 5.5 Hz)
AcO OAc
(s) 8.79 H N N
N N — H 8.30 (s)
CH_3
in $CDCl_3$

CH_3
N N
(s) 8.82 H N N — H 8.28 (s)
$AcOCH_2$ O
OAc
H 6.28 (d)
O ($J_{1,2}$ 2.5 Hz)
Ac

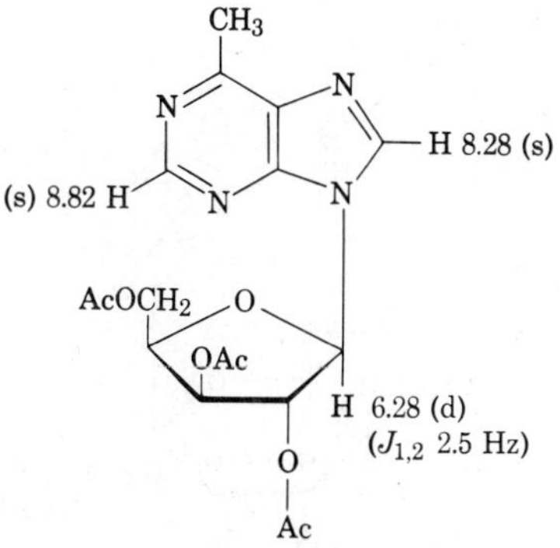

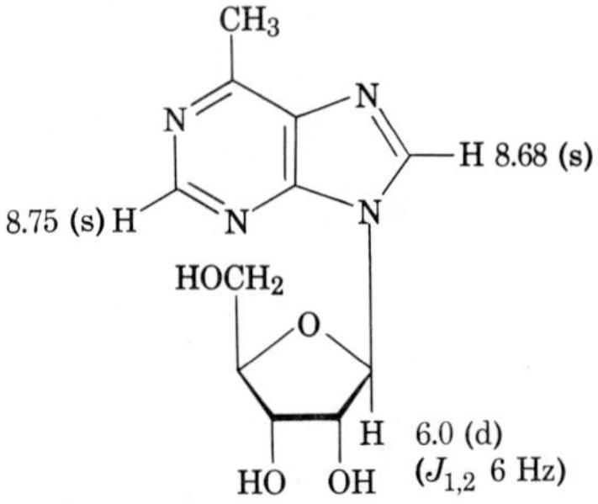

CH_3 N N H 8.68 (s) 8.75 (s) H N N $HOCH_2$ O H 6.0 (d) ($J_{1,2}$ 6 Hz) HO OH

(Ref. 99)

in methyl sulfoxide-d_6

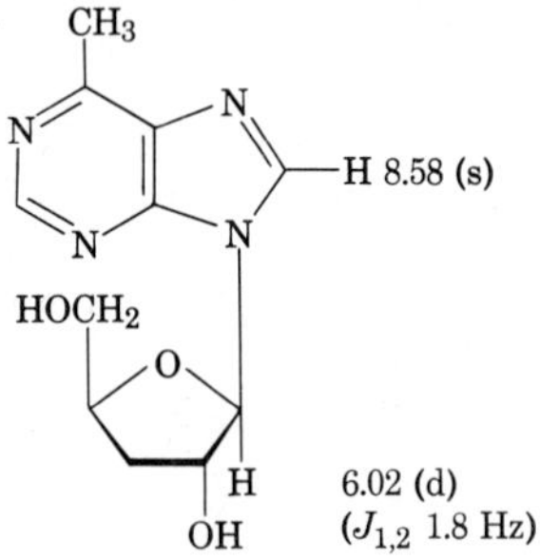

CH_3 N N H 8.58 (s) N N $HOCH_2$ O H 6.02 (d) ($J_{1,2}$ 1.8 Hz) OH

in methyl sulfoxide-d_6

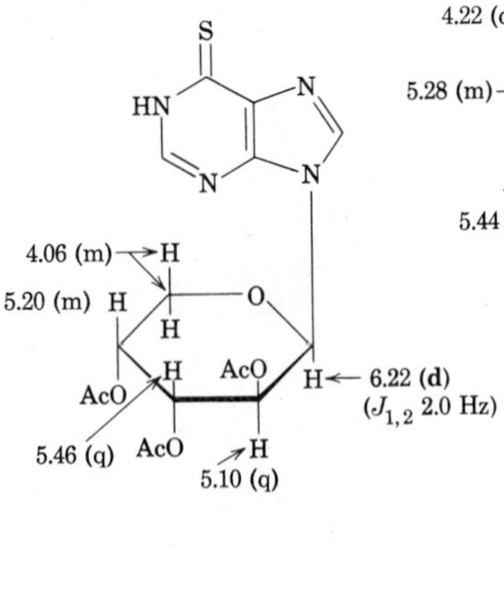

S HN N N N 4.06 (m) H 5.20 (m) H O H AcO H AcO H 6.22 (d) ($J_{1,2}$ 2.0 Hz) 5.46 (q) AcO H 5.10 (q)

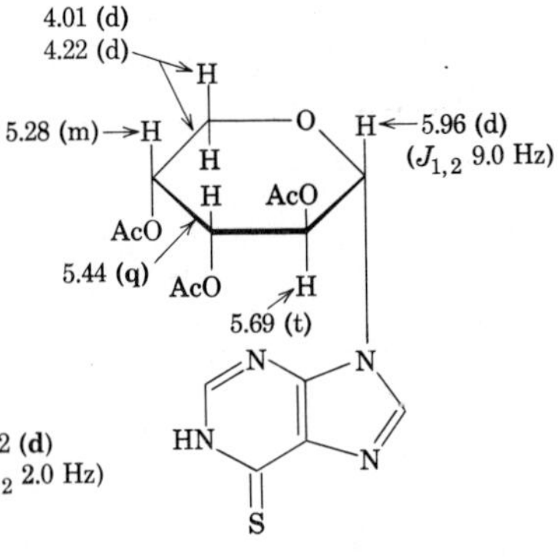

4.01 (d) 4.22 (d) H 5.28 (m) H O H 5.96 (d) ($J_{1,2}$ 9.0 Hz) H H AcO AcO 5.44 (q) AcO H 5.69 (t) N N HN N S

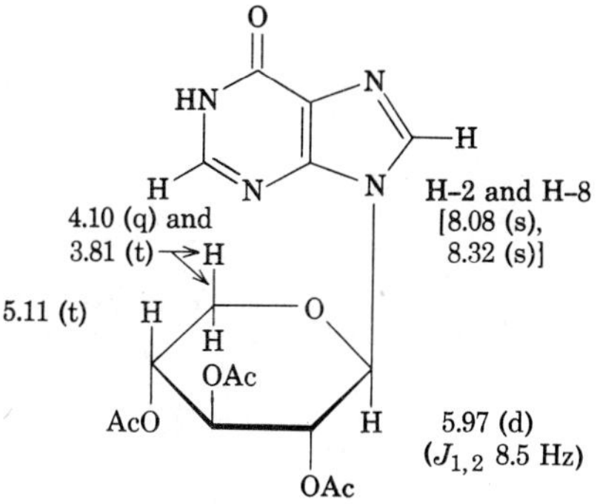

4.10 (q) and 3.81 (t) → H

5.11 (t) H

H–2 and H–8 [8.08 (s), 8.32 (s)]

5.97 (d) ($J_{1,2}$ 8.5 Hz)

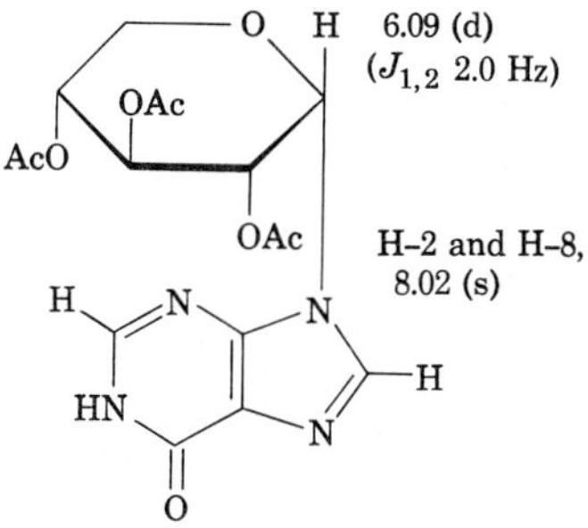

6.09 (d) ($J_{1,2}$ 2.0 Hz)

H–2 and H–8, 8.02 (s)

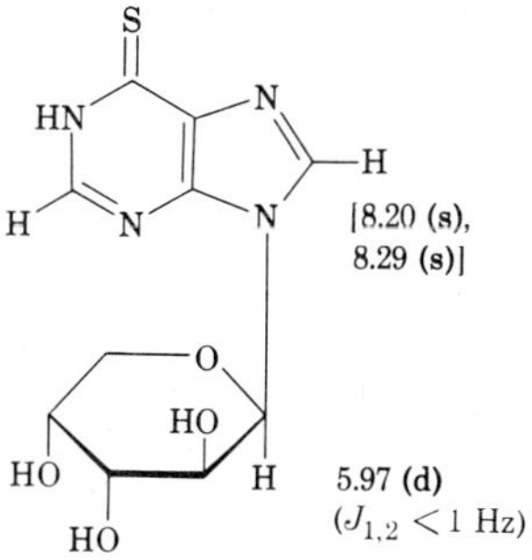

[8.20 (s), 8.29 (s)]

5.97 (d) ($J_{1,2} < 1$ Hz)

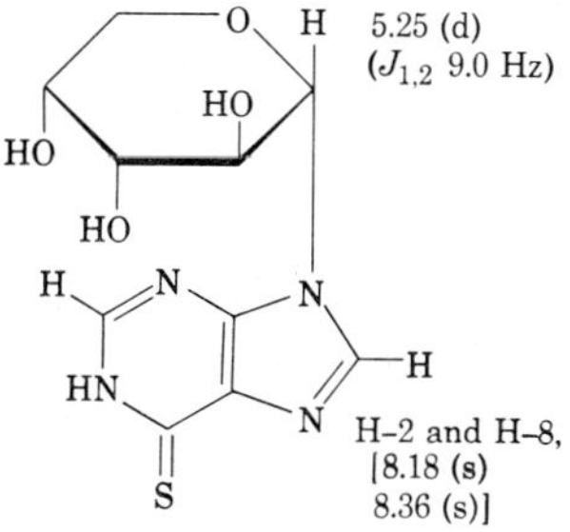

5.25 (d) ($J_{1,2}$ 9.0 Hz)

H–2 and H–8, [8.18 (s) 8.36 (s)]

In methyl sulfoxide-d_6 see Ref. 121.

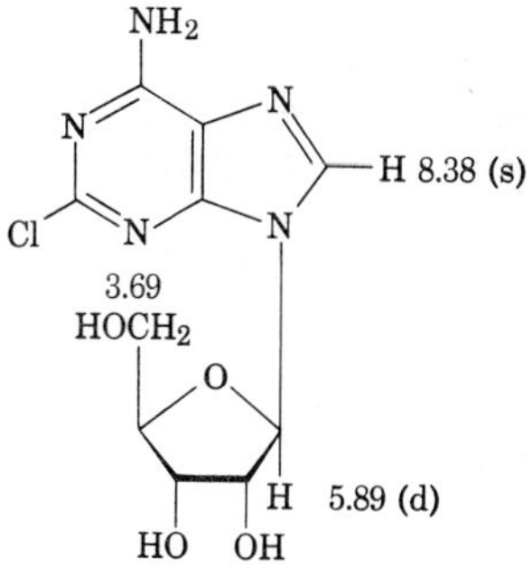

H 8.38 (s)

3.69 $HOCH_2$

H 5.89 (d)

(Ref. 255)

H 8.37 (s)

3.67 $HOCH_2$

H 5.85 (d)

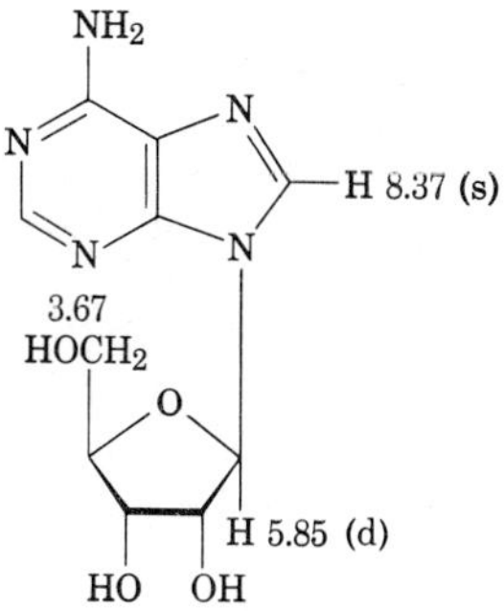

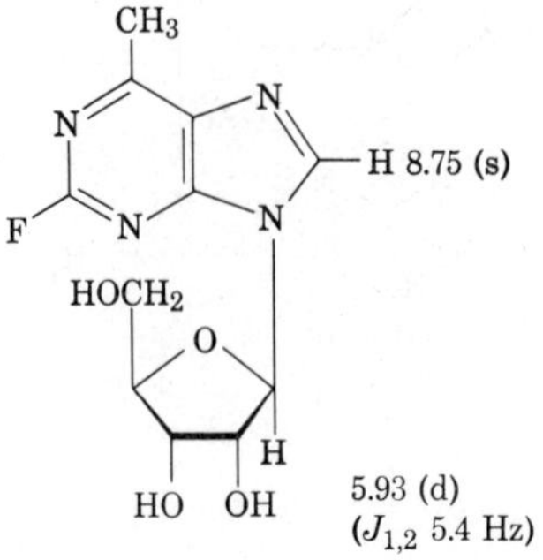

CH_3
N N
H 8.75 (s)
F N N
$HOCH_2$
O
H
HO OH
5.93 (d)
($J_{1,2}$ 5.4 Hz)

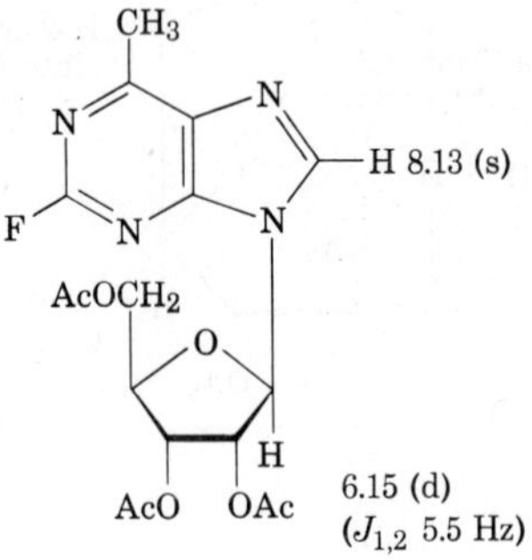

In $CDCl_3$; see Ref. 99.

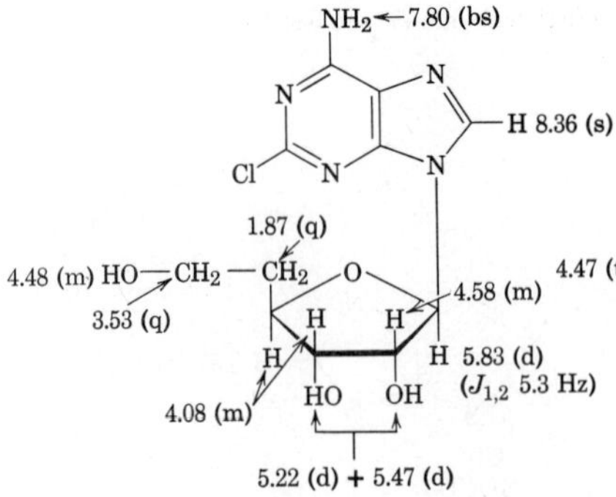

CH_3
N N
H 8.13 (s)
F N N
$AcOCH_2$
O
H
AcO OAc
6.15 (d)
($J_{1,2}$ 5.5 Hz)

NH_2 ← 7.80 (bs)
N N
H 8.36 (s)
Cl N N
1.87 (q)
4.48 (m) HO—CH_2—CH_2 O
4.58 (m)
3.53 (q)
H H
H 5.83 (d)
($J_{1,2}$ 5.3 Hz)
4.08 (m)
HO OH
5.22 (d) + 5.47 (d)

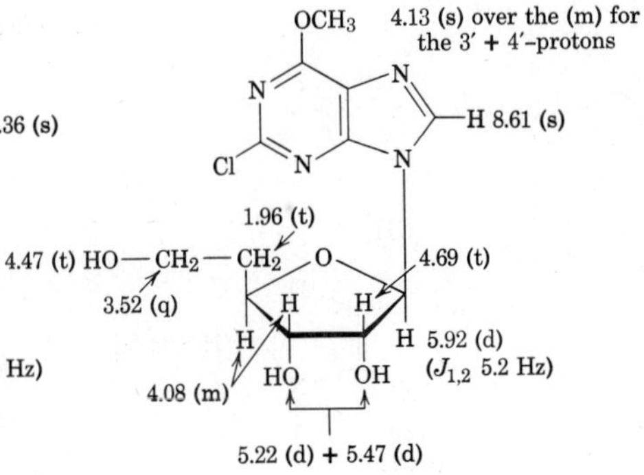

(Ref. 92) in methyl sulfoxide-d_6

OCH_3 4.13 (s) over the (m) for the 3′ + 4′-protons
N N
H 8.61 (s)
Cl N N
1.96 (t)
4.47 (t) HO—CH_2—CH_2 O
4.69 (t)
3.52 (q)
H H
H 5.92 (d)
($J_{1,2}$ 5.2 Hz)
4.08 (m)
HO OH
5.22 (d) + 5.47 (d)

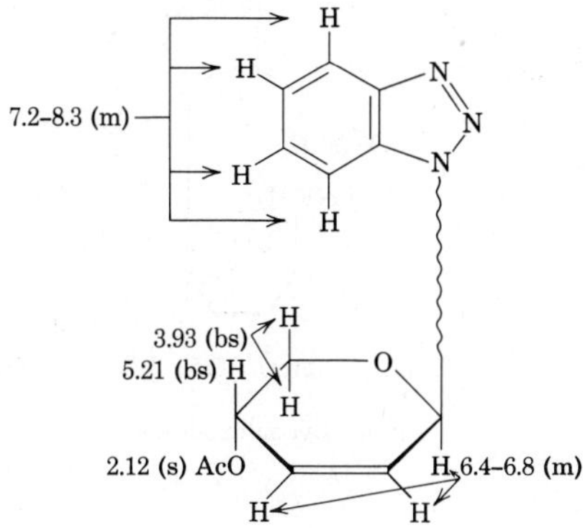

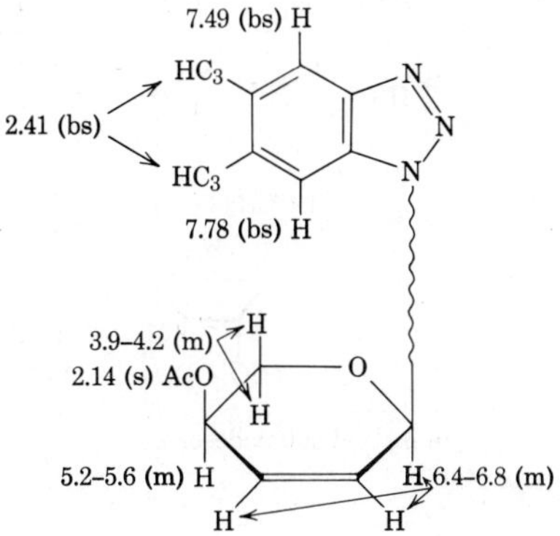

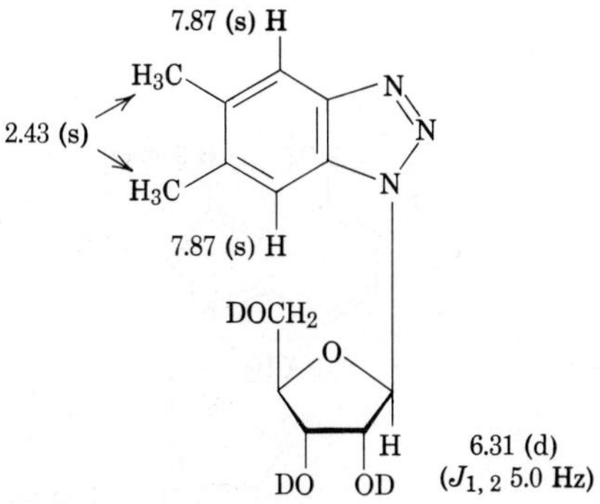

H
H
7.2–8.3 (m)
H
H
N
N
N
3.93 (bs) H
5.21 (bs) H
H
O
2.12 (s) AcO
H 6.4–6.8 (m)
H
H

7.49 (bs) H
HC_3
2.41 (bs)
HC_3
N
N
N
7.78 (bs) H
3.9–4.2 (m) H
2.14 (s) AcO
H
O
5.2–5.6 (m) H
H 6.4–6.8 (m)
H
H

7.87 (s) H
H_3C
2.43 (s)
H_3C
N
N
N
7.87 (s) H
$DOCH_2$
O
H
DO OD
6.31 (d)
($J_{1,2}$ 5.0 Hz)
in methyl sulfoxide-d_6–D_2O

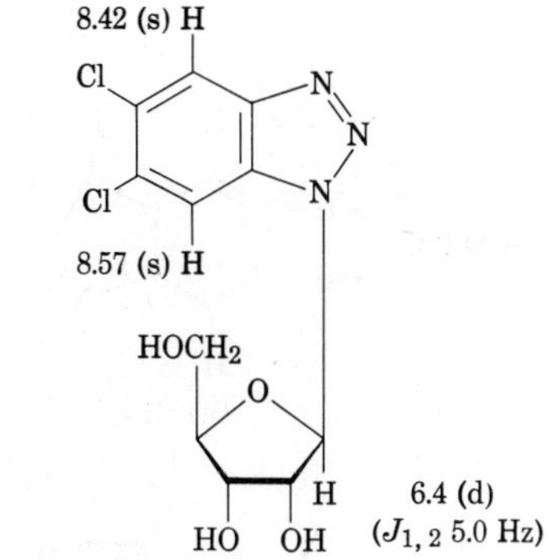

8.42 (s) H
Cl
Cl
N
N
N
8.57 (s) H
$HOCH_2$
O
H
HO OH
6.4 (d)
($J_{1,2}$ 5.0 Hz)
in methyl sulfoxide-d_6

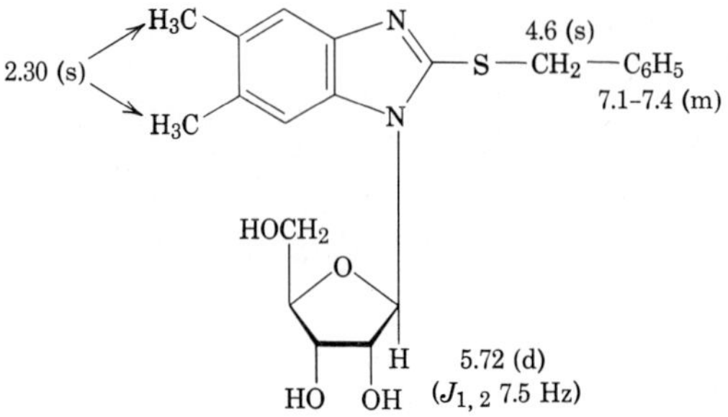

2.30 (s) → H_3C, H_3C; N; N; 4.6 (s) $S—CH_2—C_6H_5$ 7.1–7.4 (m); $HOCH_2$; O; H 5.72 (d) ($J_{1,2}$ 7.5 Hz); HO OH

in methyl sulfoxide-d_6
(Ref. 256)

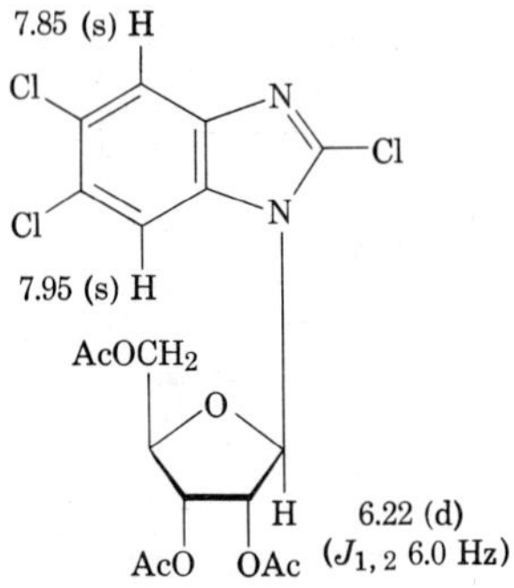

7.85 (s) H; Cl; Cl; N; N; Cl; 7.95 (s) H; $AcOCH_2$; O; H 6.22 (d) ($J_{1,2}$ 6.0 Hz); AcO OAc

in methyl sulfoxide-d_6

7.0–7.83 (m) → H, H, H, H; H 8.67 (s); N; N; $DOCH_2$; O; H 6.12 (d) ($J_{1,2}$ 3.0 Hz); DO OD

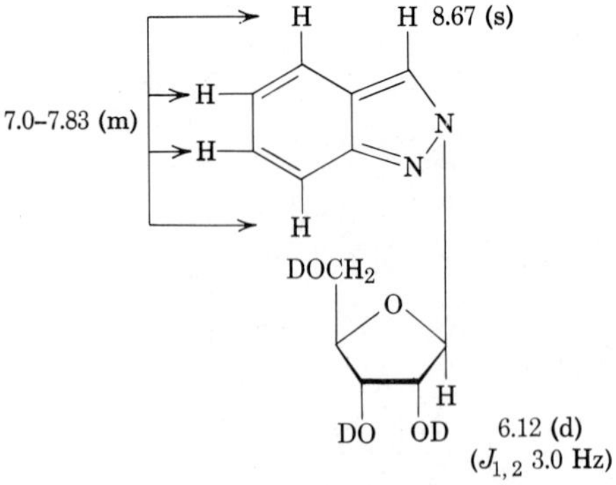

in methyl sulfoxide-d_6–D_2O

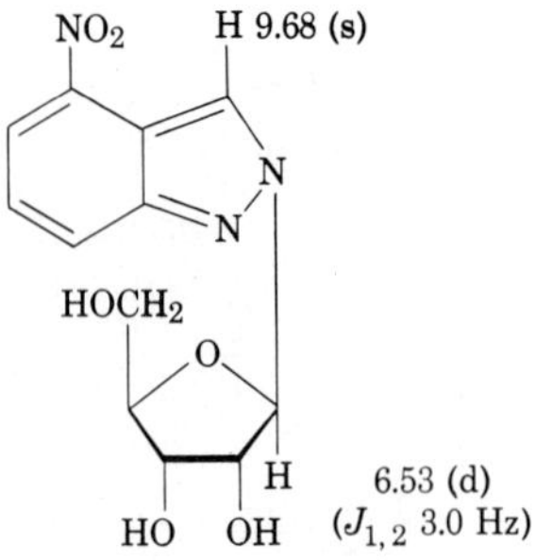

NO_2; H 9.68 (s); N; N; $HOCH_2$; O; H 6.53 (d) ($J_{1,2}$ 3.0 Hz); HO OH

in methyl sulfoxide-d_6

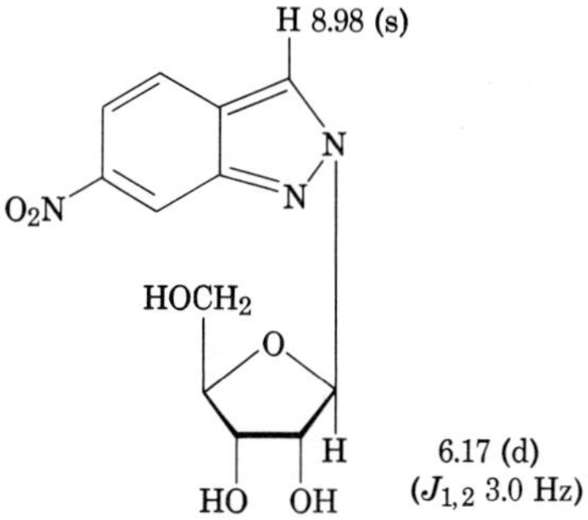

H 8.98 (s)
O_2N
N
N
$HOCH_2$
O
H
HO OH
6.17 (d)
($J_{1,2}$ 3.0 Hz)
in methyl sulfoxide-d_6

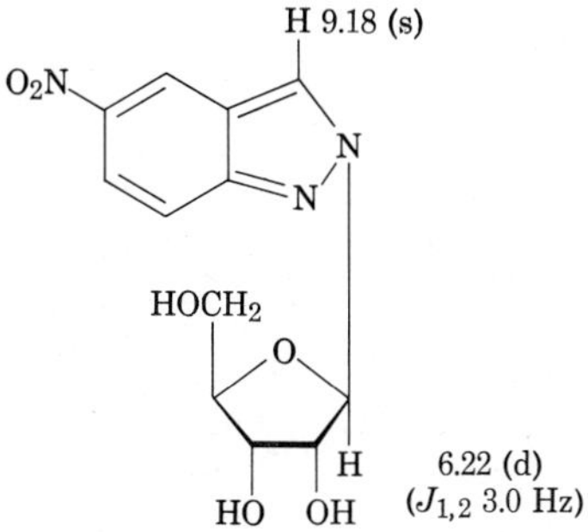

H 9.18 (s)
O_2N
N
N
$HOCH_2$
O
H
HO OH
6.22 (d)
($J_{1,2}$ 3.0 Hz)
in methyl sulfoxide-d_6

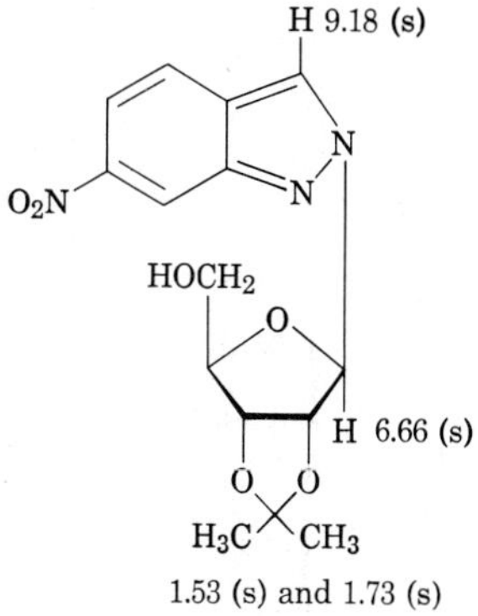

H 9.18 (s)
O_2N
N
N
$HOCH_2$
O
H 6.66 (s)
O O
H_3C CH_3
1.53 (s) and 1.73 (s)
in methyl sulfoxide-d_6

(Ref. 257)

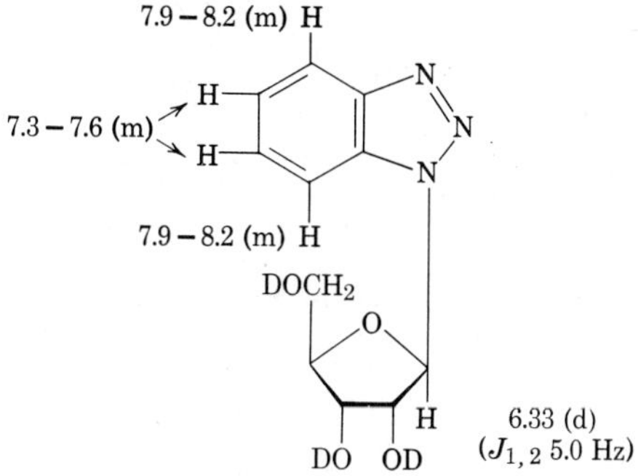

7.9 – 8.2 (m) H
7.3 – 7.6 (m)
H
H
N
N
N
7.9 – 8.2 (m) H
$DOCH_2$
O
H
DO OD
6.33 (d)
($J_{1,2}$ 5.0 Hz)
in methyl sulfoxide-d_6

*

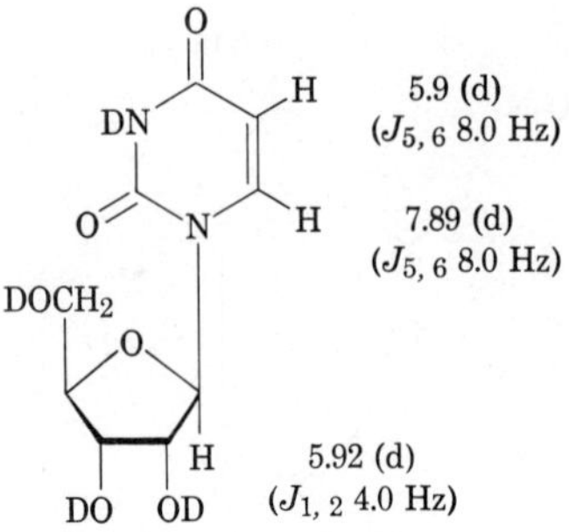

O
DN H 5.9 (d) ($J_{5,6}$ 8.0 Hz)
O N H 7.89 (d) ($J_{5,6}$ 8.0 Hz)
$DOCH_2$
O
H 5.92 (d) ($J_{1,2}$ 4.0 Hz)
DO OD

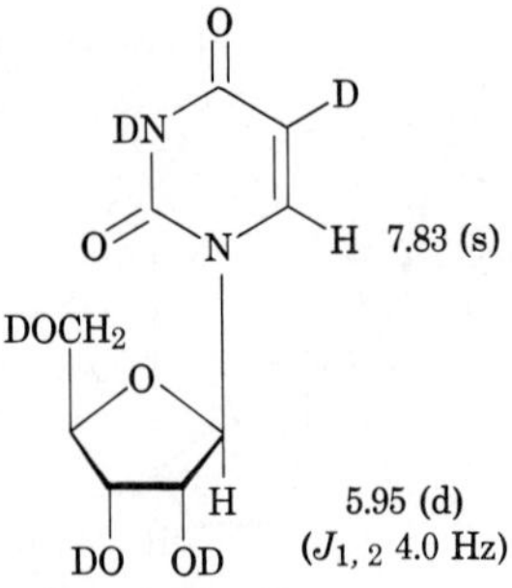

O
DN D
O N H 7.83 (s)
$DOCH_2$
O
H 5.95 (d) ($J_{1,2}$ 4.0 Hz)
DO OD

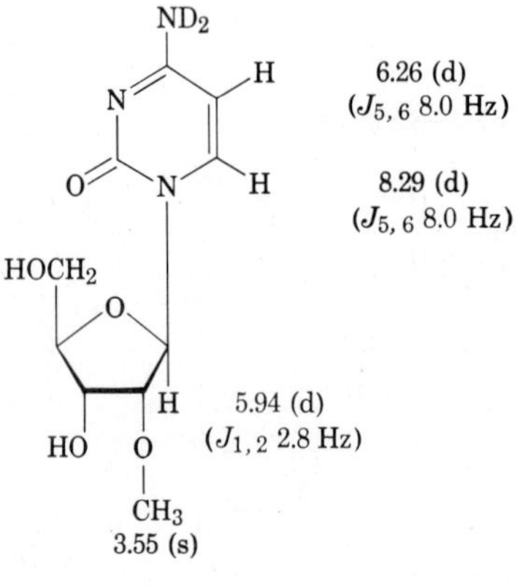

ND_2
N H 6.26 (d) ($J_{5,6}$ 8.0 Hz)
O N H 8.29 (d) ($J_{5,6}$ 8.0 Hz)
$HOCH_2$
O
H 5.94 (d) ($J_{1,2}$ 2.8 Hz)
HO O
CH_3
3.55 (s)

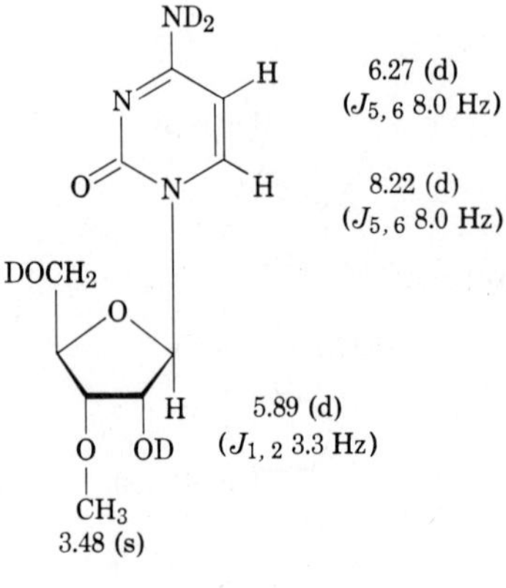

ND_2
N H 6.27 (d) ($J_{5,6}$ 8.0 Hz)
O N H 8.22 (d) ($J_{5,6}$ 8.0 Hz)
$DOCH_2$
O
H 5.89 (d) ($J_{1,2}$ 3.3 Hz)
O OD
CH_3
3.48 (s)

*In D_2O, with Tiers salt [3-(trimethylsilyl)-propanesulfonic acid sodium salt] as internal standard; see Ref. 53.

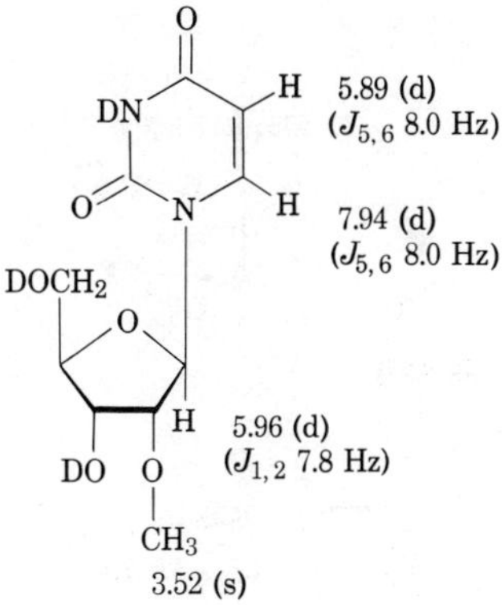

O
DN
H 5.89 (d) ($J_{5,6}$ 8.0 Hz)
O N H 7.94 (d) ($J_{5,6}$ 8.0 Hz)
DOCH$_2$
O
H 5.96 (d) ($J_{1,2}$ 7.8 Hz)
DO O
CH$_3$
3.52 (s)

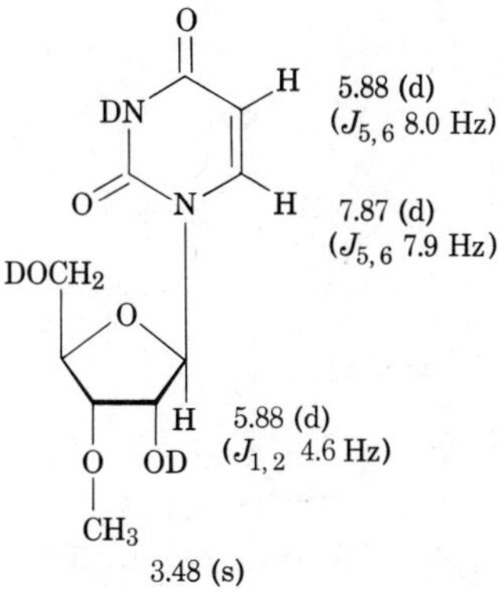

O
DN
H 5.88 (d) ($J_{5,6}$ 8.0 Hz)
O N H 7.87 (d) ($J_{5,6}$ 7.9 Hz)
DOCH$_2$
O
H 5.88 (d) ($J_{1,2}$ 4.6 Hz)
O OD
CH$_3$
3.48 (s)

(Ref. 258)

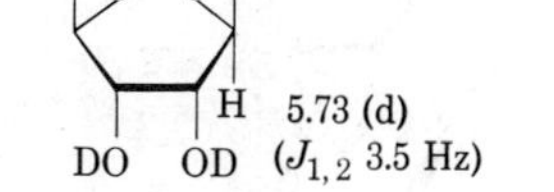

ND$_2$
H 5.88 (s)
N
O N CH$_3$ 2.37 (s)
DOCH$_2$
O
H 5.73 (d) ($J_{1,2}$ 3.5 Hz)
DO OD

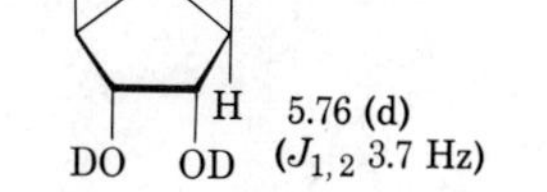

O
DN
H 5.83 (s)
O N CH$_3$ 2.42 (s)
DOCH$_2$
O
H 5.76 (d) ($J_{1,2}$ 3.7 Hz)
DO OD

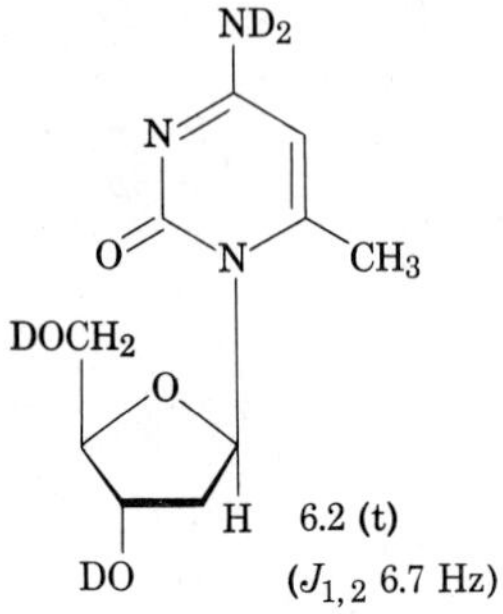

ND_2
N
O N CH_3
$DOCH_2$
O
H 6.2 (t)
DO ($J_{1,2}$ 6.7 Hz)

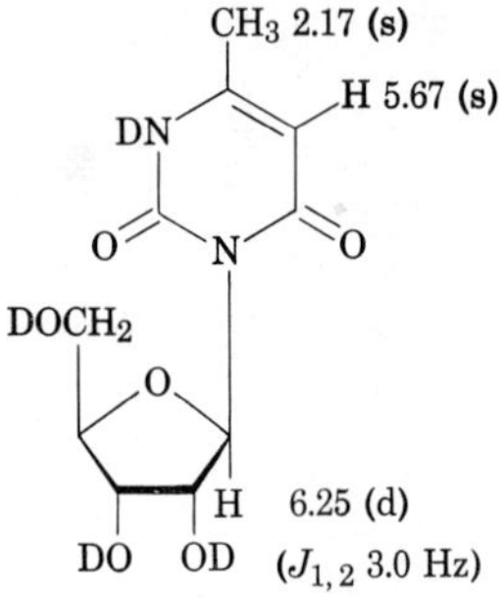

CH_3 2.17 (s)
H 5.67 (s)
DN
O N O
$DOCH_2$
O
H 6.25 (d)
DO OD ($J_{1,2}$ 3.0 Hz)

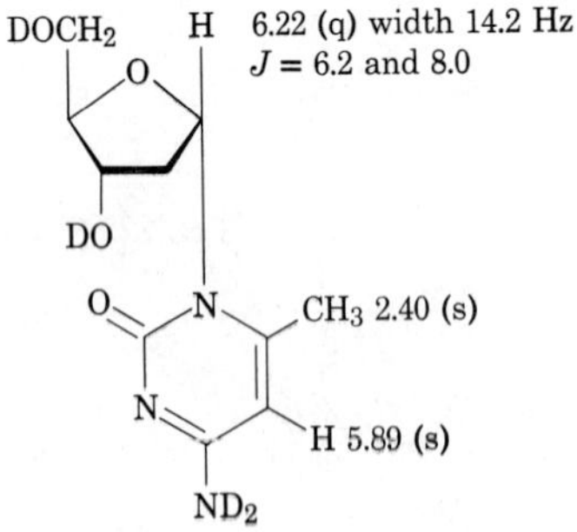

$DOCH_2$ H 6.22 (q) width 14.2 Hz
J = 6.2 and 8.0
O
DO
O N CH_3 2.40 (s)
N H 5.89 (s)
ND_2
(Ref. 61)

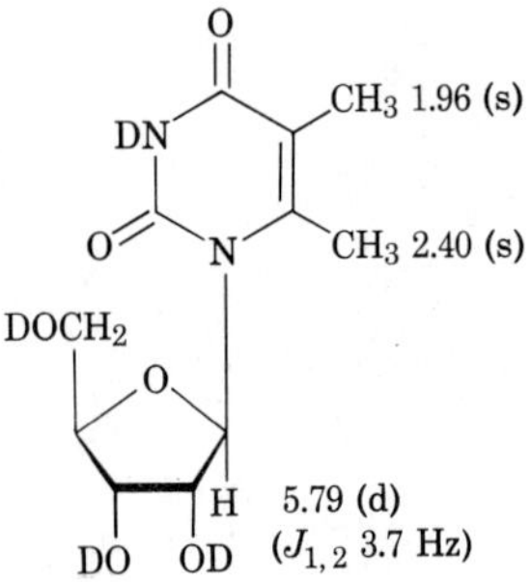

O
CH_3 1.96 (s)
DN
O N CH_3 2.40 (s)
$DOCH_2$
O
H 5.79 (d)
DO OD ($J_{1,2}$ 3.7 Hz)

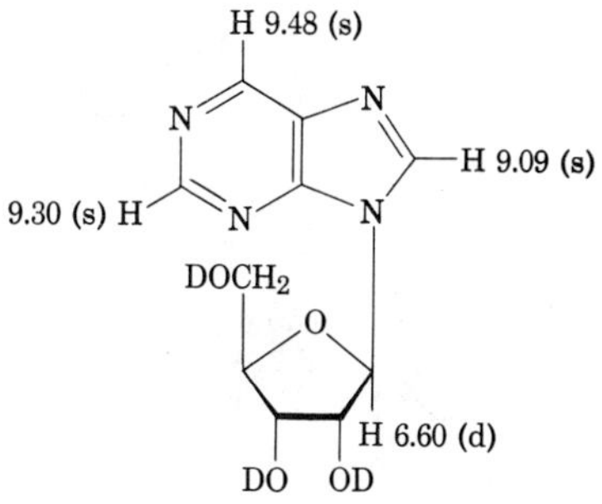

H 9.48 (s)
N N
9.30 (s) H N N —H 9.09 (s)
DOCH$_2$
O
H 6.60 (d)
DO OD

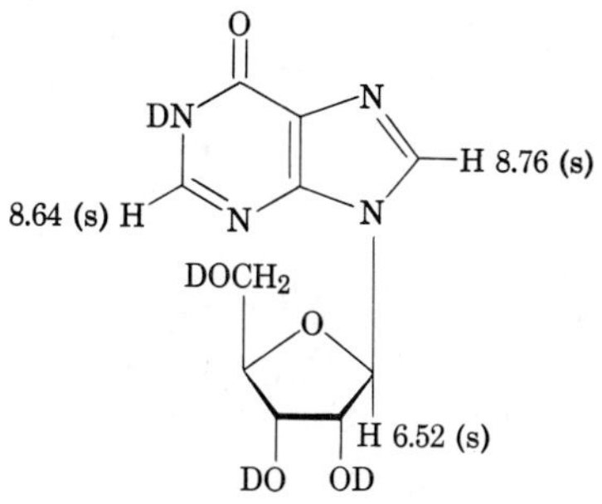

O
DN N
8.64 (s) H N N —H 8.76 (s)
DOCH$_2$
O
H 6.52 (s)
DO OD

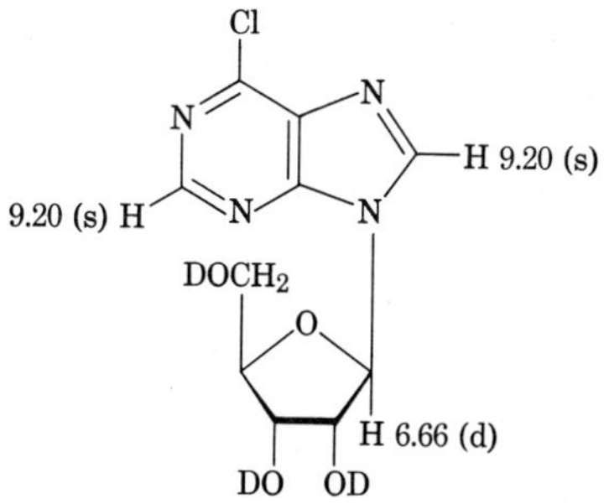

Cl
N N
9.20 (s) H N N —H 9.20 (s)
DOCH$_2$
O
H 6.66 (d)
DO OD

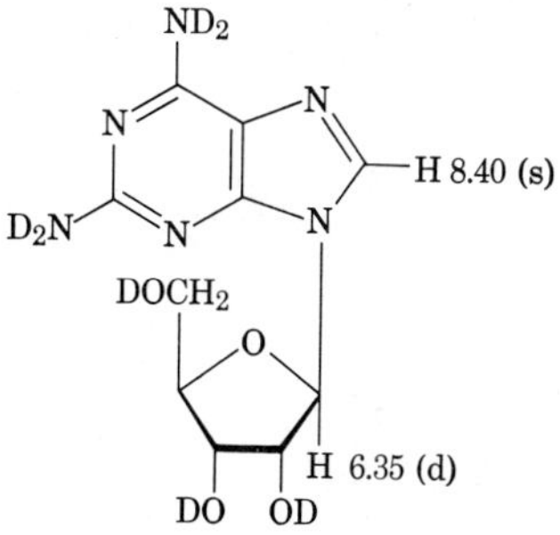

ND$_2$
N N
D$_2$N N N —H 8.40 (s)
DOCH$_2$
O
H 6.35 (d)
DO OD

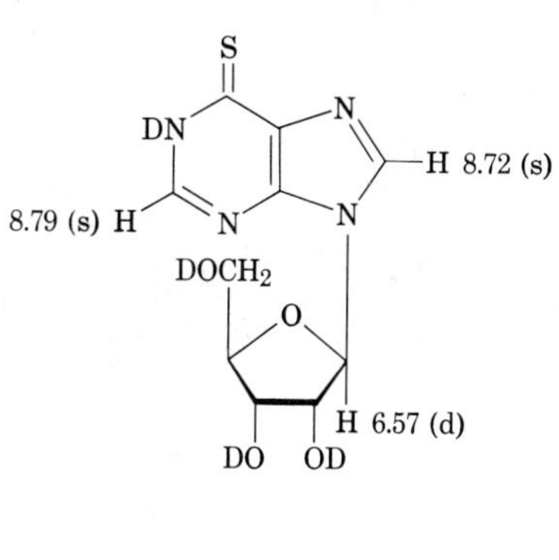

S
DN N
8.79 (s) H N N —H 8.72 (s)
DOCH$_2$
O
H 6.57 (d)
DO OD

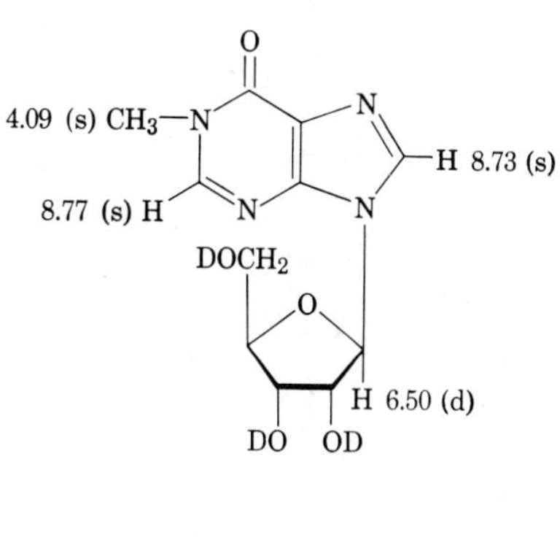

O
4.09 (s) CH$_3$—N N
8.77 (s) H N N —H 8.73 (s)
DOCH$_2$
O
H 6.50 (d)
DO OD

(Ref. 259)

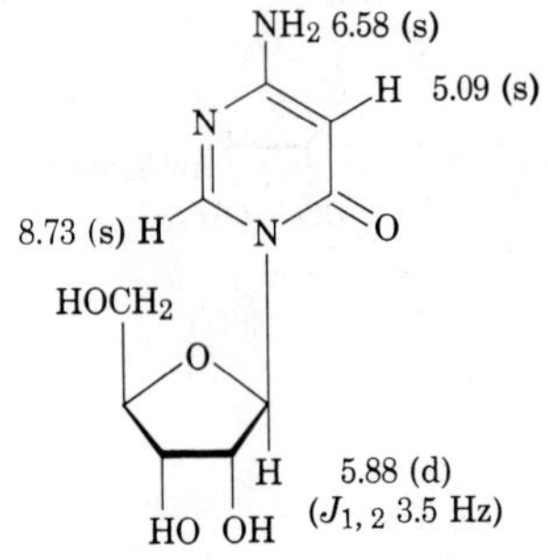
NH2 6.58 (s)
H 5.09 (s)
N
8.73 (s) H
N
O
HOCH2
O
H
5.88 (d)
(J1, 2 3.5 Hz)
HO OH

in methyl sulfoxide-d_6

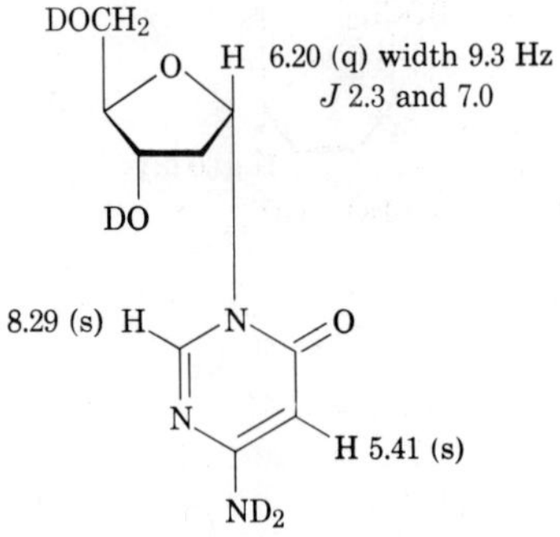
DOCH2
O
H 6.20 (q) width 9.3 Hz
J 2.3 and 7.0
DO
8.29 (s) H
N
O
N
H 5.41 (s)
ND2

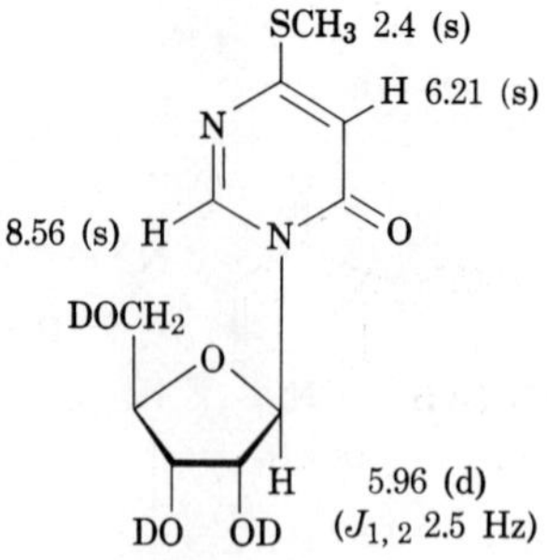
SCH3 2.4 (s)
H 6.21 (s)
N
8.56 (s) H
N
O
DOCH2
O
H
5.96 (d)
(J1, 2 2.5 Hz)
DO OD

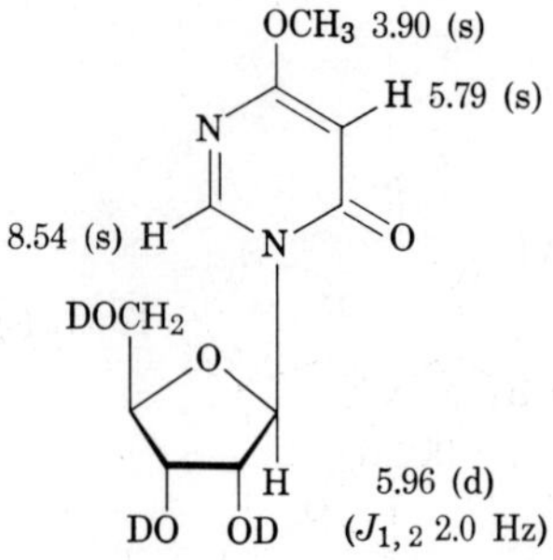
OCH3 3.90 (s)
H 5.79 (s)
N
8.54 (s) H
N
O
DOCH2
O
H
5.96 (d)
(J1, 2 2.0 Hz)
DO OD

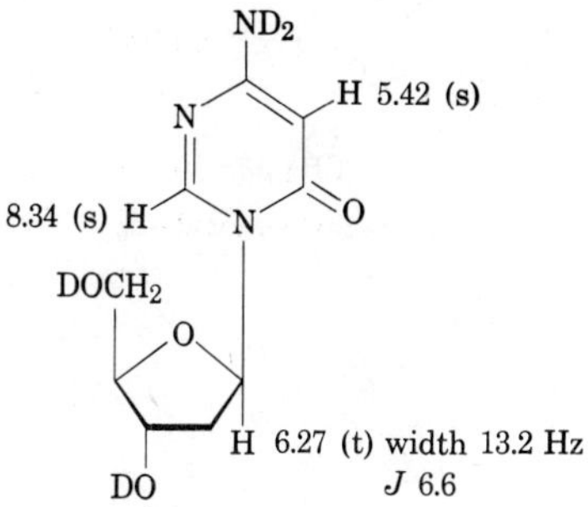

ND_2 — H 5.42 (s) — 8.34 (s) H — N — O — $DOCH_2$ — H 6.27 (t) width 13.2 Hz, J 6.6 — DO

in D_2O

(Ref. 63)

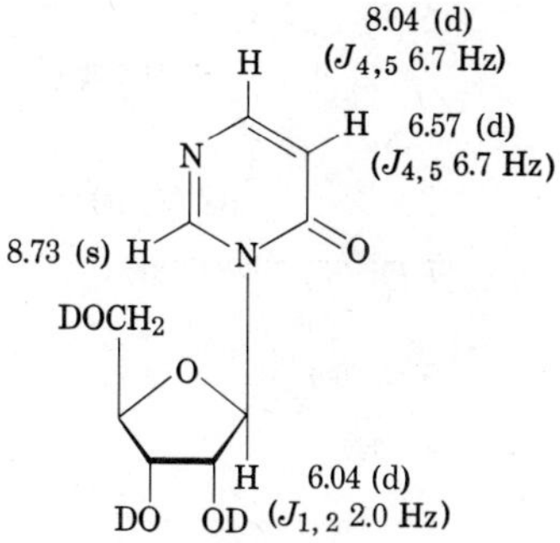

H 8.04 (d) ($J_{4,5}$ 6.7 Hz) — H 6.57 (d) ($J_{4,5}$ 6.7 Hz) — 8.73 (s) H — N — O — $DOCH_2$ — H 6.04 (d) ($J_{1,2}$ 2.0 Hz) — DO — OD

in D_2O

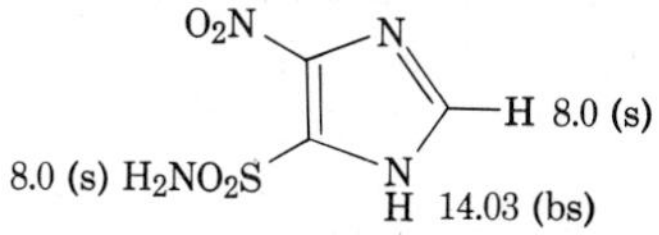

O_2N — N — H 8.0 (s) — 8.0 (s) H_2NO_2S — N H 14.03 (bs)

in methyl sulfoxide-d_6

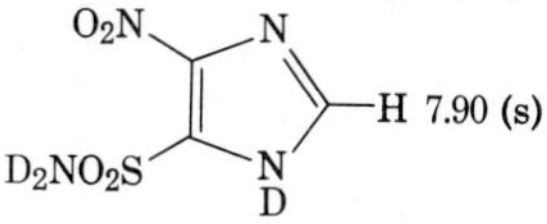

O_2N — N — H 7.90 (s) — D_2NO_2S — N D

in methyl sulfoxide-d_6—D_2O

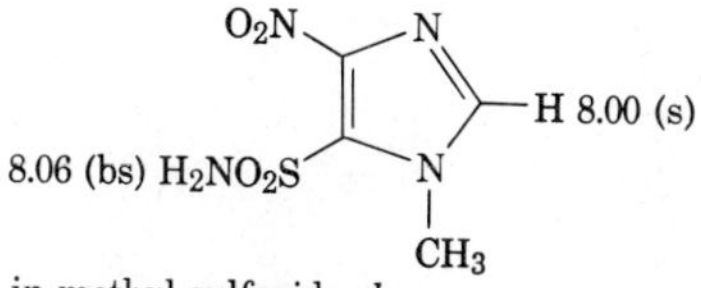

O_2N — N — H 8.00 (s) — 8.06 (bs) H_2NO_2S — N — CH_3

in methyl sulfoxide-d_6

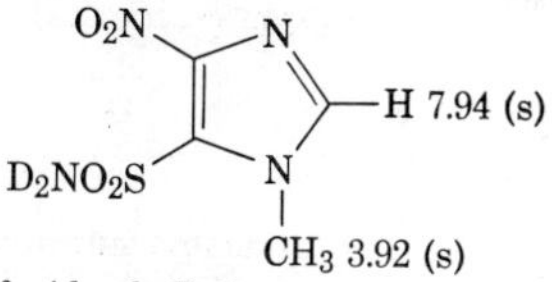

O_2N — N — H 7.94 (s) — D_2NO_2S — N — CH_3 3.92 (s)

in methyl sulfoxide-d_6-D_2O

O_2N — N — H 7.98 (s) — Cl — N — CH_3 3.75 (s)

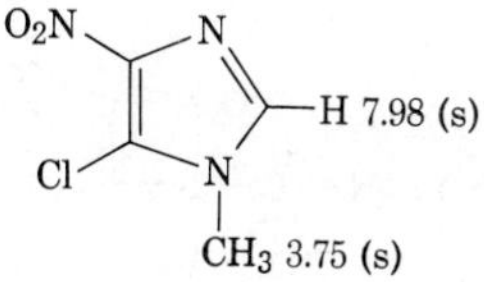

in methyl sulfoxide-d_6

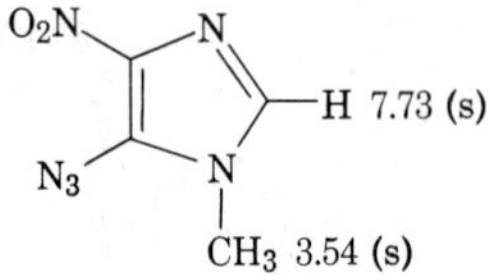

O_2N, N_3, N, N—CH_3 3.54 (s), —H 7.73 (s)

in methyl sulfoxide-d_6

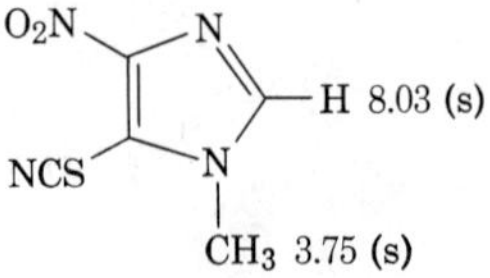

O_2N, NCS, N, N—CH_3 3.75 (s), —H 8.03 (s)

in methyl sulfoxide-d_6

(Ref. 260)

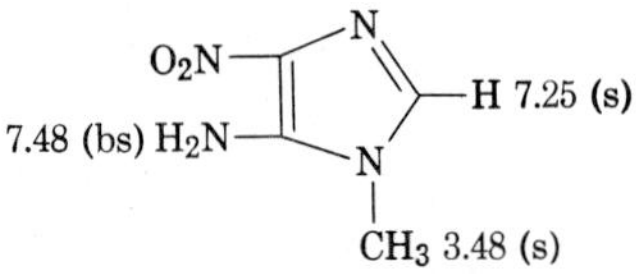

O_2N—, 7.48 (bs) H_2N—, N, N—CH_3 3.48 (s), —H 7.25 (s)

in methyl sulfoxide-d_6

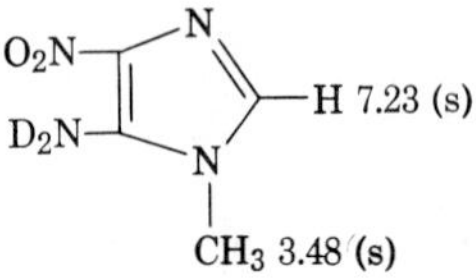

O_2N—, D_2N—, N, N—CH_3 3.48 (s), —H 7.23 (s)

in methyl sulfoxide-d_6-D_2O

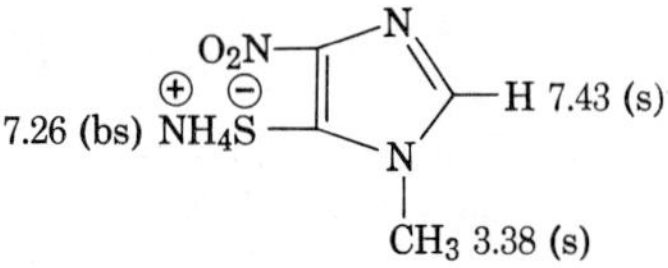

O_2N—, 7.26 (bs) $\overset{\oplus}{N}H_4\overset{\ominus}{S}$—, N, N—$CH_3$ 3.38 (s), —H 7.43 (s)

in methyl sulfoxide-d_6

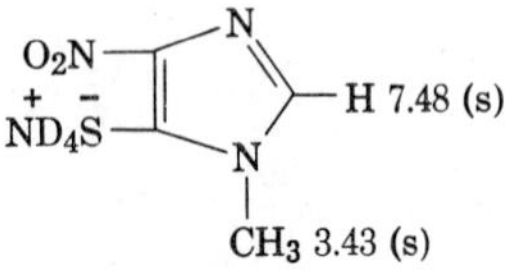

O_2N—, $\overset{+}{N}D_4\overset{-}{S}$—, N, N—$CH_3$ 3.43 (s), —H 7.48 (s)

in methyl sulfoxide-d_6′-D_2O

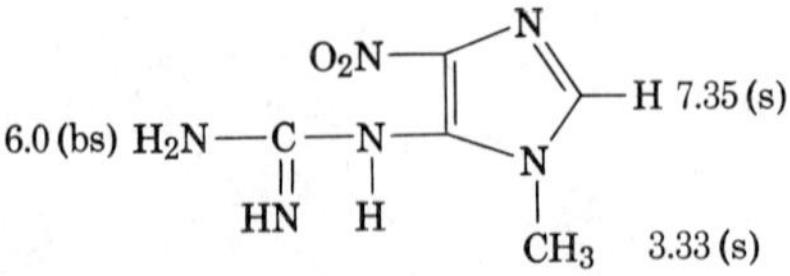

O_2N—, 6.0 (bs) H_2N—C(=HN)—N(H)—, N, N—CH_3 3.33 (s), —H 7.35 (s)

in methyl sulfoxide-d_6

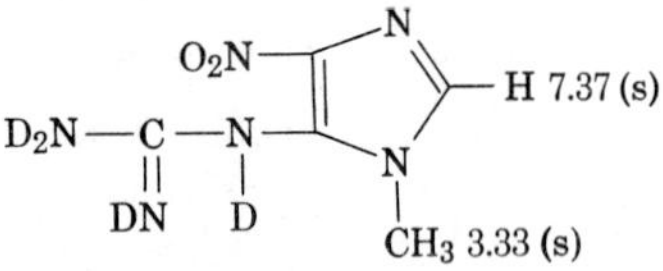

O_2N—, D_2N—C(=DN)—N(D)—, N, N—CH_3 3.33 (s), —H 7.37 (s)

in methyl sulfoxide-d_6′-D_2O

(Ref. 261)

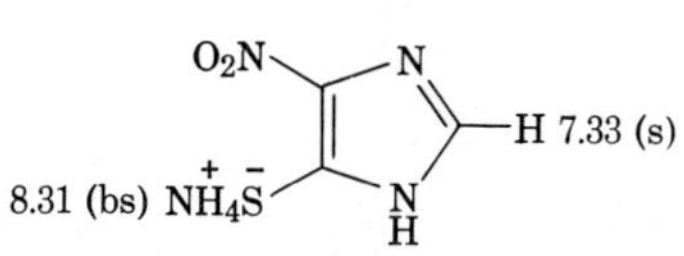

O_2N / $\overset{+}{N}H_4\overset{-}{S}$ 8.31 (bs); H 7.33 (s); N–H

in methyl sulfoxide-d_6

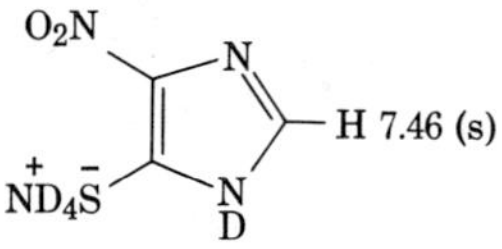

O_2N / $\overset{+}{N}D_4\overset{-}{S}$; H 7.46 (s); N–D

in methyl sulfoxide-d_6-D_2O

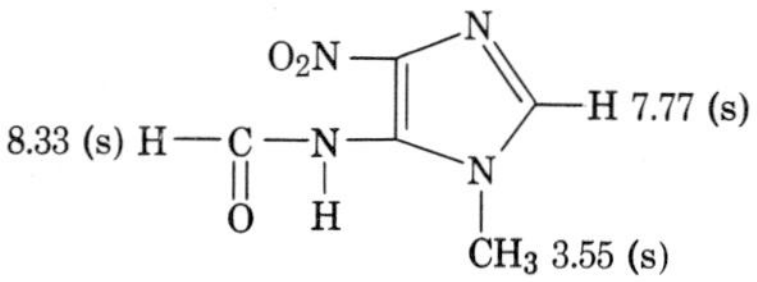

8.33 (s) H—C(=O)—N(H)— ; O_2N; H 7.77 (s); CH_3 3.55 (s)

in methyl sulfoxide-d_6

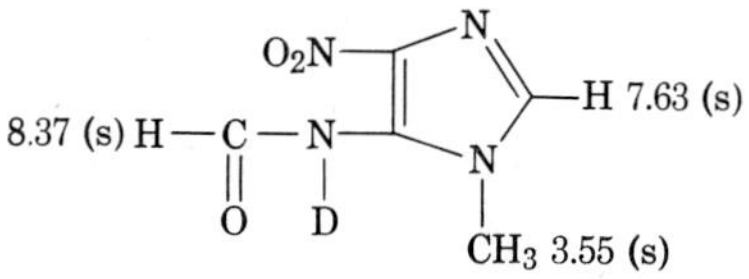

8.37 (s) H—C(=O)—N(D)— ; O_2N; H 7.63 (s); CH_3 3.55 (s)

in methyl sulfoxide-d_6'-D_2O

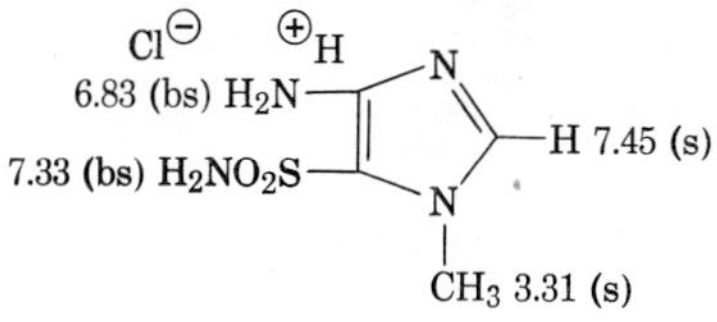

$Cl^{\ominus}$ $^{\oplus}H$ 6.83 (bs) H_2N; 7.33 (bs) H_2NO_2S; H 7.45 (s); CH_3 3.31 (s)

in methyl sulfoxide-d_6

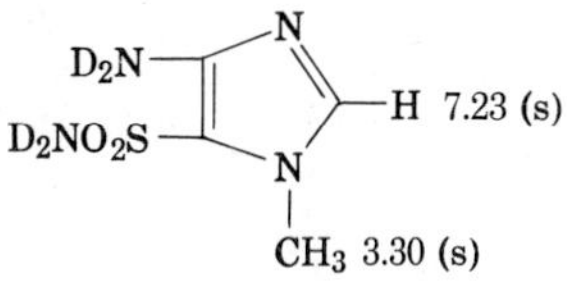

D_2N; D_2NO_2S; H 7.23 (s); CH_3 3.30 (s)

in methyl sulfoxide-d_6'-D_2O

(Ref. 262)

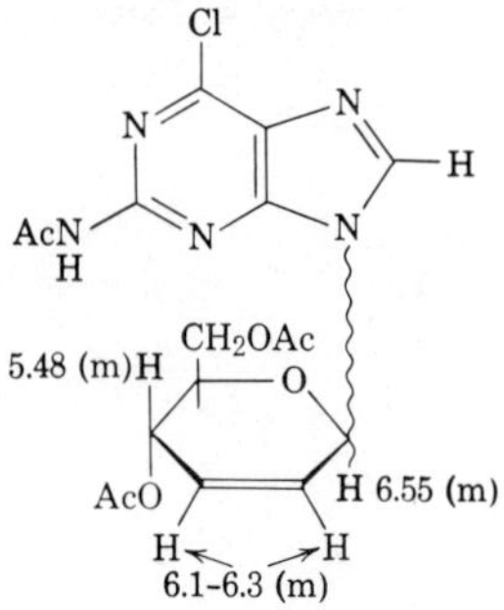

Cl, N, N, H, AcN, H, N, N, CH$_2$OAc, 5.48 (m)H, O, AcO, H 6.55 (m), H, H, 6.1–6.3 (m)

in chloroform-d

(Ref. 263)

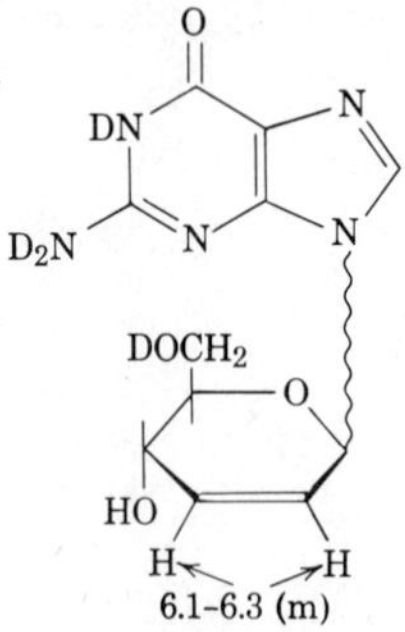

O, DN, N, D_2N, N, N, $DOCH_2$, O, HO, H, H, 6.1–6.3 (m)

in methyl sulfoxide-d_6-D_2O

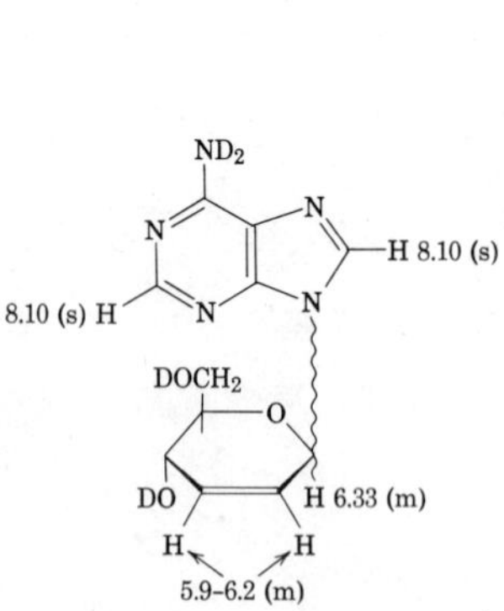

ND_2, N, N, H 8.10 (s), 8.10 (s) H, N, N, $DOCH_2$, O, DO, H 6.33 (m), H, H, 5.9–6.2 (m)

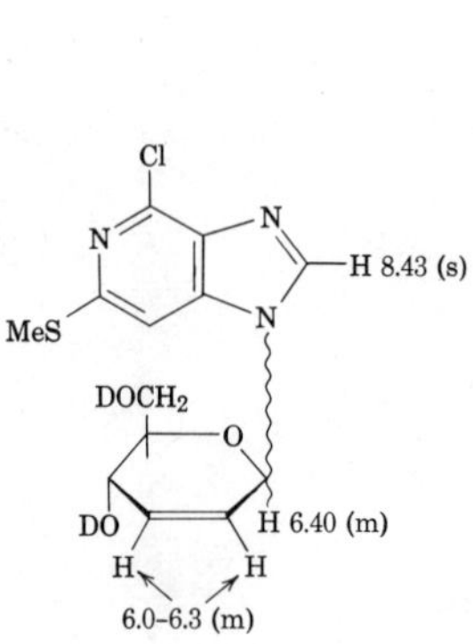

Cl, N, N, H 8.43 (s), MeS, N, $DOCH_2$, O, DO, H 6.40 (m), H, H, 6.0–6.3 (m)

in methyl sulfoxide-d_6'-D_2O

(Ref. 264)

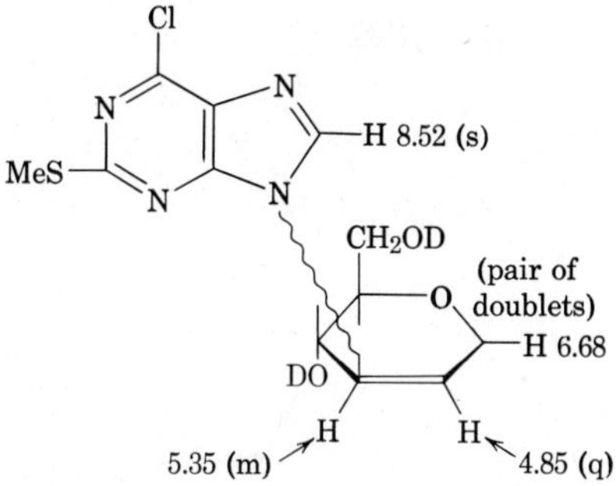

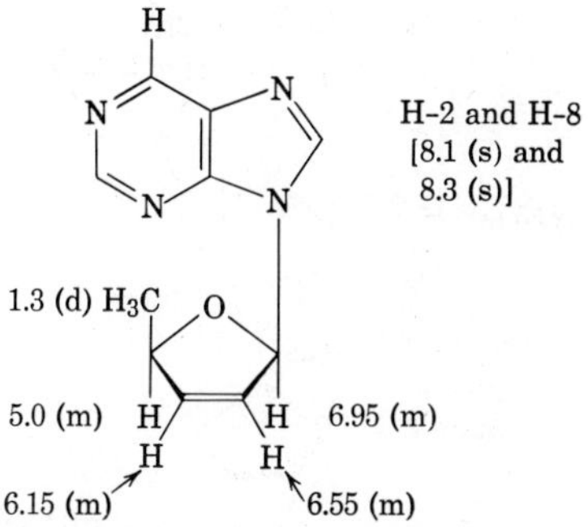

Cl
N
N
MeS
N
N
H 8.52 (s)
CH_2OD
(pair of doublets)
O
H 6.68
DO
5.35 (m) H
H 4.85 (q)

H
N
N
N
N
H-2 and H-8 [8.1 (s) and 8.3 (s)]
1.3 (d) H_3C
O
5.0 (m) H
H 6.95 (m)
6.15 (m) H
H 6.55 (m)

(Ref. 104)

in methyl sulfoxide-d_6'-D_2O

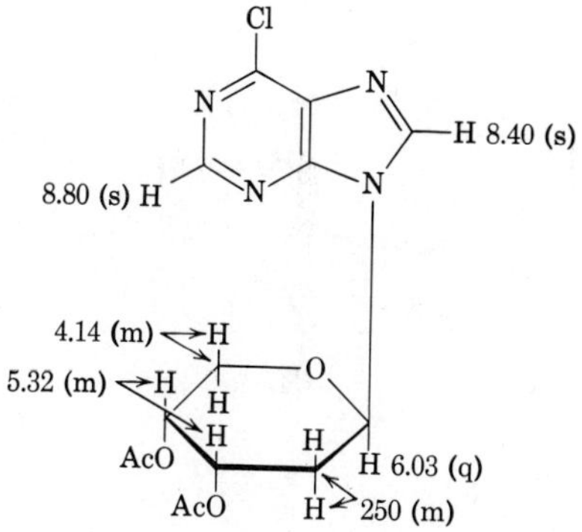

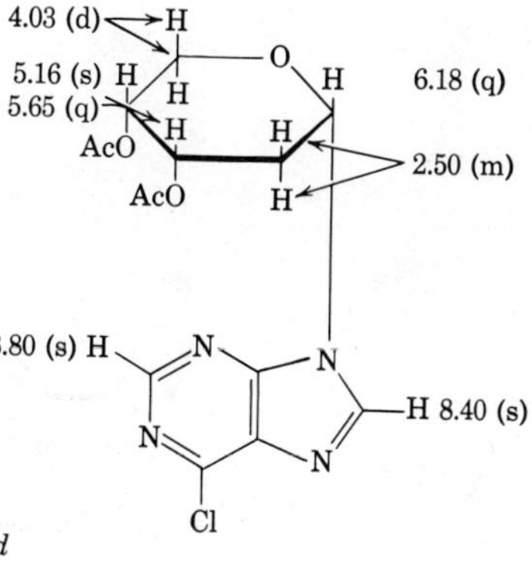

Cl
N
N
H 8.40 (s)
8.80 (s) H
N
N
4.14 (m) H
O
5.32 (m) H
H
H
H
AcO
H 6.03 (q)
AcO
H 250 (m)

4.03 (d) H
O
5.16 (s) H
H
H 6.18 (q)
5.65 (q) H
H
AcO
2.50 (m)
AcO
H
8.80 (s) H
N
N
H 8.40 (s)
N
N
Cl

(Ref. 125)

in chloroform-d

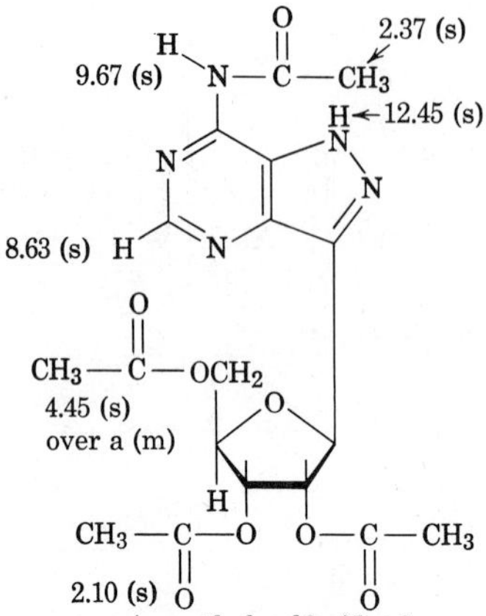

9.67 (s) H–N–C(=O)–CH$_3$ 2.37 (s)
H ← 12.45 (s)
8.63 (s) H
CH$_3$–C(=O)–OCH$_2$
4.45 (s)
over a (m)
H
CH$_3$–C(=O)–O O–C(=O)–CH$_3$
2.10 (s)

in methyl sulfoxide-d_6

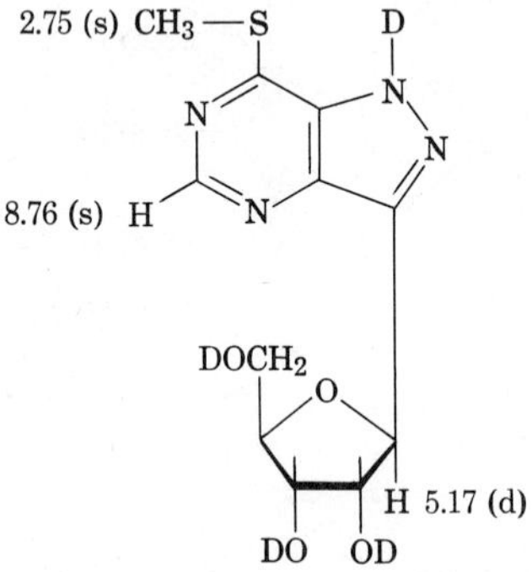

2.75 (s) CH$_3$–S D
8.76 (s) H
DOCH$_2$
H 5.17 (d)
DO OD

in methyl sulfoxide-d_6′-D_2O

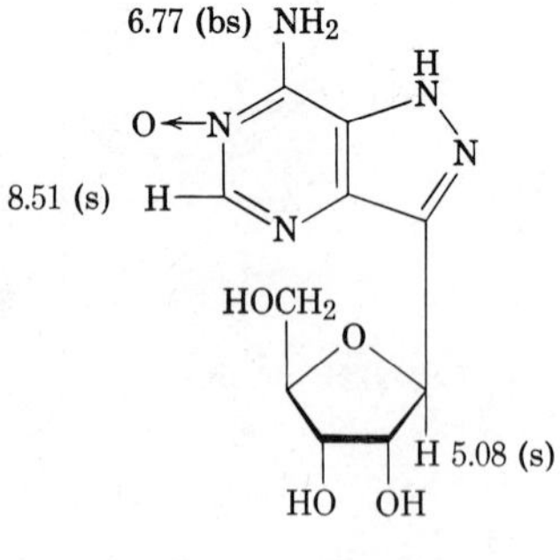

6.77 (bs) NH$_2$
O←N
8.51 (s) H
HOCH$_2$
H 5.08 (s)
HO OH

in methyl sulfoxide-d_6

CH$_3$ 4.00 (s)
NC
O_2N
CH$_3$ 2.45 (s)

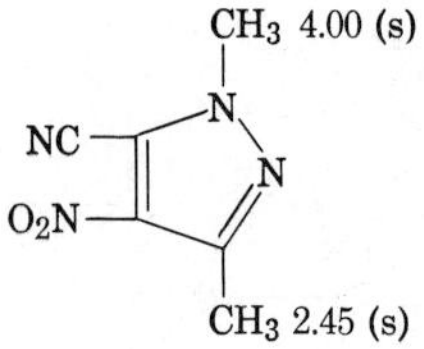

in methyl sulfoxide-d_6

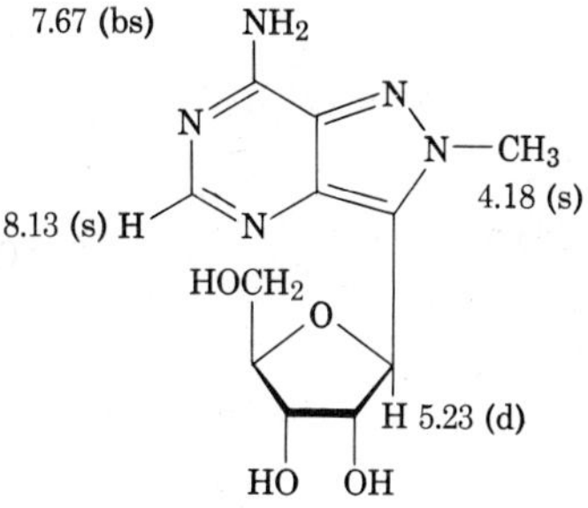

7.67 (bs) NH_2
8.13 (s) H
$N-CH_3$ 4.18 (s)
$HOCH_2$
H 5.23 (d)
HO OH

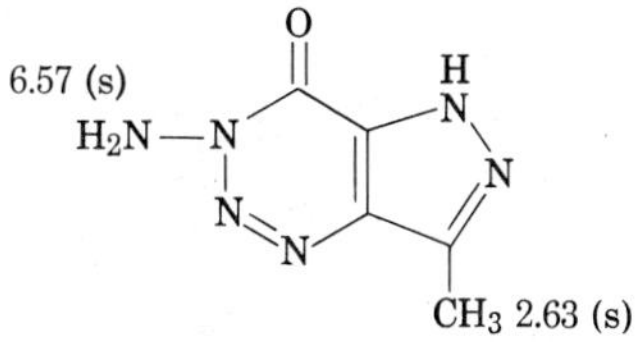

in methyl sulfoxide-d_6

(Ref. 265)

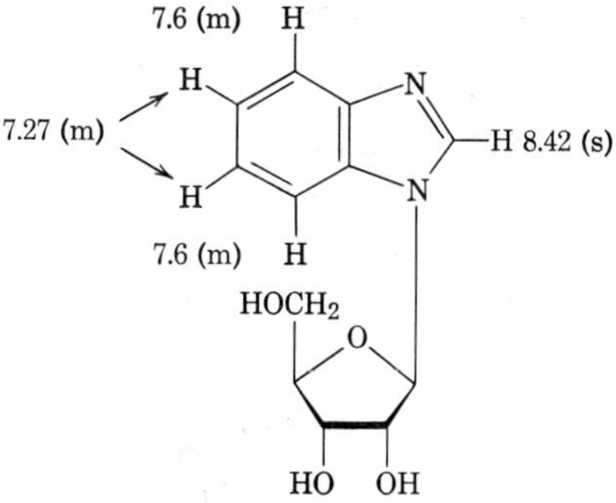

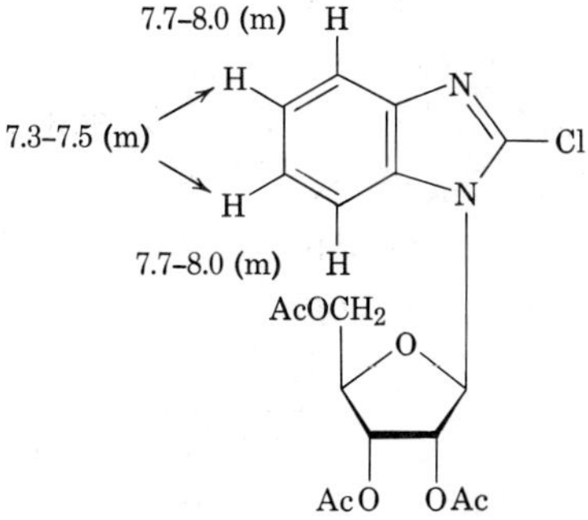

6.57 (s) H_2N-N
CH_3 2.63 (s)

in methyl sulfoxide-d_6

7.6 (m) H
7.27 (m) H H
—H 8.42 (s)
7.6 (m) H
$HOCH_2$
HO OH

7.7–8.0 (m) H
7.3–7.5 (m) H H
Cl
7.7–8.0 (m) H
$AcOCH_2$
AcO OAc

Ac = 2.0–2.2

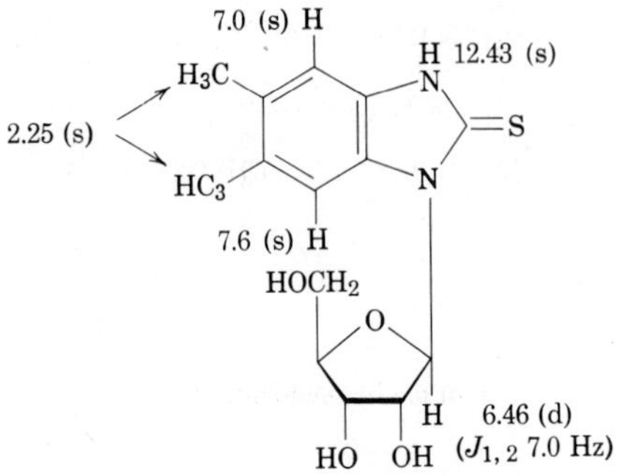

7.0 (s) H
H 12.43 (s)
H_3C
2.25 (s)
N
=S
HC_3
N
7.6 (s) H
$HOCH_2$
O
H 6.46 (d)
HO OH ($J_{1,2}$ 7.0 Hz)

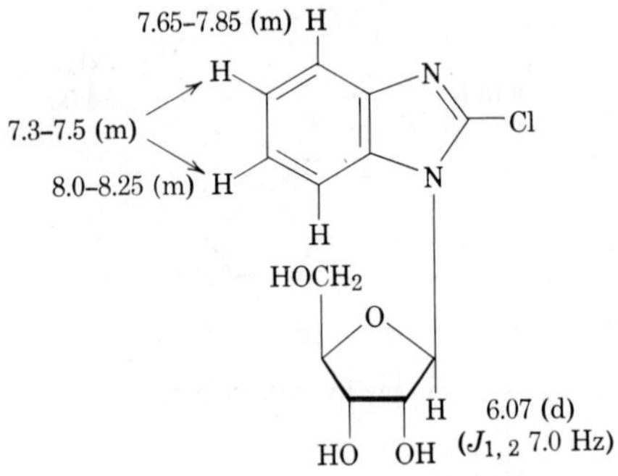

7.65–7.85 (m) H
H
N
7.3–7.5 (m)
Cl
8.0–8.25 (m) H
N
H
$HOCH_2$
O
H 6.07 (d)
HO OH ($J_{1,2}$ 7.0 Hz)

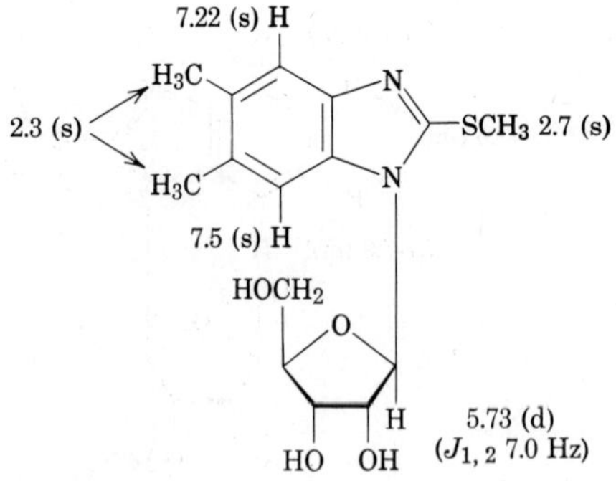

7.22 (s) H
H_3C
N
2.3 (s)
SCH_3 2.7 (s)
H_3C
N
7.5 (s) H
$HOCH_2$
O
H 5.73 (d)
HO OH ($J_{1,2}$ 7.0 Hz)

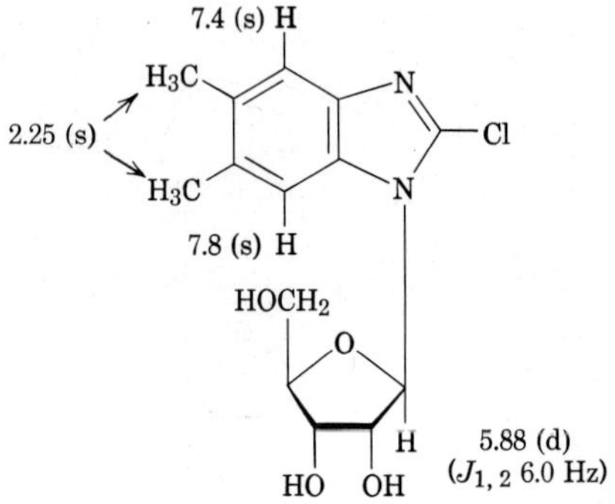

7.4 (s) H
H_3C
N
2.25 (s)
Cl
H_3C
N
7.8 (s) H
$HOCH_2$
O
H 5.88 (d)
HO OH ($J_{1,2}$ 6.0 Hz)

(Ref. 266)

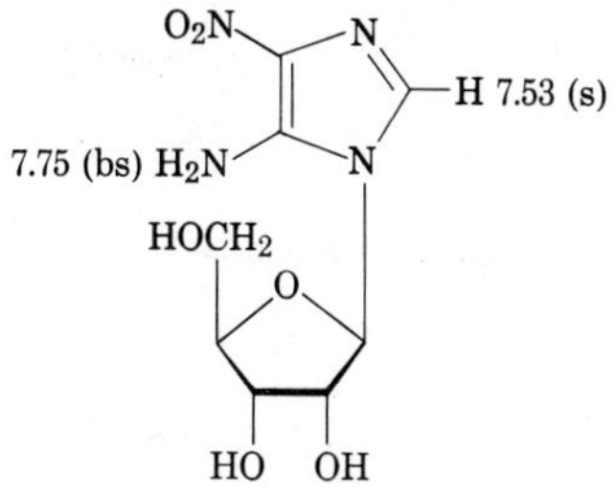

O_2N N H 7.53 (s)
7.75 (bs) H_2N N
$HOCH_2$ O
HO OH

in methyl sulfoxide-d_6

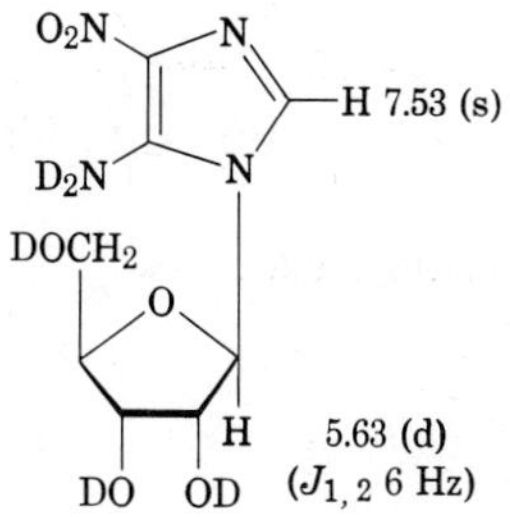

O_2N N H 7.53 (s)
D_2N N
$DOCH_2$ O
H 5.63 (d) ($J_{1,2}$ 6 Hz)
DO OD

in methyl sulfoxide-d_6-D_2O

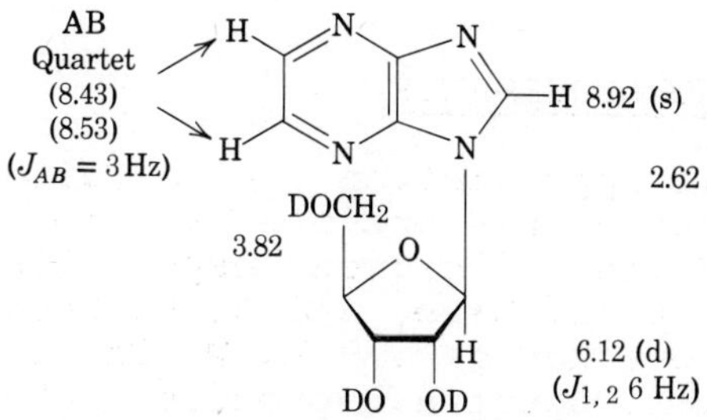

AB Quartet (8.43) (8.53) ($J_{AB} = 3$ Hz) H N N H 8.92 (s)
H N N
$DOCH_2$ 3.82 O
H 6.12 (d) ($J_{1,2}$ 6 Hz)
DO OD

in methyl sulfoxide-d_6-D_2O

2.62 (s) H_3C N N H 8.10 (s)
H_3C N N
CH_3 3.87 (s)

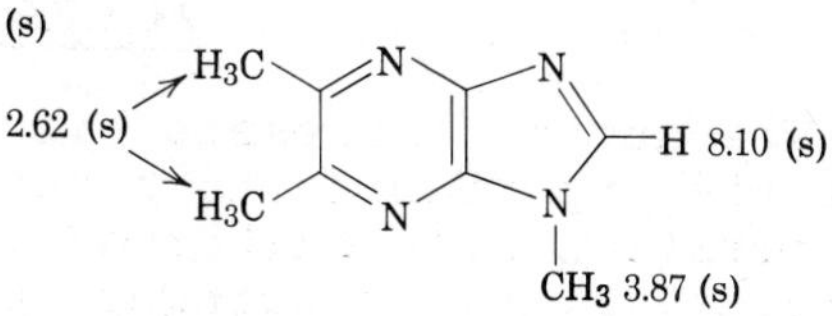

in chloroform-d

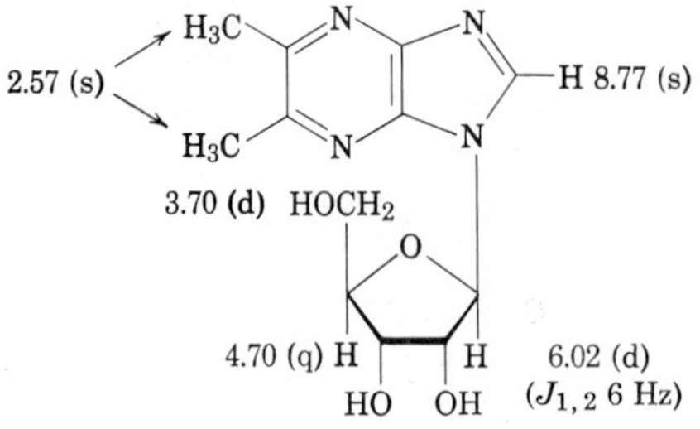

H_3C, H_3C 2.57 (s); H 8.77 (s); 3.70 (d) $HOCH_2$; 4.70 (q) H; H 6.02 (d) ($J_{1,2}$ 6 Hz); HO OH

in methyl sulfoxide-d_6

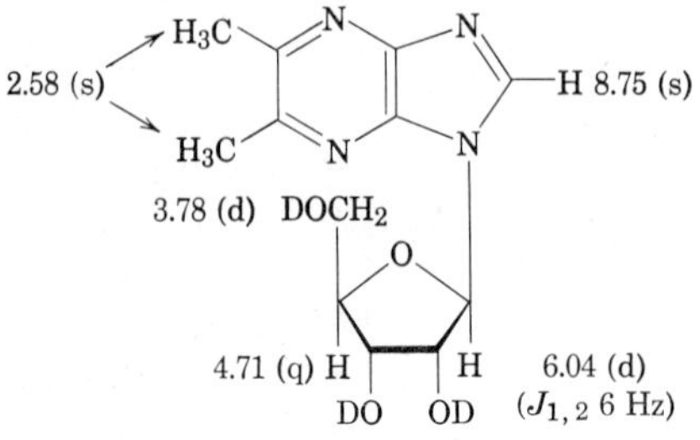

H_3C, H_3C 2.58 (s); H 8.75 (s); 3.78 (d) $DOCH_2$; 4.71 (q) H; H 6.04 (d) ($J_{1,2}$ 6 Hz); DO OD

in methyl sulfoxide-d_6-D_2O

(Ref. 267)

ACKNOWLEDGEMENTS

The author expresses his appreciation to the Editors fo for the opportunity to prepare this article, and for the cooperation of a number of his colleagues, especially R. P Panzica, E. E. Leutzinger, D. S. Wise, and A. D. Broom. Thanks are also due Ritzuko Tukunaga and R. P. Panzica for their time and for the p.m.r. spectra reproduced throughout this Chapter. He also expresses his gratitude for the encouragement of his wife Sammy and Dr. R. K. Robins, and thanks Mrs. Jeri Shamy and Miss Linda Dahle for typing this manuscript.

The foregoing acknowledgments, together with the numerous contributions throughout this article by other people to whom he is indebted, prompts the author to append a recent quotation from the "Harvard Browser": "It goes without saying that the people mentioned in the previous paragraph, having done so much already, will cheerfully share the blame for any shortcomings the book may have, so that none of the responsibilities are mine."

REFERENCES

(1) L. M. Jackman, *Applications of Nuclear Magnetic Resonance Spectroscopy in Organic Chemistry*, 2nd Ed., Pergamon Press Ltd., London, England, 1960.

(2) H. Conroy, *Advan. Org. Chem.*, 2, 65 (1960).

(3) J. D. Roberts, *Nuclear Magnetic Resonance. Applications to Problems in Organic Chemistry*, McGraw-Hill Book Company, Inc., New York, N. Y., 1959.

(4) N. S. Bhacca and D. H. Williams, *Applications of N.M.R. Spectroscopy in Organic Chemistry*, Holden-Day, Inc., San Francisco, Calif., 1964.

(5) J. R. Dyer, in *Applications of Absorption Spectroscopy of Organic Compounds*, K. L. Rinehart, ed., Prentice-Hall, Englewood Cliffs, N. J., 1965.

(6) R. H. Bible, *Interpretations of N.M.R. Spectra*, Plenum Press, New York, N. Y., 1965.

(7) I. Fleming and D. H. Williams, *Spectroscopic Methods in Organic Chemistry*, McGraw-Hill Book Co., Inc., London, England, 1966.

(8) L. R. Provost and R. V. Jardine, *J. Chem. Educ.*, 45, 675 (1968).

(9) Unpublished information and techniques personally communicated by R. P. Panzica of this Laboratory.

(10) R. J. Abraham, L. D. Hall, L. Hough, and K. A. McLaughlan, *J. Chem. Soc.*, 1962, 3699; a modification of the original equation by M. Karplus is given in *J. Chem. Phys.*, 30, 11 (1959).

(11) M. Karplus, *J. Amer. Chem. Soc.*, 85, 2870 (1963).

(12) R. J. G. Gillespie, A. Grimison, J. H. Ridd, and R. F. M. White, *J. Chem. Soc.*, 1958, 3228.

(13) J. L. Imbach, R. Jacquier, and A. Romane, *J. Heterocycl. Chem.*, 4, 451 (1967).

(14) K. J. Brunings, *J. Amer. Chem. Soc.*, 69, 205 (1947).

(15) H. B. Bensusan and M. S. R. Naidu, *Biochemistry*, 6, 12 (1967).

(16) H. Pauly and E. Arauner, *J. Prakt. Chem.*, 118, 33 (1928).

(17) E. C. Kornfeld, L. Wold, T. M. Lin, and I. H. Slater, *J. Med. Chem.*, 11, 1028 (1968); G. J. Durant, M. E. Foottit, C. R. Ganellin, J. M. Laynes, E. S. Pepper, and A. Rol, *Chem. Commun.*, 1968, 108; G. D. Gutsche and H. Voges, *J. Org. Chem.*, 32, 2685 (1967).
(18) R. G. Jones, *J. Amer. Chem. Soc.*, 71, 383 (1949).
(19) L. B. Townsend, *Chem. Rev.*, 67, 533 (1967).
(20) A. R. Katritzky and J. M. Lagowski, *The Principles of Heterocyclic Chemistry*, Academic Press Inc., New York, N. Y., 1968.
(21) A. Albert, *Heterocyclic Chemistry*, 2nd edition, Oxford University Press, New York, N. Y., 1968.
(22) M. H. Palmer, *The Structure and Reactions of Heterocyclic Compounds*, Edward Arnold Ltd., London, England, 1967.
(23) R. J. Rousseau, R. K. Robins, and L. B. Townsend, *J. Heterocycl. Chem.*, 4, 311 (1967).
(24) Unpublished data from this Laboratory.
(25) C. D. Jardetzky, *J. Amer. Chem. Soc.*, 83, 2919 (1961).
(26) R. U. Lemieux, *Can. J. Chem.*, 39, 116 (1961).
(27) L. Gatlin and J. C. Davis, Jr., *J. Amer. Chem. Soc.*, 84, 4464 (1962).
(28) M. J. Robins and R. K. Robins, *J. Amer. Chem. Soc.*, 87, 4934 (1965).
(29) A. W. Lutz and S. DeLornzo, *J. Heterocycl. Chem.*, 4, 399 (1967).
(30) P. E. Iversen and H. Lund, *Acta Chem. Scand.*, 20, 2649 (1966).
(31) J. S. G. Cox, G. Fitzmaurice, A. R. Katritzky, and G. J. T. Tiddy, *J. Chem. Soc. (B)*, 1967, 4251.
(31a) P. K. Martin, H. R. Matthews, H. Rapoport, and G. Thyagarajan, *J. Org. Chem.*, 33, 3758 (1968).
(32) R. P. Panzica and L. B. Townsend, *Tetrahedron Lett.*, 1970, 1013.
(33) B. A. Otter, E. A. Falco, and J. J. Fox, *J. Org. Chem.*, 33, 3593 (1968).
(34) G. S. Reddy, R. T. Hobgood, Jr., and J. H. Goldstein, *J. Amer. Chem. Soc.*, 84, 336 (1962).
(35) A. R. Katritzky, *Advan. Heterocycl. Chem.*, 1, 371 (1963).
(36) G. M. Kheifets, N. V. Khromov-Borisov, A. I. Koltstov, and M. V. Volkensteirn, *Tetrahedron*, 23, 1197 (1967).
(37) G. M. Kheifets and N. V. Khromov-Borisov, *Zh. Org. Khim.*, 2, 1511 (1966).
(38) J. P. Kokko, J. H. Goldstein, and L. Mandell, *J. Amer. Chem. Soc.*, 83, 2909 (1961).
(39) O. Jardetzky, P. Pappas, and N. G. Wade, *J. Amer. Chem. Soc.*, 85, 1657 (1963).

(40) H. T. Miles, *Proc. Nat. Acad. Sci.* U. S., 47, 791 (1961).
(41) D. Shugar and J. J. Fox, *Biochim. Biophys. Acta*, 9, 199 (1952).
(42) D. L. Anell, *J. Chem. Soc.*, 1961, 504.
(43) C. D. Jardetzky and O. Jardetzky, *J. Amer. Chem. Soc.*, 82, 222 (1960).
(44) H. T. Miles, *J. Amer. Chem. Soc.*, 85, 1007 (1963).
(45) L. Gatlin and J. C. Davis, Jr., *J. Amer. Chem. Soc.*, 84, 4464 (1962).
(46) L. J. Durham. A. Larsson, and P. Reichard, *Eur. J. Biochem.*, 1, 92 (1967).
(47) R. J. Cushley and S. R. Lipsky, *Tetrahedron Lett.*, 1968, 5393.
(48) J. P. Kokko, L. Mandell, and J. H. Goldstein, *J. Amer. Chem. Soc.*, 84, 1042 (1962).
(49) S. Hanessian and T. H. Haskell, *Tetrahedron Lett.*, 1964, 2451.
(50) M. E. C. Biffin, D. J. Brown, and T. C. Lee, *Aust. J. Chem.*, 20, 1041 (1967).
(51) S. Gronowitz, B. Norman, B. Getblom, B. Mathiasson, and R. Hoffmann, *Ark. Kemi*, 22, 65 (1964).
(52) T. Nishiwaki, *Tetrahedron*, 23, 2657 (1967).
(53) S. R. Heller, *Biochem. Biophys. Res. Commun.*, 32, 998 (1968).
(54) J. A. Elvidge and N. A. Zaidi, *J. Chem. Soc. (C)*, 1968, 2188.
(55) M. Prystas and F. Sorm, *Collect. Czech. Chem. Commun.*, 33, 210 (1968).
(56) M. Hirata, *Chem. Pharm. Bull.* (Tokyo), 16, 430 (1968).
(57) D. M. Brown, M. J. E. Hewlins, and P. Schell, *J. Chem. Soc. (C)*, 1968, 1925.
(58) M. W. Gray and B. G. Lane, *Biochemistry*, 7, 3441 (1968).
(59) D. J. Brown and T. C. Lee, *Aust. J. Chem.*, 21, 243 (1968).
(60) D. Lipkin, C. Cori, and M. Sano, *Tetrahedron Lett.*, 1968, 5993.
(61) M. W. Winkley and R. K. Robins, *J. Org. Chem.*, 33, 2822 (1968).
(62) C. C. Bhat, K. V. Bhat, and W. W. Zorbach, *Carbohyd. Res.*, 8, 368 (1968).
(63) M. W. Winkley and R. K. Robins, *J. Org. Chem.*, 34, 431 (1969).
(64) G. Etzold, G. Kowollik, and P. Langen, *Chem. Commun.*, 1968, 422.
(65) P. Cerutti, Y. Konda, W. R. Landis, and B. Witkop, *J. Amer. Chem. Soc.*, 90, 771 (1968).
(66) J. J. Fox, Y. Kuwada, and K. A. Watanabe, *Tetrahedron Lett.*, 1969, 6029; K. A. Watanabe, M. P.

Kotick, and J. J. Fox, *Chem. Pharm. Bull.* (Tokyo), 17, 416 (1969).
(67) A. Myles and J. J. Fox, *J. Med. Chem.*, 11, 143 (1968).
(68) F. J. Bullock and O. Jardetzky, *J. Org. Chem.*, 29, 1988 (1964).
(69) S. Matsuura and T. Gota, *Tetrahedron Lett.*, 1963, 1499.
(70) S. Matsuura and T. Gota, *J. Chem. Soc.*, 1965, 623.
(71) Unpublished data obtained by the author.
(72) W. C. Coburn, Jr., M. C. Thorpe, J. A. Montgomery, and K. Hewson, *J. Org. Chem.*, 30, 111 (1965).
(73) W. C. Coburn, Jr., M. C. Thorpe, J. A. Montgomery, and K. Hewson, *J. Org. Chem.*, 30, 114 (1965).
(74) H. C. Brown and Y. Okamoto, *J. Amer. Chem. Soc.*, 80, 4979 (1958).
(75) The p.m.r. spectra of adenosine obtained on 60-, 100-, and 220-MHz spectrophotometers are included in the last Section of this Chapter (see p. 356).
(76) L. B. Townsend, R. K. Robins, R. N. Loeppky, and N. J. Leonard, *J. Amer. Chem. Soc.*, 86, 5320 (1964).
(77) There are certain restrictions on the type of alkyl substitutent that can be used in order that the rules may still be applicable.
(78) C. P. J. Glaudemans and H. G. Fletcher, Jr., *J. Org. Chem.*, 28, 3004 (1963).
(79) K. R. Darnall and L. B. Townsend, *J. Heterocycl. Chem.*, 3, 371 (1966).
(80) L. B. Townsend and R. K. Robins, *J. Amer. Chem. Soc.*, 85, 242 (1963).
(81) D. J. Brown and P. W. Ford, *J. Chem. Soc. (C)*, 1969, 2620.
(82) F. Perini, F. A. Carey, and L. Long, Jr., *Carbohyd. Res.*, 11, 159 (1969).
(83) S. S. Danyluk and F. E. Hruska, *Biochemistry*, 7, 1038 (1968).
(84) G. K. Helmkamp and N. S. Kondo, *Biochim. Biophys. Acta*, 157, 242 (1968).
(85) A. Yamazaki, I. Kumashiro, and T. Takenishi, *J. Org. Chem.*, 33, 2583 (1968).
(86) Y. Furukawa, K. I. Imai, and M. Honjo, *Tetrahedron Lett.*, 1968, 4655.
(87) D. M. G. Martin and C. B. Reese, *J. Chem. Soc. (C)* 1968, 1731.
(88) N. J. Leonard and M. Rasmussen, *J. Org. Chem.*, 23, 2488 (1968).
(89) H. Iwamura and T. Hashizume, *J. Org. Chem.*, 33, 1796 (1968).
(90) M. Ikehara, H. Tada, and M. Kaneko, *Tetrahedron*, 24, 3489 (1968).

(91) K. H. Scheit, *Angew. Chem., Intern. Ed. Engl.*, 6, 180 (1967).
(92) J. A. Montgomery and K. Hewson, *J. Med. Chem.*, 9, 234 (1966).
(93) R. Shapiro and J. Haihmann, *Biochemistry*, 5, 2799 (1966).
(94) R. H. Hall, M. J. Robins, L. Stasiuk, and R. Thedford, *J. Amer. Chem. Soc.*, 88, 2614 (1966).
(95) E. Walton, S. R. Jenkins, R. F. Nutt, M. Zimmermann, and F. W. Holley, *J. Amer. Chem. Soc.*, 88, 4525 (1966).
(96) H. P. M. Fromageot, B. E. Griffin, C. B. Reese, and J. E. Sulston, *Tetrahedron*, 23, 2315 (1967).
(97) B. E. Griffin, M. Jarman, C. B. Reese, and J. E. Sulston, *Tetrahedron*, 23, 2301 (1967).
(98) F. W. Lichtenthaler and T. Nakagawa, *Chem. Ber.*, 100, 1833 (1967).
(99) J. A. Montgomery and K. Hewson, *J. Med. Chem.*, 11, 48 (1968).
(100) R. F. Nutt and E. Walton, *J. Med. Chem.*, 11, 151 (1968).
(101) F. W. Lichtenthaler and H. P. Albrecht, *Angew. Chem., Intern Ed. Engl.*, 7, 457 (1968).
(102) Y. Furukawa and M. Honjo, *Chem. Pharm. Bull.* (Tokyo), 16, 1076 (1968).
(103) H. Yamasaki and T. Hashizume, *Agr. Biol. Chem.* (Tokyo), 32, 1362 (1968).
(104) J. R. McCarthy, Jr., M. J. Robins, L. B. Townsend, and R. K. Robins, *J. Amer. Chem. Soc.*, 88, 1549 (1966).
(105) J. F. Gerster, B. C. Hinshaw, R. K. Robins, and L. B. Townsend, *J. Org. Chem.*, 33, 1070 (1968).
(106) R. J. Rousseau, R. K. Robins, and L. B. Townsend, *J. Amer. Chem. Soc.*, 90, 2661 (1968).
(107) R. J. Rousseau and L. B. Townsend, *J. Org. Chem.*, 33, 2828 (1968).
(108) J. A. Montgomery and H. J. Thomas, *J. Amer. Chem. Soc.*, 87, 5442 (1965).
(109) H. J. Thomas and J. A. Montgomery, *J. Org. Chem.*, 31, 1413 (1966).
(110) K. J. Ryan, H. Arzoumanian, E. M. Acton, and L. Goodman, *J. Amer. Chem. Soc.*, 86, 2503 (1964).
(111) E. J. Reist, D. F. Calkins, and L. Goodman, *J. Org. Chem.*, 32, 2538 (1967).
(112) E. J. Reist, L. V. Fisher, and L. Goodman, *J. Org. Chem.*, 33, 189 (1968).
(113) P. O. P. Ts'O, N. S. Kondo, M. P. Schweitzer, and D. P. Hollis, *Biochemistry*, 8, 997 (1969).
(114) E. Dyer and C. E. Minnier, *J. Org. Chem.*, 33, 880 (1968).

(115) E. J. Reist, D. F. Calkins, and L. Goodman, *J. Org. Chem.*, 32, 169 (1967), and references cited therein.
(116) R. U. Lemieux and D. R. Lineback, *Ann. Rev. Biochem.*, 32, 155 (1963).
(117) K. S. Pitzer and W. E. Donath, *J. Amer. Chem. Soc.*, 81, 3213 (1959).
(118) M. Karplus, *J. Chem. Phys.*, 30, 11 (1959).
(119) G. R. Revankar and L. B. Townsend, *J. Heterocycl. Chem.*, 5, 477 (1968), and references cited therein
(120) K. L. Rinehardt, Jr., W. S. Chilton, and M. Hickens *J. Amer. Chem. Soc.*, 84, 3216 (1962).
(121) A. D. Martinez and W. W. Lee, *J. Org. Chem.*, 34, 416 (1969).
(122) Y. H. Pan, R. K. Robins, and L. B. Townsend, *J. Heterocycl. Chem.*, 4, 246 (1967), and references cited therein.
(123) Z. Samek and J. Farkas, *Collect. Czech. Chem. Commun.*, 30, 2149 (1965).
(124) E. A. Davidson, *Carbohydrate Chemistry*, Holt, Rinehart, and Winston, Inc., New York, N.Y., 1967.
(125) E. E. Leutzinger, W. A. Bowles, R. K. Robins, and L. B. Townsend, *J. Amer. Chem. Soc.*, 90, 127 (1968)
(126) T. Nishimura and B. Shimizu, *Chem. Pharm. Bull.* (Tokyo), 13, 803 (1965).
(127) G. T. Rodgers and T. L. V. Ulbricht, *J. Chem. Soc.* (C), 1968, 1929.
(128) T. M. Spotswood and C. I. Tanzer, *Tetrahedron Lett.*, 1967, 911.
(129) R. H. Shoup, H. T. Miles, and E. D. Becker, *Biochem. Biophys. Res. Commun.*, 23, 194 (1966).
(130) G. T. Rodgers and T. L. V. Ulbricht, *Tetrahedron Lett.*, 1968, 1025.
(131) R. U. Lemieux, R. K. Kullnig, H. J. Bernstein, and W. G. Schneider, *J. Amer. Chem. Soc.*, 80, 6098 (1958).
(132) F. A. L. Anet, R. A. B. Bannard, and L. D. Hall, *Can. J. Chem.*, 41, 2331 (1963).
(133) F. W. Lichtenthaler and H. P. Albrecht, *Chem. Ber.* 99, 575 (1966), and references cited therein.
(134) R. U. Lemieux and J. D. Stevens, *Can. J. Chem.*, 43, 2059 (1965).
(135) R. J. Cushley, K. A. Watanabe, and J. J. Fox, *J. Amer. Chem. Soc.*, 89, 394 (1967).
(136) R. J. Cushley, S. R. Lipsky, W. J. McMurray, and J. J. Fox, *Chem. Commun.*, 1968, 1611.
(137) R. U. Lemieux and M. Hoffer, *Can. J. Chem.*, 39, 110 (1961).

(138) R. A. Long, R. K. Robins, and L. B. Townsend, *J. Org. Chem.*, 32, 2751 (1967).
(139) K. J. Ryan, E. M. Acton, and L. Goodman, *J. Org. Chem.*, 31, 1181 (1966).
(140) Personally communicated by Dr. M. W. Winkley of this Laboratory.
(141) R. A. Long, and L. B. Townsend, *Chem. Commun.*, 1970, 1087.
(142) H. P. M. Fromageot, B. E. Griffin, C. B. Reese, J. E. Sulston, and D. R. Trentham, *Tetrahedron*, 22, 705 (1966).
(143) D. P. L. Green and C. B. Reese, *Chem. Commun.*, 1968, 729.
(144) J. D. Stevens and H. G. Fletcher, Jr., *J. Org. Chem.*, 33, 1799 (1968).
(145) R. F. Nutt, M. J. Dickinson, F. W. Holly, and E. Walton, *J. Org. Chem.*, 33, 1789 (1968).
(146) S. R. Jenkins, B. Arison, and E. Walton, *J. Org. Chem.*, 33, 2490 (1968).
(147) E. J. Reist, D. F. Calkins, and L. Goodman, *J. Amer. Chem. Soc.*, 90, 3852 (1968).
(148) A. Rosenthal and M. Sprinzl, *Can. J. Chem.*, 47, 4477 (1969).
(149) R. Lenz, *J. Polym. Sci.*, 51, 247 (1961).
(150) G. R. Niaz and C. B. Reese, *Chem. Commun.*, 1969, 552.
(151) F. Keller and J. E. Bunker, *Carbohyd. Res.*, 8, 347 (1968).
(152) M. H. Freemantle and W. G. Overend, *Chem. Commun.*, 1968, 503.
(153) N. S. Bhacca and D. Horton, *J. Amer. Chem. Soc.*, 89, 5993 (1967).
(154) C. D. Jardetzky, *J. Amer. Chem. Soc.*, 83, 2919 (1961).
(155) C. D. Jardetzky, *J. Amer. Chem. Soc.*, 84, 62 (1962).
(156) F. Sweet and R. K. Brown, *Can. J. Chem.*, 46, 1481 (1968).
(157) A. P. Tulloch and A. Hill, *Can. J. Chem.*, 46, 2485 (1968).
(158) G. B. Howarth, W. A. Szarek, and J. K. N. Jones, *J. Org. Chem.*, 34, 476 (1969).
(159) I. M. Downie, J. B. Lee, T. J. Nolan, and R. M. Allen, *Tetrahedron*, 25, 2339 (1969).
(160) S. Inokawa, H. Yoshida, C.-C. Wang, and R. L. Whistler, *Bull. Chem. Soc. Jap.*, 41, 1472 (1968).
(161) A. Rosenthal and H. J. Koch, *J. Amer. Chem. Soc.*, 90, 2181 (1968).
(162) D. M. Brown and G. H. Jones, *J. Chem. Soc. (C)*, 1967, 249.

(163) J. Lehmann and H. Friebolin, *Carbohyd. Res.*, 2, 499 (1966).
(164) J. A. Zderic, *Experientia*, 20, 48 (1964).
(165) T. Kishikawa, K. Furuta, and S. Takitani, *Tetrahedron Lett.*, 1969, 1961.
(166) W. L. Salo and H. G. Fletcher, Jr., *J. Org. Chem.*, 34, 3026 (1969).
(167) B. E. Griffin and C. B. Reese, *Tetrahedron*, 25, 4057 (1969).
(168) D. Shapiro and A. J. Archer, *J. Org. Chem.*, 35, 229 (1970).
(169) J. A. Christiansen, *Acta Chem. Scand.*, 17, 2209 (1963).
(170) V. S. R. Rao, J. F. Foster, and R. L. Whistler, *J. Org. Chem.*, 28, 1730 (1963).
(171) A. S. Perlin, *Can. J. Chem.*, 44, 539 (1966).
(172) L. D. Hall, *Tetrahedron Lett.*, 1964, 1457.
(173) C. Cone and L. Hough, *Carbohyd. Res.*, 1, 1 (1965).
(174) J. D. Smith and D. B. Dunn, *Biochem. J.*, 72, 294 (1959).
(175) D. B. Dunn, *Biochem. J.*, 86, 14P (1963).
(176) H. Feldmann, D. Dutting, and H. G. Zachau, *Z. Physiol. Chem.*, 347, 236 (1966).
(177) G. R. Phillips, *Nature*, 223, 374 (1969).
(178) R. H. Hall, *Biochemistry*, 3, 876 (1964).
(179) W. E. Cohn, *Biochim. Biophys. Acta*, 32, 569 (1959)
(180) W. E. Cohn, *J. Biol. Chem.*, 235, 1488 (1960).
(181) A. M. Michelson and W. E. Cohn, *Biochemistry*, 1, 490 (1962).
(182) R. Shapiro and R. W. Chambers, *J. Amer. Chem. Soc.* 83, 3920 (1961).
(183) M. Adler and A. B. Gutman, *Science*, 130, 862 (1959).
(184) W. S. Adams, F. Davis, and M. Nakatani, *Amer. J. Med.*, 28, 726 (1960).
(185) F. E. Hruska, A. A. Grey, and I. C. P. Smith, *J. Amer. Chem. Soc.*, 92, 214 (1970).
(186) S. Aizawa, T. Hidaka, W. Otake, H. Yonehara, K. Isono, N. Igarashi, and S. Suzuki, *Agr. Biol. Chem.* (Tokyo), 29, 375 (1965).
(187) M. Hori, K. Ito, T. Takita, G. Koyama, T. Takeuchi and H. Umezawa, *J. Antibiot.* (Tokyo), 17A, 96 (1964).
(188) G. Koyama and H. Umezawa, *J. Antibiot.* (Tokyo), 18A, 175 (1965).
(189) G. Koyama, K. Maeda, H. Umezawa, and Y. Iitake, *Tetrahedron Lett.*, 1966, 597.
(190) R. K. Robins, L. B. Townsend, F. Cassidy, J. F. Gerster, A. F. Lewis, and R. L. Miller, *J. Heterocycl. Chem.*, 3, 110 (1966).

(191) L. B. Townsend and R. K. Robins, *J. Heterocycl. Chem.*, 6, 459 (1969).
(192) H. Nishimura, M. Mayama, Y. Komatsu, H. Kato, J. Shimaoka, and Y. Tanaka, *J. Antibiot.* (Tokyo), 17A, 148 (1964).
(193) K. R. Darnall, L. B. Townsend, and R. K. Robins, *Proc. Nat. Acad. Sci. U. S.*, 57, 548 (1967).
(194) Y. Nakagawa, H. Kana, Y. Tsukuda, and H. Koyama, *Tetrahedron Lett.*, 1967, 4105.
(195) K. Gerzon, R. H. Williams, M. Hoehn, M. Gorman, and D. C. DeJongh, *2nd Intern. Congr. Heterocycl. Chem., Montpelier, France (July, 1969)*, Abstract No. C-30, p. 131.
(196) R. L. Tolman, R. K. Robins, and L. B. Townsend, *J. Amer. Chem. Soc.*, 90, 524 (1968).
(197) R. L. Tolman, R. K. Robins, and L. B. Townsend, *J. Amer. Chem. Soc.*, 91, 2102 (1969), and references cited therein.
(198) J. F. Gerster, B. C. Hinshaw, R. K. Robins, and L. B. Townsend, *J. Heterocycl. Chem.*, 6, 207 (1969).
(199) B. C. Hinshaw, J. F. Gerster, R. K. Robins, and L. B. Townsend, *J. Heterocycl. Chem.*, 6, 215 (1969).
(200) L. J. Hanka, J. S. Evans, D. J. Mason, and A. Dietz, *Antimicrob. Agents Chemother.*, 1966, 619.
(201) A. Riskala and F. Sorm, *Collect. Czech. Chem. Commun.*, 29, 2060 (1964).
(202) *Chem. Eng. News* (15 Sept. 1969), p. 43.
(203) R. N. Williams, K. Gerzon, M. Hoehn, M. Gorman, and D. C. DeJongh, *Abstracts Papers Amer. Chem. Soc. Meeting*, 158, MICRO 38 (1969).
(204) K. Isono and S. Suzuki, *Tetrahedron Lett.*, 1968, 203.
(205) K. Isono and S. Suzuki, *Tetrahedron Lett.*, 1968, 1133.
(206) G. O. Morton, J. E. Lancaster, G. E. VanLear, W. Fulmor, and W. E. Meyer, *J. Amer. Chem. Soc.*, 91, 1535 (1969).
(207) J. J. Fox, Y. Kawada, K. A. Watanabe, T. Ueda, and E. B. Whipple, *Antimicrob. Agents Chemother.*, 1964, 518.
(208) C. DeBoer, A. Dietz, L. E. Johnson, T. E. Eble, and H. Hoeksema, Ger. Pat 1,101,698 (1961); *Chem. Abstr.*, 55, 27767 (1961).
(209) E. A. Kacza, N. R. Trenner, B. Arison, R. W. Walker, and K. Folkers, *Biochem. Biophys. Res. Commun.*, 14, 456 (1964).
(210) R. J. Suhadolnik, B. M. Chassy, and G. R. Waller, *Biochim. Biophys. Acta*, 179, 258 (1969).

(211) K. A. Watanabe, M. P. Kotick, and J. J. Fox, *J. Org. Chem.*, 35, 231 (1970).
(212) P. F. Wiley and F. A. MacKellar, *J. Amer. Chem. Soc.*, 92, 417 (1970).
(213) L. B. Townsend and R. K. Robins, in "CRC Handbook of Biochemistry and Selected Data for Molecular Biology," H. A. Sober (ed.), Chemical Rubber Co., Cleveland, Ohio (1970), p. G-115.
(214) R. J. Cushley, I. Wempen, and J. J. Fox, *J. Amer. Chem. Soc.*, 90, 709 (1968).
(215) R. J. Cushley, J. F. Codington, and J. J. Fox, *Can. J. Chem.*, 46, 1131 (1968).
(216) L. D. Hall and J. F. Manville, *Carbohyd. Res.*, 4, 512 (1967).
(217) P. W. Kent and J. E. G. Barnett, *Tetrahedron, Suppl.*, 7, 69 (1966).
(218) L. D. Hall and J. F. Manville, *Can. J. Chem.*, 45, 1299 (1967).
(219) T. Schaefer, S. S. Danyluk, and C. L. Bell, *Can. J. Chem.*, 47, 1507 (1969).
(220) R. D. Chambers, M. Hale, W. K. R. Musgrave, R. A. Story, and B. Iddon, *J. Chem. Soc.(C)*, 1966, 2331
(221) T. Okano and A. Takadate, *Yakugaku Zasshi*, 88, 1179 (1968).
(222) J. A. Wright, D. P. Wilson, and J. J. Fox, *J. Med. Chem.*, 13, 269 (1970).
(223) L. D. Hall, P. R. Steiner, and C. Pedersen, *Can. J. Chem.*, 48, 1155 (1970).
(224) E. F. Mooney and P. H. Winson, *Ann. Rev. N.M.R. Spectrosc.*, 2, 153 (1969).
(225) H. L. Retcofsky and R. A. Friedel, *J. Phys. Chem.* 71, 3592 (1967).
(226) H. L. Retcofsky and F. R. McDonald, *Tetrahedron Lett.*, 1968, 2575.
(227) P. C. Lauterbur, *J. Chem. Phys.*, 43, 360 (1965).
(228) R. J. Pugmire and D. M. Grant, *J. Amer. Chem. Soc.*, 90, 697 (1968).
(229) R. J. Pugmire and D. M. Grant, *J. Amer. Chem. Soc.*, 90, 4232 (1968).
(230) R. J. Pugmire, D. M. Grant, R. K. Robins, and G. W. Rhodes, *J. Amer. Chem. Soc.*, 87, 2225 (1965).
(231) A. J. Jones, M. W. Winkley, D. M. Grant, and R. K. Robins, *Proc. Nat. Acad. Sci. U. S.*, 65, 27 (1970).
(232) D. E. Dorman and J. D. Roberts, *Proc. Nat. Acad. Sci. U. S.*, 65, 19 (1970).
(233) F. J. Weigert and J. D. Roberts, *J. Amer. Chem. Soc.*, 91, 4940 (1969).

(234) W. McFarlane, *J. Chem. Soc. (B)*, 1969, 28.
(235) F. J. Weigert, M. Jautelat, and J. D. Roberts, *Proc. Nat. Acad. Sci. U. S.*, 60, 1152 (1968).
(236) W. Horsley, H. Sternlicht, and J. S. Cohen, *J. Amer. Chem. Soc.*, 92, 680 (1970).
(237) R. G. Parker and J. D. Roberts, *J. Org. Chem.*, 35, 996 (1970).
(238) B. W. Roberts, J. B. Lambert, and J. D. Roberts, *J. Amer. Chem. Soc.*, 87, 5439 (1965).
(239) E. D. Becker, H. T. Miles, and R. B. Bradley, *J. Amer. Chem. Soc.*, 87, 5575 (1965).
(240) J. A. Happe and M. Morales, *J. Amer. Chem. Soc.*, 88, 2077 (1966).
(241) J. P. Kintzinger and J. M. Lehn, *Chem. Commun.*, 1967, 660.
(242) A. K. Bose and I. Kugajevsky, *Tetrahedron*, 23, 1489 (1967).
(243) W. J. Wechter, *J. Med. Chem.*, 10, 762 (1967).
(244) W. J. Wechter, *J. Org. Chem.*, 34, 244 (1969).
(245) I. Feldman and R. P. Agarwal, *J. Amer. Chem. Soc.*, 90, 7329 (1968).
(246) Y. Inone and K. Nakanishi, *Biochim. Biophys. Acta*, 120, 311 (1966).
(247) C. C. McDonald, W. D. Phillips, and J. Penswick, *Biopolymers*, 3, 609 (1965).
(248) C. C. McDonald, W. D. Phillips, and J. Lazar, *J. Amer. Chem. Soc.*, 89, 4166 (1967).
(249) I. C. P. Smith, T. Yamane, and R. G. Shulman, *Science*, 159, 1360 (1968).
(250) I. C. P. Smith, T. Yamane, and R. G. Shulman, *Can. J. Biochem.*, 47, 48 (1969).
(251) R. A. Long, unpublished data, recorded with a Varian A-60 spectrometer with tetramethylsilane as the internal standard.
(252) A. D. Broom and R. K. Robins, *J. Org. Chem.*, 34, 1025 (1965). Varian A-60 spectrometer, methyl sulfoxide-d_6, TMS as the internal standard
(253) M. W. Winkley, unpublished data obtained with a Varian A-60 spectrometer.
(254) B. H. Rizalla, R. K. Robins, and A. D. Broom, *Biochim. Biophys. Acta*, 33, 285 (1969). Varian A-60, methyl sulfoxide-d_6, TMS as the internal standard.
(255) J. A. Montgomery and K. Hewson, *J. Heterocycl. Chem.*, 1, 213 (1964). Varian A-60, methyl sulfoxide-d_6.
(256) G. R. Revankar, personal communication. Varian A-60, TMS as internal standard.
(257) G. R. Revankar, personal communication. Varian A-60, TMS as internal standard.

(258) D. M. G. Martin, C. B. Reese, and G. F. Stephenson *Biochemistry*, 7, 1406 (1968). In D_2O, 0.1 M with respect to HCl; Varian HA-100 spectrometer.

(259) A. D. Broom, M. P. Schweizer, and P. O. P. Ts'o, *J. Am. Chem. Soc.*, 89, 3612 (1967). Varian A-60 spectrometer, in D_2O with TMS as external standard.

(260) R. P. Panzica, unpublished data obtained with a Varian 56/60 spectrometer at 60 MHz with TMS as the internal standard

(261) R. P. Panzica, unpublished data obtained with a Varian 56/60 spectrometer at 60 MHz with TMS as the internal standard.

(262) R. P. Panzica, unpublished data obtained with a Varian 56/60 spectrometer at 60 MHz with TMS as the internal standard.

(263) E. E. Leutzinger, R. K. Robins, and L. B. Townsend *Tetrahedron Lett.*, 1968, 4475.

(264) E. E. Leutzinger, unpublished data obtained with a Varian A-60 spectrometer with TMS as the internal standard.

(265) R. A. Long, unpublished data obtained with a Varian A-60 spectrometer with TMS as the internal standard.

(266) G. R. Revankar, unpublished data obtained with a Varian A-60 spectrometer in methyl sulfoxide-d_6 with TMS as the internal standard.

(267) R. P. Panzica, unpublished data obtained with a Varian 56/60 spectrometer at 60 MHz with TMS as the internal standard.

CHAPTER 8

X-Ray Methods For Determination Of Crystal Structure, As Applied To Components of Nucleic Acids*

S. T. RAO AND M. SUNDARALINGAM

DEPARTMENT OF BIOCHEMISTRY, UNIVERSITY OF WISCONSIN, MADISON, WISCONSIN

---

*Part VII of the series on "Stereochemistry of Nucleic Acids and their Constituents." For Part VI, see D. Rohrer and M. Sundaralingam, *Acta Crystallogr.*, B26, 546 (1969). Support of this work by U. S. Public Health Service Grant No. GM-17378 from the Division of General Medical Sciences, National Institutes of Health, is gratefully acknowledged. Most of the structures depicted in this article were obtained in work published up to 1968; however, at the time of submission in early 1969, a few additional structural papers became available to us, and these have been included.

I. INTRODUCTION

Single-crystal X-ray diffraction is by far the most powerful technique available for the determination of the detailed architecture of molecules of various complexitie For over fifty years, the technique has been used for making significant contributions to the advancement of chemical and physical theories. The impact of X-ray diffraction on biology was not felt until the elucidation of the structures of the genetic material 2-deoxy-D-ribonucleic acid (DNA) by Crick, Watson, and Wilkins,[1,2] the proteins hemoglobin and myoglobin by Perutz[3] and Kendrew,[4] respectively, and vitamin B_{12} by Hodgkin.[5] The theoretical and technical advances made in X-ray crystallography during the past fifteen years have resulted in a rapid growth of publications on the structure of chemical and biological molecular systems of increasing complexity. Once a rare discipline, crystallography has now become commonplace in practically every major teaching and research establishment.

Unlike some of the other spectral methods, such as infrared, ultraviolet, and nuclear magnetic resonance spectroscopy, the X-ray method cannot be employed with ease and rapidity for the determination of functional groups in a molecule, because the crystal structure has to be determined. In other words, *the entire molecular structure* is ascertained by the diffraction method, instead of that of unknown parts of a molecule. Thus, compared with the other spectral methods, the method is relatively time-consuming. However, X-ray diffraction analysis not only provides rigorous proof of a molecular structure, but also provides detailed information on the stereochemistry of the molecule and on its assemblage in a crystal lattice.

No attempt has been made in this article either to develop the theories of X-ray diffraction rigorously or to present detailed and complete structural data on the components of the nucleic acids. These subjects have been treated in various crystallographic textbooks and journals, referred to in the text. Instead, the steps involved in single-crystal structure analysis are given with particular emphasis, with illustrations of the application of the methods to the elucidation of the crystal structure of nucleic acid components.

II. SYMMETRY PROPERTIES OF CRYSTALS

1. Point and Space Groups

Crystals are three-dimensional objects exhibiting symnetries, the most important of which is the property that a crystal is composed of unit cells that repeat themselves periodically in all three directions in space. The ex-ternal forms of crystals have been studied extensively[6]; they represent the internal symmetry of the crystal (ex-cept for translations). The symmetry of the shape of the crystal can be described by a combination of symmetry elements known as point groups. Several symmetry elements are possible for an object, namely, rotation axes, mirror planes, and center of inversion, or a combination of these. Of the innumerable symmetry elements that could be generated for a crystal, only a restricted number are of in-terest in crystallography; this situation arises from the nteraction between lattice periodicity and axis of ro-ation.

Consider two points, A and B, in a lattice (see Fig. 1). hatever the property is at A, it should also be present t B. Consider a rotation axis of order n at A; this orresponds to a rotation by an angle $\alpha = 360°/n$, which akes the point B to B'. Similarly, applying the rotation xis at B, A' is generated from A. Because A' and B' re generated points, they should also be lattice points. hus, $A'B' = m\alpha$, where m is an integer. From Fig. 1, $'B' = A(1-2 \cos \alpha)$. If a rotation axis of order n is to onform to lattice repetition, then

$$\cos \alpha = m'/2,$$

nere m' is an integer. Because $-1 \leq \cos \alpha \leq +1$, only ive values of m' are possible, namely, -2, -1, 0, 1, and . The corresponding values of α are 180, 120, 90, 60,

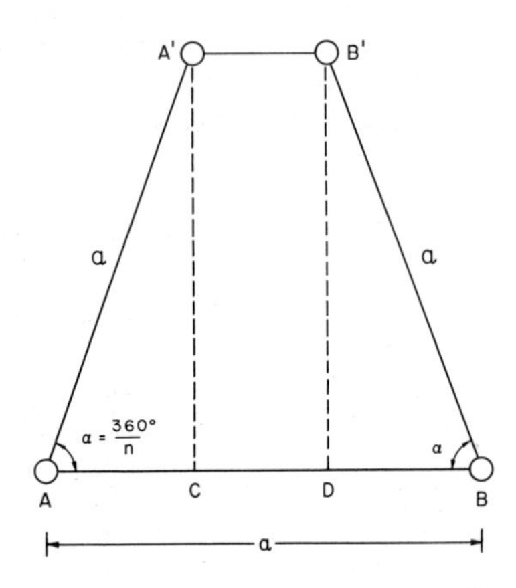

Fig. 1. The Effect of Lattice Translation on Rotation Axis.

and 0°, respectively. These angles represent rotation axes of orders 2, 3, 4, 6, and 1, respectively. Thus, no other rotation axes (five-fold, seven-fold, and higher axes of rotation) can be accommodated by crystals. A combination of the five rotation-axes, the mirror plane, and the center of inversion gives the 32 crystallographic point-groups that describe the possible symmetries of a crystal.[7,8]

A crystal lattice belongs to one of the fourteen space or Bravais lattices, classified into primitive *P*, *C*-face centered *C*, all faces-centered *F*, and body-centered *I*. The lattice type is determined by the components of the translational symmetries in the unit cell. The translational repeats within a unit cell are fractions of the cell edges, except for the *P* lattice (see Table I).

In space lattices, in addition to the point-group symmetries, symmetry elements having a translational component may be present. For example, application of a rotation about a two-fold axis parallel to *b*, followed by a translation of $b/2$ along *b*, gives rise to a two-fold screw-axis denoted by 2_1. Similarly, an operation of reflection across a plane *ac* can be followed by half a translation along the *a* or *c* axis, or both, yielding,

TABLE I

Translation Components for Different Types of Lattice

Lattice type	Symbol	Translations
Primitive	P	none
c-Face-centered	C	$(a + b)/2$
All faces centered	F	$(a + b)/2$, $(b + c)/2$, $(c + a)/2$
Body-centered	I	$(a + b + c)/2$

respectively, an a, c, or n glide-plane. The 230 space groups are generated when these translational elements of symmetry are combined with the 32 point-groups and the 14 Bravais lattices.[7,8] The space groups are assigned to seven crystal systems; their relationship to the lattice type, point groups, and Laue groups are listed in Table II. The 11 Laue groups are a subset of the 32 point-groups that possess the operation of a center of inversion. The Laue groups are important when the symmetry of X-ray reflections from crystals is considered.

Many symmetry elements in a crystal lattice repeat themselves at every half of the translation along the axes. Consider a one-dimensional lattice (see Fig. 2) in which A and B are two consecutive lattice-points and P a general point on this line. Suppose that a center of inversion is introduced at A; then the point Q is generated

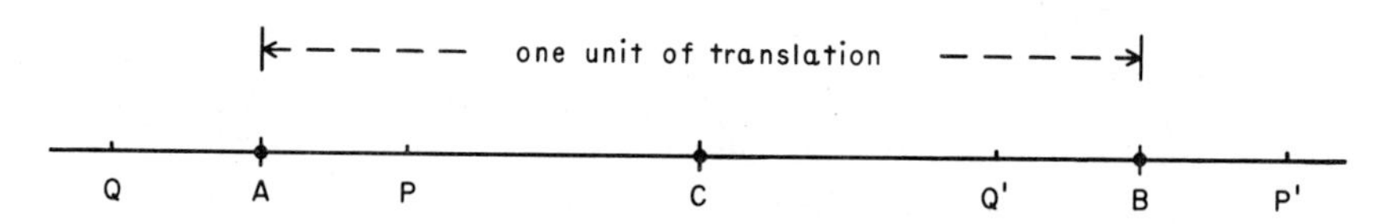

Fig. 2. The Effect of Lattice Translation on Symmetry Elements.

from P. P' and Q', generated by lattice translations of P and Q, respectively, are related by the center of inversion at B. However, the points P and Q' are now related by a center at C, which is midway between A and B. The points P and Q are the same; if P has a coordinate x, then Q would have a coordinate $-x$. The values x and $-x$ are called the equivalent points. The dispositions of the symmetry elements and the equivalent-

TABLE II

Unit-cell Parameters for the Seven Crystal Systems, and Other Details

System	Restrictions on cell-edges and interaxial angles[a]	Point-group symmetry	Laue-group symmetry	Lattice type	Number of		
					Point groups	Laue groups	Space groups
Triclinic	$a \neq b \neq c$ $\alpha \neq \beta \neq \gamma \neq 90°$	$1, \bar{1}$	$\bar{1}$	P	2	1	2
Monoclinic	$a \neq b \neq c$ $\alpha = \gamma = 90°$, $\beta \neq 90°$	$2, m, 2/m$	$2/m$	P, C	3	1	13
Orthorhombic	$a \neq b \neq c$ $\alpha = \beta = \gamma = 90°$	222, $mm2$, mmm	mmm	P, C, I, F	3	1	59
Tetragonal	$a = b \neq c$ $\alpha = \beta = \gamma = 90°$	$4, \bar{4}, 4/m, 422, 4mm, \bar{4}2m, 4/mmm$	$4/m$, $4/mmm$	P, I	7	2	68

TABLE II (continued)

Hexagonal	$a = b \neq c$	$6, \bar{6}, 6/m,$	$6/m, 6/mmm$	P	12	4	52
	$\alpha = \beta = 90°,$ $\gamma = 120°$	$622, 6mm,$ $\bar{6}m2, 6/mmm$					
Trigonal	$a = b = c$	$3, \bar{3},$	$\bar{3}, \bar{3}m$	R			
	$\alpha = \beta = \gamma \neq 90°$	$32, 3m, \bar{3}m$					
Cubic	$a = b = c$	$23, m3,$	$m3, m3m$	P, I, F	5	2	36
	$\alpha = \beta = \gamma = 90°$	$432, \bar{4}3m,$ $m3m$					
			Totals	14	32	11	230

[a]It is possible that, of the quantities not required by symmetry to be equal, two may accidentally be identical within the limits of observational error.

Monoclinic 2/*m* P 1 2_1/c 1 No. 14 P 2_1/C C_{2h}^5

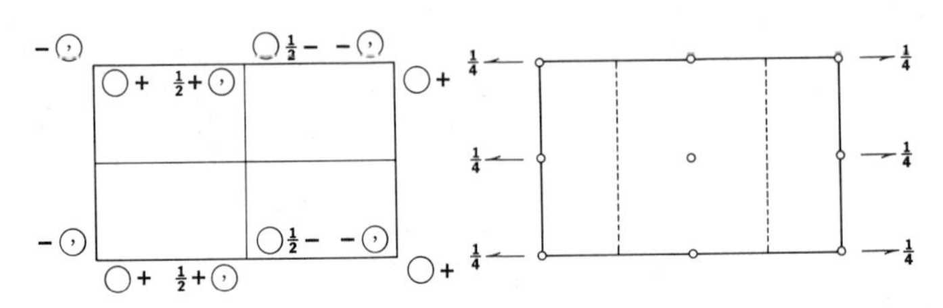

Origin at $\bar{1}$; unique axis *b* 2nd SETTING

Number of positions, Wyckoff notation, and point symmetry	Co-ordinates of equivalent positions	Conditions limiting possible
		General
4 *e* 1	x,y,z; $\bar{x},\bar{y},\bar{z}$; $\bar{x},\frac{1}{2}+y,\frac{1}{2}-z$; $x,\frac{1}{2}-y,\frac{1}{2}+z$.	*hkl*: No conditions *h0l*: $l=2n$ *0k0*: $k=2n$
		Special: as above, plus
2 *d* $\bar{1}$	$\frac{1}{2},0,\frac{1}{2}$; $\frac{1}{2},\frac{1}{2},0$.	*hkl*: $k+l=2n$
2 *c* $\bar{1}$	$0,0,\frac{1}{2}$; $0,\frac{1}{2},0$.	*hkl*: $k+l=2n$
2 *b* $\bar{1}$	$\frac{1}{2},0,0$; $\frac{1}{2},\frac{1}{2},\frac{1}{2}$.	*hkl*: $k+l=2n$
2 *a* $\bar{1}$	$0,0,0$; $0,\frac{1}{2},\frac{1}{2}$.	*hkl*: $k+l=2n$

Symmetry of special projections

(001) *pgm*; $a'=a, b'=b$ (100) *pgg*; $b'=b, c'=c$ (010) *p2*; $c'=c/2, a'=a$

Orthorhombic 222 P $2_1 2_1 2_1$ No. 19 P $2_1 2_1 2_1$ D_2^4

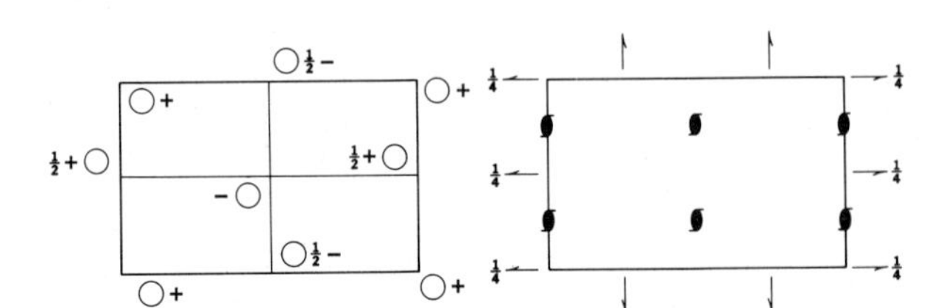

Origin halfway between three pairs of non-intersecting screw axes

Number of positions, Wyckoff notation, and point symmetry	Co-ordinates of equivalent positions	Conditions limiting possible reflections
4 *a* 1	x,y,z; $\frac{1}{2}-x,\bar{y},\frac{1}{2}+z$; $\frac{1}{2}+x,\frac{1}{2}-y,\bar{z}$; $\bar{x},\frac{1}{2}+y,\frac{1}{2}-z$.	*hkl*, *0kl*, *h0l*, *hk0*: No conditions *h00*: $h=2n$ *0k0*: $k=2n$ *00l*: $l=2n$

Symmetry of special projections

(001) *pgg*; $a'=a, b'=b$ (100) *pgg*; $b'=b, c'=c$ (010) *pgg*; $c'=c, a'=a$

Fig. 3. Representative Diagrams for the Space Groups (*a*) $P2_1/c$ and (*b*) $P2_12_12_1$. (Reproduced, by permission of the International Union of Crystallography, from *International Tables for X-ray Crystallography*, Vol. I., Kynoch Press, Birmingham, England, 1965.)

points of the 230 space-groups have been tabulated.[8] The illustrations for two common space-groups are reproduced in Figs. 3a and 3b. In Fig. 3a, the top line contains information about the crystal system, the point group, and the space group. The symbol 2_1 stands for a two-fold screw axis parallel to the *b* axis, and the symbol *c*-glide perpendicular to *b*. The sketch on the right (Fig. 3a) represents the disposition of the symmetry elements in the unit cell, namely, screw axes, glide planes, and centers of inversion. The sketch on the left represents the effect of the symmetry elements on a molecule, denoted by a circle, placed in the cell. The open circles bear an enantiomorphic relationship to those containing a comma. The origin here is so chosen as to coincide with the center of inversion and the equivalent-points are given below. There are four such points, and the multiplicity of the space group is four. The unique portion of the unit cell that generates the rest of the unit cell by symmetry elements of the space group is called the asymmetric unit. In this case, the asymmetric unit is one-quarter of the volume of the unit cell. To the right of Figure 3a are listed certain conditions, as to the X-ray reflections, that arise from the symmetry elements of the space group, and this aspect will be dealt with later.

A similar illustration for the orthorhombic space group $p2_12_12_1$ is given in Fig. 3b. Here, the origin is chosen at the top left-hand corner of the sketch, midway between the three pairs of non-intersecting screw-axes. There is no center of inversion here, and the space-group is non-centrosymmetric. Crystals belonging to such space-groups are generally optically active. In the triclinic, monoclinic, and orthorhombic systems, optically active crystals belong to the point-groups 1, 2, and (222) (see Table II).

2. Lattice Planes and Miller Indices

It is convenient to imagine the lattice to be composed of a series of planes. Several such planes for a projection of the lattice on the *ab* plane are illustrated in Fig. 4. Consider the set of planes, labelled (120), that make intercepts of 4 units along *a*, 2 units along *b*, and an infinity of units along *c*. Taking the indices to be inversely proportional to the corresponding intercepts, namely, 1/4, 1/2, and 0, and clearing the fractions, we have (120). These are called the Miller indices of the planes. Negative indices result if the corresponding intercepts are negative. The indices contain no common factors among them.

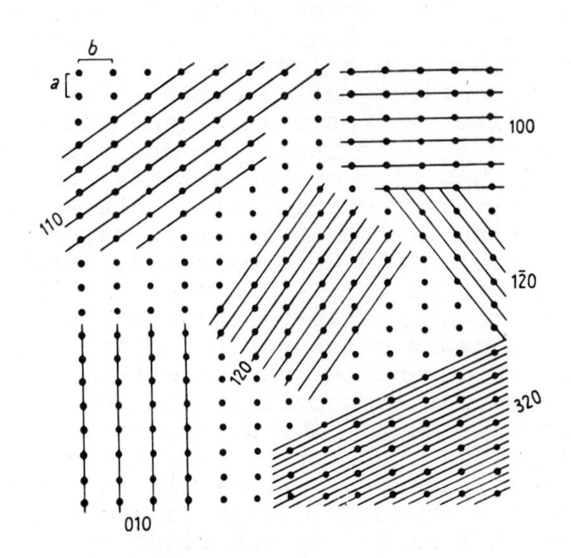

Fig. 4. Lattice Planes having Different Miller Indices.

III. GENERATION OF X-RAYS

Like visible light, X-rays are electromagnetic radiation, but they have wavelengths in the range of 0.1 to 100 Å (10 pm-10 nm). In the study of the diffraction of X-rays from crystals, we are interested in wavelengths of ~1 Å, a distance of the order of interatomic distances. When electrons accelerated by a suitable electric field are allowed to strike a target of an element such as copper or molybdenum, X-rays are produced.[9] The spectrum of wavelengths generated, shown in Fig. 5a, consists of a background of white radiation, and the characteristic $K\alpha$ and $K\beta$ lines corresponding to electronic transitions from the L and M shells to the K shell. The $K\alpha$ line is a doublet for which the intensity of the shorter wavelength component $K\alpha_1$ is roughly double that of $K\alpha_2$, the longer wavelength component. The $K\beta$ line is relatively weak, with only about 15–20% of the intensity of the $K\alpha$ line. In practice, therefore, the $K\alpha$ line is used for measurements of intensity, but the $K\beta$ line may be used for accurate determination of cell constants. In single-crystal analysis, we are interested in monochromatic X-rays, that is, X-rays of a specific wavelength. Thus, we have to single out the strong $K\alpha$ line from the spectrum and, especially, suppress the $K\beta$ line, which is very close to it; this is usually done by means of a filter.[9] The element

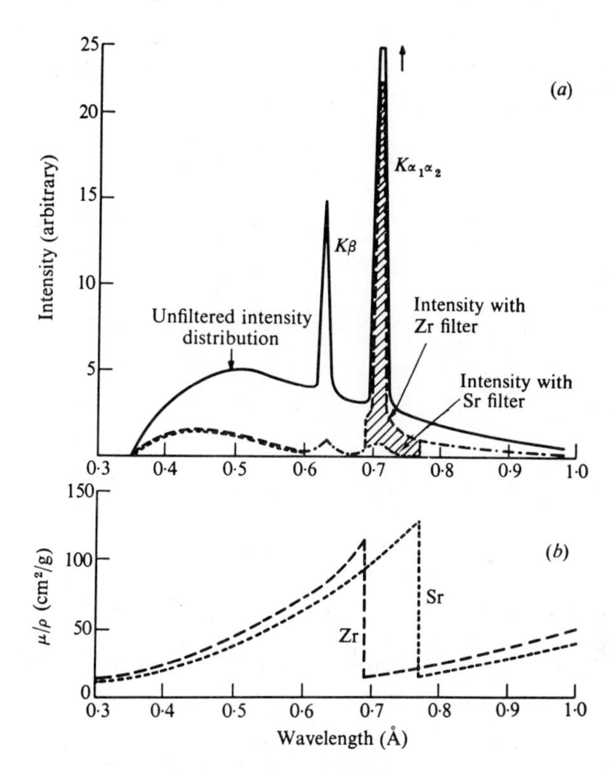

Fig. 5. (a) Intensity Distribution of X-Rays from a Molybdenum Target. (b) The Absorption Edges for Zirconium and Strontium, the α- and β-Filter Materials. [The shaded area in (a) is the band of wavelengths isolated by the balanced filters.] (Reproduced by permission of the International Union of Crystallography from *International Tables for X-ray Crystallography*, Vol. III, Kynoch Press, Birmingham, England, 1962)

usually chosen for the filter material has an atomic number higher by unity than that of the target material, and its absorption-edge lies between the $K\alpha$ and $K\beta$ lines (see Fig. 5b), preferentially absorbing the $K\beta$ line. A list of target and filter materials commonly used is given in Table III.

The effect of the white radiation is to cause an uneven background for the $K\alpha$ peak, and the use of β-filters, alone, does not correct for this effect. In diffracto-

TABLE III

Data for Some Target Materials Commonly Used

	Cr	Fe	Cu	Mo
Atomic number (Z)	24	26	29	42
α_1, pm	228.96	193.60	154.05	70.926
α_2, pm	229.35	193.99	154.43	71.354
$\bar{\alpha}$, pm (weighted mean)	229.09	193.73	154.18	71.069
β_1, pm	208.48	175.65	139.22	63.225
β-Filter	V	Mn	Ni	Nb
α-Filter	Ti	Cr	Co	Y
Normal operating voltage (kV)	30–40	35–45	35–45	50–55
Current [full-wave or half-wave rectified (mA)]	10	10	20	20

metric data, to diminish the effect of background, a pair of filters, α and β, referred to as balanced filters,[10] is employed. The function of the α-filter, which is composed of a material of atomic number one or two numbers less than that of the target material, is to eliminate the $K\alpha$ peak. When the filters are properly balanced, the difference in the intensity transmitted by the two filters, represented by the shaded curve in Fig. 5a, is the intensity due to the $K\alpha$ line alone.

IV. X-RAY DIFFRACTION FROM CRYSTAL LATTICES

In this Section, the various concepts associated with the diffraction of X-rays from crystal lattices will be developed.

1. Bragg's Law and the Reciprocal Lattice

When a monochromatic beam of light is shone onto a grating, the diffracted beam builds up in specific directions, depending on the spacing of the grating and the wavelength of light used. Similarly, when X-rays are incident on a crystal, which may be considered to be a three-dimensional grating of unit cells, the diffracted beams build up for certain directions only. The relation between the angles of diffraction and the periodicity of the lattice in terms of the wavelengths of X-rays is known as Bragg's law.[11]

Consider a beam of monochromatic X-rays of wavelength λ, incident on the two-dimensional crystal at an angle θ with the set of planes having Miller indices (11) with an interplanar spacing d (see Fig. 6a). The diffracted beams are considered when they make the same angle θ with the planes. The "reflections" from two consecutive planes will generally interfere with each other. If the path difference between the two diffracted beams is an integral multiple of the wavelength, the beams will reinforce each other and produce a diffracted beam in that direction. The glancing angle θ is increased from zero upwards until, at a particular value of θ, the path difference $2d \sin \theta$ is equal to λ. Therefore, the condition for a reflection to occur is

$$2d \sin \theta = \lambda,$$

and this relation is known as Bragg's law. When the glancing angle is increased to a value shown by the dotted lines in Fig. 6a, the path difference becomes equal to 2λ. The relation now is

$$2d \sin \theta = 2\lambda,$$

or

$$2(d/2) \sin \theta = \lambda.$$

Thus, a second-order reflection from the set of planes (11) may be imagined as being the first-order reflection from a set of planes having half the spacing, or having indices (22). It is seen, therefore, that, in X-ray

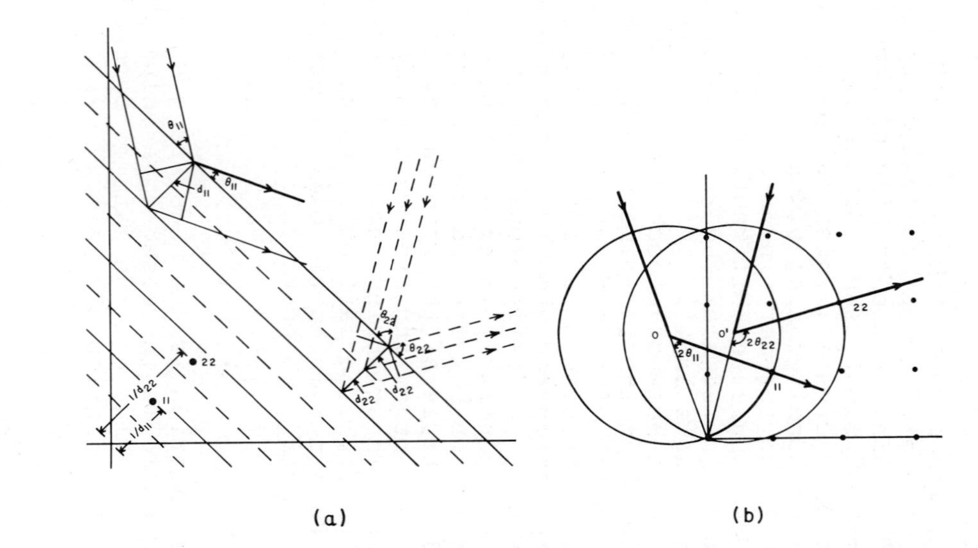

Fig. 6. (a) The Condition for X-Ray Reflection to Occur from Lattice Planes. (b) Reflecting Condition in Terms of the Reciprocal Lattice.

diffraction, the Miller indices are modified to accommodate the order of the reflection. For a given wavelength, there is a lower limit to the interplanar spacing *d* from which reflections could occur; this may be deduced from Bragg's law. For a glancing angle of $\theta = 90°$, the *d* spacing is a minimum, equal to $\lambda/2$. Thus, the resolution that can be obtained with a certain wavelength is equal to half of the wavelength. For CuK_{α} radiation, the wavelength is 1.5418 Å, and the limit of resolution is thus 0.7709 Å. As the normal distances between atoms are much larger than this value, atomic resolution can be realized with this radiation.

It is conceptually elegant to conceive of Bragg's law in terms of the reciprocal lattice, because of the inverse relationship between the interplanar spacing and the glancing angle. The reciprocal lattice-point of the set of planes (11) is at a distance of $1/d$ from the origin, along the line normal to the plane. The reciprocal lattice-point (22) would lie along the same line at twice the distance. Thus, the entire, reciprocal lattice-points can be constructed from all of the planes of the direct lattice, as shown in Fig. 6b. The relationship between the dimensions of the direct-lattice parameters and the reciprocal-lattice parameters is given in Table IV.

In Fig. 6b, *OA* is the incident X-ray beam. With *O* as the center, a sphere having a radius of $1/\lambda$ is so constructed that it passes through the origin of the reciprocal lattice, and is called the sphere of reflection. The reciprocal lattice-point (11) lies on this sphere, and the line joining the center of the sphere *O* to the reciprocal lattice-point (11) gives the direction of the reflected beam. Thus, to generate a reflection, the incident-beam angle should be so adjusted that the sphere of reflection passes through the corresponding, reciprocal lattice-point.

2. Atomic-Scattering Factors, Structure Factors, and Friedel's Law

The diffracted-beam intensity from a set of planes is composed of the sum of the scattering of the individual atoms in the unit cell. When X-rays are incident on the electron-density distribution of an atom, scattering results (see Fig. 7a). When the wavelength of the X-rays is far removed from the absorption-edges of the atoms in the structure, the scattering is said to be normal, and it is this type of scattering that commonly occurs. Consider the scattering from two points *A* and *B* in the electron-density distribution of an atom. When the scattering takes place in the forward direction, namely, $\theta = 0$,

TABLE IV

Relation Between Direct and Reciprocal Lattice-constants for a General (Triclinic) System[a]

$$V = abc\sqrt{1 - \cos^2\alpha - \cos^2\beta - \cos^2\gamma + 2\cos\alpha\cos\beta\cos\gamma}$$

$$V^* = a^*b^*c^*\sqrt{1 - \cos^2\alpha^* - \cos^2\beta^* - \cos^2\gamma^* + 2\cos\alpha^*\cos\beta^*\cos\gamma^*}$$

$$V^* = \frac{1}{V}$$

$$a^* = bc\sin\alpha/V \qquad a = b^*c^*\sin\alpha^*/V^*$$

$$b^* = ca\sin\beta/V \qquad b = c^*a^*\sin\beta^*/V^*$$

$$c^* = ab\sin\gamma/V \qquad c = a^*b^*\sin\gamma^*/V^*$$

$$\cos\alpha^* = \frac{\cos\beta\cos\gamma - \cos\alpha}{\sin\beta\sin\gamma} \qquad \cos\alpha = \frac{\cos\beta^*\cos\gamma^* - \cos\alpha^*}{\sin\beta^*\sin\gamma^*}$$

$$\cos\beta^* = \frac{\cos\gamma\cos\alpha - \cos\beta}{\sin\gamma\sin\alpha} \qquad \cos\beta = \frac{\cos\gamma^*\cos\alpha^* - \cos\beta^*}{\sin\gamma^*\sin\alpha^*}$$

$$\cos\gamma^* = \frac{\cos\alpha\cos\beta - \cos\gamma}{\sin\alpha\sin\beta} \qquad \cos\gamma = \frac{\cos\alpha^*\cos\beta^* - \cos\gamma^*}{\sin\alpha^*\sin\beta^*}$$

[a] The corresponding relations for all the other crystal systems of symmetry higher than triclinic may be derived by using the appropriate restrictions on the unit-cell parameters, given in Table II.

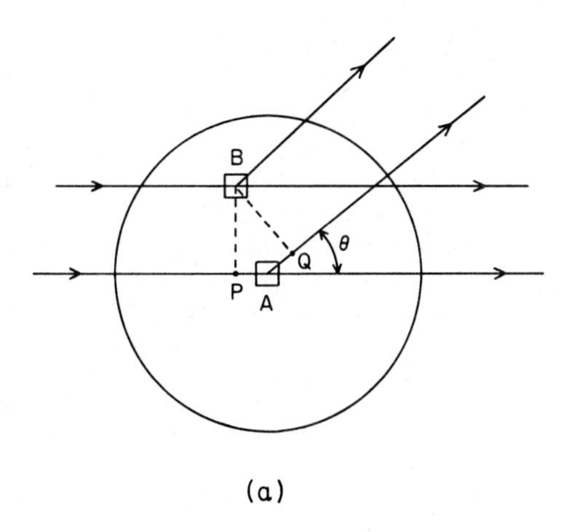

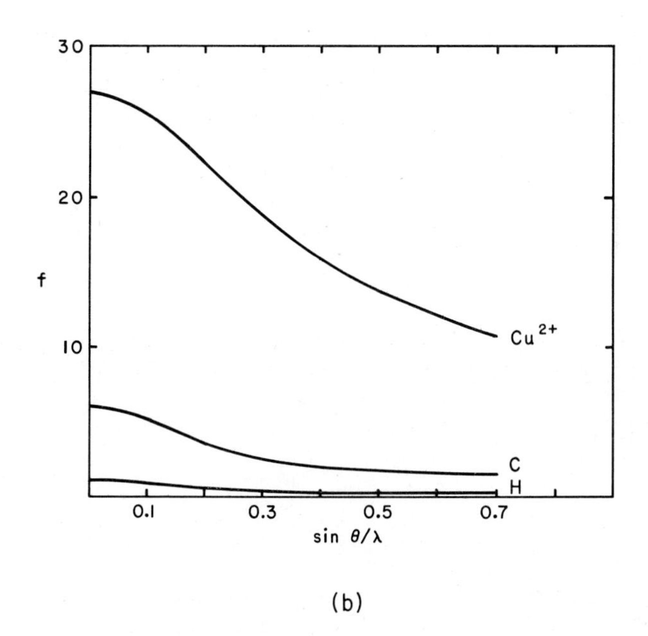

Fig. 7. (a) The Path Difference for Scattering by an Atom at an Angle $\theta > 0$. [The presence of some destructive interference between the constituent waves, *e.g.*, emerging from A and B, lessens the scattered amplitude in comparison to the scattered amplitude at $\theta = 0$. Thus, the scattering from an atom diminishes with increase in angle θ.] (b) The Scattering-factor Curves for $Copper^{2+}$, Carbon, and Hydrogen as a Function of $\sin \theta/\lambda$.

there is no path-difference between the two beams diffracted from the points A and B, and they reinforce each other completely. When the scattering takes place at an angle θ, the scattered waves from A and B have a path-difference; this results in some destructive interference, and a consequent diminution in intensity. The ratio of the amplitude scattered in a certain direction by an atom to that of a free electron is called the atomic scattering-factor, f. Clearly, this value for the forward direction is equal to the atomic number Z of the atom, and the value decreases progressively at higher angles. The angular dependences of the atomic scattering factors for Cu^{2+}, carbon, and hydrogen are shown in Fig. 7b.

The atomic coordinates are commonly given in terms of the fraction of the unit cell-edges. The diffracted amplitude of the $(hk\ell)$ reflection from an atom at x_j, y_j, z_j is given by

$$f_j \exp(i\alpha_j),$$

where $i = \sqrt{-1}$, $\alpha_j = 2\pi(hx_j + ky_j + \ell z_j)$, and f_j is the scattering-factor of the atom corresponding to the Bragg angle θ. When there are N atoms in the unit cell, the total diffracted amplitude is given by

$$\begin{aligned} F &= \Sigma f_j \exp(i\alpha_j) \\ &= \Sigma f_j \cos \alpha_j + i\Sigma f_j \sin \alpha_j \\ &= A + iB. \end{aligned}$$

The value F is called the structure-factor, and is a complex quantity having real and imaginary parts, namely, A and B. It is possible to put F in the form $F = |F| \exp i\alpha$; therefore,

$$|F|^2 = A^2 + B^2,$$

$$\text{and } \alpha = \tan^{-1} B/A.$$

$|F|$ is called the structure amplitude, and α, the phase of the reflection $(hk\ell)$. The structure-factor can be constructed on the argand diagram, and the effect of the jth atom can be represented by a vector of length f_j that makes an angle of a_j with the real axis. The effect of several atoms is the vectorial addition of the individual atomic vectors, and this results in the structure factor F (see Fig. 8a).

For a crystal having a center of symmetry, there is, for every atom at x_j, y_j, z_j, a corresponding atom at

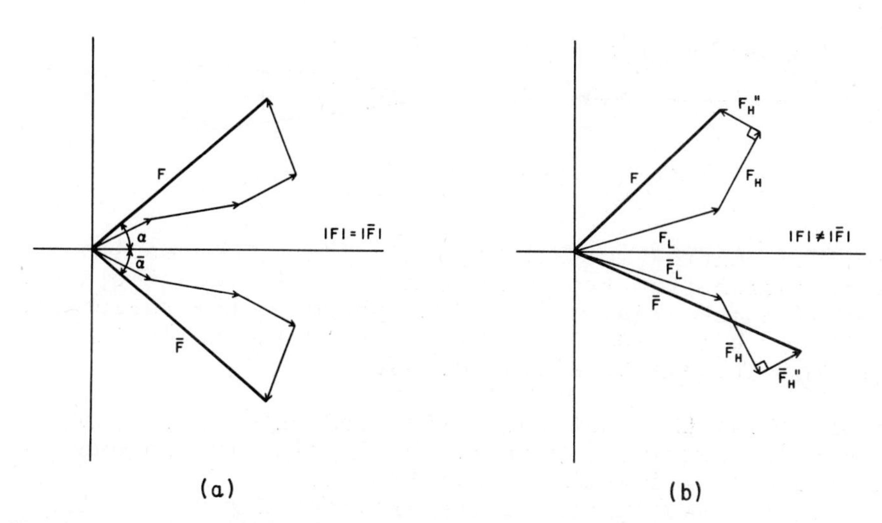

Fig. 8. Noncentrosymmetric Crystal. [(a) Friedel's law for normal scattering. (b) Invalidation of Friedel's law when anomalous scattering occurs.]

$-x_j, -y_j, -z_j$. Such a pair of atoms produces a structure-factor of value

$$F = f_1 \exp(i\alpha_1) + f_1 \exp(-i\alpha_1)$$

$$= 2f_1 \cos \alpha_1.$$

There is no imaginary component, and the structure-factors are real. Thus, the phase-angle of a reflection from a centrosymmetric crystal is restricted to two possible values, namely, 0° or 180° (see Fig. 9a).

When the inverse reflection from the planes $(\bar{h}\bar{k}\bar{\ell})$ is considered,

$$\bar{F} = \Sigma f_j \exp(-i\alpha_j)$$

$$= A - iB.$$

Thus, $$|\bar{F}|^2 = A^2 + B^2.$$

Therefore, the structure-amplitudes of inverse reflections are the same. Because the structure-amplitudes are related to the intensities of the reflections, the intensities of inverse reflections are the same for normal scattering; this relationship is known as Friedel's law.

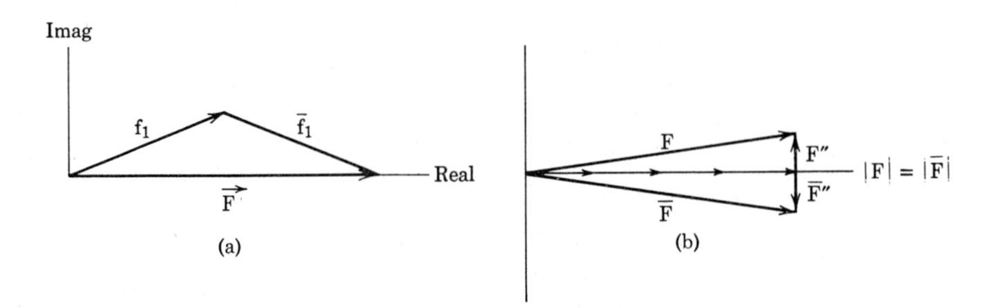

Fig. 9. Centrosymmetric Crystal. [(a) The resultant contribution from two centrosymmetrically related atoms lies along the real axis. (The phase of the reflection is, thus, 0 or 180°.) (b) Friedel's law holds, even when anomalous scattering occurs.]

It is noteworthy that the diffracted intensities have a center of inversion, even if the crystal itself does not; this result is illustrated in Fig. 8a.

When the wavelength of the incident X-rays is close to, and on the shorter-wavelength side of, the absorption-edge of any one (or more) atoms in the cell, then, in addition to the normal scattering from these atoms, a small, but significant, component of scattering, having a phase-lead of 90°, takes place; this is referred to as anomalous scattering. The normal scattering is also modified to some extent, and the scattering-factor for such anomalously scattering atoms can be put as

$$f_a = f + \Delta f' + i\Delta f'',$$

where values $\Delta f'$ and $\Delta f''$ are anomalous-dispersion corrections;[9] they are listed in Table V for a few elements, for CuK_α and MoK_α radiation. The values of the dispersion corrections are essentially independent of the angle of scattering, as they arise from the inner core of electrons that have a much smaller spatial spread.

The anomalous scattering leads to a difference in intensity between inverse reflections in a noncentrosymmetric crystal. Consequently, Friedel's law is violated, and $|F|^2 \neq |\bar{F}|^2$ (see Fig. 8b). It is this distinction in inverse reflections that makes possible the determination of the absolute configuration of molecules by X-ray diffraction methods; this aspect is considered later (see p. 511). In the centrosymmetric structure, $|F|^2 = |\bar{F}|^2$, and Friedel's law always holds (see Fig. 9b). This result is to be expected because the planes $(\bar{h}\bar{k}\bar{\ell})$ and $(\bar{h}\bar{k}\bar{\ell})$ are related by the center of symmetry present in the crystal.

TABLE V

Typical Dispersion Corrections for CuK_α and MoK_α Radiations

	CuK_α		MoK_α	
Element	$\Delta f'$	$\Delta f''$	$\Delta f'$	$\Delta f''$
O	0.0	0.1	0.0	0.0
P	0.2	0.5	0.1	0.2
Co	-2.2	3.9	0.4	1.1
Br	-0.9	1.5	-0.3	2.6
Ba	-2.1	8.9	-0.4	3.0
Hg	-5.0	9.0	-2.6	10.6

3. Systematic Absences

Whenever symmetry elements involving components of translation are present in the crystal, they manifest themselves in producing characteristic absences from the diffracted spectra. The situation for a two-fold screw-axis is illustrated in Fig. 10. Consider the reflection 010. The (010) planes are one unit apart, and are perpendicular to b. The atoms in these planes scatter in phase; however, the presence of the two-fold screw-axis generates an identical number of atoms in a plane midway between the (010) planes. The scattering from these atoms is out of phase by 180°. Thus, the net intensity of the reflection (010) is zero, and the reflection is absent from the recorded spectra. Similarly, it can be shown that the reflections (030), (050),... are also absent, or, in general, that (0k0)-type reflections are absent when $k = 2n + 1$. This result can be derived by using for the structure factor the expression

$$F = \sum_{1}^{N} f_j \exp 2\pi i(hx_j + ky_j + \ell z_j).$$

For every atom at x_j, y_j, z_j, there is an atom at $\bar{x}_j, \frac{1}{2} + y_j + \bar{z}_j$. Thus,

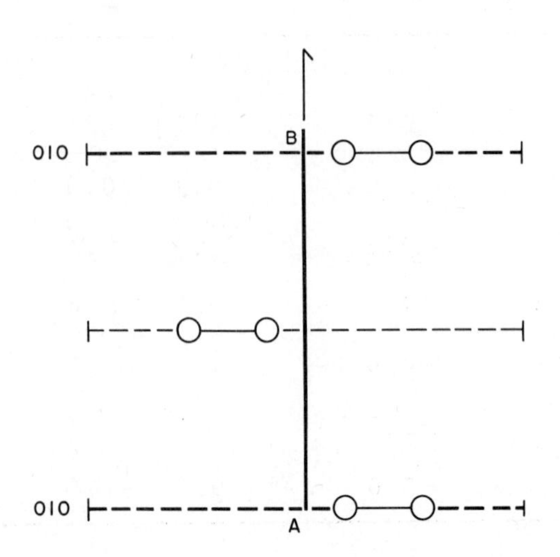

Fig. 10. The Reflection 010 from the Planes 010 in the Presence of a Two-fold Screw-axis parallel to the *b*-Axis.

$$F = \sum_{1}^{N/2} f_j [\exp 2\pi i(hx_j + ky_j + \ell z_j) + \exp 2\pi i(-hx_j + k(y_j + \tfrac{1}{2}) - \ell z_j)].$$

Consider the reflections 0k0. Here,

$$F = \Sigma f_j \exp 2\pi iky_j [1 + \exp \pi ik]$$
$$= \Sigma f_j \exp 2\pi iky_j [1 + (-1)^k].$$

When $k = 2n + 1$, $(-1)^k = -1$, and $F = 0$. Such absences are called systematic, or space-group, absences. The absences due to different types of symmetry elements are given in Table VI.

4. Perfect and Mosaic Crystals

A crystal is said to be ideally perfect if the lattice periodicity is strictly maintained throughout the volume of the crystal. The intensities of the diffracted beams from such crystals are considerably diminished by multiple

TABLE VI

Systematic Absences and Harker Sections Arising from Some Symmetry Elements

Symmetry element	Symbol	Absent reflections	Harker sections in the Patterson function
Center of inversion	$\bar{1}$	none	none
Two-fold axis, parallel to *b* axis	2	none	plane at $V = 0$
Two-fold screw, parallel to *b* axis	2_1	0k0, $k = 2n + 1$	plane at $V = 1/2$
Three-fold screw, parallel to *c* axis	3_1	00ℓ, $\ell \neq 3n$	plane at $W = 1/3$
Four-fold screw, parallel to *c* axis	4_1, 4_3	00ℓ, $\ell \neq 4n$	plane at $W = 1/4$
	4_2	00ℓ, $\ell = 2n + 1$	plane at $W = 1/2$
Six-fold screw, parallel to *c* axis	6_1, 6_5	00ℓ, $\ell \neq 6n$	plane at $W = 1/6$
	6_2, 6_4	00ℓ, $\ell \neq 3n$	plane at $W = 1/3$
	6_3	00ℓ, $\ell = 2n + 1$	plane at $W = 1/2$
Glide-plane, perpendicular to *b*, translation $c/2$	*c*	$h0\ell$, $\ell = 2n + 1$	line at $V = 0$, $W = 1/2$
Glide-plane, perpendicular to *b*, translation $(a + c)/2$	*n*	$h0\ell$, $h + \ell = 2n + 1$	line at $V = 1/2$ and $W = 1/2$

TABLE VI (continued)

Body-centered lattice, translation $(a + b + c)/2$	*I*	hkℓ, h + k + ℓ = $2n + 1$
Face-centered lattice, translations $(a + b)/2$, $(b + c)/2$, $(c + a)/2$	*F*	hkℓ, h + k = $2n + 1$ or k + ℓ = $2n + 1$ or ℓ + h = $2n + 1$
Lattice centered on *ab* plane	*C*	hkℓ, h + k = $2n + 1$

reflections, and the intensity of reflection I can be expressed as $I \alpha |F|$. Moreover, the reflections from such crystals should be ideally sharp.

In practice, very few ideally perfect crystals are encountered. Most crystals of interest are mosaic, being composed of several small blocks that are slightly misoriented with reference to each other (see Fig. 11).

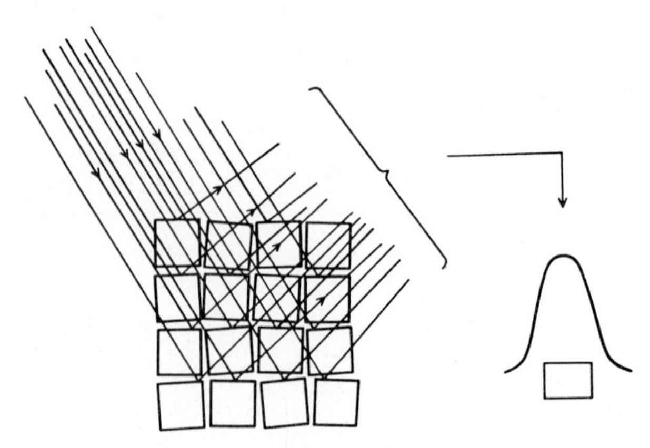

Fig. 11. Mosaic Spread in the Crystal, Giving Rise to the Spread of the X-ray Reflections.

The diffracted beam from such crystals has an angular width attributable to the fact that different blocks reflect at slightly different angles, thus producing a mosaic spread in the reflection. Almost all organic crystals are mosaic, and, for such crystals, $I \alpha |F|^2$. Sometimes, mosaicity is induced in the crystal under study by subjecting it to mechanical shock, such as grinding, or to thermal shock by repeatedly dipping it into and withdrawing it from liquid air or nitrogen.

Despite such treatment, some low-angle, strong reflections do suffer diminution in intensity; this is called *secondary extinction*, and it can be detected at the end of a structure analysis. Usually, there are only a few reflections that suffer from such secondary extinction effects.

5. Weighted Reciprocal-lattice

If a weight is attached to the reciprocal lattice-point $hk\ell$, proportional to the observed intensity, a weighted reciprocal-lattice is obtained. The symmetry of the weighted reciprocal-lattice is the same as the point-group symmetry of the crystal. When Friedel's law holds, an inversion center is added to this symmetry, and the effective symmetry is that of the Laue group of the crystal. The reflections that are related to the others in the reciprocal lattice by the Laue group-symmetry are called equivalent reflections; they are tabulated in Table VII for triclinic, monoclinic, and orthorhombic systems. In practice, it is sufficient to measure the intensities of an independent set of reflections.

TABLE VII

Equivalent and Independent Reflections in Triclinic, Monoclinic, and Orthorhombic Systems, when Friedel's Law is Obeyed

System	Laue symmetry	Equivalent reflections	Independent reflections	Fraction of total number of reflections
Triclinic	$\bar{1}$	$[hk\ell, \bar{h}\bar{k}\bar{\ell}]$	$hk\ell$	0.5
		$[\bar{h}k\ell, h\bar{k}\bar{\ell}]$	$\bar{h}k\ell$	
		$[h\bar{k}\ell, \bar{h}k\bar{\ell}]$	$h\bar{k}\ell$	
		$[\bar{h}\bar{k}\ell, hk\bar{\ell}]$	$\bar{h}\bar{k}\ell$	

TABLE VII (continued)

Monoclinic (b-Axis unique)	2/*m*	[$hk\ell$, $h\bar{k}\ell$, $hk\ell$ $hk\bar{\ell}$, $\bar{h}\bar{k}\bar{\ell}$] [$\bar{h}k\ell$, $hk\ell$, $\bar{h}k\ell$ $\bar{h}k\bar{\ell}$, $hk\ell$]	0.25
Orthorhombic	*mmm*	[$hk\ell$, $\bar{h}k\ell$, $hk\ell$ $hk\ell$, $\bar{h}\bar{k}\ell$, $hk\ell$, $\bar{h}k\ell$, $hk\ell$, $\bar{h}\bar{k}\bar{\ell}$]	0.125

V. EXPERIMENTAL PROCEDURES

In this Section, a very brief account is given of the experimental procedures that lead to determination of the unit-cell dimensions and the space-group, and to collection of the intensity data.

1. Crystal Growth and Mounting

For this kind of work, the crystals are generally grown by dissolving the compound in a suitable solvent, usually water or aqueous ethyl alcohol mixture, and allowing the solution to evaporate slowly. In a successful crystallization, crystals of various sizes and shapes are deposited as the solvent evaporates.

Crystals 100–200 μm thick and 200–400 μm long are usually adequate for X-ray diffraction analysis. If the crystals are too large, a fragment having the desired size may be cut, and properly shaped.

The crystal is mounted at the tip of a glass fiber with epoxy resin; the fiber is placed in the hollow of a brass pin, and is held firmly by means of sealing wax (see Fig. 12). One of the prominent faces of the crystal is usually parallel to the axis of the fiber, which generally represents a principal axis of the crystal. The brass pin bearing the crystal is transferred to a goniometer head (see Fig. 13) that is provided with two arcs for angular movements at right angles to each other. Two translational devices parallel to the arcs are also provided, and a third device permits adjustment of the height of the crystal. By the use of the arcs and the translational devices, the crystal may be so aligned optically that the crystal-axis is coincident with that of the goniometer. The final adjustments are made after examination of preliminary, X-ray photographs.[12]

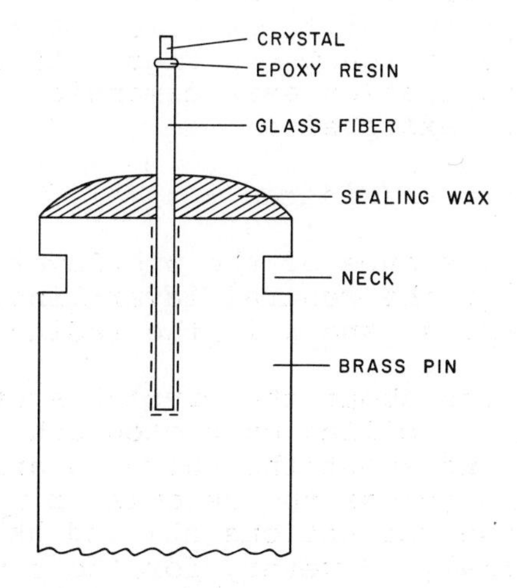

Fig. 12. The Crystal Mounted on a Brass Pin.

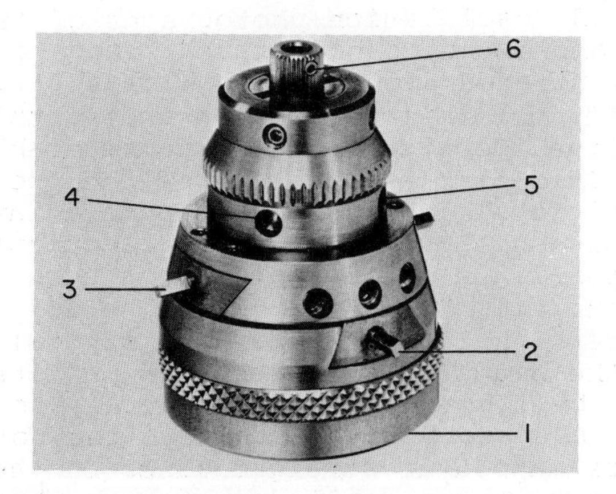

Fig. 13. Goniometer Head. [(1) Base; (2) and (3), arcual movements; (4), (5), and (6), translational adjustments. (Courtesy of Electronics and Alloys, Inc.)]

2. Preliminary Crystal Data

Fig. 14 shows the X-ray beam that is incident normal to the crystal mounted about the *b* axis. On oscillating the crystal about this axis, the reciprocal planes (h0l), (h1ℓ), (h21), (h$\bar{1}$ℓ), (h$\bar{2}$1), *etc.*, intersect the sphere of reflection in a series of circles. The diffracted beams are developed in a series of cones that intersect a cylindrical photographic film wrapped circularly around the crystal. The circles correspond to the layer-lines in the unwrapped oscillation photograph. The oscillation photographs about two different axes, *b* and *c*, for adenosine 3'-phosphate are shown in Figs. 15a and 15b. From Fig. 15a, the oscillation-axis dimension can be obtained by the use of the expression

$$b = n\lambda/\sin[\tan^{-1}(y_n/R)],$$

where y_n is the distance of the *n*th layer-line from the zero layer-line or the central layer-line, λ is the wavelength of the X-rays, and *R* is the radius of the cylindrical camera.

Some information about the crystal system can also be obtained from the oscillation photograph. In a monoclinic crystal oscillating about the unique *b* axis, there will be a mirror plane across the zero-layer line, because the intensities of the reflections hkl and hk$\bar{l}$ are the same (see Fig. 15a). However, for the crystal mounted about the (non-unique) *c* axis, there is no mirror plane, because the intensities of the reflections hkl and hk$\bar{l}$ are not the same (see Fig. 15b). The symmetries that are observed in the oscillation photographs of triclinic, monoclinic, and orthorhombic crystals mounted about the three different axes are given in Table VIII.

It is difficult to identify the individual reflections in a layer-line, because there is considerable overlap of the spots. This overlap arises from the projection of a plane of reciprocal lattice-points onto a layer-line.

In the Weissenberg method, the reflections on the individual layer-lines are isolated by a screen, and recorded on a moving film (see Fig. 16). The zero-layer Weissenberg photographs h0ℓ and hk0 for the crystal oscillating about the *b* and *c* axes are shown in Figs. 17a and 17b. In the h0ℓ photograph, the central, reciprocal-lattice rows h00 and 00ℓ pass through the rotation axis, and are mapped as straight lines on the film. The noncentral rows, such as h01 do not pass through the rotation axis, but, instead, lie on curves known as festoons. Thus, the reciprocal lattice is recorded on the film in a distorted way.

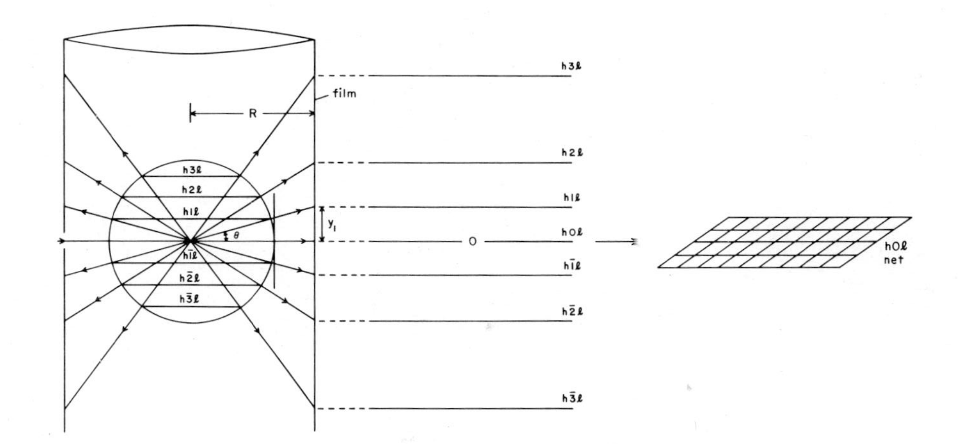

Fig. 14. The Geometry of Recording Reflections on Layer-lines in an Oscillation Photograph. [Each layer-line comprises reflections from one reciprocal, lattice-plane.]

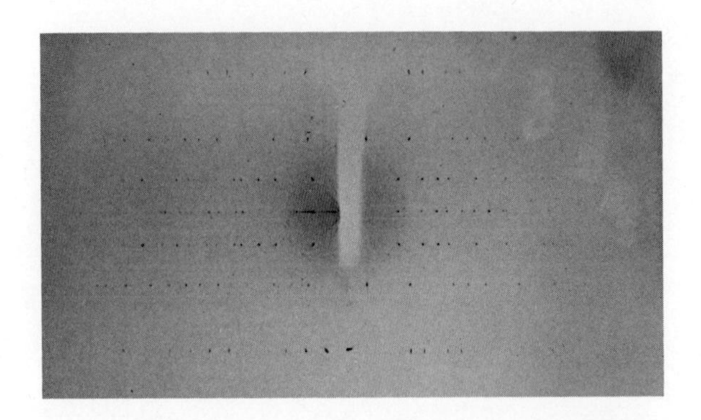

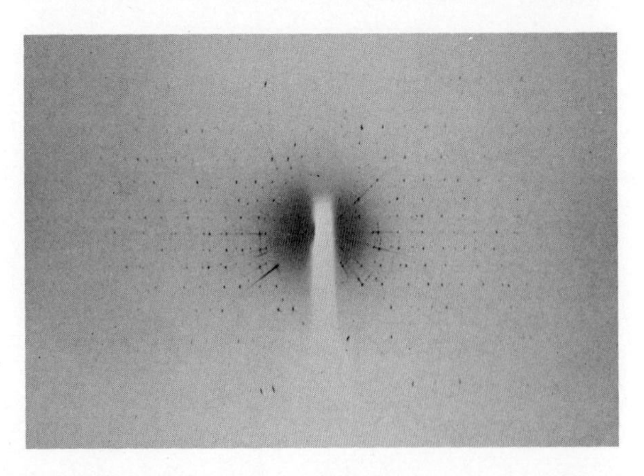

Fig. 15. Oscillation Photographs of Adenosine 3'-Phosphate. [(a) About the unique *b*-axis, and (b), about the nonunique *c*-axis.]

The standard diameter of the camera is chosen to be 57.3 mm, so that the distance *y* of a reflection from the trace of the direct beam 00 gives the Bragg angle θ (in degrees). If θ_h are the angles for the h00 reflections, the mean value of a^* (see Table IV) is given by the relation

$$a^* = \frac{2}{\lambda} \left\langle\left(\frac{\sin\, \theta_h}{h} \right)\right\rangle.$$

TABLE VIII

Symmetries of Oscillation Photographs and Weissenberg Photographs

System	Oscillation axis	Symmetry in oscillation photograph	Symmetry in the upper half of the Weissenberg photograph	
			Zero level	First level
Triclinic	*a*, *b*, or *c*	none	none	none
Monoclinic	*b*	mirror	none	none
	a (or *c*)	none	mirror about *b** and *c** (*a**)	mirror about *c** (*a**)
Orthorhombic	*a* (or *b* or *c*)	mirror	mirrors about both axes	mirrors about both axes

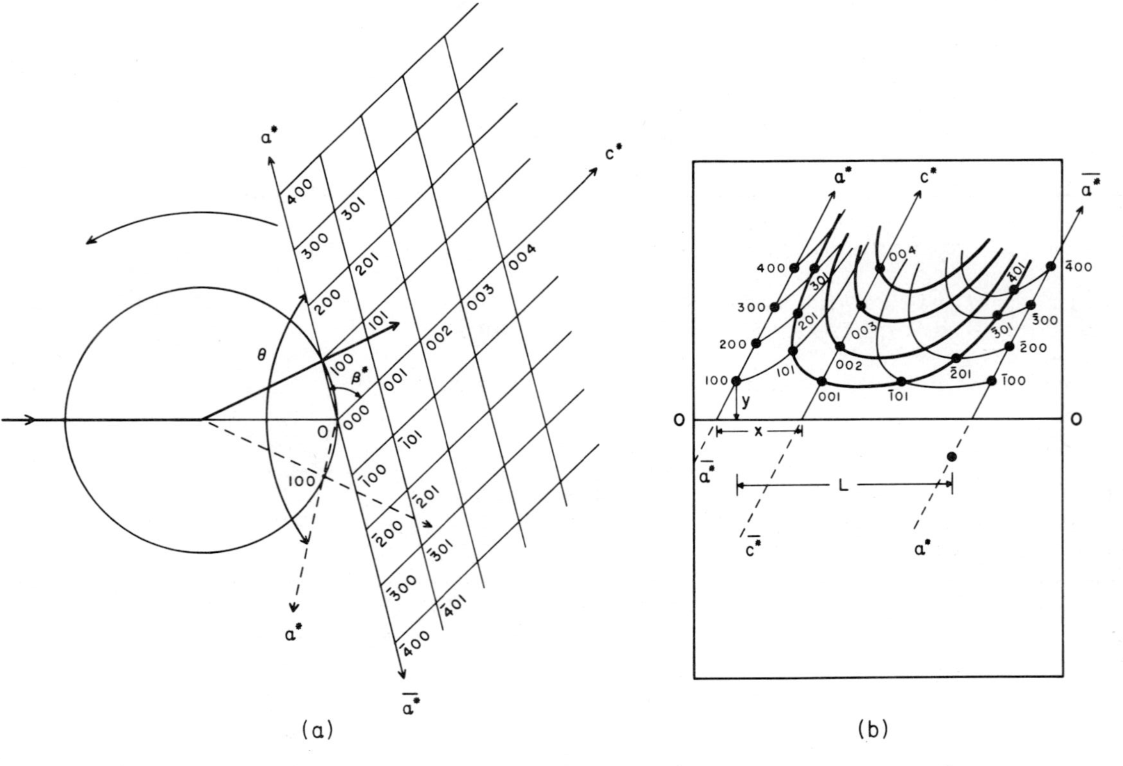

Fig. 16. The Geometry of Recording Reflections in a One-layer Line on a Moving Film (Weissenberg) Photograph. [As the crystal rotates counterclockwise, the film is translated from right to left. The central rows of the reciprocal lattice are recorded on a straight line, and the noncentral rows, on curves known as festoons.]

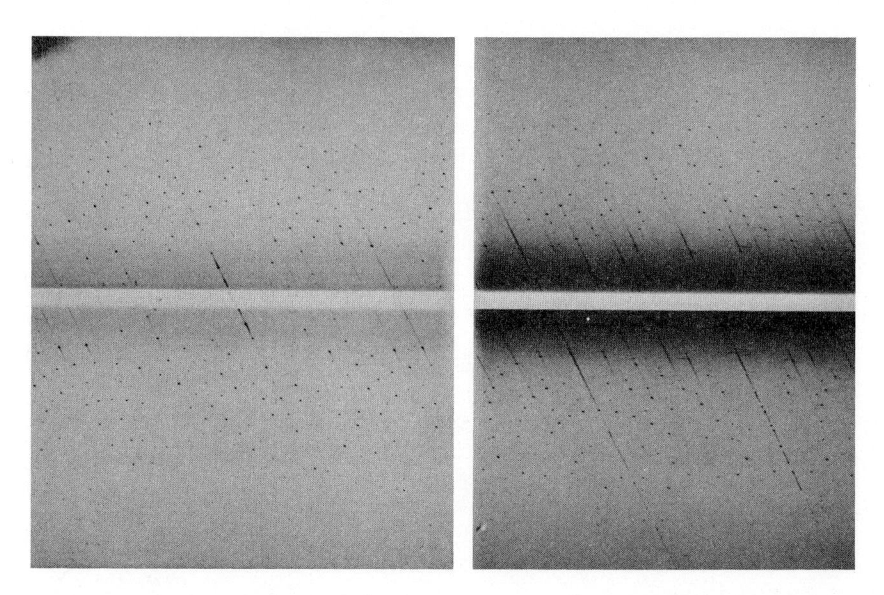

Fig. 17. Zero-layer Weissenberg Photographs for Adenosine 3'-Phosphate. [For the crystal rotating about (a) the unique *b*-axis, and (b) the non-unique *c*-axis.]

Similarly, the mean value of c^* can be determined from the 00ℓ reflections. The angle $\beta^* = 2x$, where x is the horizontal distance (in mm) between the a^* and c^* axes on the film (see Fig. 16b). This relation arises because the camera is usually so designed that, for every 2° of rotation of the crystal, the film is translated by 1 mm. The Weissenberg photograph may also be used for obtaining information concerning the crystal system (see Table VIII). For a zero-layer Weissenberg photograph of a monoclinic crystal oscillating about the unique b axis, there is no mirror plane across either the a^* or the c^* axis (see Fig. 17a). However for the c-axis oscillation, the zero-layer Weissenberg photograph shows mirror planes across both the a^* and the b^* axes, and the angle between these axes is 90° (see Fig. 17b).

The reflections corresponding to higher layers $h1\ell$ are usually recorded by the equi-inclination Weissenberg technique (see Fig. 18). Here, the camera carrying the crystal is rotated through an angle μ related to the dimension of the axis of rotation b by

$$\mu = \sin^{-1} (b/\lambda).$$

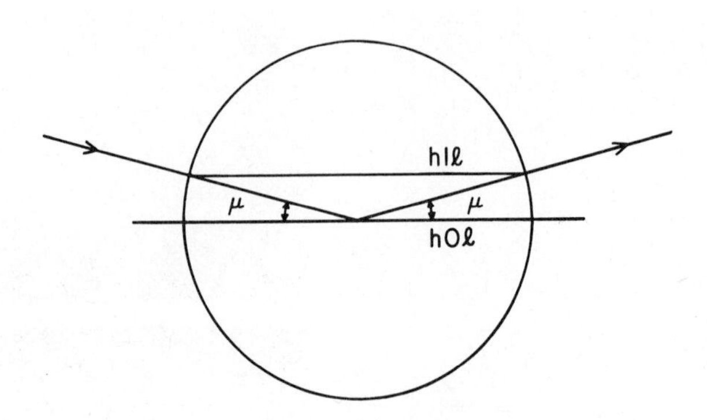

Fig. 18. Geometry of the Equi-inclination Method of Recording Upper-level Reflections in a Weissenberg Photograph.

The incident beam and the diffracted beam are equally inclined at μ to the reciprocal lattice-plane (h1ℓ) that is being recorded.

The presence of certain symmetry elements can give rise to systematic absences (see Table VI) which, together with a knowledge of the crystal system, lead to the assignment of the space-group of the crystal. It is sometimes possible to have more than one space-group for a given set of systematically absent reflections (see Table IX). The choice of the correct space-group may be made after study of the intensity distribution and the structure analysis.

One other way of recording the reciprocal lattice is to use the precession method.[13,14] In this method, a principal axis of the crystal is usually allowed to precess about the incident beam, and the reciprocal lattice-planes perpendicular to this axis can be recorded in an undistorted way (see Fig. 19). If y_h is the distance (in cm) of the reflection (h00) from the center of the film,

$$a^* = y_h/hL\lambda,$$

where L is the distance (in cm) between the crystal and the film, then by using this expression, the reciprocal lattice parameter, a^*, can be calculated.

In accurate determinations of the cell constants, the film is usually calibrated by recording the reflections from a sample, such as aluminum powder, having known unit-cell dimensions. In the diffractometric methods,

TABLE IX

Space Groups that Cannot Be Distinguished on the Basis of Systematic Absences Alone

System	Space Groups
Triclinic	$P1$, $P\bar{1}$
Monoclinic	$P2$, Pm, $P2/m$ $P2_1$,$P2_1/m$ Pc, $P2/c$ $C2$,Cm,$C2/m$ Cc,$C2/c$
Orthorhombic	$P222$,$Pmm2$,$Pmmm$ $C222$,$Cmm2$,$Cmmm$ $F222$,$Fmm2$,$Fmmm$ $I222$,$I2_12_12_1$,$Imm2$,$Immm$ $Pmc2_1$,$Pma2$,$Pmma$ $Pcc2$,$Pccm$ $Pca2_1$,$Pbcm$ $Pnc2$,$Pmna$ $Pmn2_1$,$Pmmn$ $Pba2$,$Pbam$ $Pna2_1$,$Pnma$ $Pnn2$,$Pnnm$ $Cmc2_1$,$Cmcm$ $Ccc2$,$Cccm$ $Iba2$,$Ibam$ $Ima2$,$Imma$

accurate cell-constants are obtained; it is usually necessary to determine these before the intensity data are collected.

3. Determination of Molecular Weight

By use of the reciprocal cell-constants obtained in the foregoing way, the unit-cell dimensions and the volume (V) of the unit cell can be calculated by means of the relationships given in Table IV. The density, ρ, of the crystal may be determined by flotation in a suitable mixture of liquids. The liquids commonly employed are benzene, chloroform, carbon tetrachloride, cyclohexane, and bromoform. The molecular weight, M, of the compound in atomic-mass units can be calculated from the relation

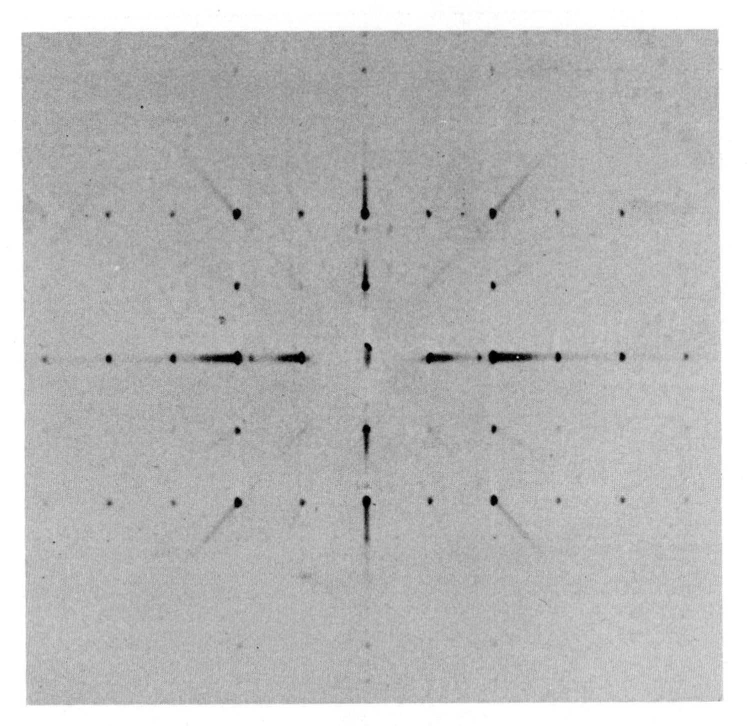

Fig. 19. Precession Photograph of the b^*c^* Plane for $(NH_4)_2MnCl_5 \cdot H_2O$. [Photograph kindly supplied by Mr. E. Putkey.]

$$\rho = Z \times M \times 1.667/V,$$

where Z is the number of molecules in the unit cell and V is given in $\AA^3$. Alternatively, if the molecular weight of the compound is known, the density can be calculated. The unit-cell dimensions, space-groups, and densities (calculated and observed) for all of the known crystals of the bases, paired bases, nucleosides, paired nucleosides, and nucleotides of the nucleic acids are listed in Tables X, XI, XII, and XIII.

4. Resolution and Absorption

The choice of radiation is made in relation to the ultimate accuracy of the analysis and the effect of absorption. The two radiations commonly used are CuK_α

TABLE X

Crystal Data for Purine and Pyrimidine Bases

Base	Crystal[a] system	Space group	Unit-cell constants a (Å) b c	α (°) β γ	Z[b]	D,calc.; D,obs.[c] (g/cm³)	Vol.[d] Mol. Wt.	Data[e]; Method	R[f]	σ(ℓ) pm[g] σ(θ)°	Ref.
Adenine, hydrochloride hemihydrate	*M*	$P2/c$	8.77	90	4	1.596	751.6	250	0.061	1	15, 16
			4.83	114°15'			180.6	*C*			
			19.46	90							
9-methyl-	*M*	$P2_1/c$	7.67 (0.03)	90	4	1.471	663.6	385	0.091	1	17
			12.24 (0.04)	123°26'			149.2	*P*		0.8	
			8.47 (0.03)	90							
dihydrobromide	*O*	$Pna2_1$	17.59 (0.006)	90	4	2.03	1008.6	(h0ℓ) 130	0.07	10	18
			4.88 (0.02)	90		2.06	311.0	(hk0) 88	0.11	6.0	
			11.75 (0.04)	90				*P*			

TABLE X (continued)

Base	Crystal[a] system	Space group	Unit-cell constants a (Å) b c	α (°) β γ	Z[b]	D,calc.; D.obs[c] (g/cm³)	Vol.[d] Mol. Wt.	Data[e]; Method	R[f]	σ(ℓ) pm[g] σ(θ)°	Ref.
Cytosine	*O*	$P2_12_12_1$	13.041 (0.002)	90	4	1.562	472.3	613	0.085	0.3	19
			9.494 (0.001)	90			111.1	*P*		0.6	
			3.815 (0.001)	90							
mono-hydrate	*M*	$P2_1/c$	7.801 (0.005)	90	4	1.473	581.7	1280	0.11	0.4	20
			9.844 (0.007)	99°42' (4')		1.478	129.1	*P*		0.4	
			7.683 (0.006)	90							
1-methyl-	*T*	$P\bar{1}$	6.09	95.47	2	1.455	298.3	830	0.115		21
			6.82	84.48		1.450	125.2	*P*			
			7.01	100.0							
1-methyl-, hydro-bromide	*O*	*Pnma*	12.98 (0.04)	90	4	1.87	779.4	349	0.12	4	64, 65
			6.80 (0.02)	90		1.85	206.1	*P*		2.0	
			8.83 (0.03)	90							

Cytosine-5-acetic acid (4-Amino-1,2-dihydro-2-oxo-5-pyrimidine-acetic acid)	*M*	$P2_1/c$	10.930 (0.006) 5.016 (0.008) 14.042 (0.005)	90 114.69 (0.03) 90	4	1.606 1.611 $CHBr_3$-CCl_4	699.9 171.2	1258 *P*	0.096	0.6 0.3	22
Guanine, monohydrate	*M*	$P2_1/n$	16.51 11.28 3.65	90 96.8 90	4	1.664	675.0 169.2				23
hydrochloride, monohydrate	*M*	$P2_1/a$	19.29 (0.02) 9.94 (0.02) 4.494 (0.005)	90 107°54' (2') 90	4	1.665 1.662	820.0 205.6	340 *P*	0.17	4	24
hydrochloride, dihydrate	*M*	$P2_1/a$	14.69 (0.01) 13.40 (0.01) 4.849 (0.005)	90 93.8 90	4	1.569 1.562	950.6 223.6	1600 *P*	0.073	0.5 0.3	25, 26

TABLE X (continued)

Base	Crystal[a] system	Space group	Unit-cell constants a (Å) b c	α (°) β γ	Z^b	D,calc.; D,obs[c] (g/cm^3)	Vol.[d] Mol Wt.	Data[e]; Method	R^f	$\sigma(\ell)$ pm[g] $\sigma(\theta)$°	Ref.
9-methyl-, hydrobromide	*M*	$P2_1/b$	4.54 (0.01)	90	4	1.940	846.6	783	0.077	3	27
			17.46 (0.01)	90		1.933	246.1	*P*		2.0	
			10.68 (0.08)	90							
Thymine, monohydrate	*M*	$P2_1/c$	6.077 (0.003)	90	4	1.486	644.30	1068	0.078	1	28
			27.862 (0.007)	94°19' (3')			144.1	*P*		0.5	
			3.816 (0.002)	90							
1-methyl-	*M*	$P2_1/c$	7.351 (0.004)	90	4	1.379	675.7	1166	0.072	0.4	29
			12.091 (0.006)	89°58' (3')		1.381	140.2	*P*		0.2	
			7.602 (0.004)	90		C_6H_6-$CHCl_3$					

Uracil	*M*	*P*2/*a*	11.938 (0.001)	90	4	1.606	463.4	1163	0.045	0.3	30, 31
			12.376 (0.0009)	120°54' (0.4')		1.617	112.1	*C*		0.2	
			3.6552 (0.0003)	90							
5-ethyl-6-methyl-	*T*	$P\bar{1}$	7.478 (0.001)	99.39 (0.02)	2	1.297	445.1	1368	0.080	1	34
			11.803 (0.001)	107.65 (0.01)		1.27	154.2	*P*		1.0	
			4.776 (0.001)	92.14 (0.01)		*p*-dioxane-$CHCl_3$					
1-methyl-	*O*	*Ibam*	13.22	90	8	1.525	1098.3	517	0.11	0.8	32
			13.25	90			126.1	*P*		0.6	
			6.27	90							
hydrobromide	*O*	*Pnma*	13.24 (0.03)	90	4	1.834	754.0	226	0.17	5	33
			6.82 (0.04)	90		1.805	207.1	*P*		2.0	
			8.35 (0.04)	90							
5-nitro-, monohydrate	*M*	$P2_1/c$	5.137 (0.005)	90	4	1.808	642 (643.2)	1210	0.062	0.6	35
			21.956 (0.006)	143°30' (5')		1.79	175.1	*P*		0.4	
			9.587 (0.007)	90							

TABLE X (continued)

[a] T = triclinic, M = monoclinic, and O = orthorhombic. [b] The number of molecules in the unit cell. [c] The solution or mixture utilized in the density measurements is given wherever it is known. [d] Volume of the unit cell in $Å^3$. [e] The total number of X-ray reflections utilized in the final refinement of the structure. P, by the photographic method, and C, by the counter method of measurement of intensities. [f] Agreement index (R) is defined in Equation 1. [g] Estimated standard-deviations in bond distances $\sigma(\ell)$ (in pm) and bond angles $\sigma(\theta)$ (in degrees).

TABLE XI

Crystal Data for Nucleosides[a]

Nucleoside	Crystal system	Space group	Unit-cell constants a, b, c (Å)	α, β, γ (°)	Z	D, calc.; D, obs. (g/cm³)	Vol.; Mol. Wt.	Data; Method	R	$\sigma(\ell)$ pm; $\sigma(\theta)$°	Ref.
Adenosine, 2'-deoxy-, hydrate	*M*	$P2_1$	16.060 (0.007) 7.866 (0.003) 4.700 (0.002)	90 96°4' (1') 90	2	1.514 1.510	590.4 269.3	1400 *P*	0.078	1 1.2	36
(Guanosine)$_2$ dihydrate	*M*	$P2_1$	17.518 11.502 6.658	90 98.17 90	2	1.597	1328.0 638.6				23
(Inosine)$_2$ dihydrate	*M*	$P2_1$	17.573 11.278 6.654	90 98.23 90	2	1.548	1305.2 608.5				23
Cytidine	*O*	$P2_12_12_1$	13.991 (0.002) 14.786 (0.002) 5.116 (0.001)	90 90 90	4	1.526	1058.4 243.2	1195 *P*	0.080	0.7 0.5	37

TABLE XI (continued)

Nucleoside	Crystal system	Space group	Unit-cell constants a (Å) b c	α (°) β γ	Z	D, calc.; D,obs. g/cm^3	Vol; Mol. Wt.	Data; Method	R	$\sigma(\ell)$ pm $\sigma(\theta)$°	Ref.
2'-deoxy-, hydro-chloride	*M*	$P2_1$	6.5707 (0.0003) 17.659 (0.001) 5.1248 (0.0002)	90 108.09 90	2	1.551 1.548	564.6 263.7				38
Thymidine	*O*	$P2_12_12_1$	4.86 13.91 16.32	90 90 90	4	1.479	1087.4 242.2	1203 *C*	0.08		39
5'-bromo-5'-deoxy-	*M*	$P2$	25.50 5.50 8.95	90 90 108 (1)	4	1.703	1193.7 306.1				40
	O	$P2_12_12_1$	23.10 9.55 5.42	90 90 90	4	1.700	1195.7 306.1				40
5'-bromo-2',3',5'-trideoxy-	*O*	$P2_12_12_1$	14.78 4.89 15.17	90 90 90	4	1.661	1096.4 274.1				41

Uridine, 5-bromo-	*M*	$P2_1$	7.725 (0.006)	90	2	1.842	584.3	1221	0.07	2	42
			5.813 (0.004)	101.2			324.1	*P*		1.0	
			13.264 (0.010)	90							
Unimolar complex with methyl sulf- oxide	*M*	$P2_1$	13.65 (0.01)	90	2	1.675	795.0				43
			4.820 (0.005)	91.8 (0.1)			401.1				
			12.09 (0.01)	90							
2'-deoxy-	*M*	$P2_1$	9.150 (0.008)	90	2	1.890	541.2	1312	0.074	2	42
			5.142 (0.004)	108.2 (0.1)			308.1	*P*		1.0	
			12.108 (0.009)	90							
5-fluoro	*O*	$P2_12_12_1$	19.39 (0.03)	90	4	1.562	1051.2	1187	0.100	1.1	45
			11.76 (0.02)	90			247.2	*C*		0.7	
			4.61 (0.01)	90							

TABLE XI (continued)

Base	Crystal system	Space group	Unit-cell constants a, b, c (Å)	α, β, γ (°)	Z	D,calc.; D,obs. (g/cm^3)	Vol.; Mol. Wt.	Data; Method	R	$\sigma(\ell)$ pm $\sigma(\theta)$°	Ref.
5-iodo-	*T*	*P*1	4.98 (0.01) 6.83 (0.01) 9.60 (0.02) (0.02)	101°40' (5') 109°18' (5') 98°20' (5')	1	2.008	356.3 355.1	1434 *C*	0.054	2.8 1.9	46
5-methyl-, hemihydrate	*O*	$P2_12_12_1$	14.026 (0.003) 17.302 (0.001) 4.861 (0.001)	90 90 90	4	1.510	1179.7 268.2				44

[a]For an explanation of the column headings, see the footnotes to Table X.

TABLE XII

Crystal Data for Some Nucleotides[a]

Nucleotide	Crystal system	Space group	Unit-cell constants a, b, c (Å)	α, β, γ (°)	Z	D,calc.; D,obs. (g/cm^3)	Vol.; Mol. Wt.	Data; method	R	$\sigma(\ell)$ pm; $\sigma(\theta)$°	Ref.
Adenosine 3'-phosphate, dihydrate	*M*	$P2_1$	9.939 (0.005) 6.343 (0.002) 11.896 (0.005)	90 92°13' (3') 90	2	1.698	749.4 383.3	1411 *P*	0.043	0.6 0.3	50
Adenosine 5'-phosphate, monohydrate	*M*	$P2_1$	12.77 (0.02) 11.82 (0.02) 4.882 (0.01)	90 92°24' (5') 90	2	1.647	736.4 365.2	1197 *P*	0.068	1.3 0.8	51, 52
Cytidine 3'-phosphate	*O*	$P2_12_12_1$	8.778 (0.001) 21.649 (0.003) 6.847 (0.001)	90 90 90	4	1.64 1.66	1301.2 323.2	1196 *P*	0.045	0.8 0.6	47, 48

TABLE XII (continued)

Nucleotide	Crystal system	Space group	Unit-cell constants *a* (Å) *b* *c*	α (°) β γ	Z	D,calc.; D,obs. (g/cm^3)	Vol.; mol. wt.	Data; method	R	σ(ℓ) pm σ(θ)°	Ref.
Cytidine 3'-phosphate	*M*	$P2_1$	5.987 (0.001)	90	2	1.723	621.7	1332	0.073	0.9	49
			17.040 (0.003)	114.0 (0.01)		1.73	323.2	*P*		1.0	
			6.670 (0.001)	90							
Inosine 5'-phosphate, barium salt, hexahydrate	*O*	$C222_1$	21.45	90	8	1.928	4128.9				58
			8.85	90		1.935	484.6				
			21.75	90							
Inosine 5'-phosphate, sodium salt, octahydrate (found: 6 H_2O)	*O*	$C222_1$	8.682 (0.003)	90	8	1.547	4417.6	1400	0.10	2	56, 57
			23.216 (0.009)	90		1.576	514.4	*C*		1.3	
			21.917 (0.007)	90		$CHCl_3$-$CHBr_3$					

Inosine 5'-phosphate, disodium salt, 7.5 H_2O	*O*	*C*222_1	23.06 8.64 21.92	90 90 90	8	1.616 1.62	4367.3 393.2				58
Thymidine 5'-phosphate, calcium salt, octahydrate	*M*	*P*2_1	14.40 (0.02) 6.87 (0.01) 9.81 (0.01)	90 90°58' (3') 90	2 Freon-113	1.61 1.60	970.4 630.4	1575 *P*	0.116	2 1.4	53, 54
Uridine 5'-phosphate, barium salt, 7 H_2O	*O*	*C*222_1	21.11 (0.02) 9.06 (0.03) 20.98 (0.02)	90 90 90	8	2.05 2.05 $CHCl_3$-$CHBr_3$	4012.6	2502 *P*	0.098	3 2.0	55
Dinucleotide											
Adenylyl-(2'→5')-uridine, tetrahydrate	*M*	*P*2_1	9.783 (0.006) 16.013 (0.007) 8.718 (0.007)	90 90.79 (0.05) 90	2	1.569 1.56 $CHBr_3$-CCl_4	1365.6 663.5	1786 *P*	0.088	12 0.7	59

TABLE XII (continued)

Nucleotide	Crystal system	Space group	Unit-cell constants a, b, c (Å)	α, β, γ (°)	Z	D,calc.; D,obs. (g/cm3)	Vol.; mol. wt.	Data; method	R	$\sigma(\ell)$ pm; $\sigma(\theta)$°	Ref.
5'-*O*-Phosphono-thymidylyl-(3'→5')-thymidine, dodecahydrate	*O*	$P2_12_12_1$	16.06 (0.04) 15.13 (0.04) 15.65 (0.04)	90 90 90	4	1.587 1.588	3754.2 860.7	953 *c*			60

[a]For an explanation of the column headings, see the footnotes to Table X.

TABLE XIII

Crystal Data for Paired Bases and Paired Nucleosides[a]

Paired bases	Crystal system	Space group	Unit-cell constants a b c (Å)	α β γ (°)	Z	D,calc.; D,obs. (g/cm³)	Vol.; mol. wt.	Data; method	R	σ(ℓ) pm σ(θ)°	Ref.
Adenine, 9-methyl-: thymine, 1-methyl-	*M*	$P2_1/m$	8.24 (0.03) 6.51 (0.03) 12.75 (0.05)	90 106°48' (10') 90	2	1.468 1.433	654.7 289.3	1361 *P*	0.081	0.5	61, 62
Adenine, 9-ethyl-: uracil, 5-bromo-1-methyl-	*T*	$P\bar{1}$	7.91 12.56 8.68	104.3 72.8 117.4	2	1.187	1030.2 368.2	3064 *P*	0.13 disordered	2.5 1	63a
Adenine, 9-ethyl-: uracil, 5-fluoro-1-methyl-	*T*	$P\bar{1}$	7.82 8.39 12.38	105.28 62.18 104.40	2	1.487 1.483	685.4 307.3	1264 *P*	0.18	3	73

TABLE XIII (continued)

Paired Bases	Crystal system	Space group	Unit-cell constants a (Å) b c	α (°) β γ	Z	D, calc.; D, obs. (g/cm^3)	Vol.; mol. wt.	Data; method	R	$\sigma(\ell)$ pm $\sigma(\theta)$°	Ref.
Adenine, 9-ethyl-: uracil, 1-methyl-	*T*	$P\bar{1}$	7.74 8.39 12.33	107.84 63.75 105.25	2	1.003	957.4 289.3	2700 *P*	0.041	0.8 0.2	63
Adenine, 9-methyl-: uracil, 5-bromo-1-methyl-	*T*	$P\bar{1}$	9.26 10.78 7.4	79 77 101	2	1.82 1.79	777.2 354.2	1120 *P*	0.11	2	72
Adenosine: uridine, 5-bromo-	*O*	$P22_12_1$	4.80 (0.01) 15.19 (0.01) 31.76 (0.03)	90 90 90	4	1.699 1.706	2315.7 592.4	2511 *P*	0.14	3 2.0	76, 77
Cytosine: uracil, 5-fluoro-, monohydrate	*T*	$P\bar{1}$	4.29 9.29 15.14	111.83 98.02 101.30	2	1.56 1.59	654.2 259.2	--	--	--	71a

Guanine, 9-ethyl-: cytosine, 5-bromo- 1-methyl-	*M*	$P2_1/c$	8.38 (0.01) 13.16 (0.01) 14.51 (0.01)	90 97°30' 90	4	1.602 1.628	1646.7 384.3	2200 *P*	0.20	--	68
Guanine, 9-ethyl-: cytosine, 5-fluoro- 1-methyl-	*T*	$P\bar{1}$	8.745 (0.008) 11.227 (0.010) 7.513 (0.009)	109°1' (5') 84°58' (5') 90°59' (5')	2	1.374	778.7 322.3	1900 *P*	0.13	0.9 0.5	66, 67
Guanine, 9-ethyl-: cytosine, 1-methyl-	*T*	$P\bar{1}$	8.838 (0.008) 11.106 (0.010) 7.391 (0.006)	107°49' 5' 87°3' (5') 91°27' (5')	2	1.330	759.8 304.32	1976 *P*	0.11	0.7 0.4	66, 67
Guan- osine,2'- deoxy-: cytidine, 5-bromo- 2'-deoxy-	*O*	$P22_12_1$	5.14 (0.02) 19.11 (0.02) 23.66 (0.03)	90 90 90	4	1.64 1.63	2324.0 481.3	1350 *P*	14	0.05 3.0	74, 75

TABLE XIII (continued)

Paired Bases	Crystal system	Space group	Unit-cell a (Å) b c	con-stants α (°) β γ	Z	D,calc.; D,obs. (g/cm^3)	Vol.; mol. wt.	Data; meth-od	R	σ(ℓ) pm σ(θ)°	Ref.
Hypo-xanthine, 9-ethyl-: uracil, 5-fluoro-	*M*	$P2_1/c$	4.656 15.276 17.807	90 90°48' 90	4	1.54 1.55	1266.4 296.3	1464 *C*	5.5	--	69
9*H*-Purine, 2-amino-9-ethyl-: uracil, 5-fluoro-1-methyl-	*M*	$P2_1/c$	8.56 (0.01) 21.15 (0.015) 7.63 (0.01)	90 97°48' 90	4	1.453 --	1404 307	2211 *P*	18 --	--	71
9*H*-Purine, 2,6-diamino-9-ethyl-: (uracil, 5-iodo-1-methyl-)$_2$	*M*	$P2_1/c$	9.125 (0.010) 12.605 (0.015) 20.392 (0.025)	90 96°20' 90	2	--	2331.4 682.2	1195 *P*	16 --	--	70

[a]For the explanation of the column headings, see the footnotes to Table X.

and MoK_α, which are adequate for single-crystal work, and the wavelength of the X-rays is related to the resolution obtained. The electron-density maps, calculated for a molecule by using reflections up to a certain value of d spacing, are illustrated in Fig. 20.

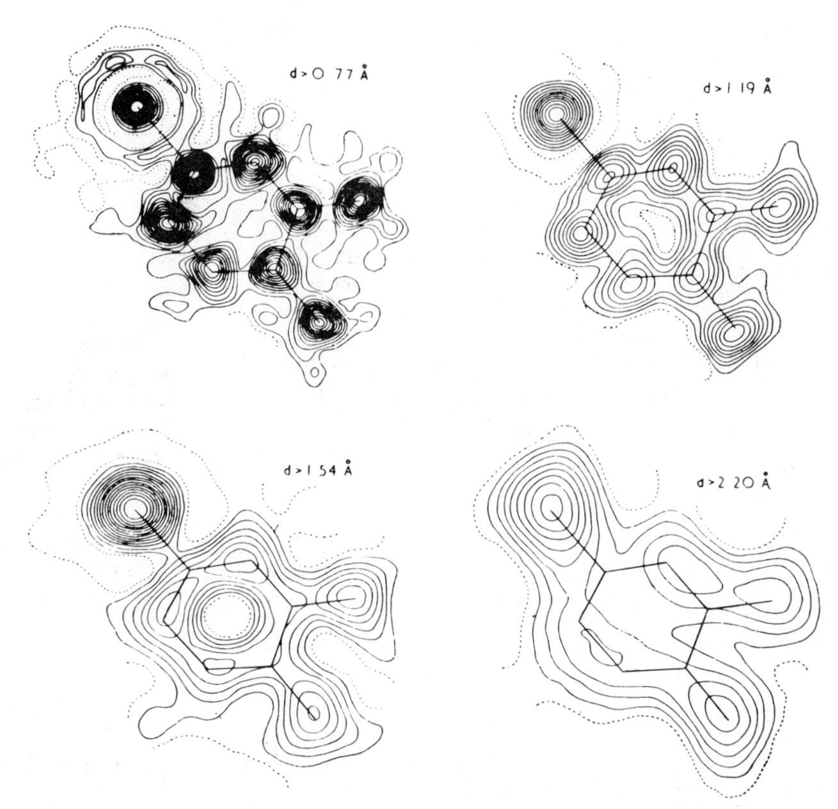

Fig. 20. The Image of a Molecule of 4,5-Diamino-2-chloropyrimidine when Data having Decreasing Resolution (Increasing d-Spacings) are Used in the Calculation of the Electron Density. [Reproduced by permission of the copyright owner from J. C. D. Brand and J. A. Speakman, *Molecular Structure, The Physical Approach*, Edward Arnold (Publishers) Ltd., London, England, 1961.]

It may be seen that the clarity of the detail increases as the minimum d value decreases, that is, as the resolution increases.

The absorption effects are connected with the atomic species composing the crystal and with the actual thickness of the crystal. Both the intensity and the absorption are dependent on the thickness, t, of the crystal. An optical crystal thickness that compromises between these two effects is given by

$$t \approx 2/\mu ,$$

where μ is the linear absorption coefficient of the crystal. The value μ is given by

$$\mu = \frac{n}{V} \sum_i \mu_a^i ,$$

where n is the multiplicity of the unit cell of volume V, μ_a^i is the atomic absorption coefficient of the various atoms, and the summation is carried over the atoms in one molecule or one asymmetric unit. The values of μ_a^i are listed in Ref. 9 for various elements, and a sample is given in Table XIV. The value of μ for cytidine 3'-phosphate can be calculated as follows. The chemical formula is $C_9H_{14}N_3O_8P$, $n = 4$, and $V = 1308.1$ Å^3. By use of the appropriate values of μ_a^i from Table XIV, the values $\mu = 21.5$ cm^{-1} for CuK_α and $\mu = 2.4$ cm^{-1} for MoK_α are obtained. The corresponding values of the optimal thickness are 0.95 and 7.25 mm, respectively. Thus, larger crystals may be employed when MoK_α radiation is used, and it is usually necessary to produce strong, diffracted beams. In practice, the crystal chosen is usually smaller than the optimal size, and its thickness is lessened if there are heavy atoms in the structure.

5. Methods for Collection of Data

The collection of intensity data constitutes the major experimental part of any crystal-structure analysis. The selection of a suitable single crystal and the choice of radiation have to be made before the collection of the data is begun. The data are collected in one of two ways—photographic or diffractometric.

In the photographic method of data collection, the reflections are usually recorded by the Weissenberg technique, in which the blackening of the film is a measure of the intensity of the diffracted beam. However, for strong reflections, the linear response of the film is lost, and, usually, a pack of films (called multiple films[78]) is used, so that the strong reflections are attenuated by the preceding films and are recorded in the linear range of the succeeding film. The intensities

TABLE XIV

Atomic Absorption Coefficients [a] for CuK_α and MoK_α Radiation for a Few Elements

Element	Atomic absorption coefficient ($\times 10^{23}$ cm^2)	
	CuK_α	MoK_α
H	0.073	0.064
C	9.17	1.25
N	17.50	2.13
O	30.50	3.49
P	381	40.6
Cl	621	67.1
Cu	558	537
Br	1320	1060
I	6200	781
Ba	7510	993
Hg	7180	3900

[a] Adapted from Ref 9.

of the reflections are measured visually by using a calibrated intensity-scale, and the data are generally collected about two different axes of the crystal, so that a large proportion of the reflections accessible to the radiation is recorded. It is also possible to use the precession method for collecting the data. Sometimes, the spots are linearly integrated, or tangentially and radially integrated, on the film by moving the film suitably during the exposure; it is then possible to use a densitometer that determines the intensities of the reflections by the attenuation of a light beam passing through the spots. In the integrated photographic methods, the accuracy of the intensity data can be comparable to that obtained with diffractometers.

In the diffractometric method of data collection, the diffracted X-ray beam is allowed to pass through a counter which records the intensity of the diffracted beam. The most commonly used counters are Geiger, proportional, and scintillation counters; the last type is widely used nowadays. For geometrical convenience, the counter is usually constrained to move in one plane (generally, the horizontal plane), and the reflections from the crystal are brought to this plane as will be described. A general, reciprocal lattice-point, P, will have to be rotated by an angle, χ, to bring it to the basal plane, and then be rotated through an angle ϕ to bring it onto the sphere of reflection. The counter is then moved to the appropriate 2θ value in order that this reflection may be recorded (Fig. 21).

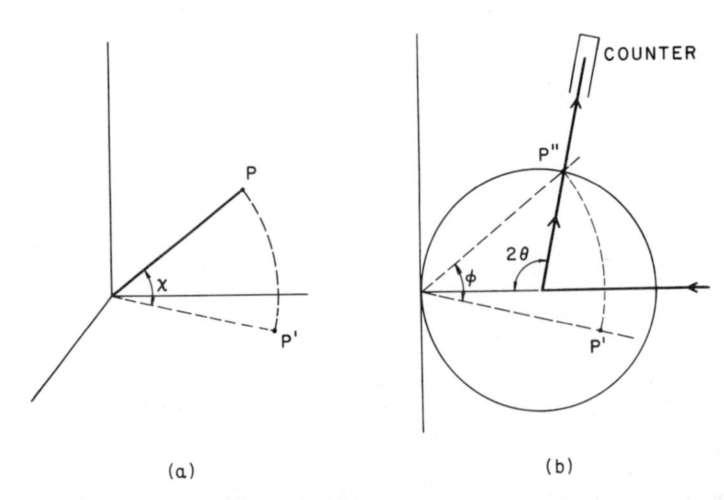

Fig. 21. The Process of Recording a General Reflection Represented by the Reciprocal Lattice-point P with a Counter Constrained to Move in a Horizontal Plane.

In commercially available diffractometers, these angles can be set to the desired values by independently setting the appropriate circles. Also, the plane containing the normal to the reflecting planes, called the ω-circle, can be independently changed. Thus, these diffractometers are all "four-circle" instruments.

The procedure for aligning the crystal is described in detail in crystallographic manuals.[12,79] Once the crystal has been perfectly aligned, the values of χ,

ϕ, 2θ for a reflection ($hk\ell$) can be calculated from the expressions $\sin\theta = 0.5\lambda[h^2a^{*2}+k^2b^{*2}+\ell^2c^{*2}+2hka^*b^* \cos\gamma^* +2k\ell b^*c^* \cos\alpha^*+2\ell hc^*a^* \cos\beta^*]$, and $\chi = \sin^{-1}(\zeta/2 \sin\theta)$, where ζ is the height of the reciprocal lattice-point above the basal plane. The angle ϕ measures the azimuth of the reciprocal lattice-vector projected onto the basal plane. In semi-automatic diffractometers, this information is punched-out on cards or on paper tape. The diffractometer reads-in such information, sets the proper angles, and determines the intensity of the reflection, which is punched out for further processing. In automatic machines, the settings are made on line, by means of a small computer.

There are generally two methods by which the intensities of the reflections may be recorded. In the *peak-height method*, the incident X-ray beam is made divergent to $\sim 8°$ (or more), so that the reflection occurs over a wide angular range and consists of a plateau. The counter is brought to the midpoint of this plateau, and counts are taken for a fixed time. The background on each side of the plateau is also recorded, by changing ω.

In the *scan method*, the crystal is usually swept through the reflecting position, and the integrated intensity of the reflection is recorded. Either of two methods is generally used here. In the first, called the ω-scan method, the counter is held stationary and the crystal is moved through the reflecting position by changing the angle ω; this is also known as the *moving-crystal, stationary-counter method*. In the second, called the *moving-crystal, moving-counter method*, or the θ-2θ scan-mode, the crystal is swept through the reflecting position, the counter is moved at twice the rate of rotation of the crystal, and the integrated intensity is recorded. In both procedures, the background of the reflection at both ends of the scan-range is measured. The effective intensity of the reflection is the net count above the background, represented by the shaded area in Fig. 22, and is given by the expression

$$I = I_{net} - (B_1 + B_2)t',$$

where B_1 and B_2 are the background counts on the two sides of the peak, and t' takes into account the different times required for measurement of the scans and backgrounds. A set of such values for all of the reflections constitutes the raw-intensity data; these are subsequently processed to obtain the structure amplitudes.

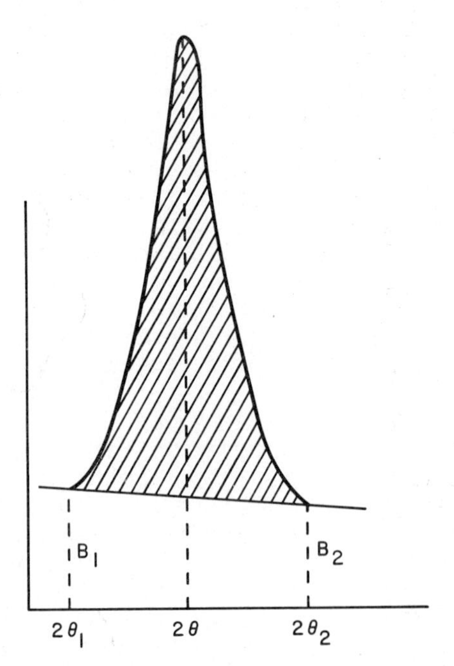

Fig. 22. Trace of a Reflection. [The shaded area represents the net, integrated intensity above the background.]

For a nucleotide, the data, consisting of about 2,000 reflections, can be collected in 5-7 days by use of semi-automatic machines and the scan mode. With the PDP 8-I computer-controlled diffractometer, the data collection may be completed in ∿2-3 days. The collection of photographic data usually requires several weeks. Moreover, photographic intensities are generally less accurate than the diffractometric intensities.

VI. PROCESSING OF INTENSITY DATA

The raw data that have been collected have to be processed for several factors before a set of structure amplitudes can be obtained. These correction factors will be discussed next.

The beam of X-rays incident on the crystal is not polarized; that is, the electric vector parallel to the crystal planes is the same as that perpendicular to them. However, on reflection, the component parallel to the planes changes in a way dependent on the angle of incidence, whereas the perpendicular component is not affected. The dependence of the intensity of the reflected beam on the angle is given by

$$p = (1 + \cos^2 2\theta)/2,$$

where p is the polarization factor.

As the reciprocal lattice is swept through the sphere of reflection, the reflections closer to the axis of rotation take a longer time to cross the sphere than do those farther away from it. Thus, the planes have different opportunities to reflect the incident beam (see Fig. 23). The Lorentz factor, L, for a reflection on a plane that is normal to the incident beam and that passes through the origin is given by

$$L = 1/\sin 2\theta \ .$$

Because, in the Picker diffractometer, this reflection is always brought into the basal plane, this Lorentz factor applies to all of the reflections, but, in any other geometry of recording reflections, it has to be modified.

The total path-length of the incident and diffracted beam is different for the different directions in an irregular crystal, and this causes uneven absorption for different reflections. The paths for two different reflections through an aspherical and a spherical crystal are shown in Fig. 24. It is seen that the two paths are different for an aspherical crystal, whereas they are the same for a spherical crystal. Therefore, the absorption factor (difference in path length) for an irregular crystal varies from reflection to reflection, and this variation can be allowed for only by a tedious calculation.[79a] Usually, the crystal size is so chosen (as discussed on p. 454) that this effect is not great and can be ignored. Where the data are collected photographically by the Weissenberg technique, a correction is usually made for variations in the shape of the spot.[80,81]

The combined corrections for the foregoing factors are termed data reduction or data processing. The corrected intensities, I_{corr}, are, therefore, related to the raw data in the following way:

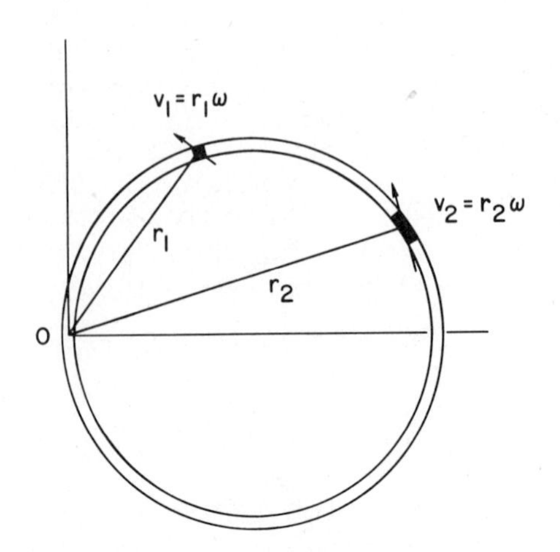

Fig. 23. Diagram Illustrating the Lorentz Factor. [The two reciprocal lattice-points 1 and 2 sweep through the sphere of reflection at different speeds. Point 1 is closer to the origin and moves more slowly than point 2, which is farther away.]

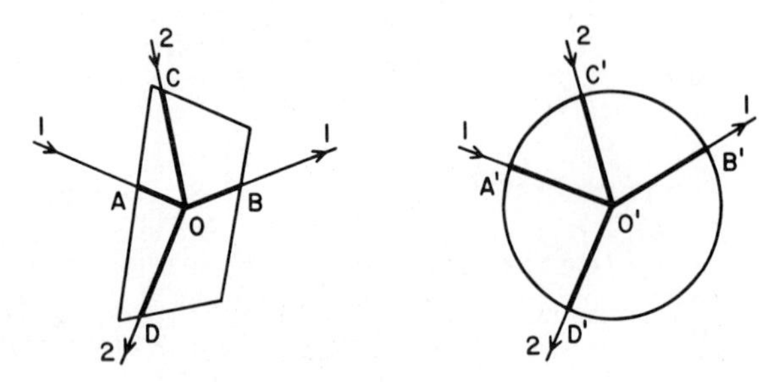

Fig. 24. Absorption Effects for Two Reflections. [(a) An irregularly shaped crystal, and (b), a spherical crystal.]

$$|\underline{F}|^2 \,\alpha\, I_{corr} = I_{obs}/Lp \, .$$

Thus, a set of structure amplitudes can be derived.

In photographic work, data are usually collected about two axes, at least; thus, certain numbers of reflections are then recorded in both sets of data (see Fig. 25). By using these common reflections, all of the data may be placed on the same relative scale.[82] (Data collected diffractometrically are already on a common scale.)

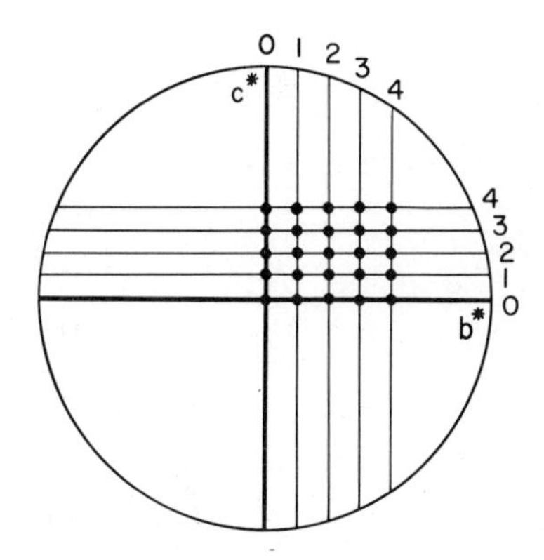

Fig. 25. Correlation of Data Collected about Two Axes (b^* and c^*). [The dots represent the reflections observed in both sets, and are used to cross-correlate the data.]

To place the data on an approximately absolute scale, they are so adjusted that, numerically, they match the scattering from the atoms in a unit cell. If the atoms are assumed to be randomly distributed in the cell, the average intensity $\langle I' \rangle$ at any angle is independent of the actual positions of the atoms, and may be expressed as[83]

$$\langle I \rangle \propto \langle I' \rangle = \exp\left(-2B\frac{\sin^2\theta}{\lambda^2}\right) \sum_1^N f_j^2,$$

where f_j is the scattering factor of atom j at the angle θ, and B is the parameter describing the average, isotropic, thermal motions of the atoms. Actually, B is related to the mean-square amplitude of vibration, $\overline{u^2}$, by relationship $B = 8\pi^2\overline{u^2}$. The proportionality may be converted into an equality by introducing a constant k, referred to as a scale-factor. This gives

$$k\langle I \rangle = \exp\left(-2B\frac{\sin^2\theta}{\lambda^2}\right) \sum f_j^2.$$

By taking logarithms and rearranging,

$$\ln \frac{\Sigma f_j^2}{<I>} = 2B \frac{\sin^2 \theta}{\lambda^2} + \ln k,$$

and this is an expression of the form $y = mx + c$. Thus, a plot, called the Wilson plot, of $\ln \frac{\Sigma f_j^2}{<I>}$ against $\sin^2 \theta/\lambda^2$ gives a straight line, and the slope and intercept determine B and k, the mean thermal parameter and the scale-factor, respectively. Such a plot for an actual case is shown in Fig. 26.

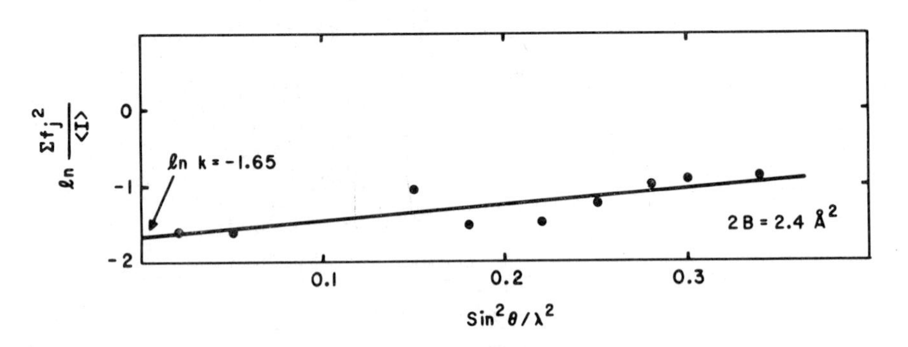

Fig. 26. Wilson Plot for Hydrouracil.[118]

Even though the average intensity is independent of the actual distribution of atoms, the distribution of the intensities about the average is markedly dependent on the presence or absence of a center of symmetry in the distribution of atoms.[84] This average intensity is usually given in terms of two variables, z and E, where z is the ratio of the intensity of a reflection to the average intensity at that angle, and E is the square root of z. These parameters are related by

$$z = E^2 = \left[\frac{I}{<I>}\right]^{0.5}$$

where z is the normalized intensity and E is the normalized structure-amplitude. The overall average involving the values of E are different for the two cases.[84] Also, the distribution of the E values for values above unity (that is, reflections having intensities above average) is different for the two cases. The theoretically expected values for centrosymmetric and noncentrosymmetric cases are tabulated in Table XV,[85] together with the experimentally observed values. It should be noted that the distribution of the E values does not depend on the actual number of atoms in the unit cell; thus, the distribution would be essentially identical for any two centrosymmetric or any two noncentrosymmetric structures.

The distribution of weak reflections (that is, those having an E value less than unity) is also affected by the presence of a center of symmetry. The distribution is evaluated[86] in terms of a cumulative function $N(z)$, the fraction of the reflections having a normalized intensity less than a specified value of z. The plots of this function for both cases are shown in Fig. 27, together with the actual points obtained from two crystals.

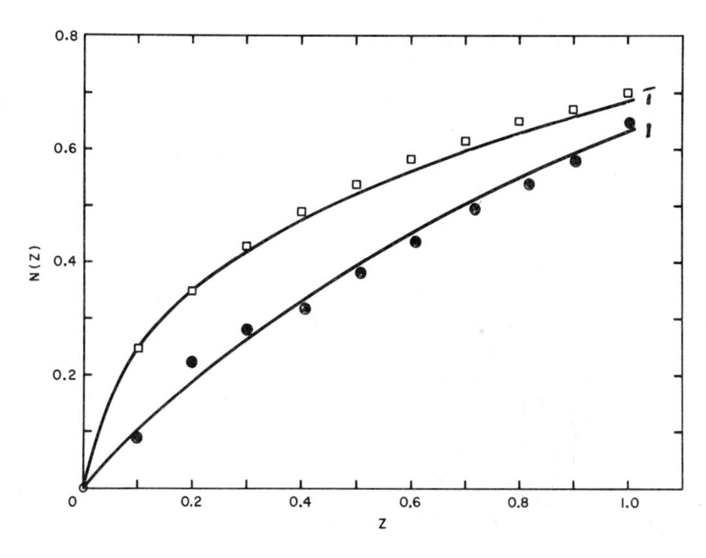

Fig. 27. The Cumulative Function $N(z)$. [The lower, solid curve represents the theoretical distribution for a noncentrosymmetric crystal, and the upper curve, that for a centrosymmetric crystal. Circles represent the experimental points for a crystal of noncentrosymmetric N-acetyladenosine, and the squares, those for a crystal of centrosymmetric hydrouracil.[118]]

TABLE XV

Intensity Distribution for Centrosymmetric ($\bar{1}$) and Noncentrosymmetric (1) Crystals

	Theoretical		Experimental	
	1	$\bar{1}$	1[a]	$\bar{1}$[b]
$E>3$	0.000	0.004	0.000	0.002
$E>2$	0.018	0.045	0.017	0.040
$E>1$	0.368	0.319	0.337	0.321
$<\|E\|>$	0.886	0.798	0.879	0.853
$<E^2>$	1.000	1.000	0.960	1.006
$<\|E^2-1\|>$	0.736	0.968	0.693	0.873

[a]For *N*-acetyladenosine. [b]For hydrouracil.

This type of analysis of data is especially valuable in certain instances where the systematic absences, alone, are unable to provide a decision between a pair of space-groups. Usually, in such instances, one space-group has a center of symmetry and the other does not. Some of these ambiguous pairs of space-groups are listed in Table IX.

VII. STRUCTURE DETERMINATION

1. The Phase Problem

The two fundamental relations that are central to determination of a crystal structure are the expressions for the structure factor F and the electron density ρ, given by

$$F = \sum_{j=1}^{N} f_j \exp 2\pi i(hx_j + ky_j + \ell z_j)$$

$$\text{and } \rho(x,y,z) = \frac{1}{V} \sum_{h,k,\ell} F \exp - 2\pi(hx + ky + \ell z).$$

These expressions are Fourier transforms of one another. To construct the image of structure, the electron-density distribution has to be calculated. This calculation requires a knowledge of the amplitude and phase of the reflections, and, in practice, the amplitude is known (from the intensity observed), but the phase is not known. The task of a crystal-structure analyst is to supply this missing information. The problem is called the phase problem (in crystallography), and several ways of solving it have been developed, some of which will be discussed next.

2. Patterson Methods

Patterson[87] showed that a Fourier synthesis of the type

$$P(u,v,w) = \frac{1}{V} \Sigma |F|^2 \exp -2\pi i(hu + kv + \ell w),$$

computed with the intensities as coefficients, is a vector-density map containing maxima at points corresponding to all interatomic (intra- and inter-molecular) vectors in the unit cell. According to Friedel's law, $|F|^2 = |\overline{F}|^2$, and use of this simplifies the previous expression to

$$P(u,v,w) = \frac{2}{V} \Sigma |F|^2 \cos 2\pi(hu + kv + \ell w),$$

where the summation is made over half of the reciprocal space, and the function is centrosymmetric. The heights of maxima in this function are proportional to the product of the atomic numbers of the atoms involved, and, if there are N atoms in the unit cell, there are N^2 vectors between them. Of these vectors, N are self-interactions that coincide with the origin. Because the Patterson function has a center of symmetry, there are $(N^2-N)/2$ independent peaks.

As an example, the Patterson function for a hypothetical, 4-atom problem (see Fig. 28a) is shown in Fig. 28b. The corresponding atomic positions and interatomic vectors are depicted in Figs. 28c and 28d, respectively. From Fig. 28d, it may be seen that there are 16 interactions, and that four of these lie at the origin. Thus, the vector-set is a combination of four images of the structure, each atom being successively placed at the origin. The problem is to extract one image of the structure from this composite set.

Wrinch[88] pointed out that, in principle, this extraction can be accomplished by performing certain

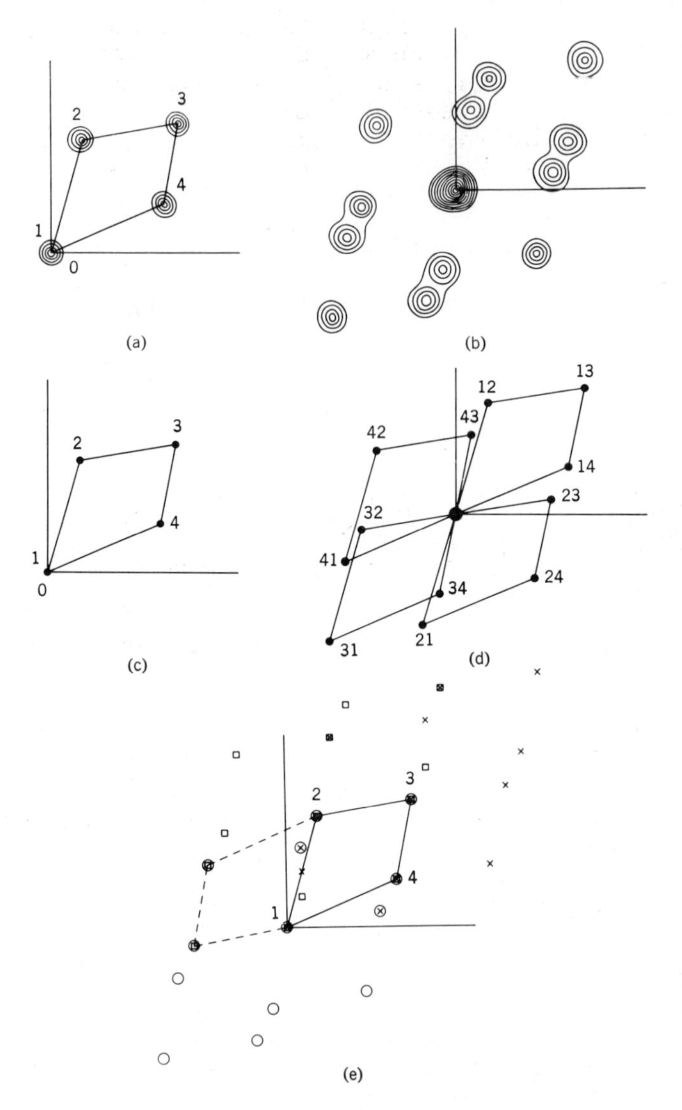

Fig. 28. Illustration of the Patterson Superposition Method for a Noncentrosymmetric Crystal Composed of Four Atoms in the Unit Cell (a) and (c). [(b) and (d) represent the Patterson map of the structure. (e) The original Patterson map is denoted by circles. An identical Patterson map shifted through the vector 12 is marked with squares, and a third Patterson map, shifted through the vector 13, is marked with crosses. The four points at which coincidences occur between the three maps represent the image of structure (c).]

geometric operations on the vector-set. The principle of the method is illustrated in Fig. 28e, where the termini of the vectors in the Patterson map are denoted by open circles. Suppose that a second Patterson map is placed on the first, and that it is so translated that its origin coincides with the end of the vector passing through atoms 1 and 2 (vector 12), the axes of the maps being kept parallel to each other. The peaks corresponding to the second Patterson map are denoted by squares, and may be seen that there are coincidences at only six points, which are represented by two quadrilateral figures. Thus, two images have been isolated from the original Patterson map, which was comprised of four images. These two images are related to each other through a center of inversion at the midpoint of vector 12, and to separate the two images, a third copy of the Patterson map is placed with its origin at the end of vector 13, the peaks being marked by crosses. There are now only four coincidences involving *all* of the Patterson maps, and these correspond to the atomic positions of the structure. This method of geometrical manipulation of the Patterson map is called image seeking or superposition; it has been extensively studied by Buerger.[89]

The procedure is markedly simplified if the structure is centrosymmetric. Consider a molecule with three atoms 1, 2, and 3 (Fig. 29a) which are related to the atoms 1',2', and 3', respectively, through a center of symmetry at the origin. The vector-set of this structure is shown in Fig. 29b; in Fig. 29c, the original Patterson map is shown by open circles, and the Patterson map displaced by vector 11' is shown by squares. The six points at which coincidences occur correspond to the six atoms of the structure, the origin being at the midpoint of vector 11'. Thus, in a centrosymmetric crystal, the structure is extracted by shifting the Patterson map by a vector between atoms related by the center. For a noncentrosymmetric crystal, the Patterson map is shifted through two vectors (12 and 13) having one atom (1) in common.

From the preceding discussion, it might appear that there is a simple way in which the structure can be obtained from the intensities observed without the phases having been ascertained, but this conclusion is quite misleading because, for a structure containing atoms of nearly the same atomic numbers, it is not easy to locate the appropriate vectors just discussed. If inappropriate vectors are used for the superposition (such as vectors 12 and 34 in the first example),

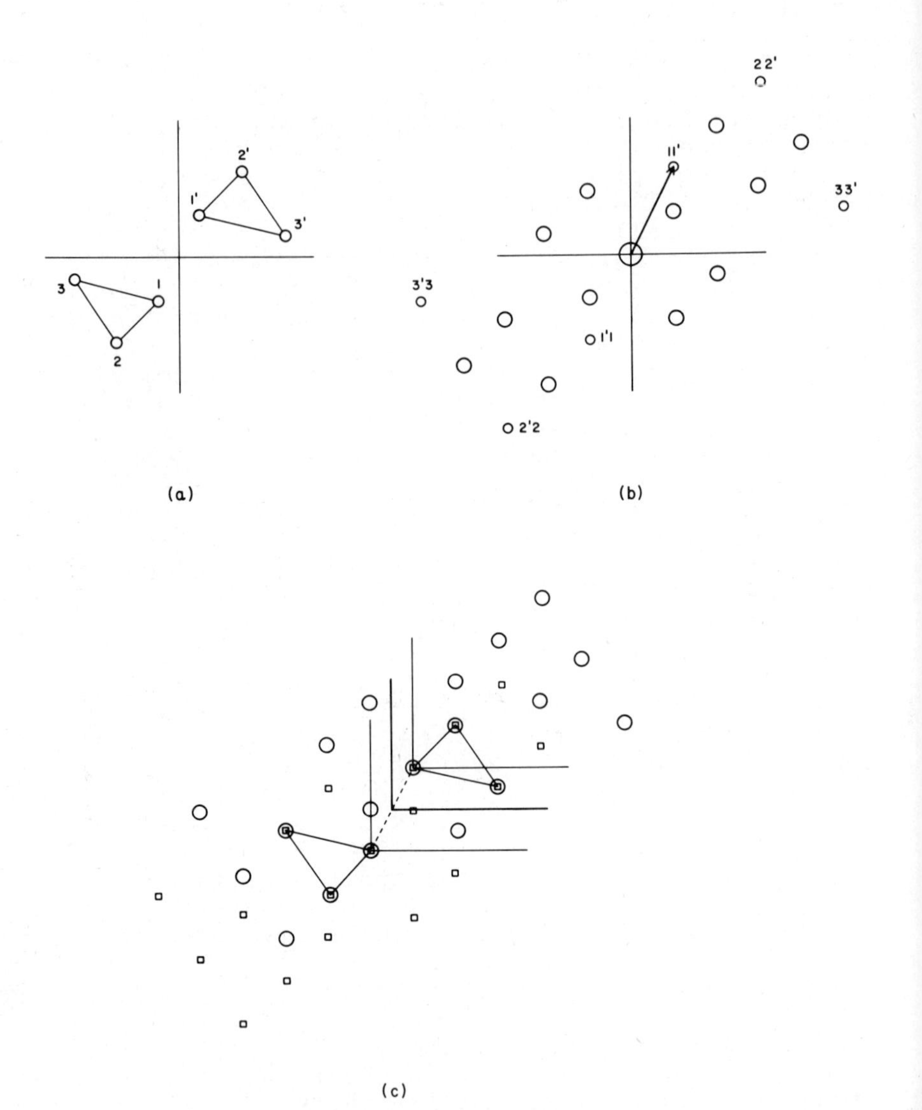

Fig. 29. Illustration of the Patterson Superposition Method for a Centrosymmetric Crystal. [(a) Six atoms in the unit cell, related by a center of inversion at the origin. (b) Patterson map of the structure shown in (a). (c) represents identical Patterson maps, indicated by circles and squares displaced through the vector 11'. It may be seen that the six points at which superposition occurs represent the structure provided, where the origin is chosen at the midpoint of the shift-vector 11'.]

multiple images result, and these cannot be interpreted meaningfully. If a structure contains heavy atoms, the vectors between them will be prominent in the Patterson map and are readily identifiable. By employing these vectors, the superposition method can be used for finding the other atoms. Alternatively, the approximate values of the phases given by the heavy atoms can be calculated, and the structure can be developed by Fourier methods, discussed in detail later (see p. 479).

An interesting situation occurs when a heavy atom is situated at the origin. The Patterson function for the mercuric chloride complex of uracil,[90] namely, $HgCl_2\cdot(uracil)_2$, is shown in Fig. 30. Here, the mercury atom

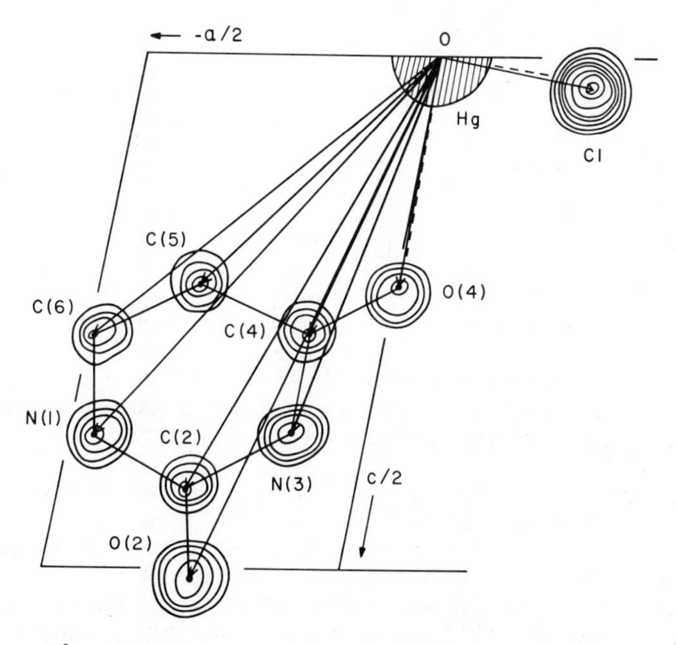

Fig. 30. The Patterson map of $HgCl_2\cdot(uracil)_2$, with Hg at the Origin.[90] [The peaks represent the actual atomic positions.]

is at the origin, which is also a center of inversion. The heights of the various interactions involving mercury are Hg-Hg = 80 x 80 = 6,400, Hg-Cl = 1,360, Hg-O = 640, Hg-N = 560, and Hg-C = 480. The largest interactions are between mercury and the remaining atoms of the structure, except for the Hg -Hg interaction at the origin. Thus, the prominent peaks in the Patterson map indicate the actual atomic sites of the other atoms.

a. Harker Sections.- When the atoms in the unit cell are related by a symmetry element other than a center of inversion, the interactions between symmetry-related atoms fall on certain planes or lines, called Harker sections or lines.[91] For three atoms in the cell related to three others by a two-fold screw-axis parallel to *b*, the situation is illustrated in Figs. 31a and 31b.

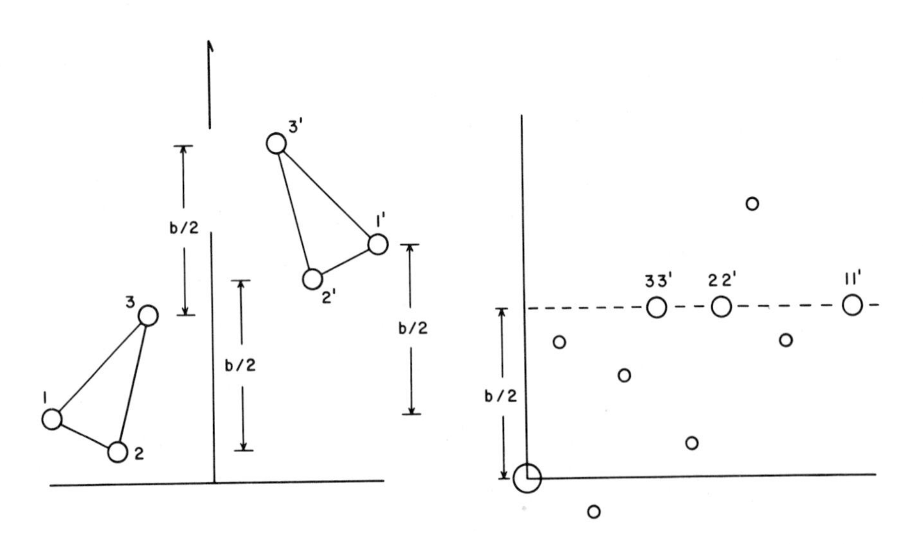

Fig. 31. Harker Section Caused by a Two-fold Screw-axis Parallel to the *b*-Axis.

The interactions (11'), (22'), and (33') between the three pairs of atoms related by the symmetry lie on a line at $y = b/2$. The other (non-Harker) interactions fall, in general, at different values of y. The Harker sections for various symmetry-elements are given in Table VI (see p. 421). In principle, an examination of such sections should give sufficient information about the positions of the atoms. However, in practice, the interpretation of the Harker sections is complicated by the fact that a large number of non-Harker interactions occur on the sections because of the presence of atoms that are separated by a distance equal to the translational component of the symmetry element.

The crystal structure of the copper complex of guanine,[92] namely, $CuCl_3 \cdot (guanine)_2 \cdot H_2O$, belongs to the space group $C2/c$, and the Harker section at $v = 0$ is shown in Fig. 32. In a normal situation, the highest peaks

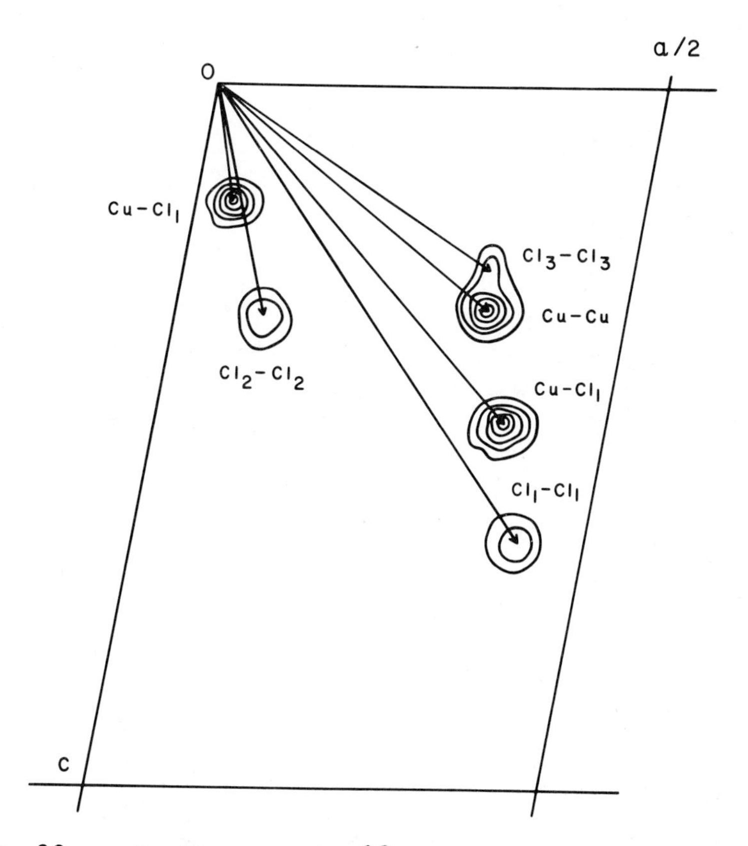

Fig. 32. Harker Section[92] at $v = 0$ for $CuCl_3 \cdot (guanine)_2$. [The Cu and Cl atoms accidentally have the same y coordinate, and the non-Harker Cu-Cl interactions, which have a vector density higher than that of the chlorine atoms, also lie on this section.]

in this section should be due to Cu-Cu and to the three Cl-Cl vectors. However, the copper atom and one of the chlorine atoms happen to have the same y coordinates, and this section also contains strong peaks due to Cu-Cl non-Harker interactions. This situation having been recognized, the x and y coordinates of the copper atom and the three chlorine atoms can be deduced. It is, therefore, necessary to exercise caution when Harker sections are interpreted.

b. Planar Groups.- Several crystal structures of bases and nucleotides have been determined by extensive use of *the Patterson function*. In this Section, the different methods are briefly described, with specific examples.

If a planar group of atoms forms a fragment of the structure, for example, a pyrimidine ring, it will have its own planar vector-set (see Fig. 33a). A tilting of

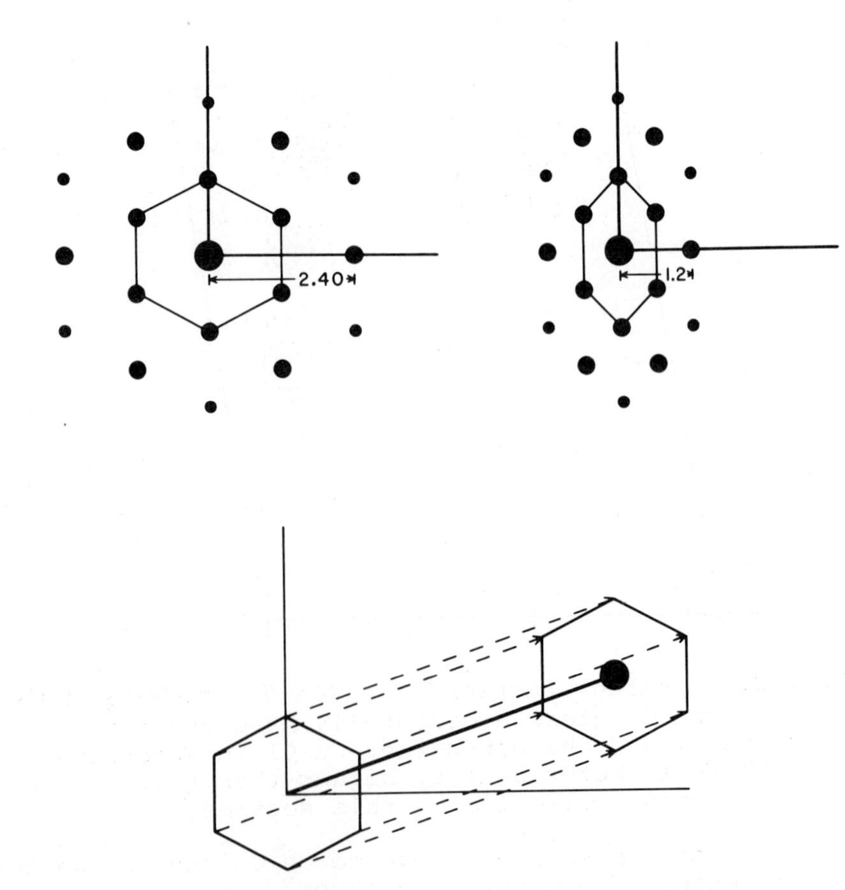

Fig. 33. (a) This Represents the Patterson Map of a Six-membered Planar Ring Parallel to the Plane of the Paper. (b) The Ring is Tilted at an Angle to the Paper (c) This Denotes the Clustering of the Interactions Between the Corresponding Atoms in the Two Rings which Gives Rise to the "Center-to-center" Vector.

the ring will result in a corresponding tilt of the vector set, and, for simplicity, the tilted vector-set for a benzene ring is shown, in projection, in Fig. 33b. If the set of vectors belonging to such a fragment could be identified in the Patterson map, the positions of all of the atoms in this group could be determined, as will be illustrated. The positions of the remaining atoms could then be established by superposition methods.

The vector-set of the planar group of atoms will lie in the plane through the origin of the Patterson function that has the highest concentration of peaks.[93] To determine the orientation of the planar group, the integral of the Patterson map, namely,

$$I(\theta,\phi) = \int_{\text{disc}} P(u,v,w)\,ds,$$

is calculated over a disc of radius R, at an orientation defined by the angles (θ,ϕ) made by the normal to the disc (see Fig. 34). This integral can be written in

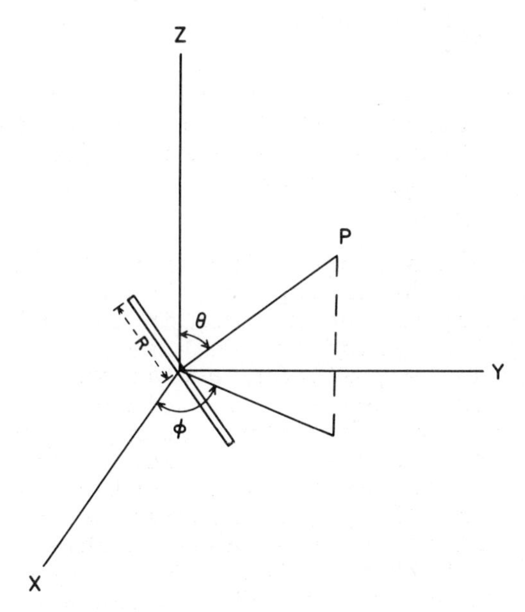

Fig. 34. The Angles θ and φ Define the Normal to the Disc of Radius R.

terms of the Fourier transforms as

$$I(\theta,\phi) = \sum_{h,k,\ell} |F|^2\, 2\pi R^2 \frac{J_1(2\pi RS)}{2\Pi RS},$$

where S is the perpendicular distance of the reciprocal lattice-point of a reflection to the normal to the disc, and J_1 is the Bessel function of first order. The values of I are found for all possible orientations of the disc, and the angles θ,ϕ at which a maximum value occurs indicate the orientation of the plane in the crystal. Such a map, obtained for the complex of 9-ethylguanine with 1-methylcytosine,[67] is shown in Fig. 35. The two bases forming

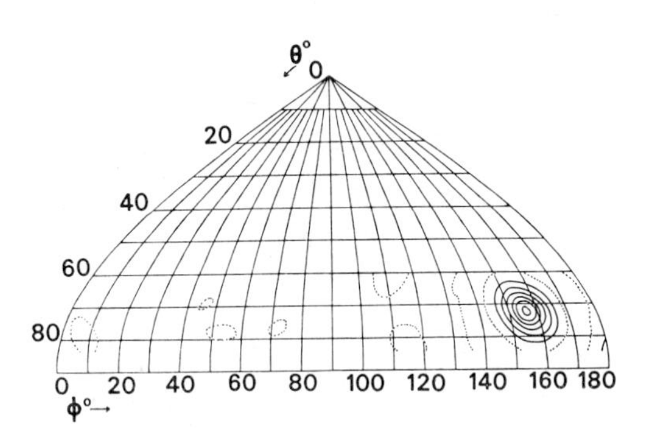

Fig. 35. The Map Obtained by Integrating the Patterson Function over a Disc, at Different Orientations of a Complex, namely, 9-Ethylguanine: 1-Methylcytosine. [Reproduced from E. J. O'Brien, *Acta Crystallogr.*, 23, 92 (1967).]

the base-pair have a linear dimension of R=1 nm (10 Å), over which the Patterson map is integrated. The angles θ,ϕ, corresponding to the maximum in the map, describe the orientation of the normal to the base-pair. Moreover, the reflection 112 is very strong, and it was therefore suspected that the base-pairs lie close to this plane. It may be pointed out that the presence of unusually strong reflections in a structure leads often to a correct guess as to the orientation of the molecules in the crystal.[94] From a knowledge of the orientation of the base, the relative position of the bases in this plane can be determined. The Patterson function computed on the $(11\bar{2})$ plane is reproduced in Fig. 36b. Because the rings are separated by a distance, the vectors between some of the atoms in the rings are parallel, and are superimposed, forming what is called the center-to-center vector (see Fig. 33c). The distribution of the base-pairs and the center-to-center vector are shown in Fig. 36a, and the corresponding vector in the Patterson map is shown in Fig. 36b; thus, the positions of all of the atoms in the

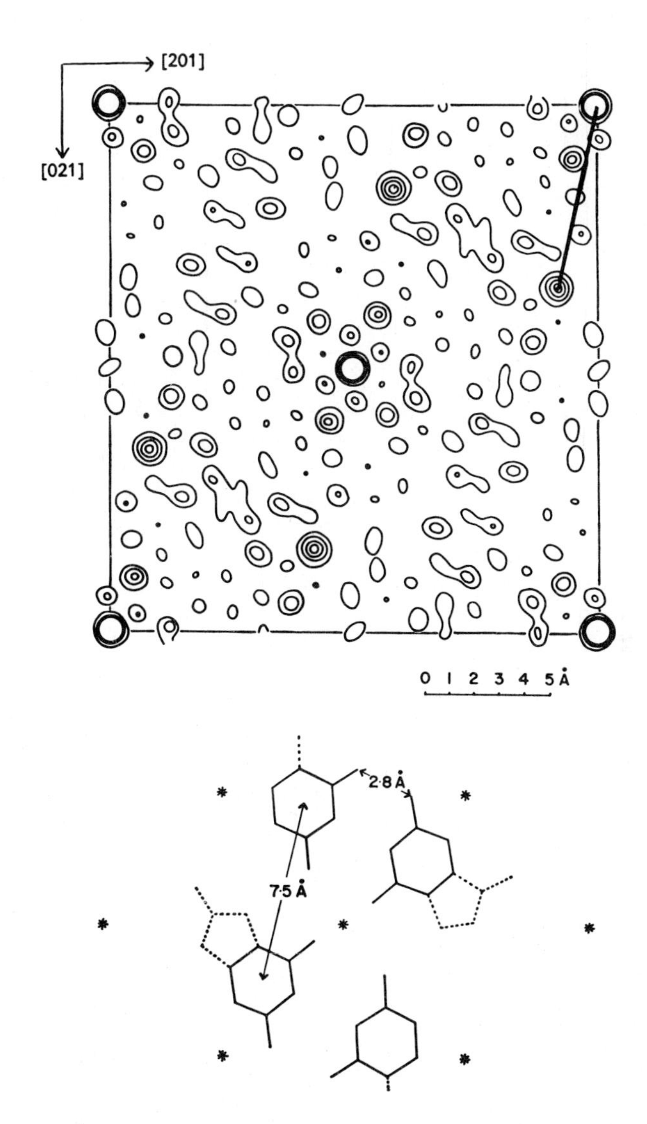

Fig. 36. The Bottom Part Describes the Proposed Arrangement of the Base-pair 9-Ethylguanine: 1-Methylcytosine. [The corresponding center-to-center vector is identified in the Patterson map shown on top. Reproduced from E. J. O'Brien, *Acta Crystallogr.*, 23, 92 (1967).]

base-pair have been ascertained. For nucleosides, the positions of the atoms in the base are used in superposition methods in order to determine the location of the atoms of the sugar, or, alternatively, the known part of the structure (the base) can be used for "developing" the remaining atoms by Fourier methods, discussed later (see p. 502). A general procedure based on similar principles has been developed by Nordman and co-workers for solving the crystal structures of several natural products.[95,96]

There is another way in which the Patterson function may be used. Sometimes, from considerations of crystal packing and hydrogen bonding, a model can be proposed for the structure. The correctness of these models may be checked by comparing the calculated Patterson function of the model with that obtained from experimentally observed intensities. This method has been elegantly used in the crystal-structure analysis of cytosine monohydrate.[20] The model proposed for hydrogen-bonded cytosine bases is shown in Fig. 37a, the Patterson function from the observed intensities, in Fig. 37b, and the Patterson function calculated for the model, in Fig. 37c. The close agreement between the Patterson maps observed and calculated indicates that the model proposed is correct. The validity of the model was further checked by structure-factor, Fourier calculations.

c. Sharpened, Patterson Function. — A standard procedure widely used nowadays is computation of a "sharpened" Patterson function, the need for which is clear from the following. Consider a structure having 40 atoms in a unit cell of volume 1,000 $\mathring{A}^3$. The atoms are well resolved in space, there being one atom in every 25 $\mathring{A}^3$. In the corresponding Patterson map, there would be 40 x 39 = 1,560 peaks in the same volume, that is, about 1.5 peaks per cubic Ångstrom. The resolution of the peaks in the Patterson map is quite poor, and the peaks are broadened because of the thermal motions and the quantum-mechanical spread of the scattering matter. The usual procedure is to obtain a Patterson function corresponding to point-atoms at rest by using the normalized intensities, z, (instead of the observed intensities) as coefficients in the calculations. A disadvantage of this method is that, because, from point-atoms, there is no decrease in intensity with angle, the use of only a limited number of terms in the calculation produces spurious peaks due to errors attributable to series-termination. Yet another way of sharpening the Patterson map is to use, as coefficients,[97] the intensities multiplied by $\sin^2 \theta$; this is equivalent to using the gradient of the electron density. This method has been used successfully in solution of the crystal structure of cellobiose.[11] In structures contain-

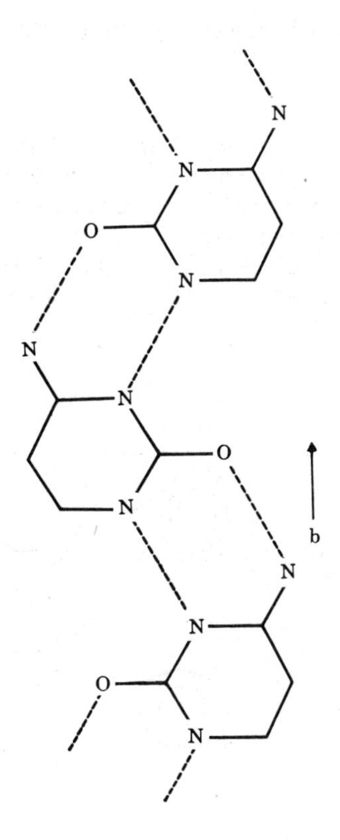

Fig. 37. (a) Proposed Arrangement of Cytosine Molecules Hydrogen-bonded Along the *b*-Axis. (b) On the Left is shown the Actual Patterson map, and, on the Right, the Calculated Patterson Function for the Orientation Proposed. [Reproduced from G. A. Jeffrey and Y. Kinoshita, *Acta Crystallogr.*, 16, 20 (1963).]

ing moderately heavy atoms, sharpening brings out the vectors between the heavy atoms in the Patterson map. For example, in the sodium salt of inosine 5'-phosphate, the section of the Patterson map at $U = 0$, calculated from the intensities only, is shown in Fig. 38a, and that "sharpened" with $\sin^2 \theta$ is shown in Fig. 38b. In the unsharp-

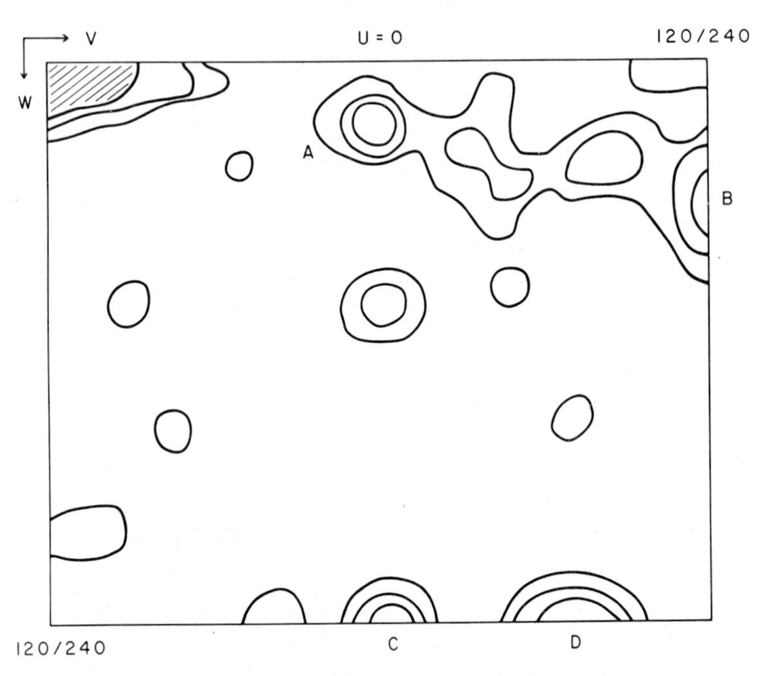

Fig. 38(a). Harker Section, at $U = 0$, for the Sodium Salt of Inosine 5'-Phosphate.[57] [Unsharpened. One of the four peaks A, B, C, or D can represent the P/P vector.]

ened map, there are four peaks (*A*, *B*, *C*, and *D*) of about equal height, and it is difficult to identify the vector between the phosphorus atoms. However, in the sharpened diagram, the height of peak *A* is considerably greater than that of the other three, and this peak is attributed to the P—P vector (see p. 483).

The Patterson methods have proved exceedingly useful in analysis of the structure of nucleic acid components, even though its use requires a certain ingenuity. Except where the phases are directly determined (as discussed later, see p. 493), the Patterson function is invariably the starting point in any determination of structure.

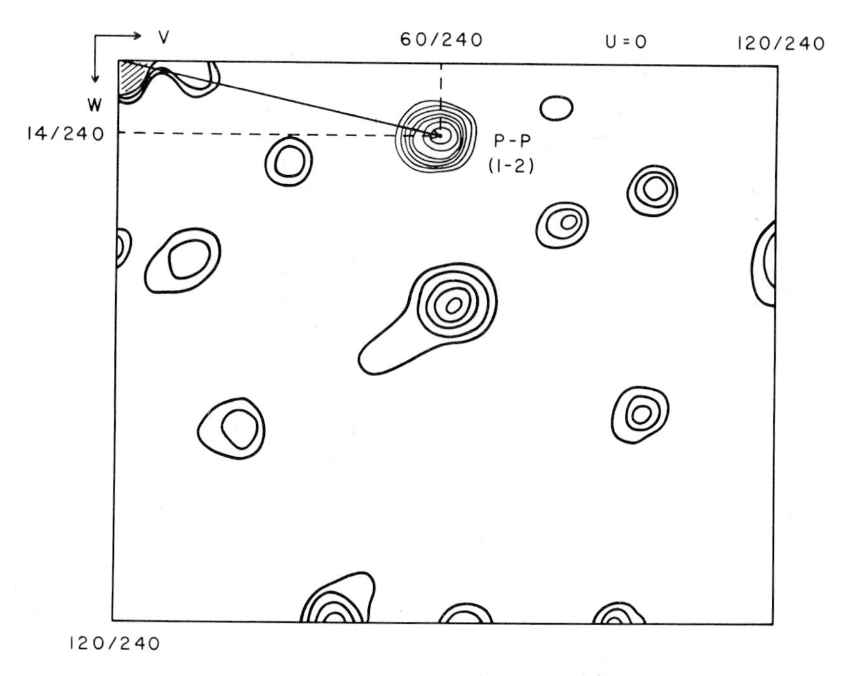

Fig. 38(b). Harker Section, at $U = 0$, for the Sodium Salt of Inosine 5'-Phosphate.[57] [Patterson section whose coefficients are sharpened by the function $\sin^2\theta$ shows A as the largest peak that was identified with the P/P vector.]

3. The Heavy-atom Method

For solving crystal structures,[98] the heavy-atom method is a popular, useful method based on the principle that, on the average, the phases of reflections are biased towards those of the heavy atoms. The heavy atoms dominate the scattering, whereas the light atoms have smaller scattering-factors. A hypothetical example, consisting of one heavy atom and four light atoms, is illustrated in Fig. 39a. The approximate phases from the contribution of the heavy atom are combined with the observed amplitude of the structure in calculating the electron-density function, which reveals either all or part of the light atoms; the latter atoms are included in a second set of structure-factor calculations, yielding better-fitting phases,

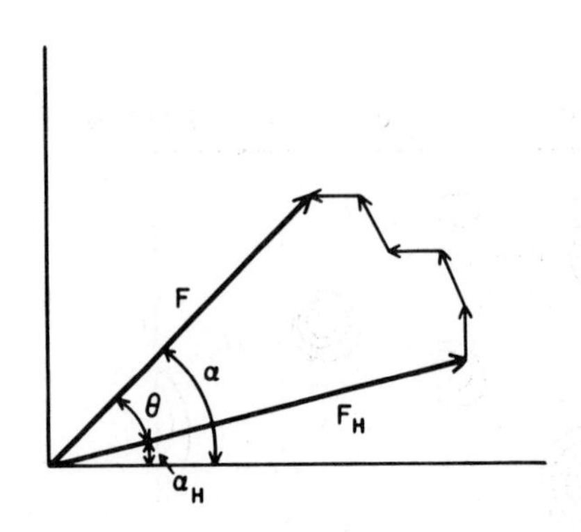

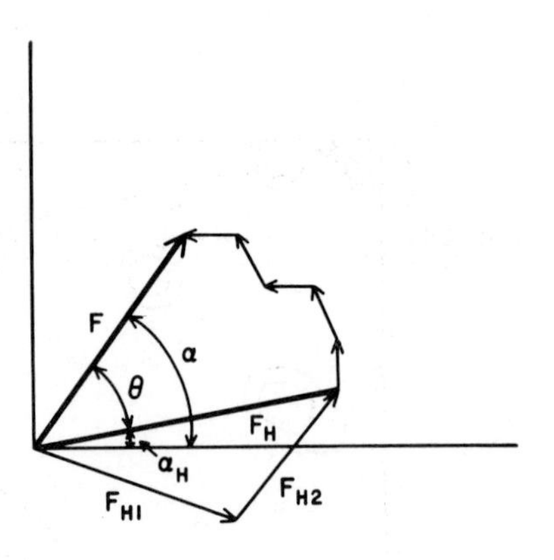

Error $\theta = \alpha - \alpha_H$

Fig. 39. (a) The Phase from the Heavy Atom is Close to the Actual Phase. [The heavier the atom, the longer is the vector F_H and the smaller the error θ.] (b) When Two Heavy Atoms, Each Half as Heavy as the Original Atom, are Present, the Length of the Resultant Vector F_H is Smaller and the Error in the Phase is Larger than in the Previous Case when the Scattering is from One Heavy Atom of Double the Atomic Number.

and the process is repeated until all of the atoms in the structure have been located.

Should two heavy atoms be present in a molecule of the crystal, each having half the atomic number of a single heavy atom (for example, two chlorine atoms instead of one bromine atom), the phases obtained from the two light atoms have a larger error, because the two vectors from the two light atoms produce a resultant vector having a length shorter than that of the heavy atom (see Fig. 39b).

These ideas may be expressed mathematically in terms of a parameter σ^2 that is defined by the expression

$$\sigma^2 = \Sigma z_H{}^2/(\Sigma z_H{}^2 + \Sigma z_L{}^2),$$

where z_H is the atomic number of the heavy atoms, z_L is the atomic number of the light atoms, and the summation is performed over the atoms contained in one asymmetric unit. For example, the formula of cytidine 3'-phosphate, is $C_9H_{14}N_3O_8P$, and, by using the atomic numbers for the fferent atoms, the σ^2 value for the phosphorus atom is lculated to be 0.184. The σ^2 values compiled for the avy-atom contribution and the phosphate-group contribu-on for a number of nucleotides and related compounds e tabulated in Table XVI. Theoretical calculations

TABLE XVI

Values of σ^2 for Some Nucleotides

me	Formula	Heavy atom	σ^2	References
enosine -phos-ate, di-drate	$C_{10}H_{14}N_5O_7P \cdot 2\ H_2O$	P	0.158	50
		PO_4	0.339	
enosine -phos-ate, nohy-ate	$C_{10}H_{14}N_5O_7P \cdot H_2O$	P	0.166	51, 52
		PO_4	0.354	
enylyl- '→5')-idine, trahy-ate	$C_{19}H_{24}N_7O_{12}P \cdot 4\ H_2O$	P	0.098	59
		PO_4	0.209	
tidine -phos-ate	$C_9H_{14}N_3O_8P$	P	0.184	47, 48
		PO_4	0.394	
ctor la	$C_{46}H_{66}CoO_9N_{11} \cdot 11\ H_2O$	Co	0.171	99
osine -phos-ate, so-um salt, tahy-ate	$C_{10}H_{12}N_4NaO_8P \cdot 8\ H_2O$	P	0.116	56, 57
		P, Na	0.179	
		PO_4	0.248	
idine -phos-ate, rium lt, nona-drate	$C_9H_{11}BaN_2O_9P \cdot 9\ H_2O$	Ba	0.634	55
		Ba, P	0.680	

have been made that relate the average error in the phase angle to the σ^2 value.[100-103] The fractional number of reflections that have a phase error of less than 90°, as a function of the value of σ^2, are listed in Table XVII. The same Table also provides the number of reflections

TABLE XVII

Errors in the Signs and Phases as Determined by the Heavy Atoms, for Various Values of σ^2

σ^2	P+ % of reflections having signs agreeing with results from use of heavy atoms	N (90) % of reflections having phases differing by less than 90° from the heavy-atom phases
0.0	50.0	50.0
0.2	64.8	72.0
0.4	71.8	81.5
0.5	75.0	84.0
0.6	78.2	88.5
0.8	85.2	94.5
1.0	100.0	100.0

that would have the sign agreeing with that of the heavy atoms in the centrosymmetric crystal. It may be seen that, the larger the value of σ^2, the better fitting are the phases, and for such compounds, the rest of the atoms in the structure can be located in the heavy-atom, phased, electron-density distribution. When the value of σ^2 is very high, however, the accuracy with which the other atoms may be located decreases rapidly. For small values of σ^2, the phases from only the heavy atom possess large errors, and the structure may be obtained by structure-factor, electron-density iterations, more and more atoms being located at each stage. The accuracy with which the light-atom coordinates are defined increases with decreasing values of σ^2.

Usually, the progress of the determination of structure is followed by the R value, a measure of the agreement between the observed (F) and calculated (F_c) structure amplitudes, defined as

$$R = \Sigma \Big| |F| - |F_c| \Big| / \Sigma |F|, \qquad (1)$$

where the summation is made over all of the reflections. If the progress in the structure determination is satisfactory, the R value steadily drops.

The actual steps involved in practical application of the method will be illustrated for the sodium salt of inosine 5'-phosphate, in which the heavy atom is the phosphorus atom, the σ^2 value is 0.116 (see Table XVI), and the crystal belongs to the space group $C222_1$. The four equivalent positions in the unit cell are

1	x	y	z
2	x	$-y$	$-z$
3	$-x$	$-y$	$\frac{1}{2}+z$
4	$-x$	y	$\frac{1}{2}-z$

These are related to another four positions by the lattice centering. The vectors between the equivalent positions are

1–2	0	$2y$	$2x$
1–3	$2x$	$2y$	$\frac{1}{2}$
1–4	$2x$	0	$\frac{1}{2}+2z$

The vector 1–2 lies on the Harker section at $U = 0$, 1–3 lies on $W = 1/2$, and 1–4 lies on $V = 0$. The Harker section at $U = 0$ has already been shown in Fig. 38b, and the section of the sharpened Patterson function at $U = 56/240$ is shown in Fig. 40, in which the interactions due to the phosphorus atoms are indicated. The coordinates corresponding to these interactions are (0,16,14/240), (56,60,120/240) and (56,0,134/240), respectively. From these, the coordinates of the phosphorus atom can be deduced to be

$$x = 28/240 = 0.117$$

$$y = 30/240 = 0.125$$

$$z = 7/240 = 0.029.$$

By using these coordinates for the phosphorus atom, the structure factors for all of the reflections were calculated, giving an R value of 0.56. If the structure-factor from the phosphorus atom is much smaller than the observed structure-amplitude for a given reflection, the

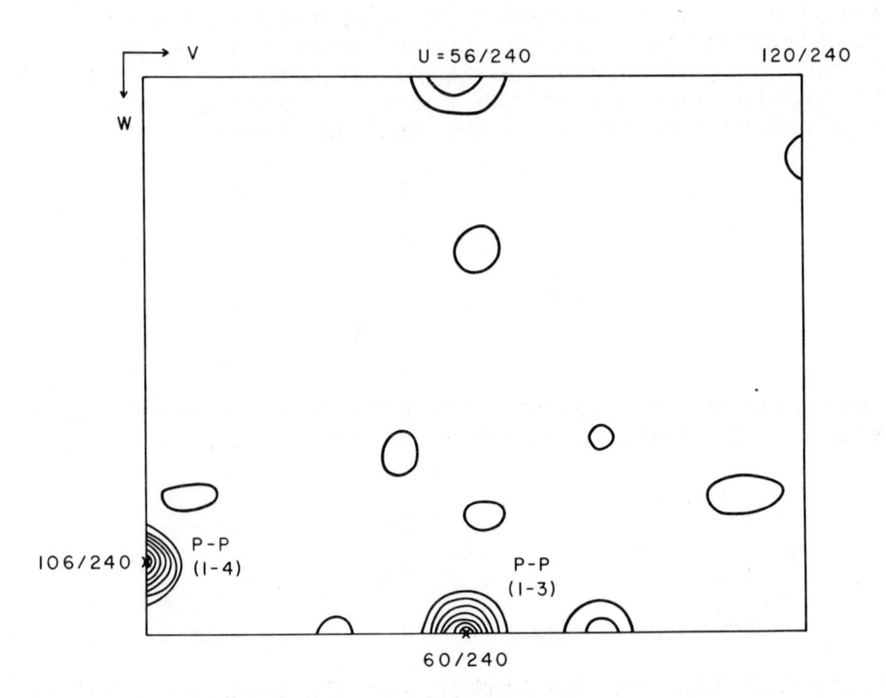

Fig. 40. Gradient-sharpened Patterson Section, U = 56/240, for the Sodium Salt of Inosine 5'-Phosphate.[57] [The two P/P vectors are shown.]

phosphorus atom does not contribute significantly to this reflection, and the phase is not reliable. Therefore, all reflections in which the ratio of F_c/F is less than 0.33 are ignored in calculation of the electron-density distribution. Alternatively, it is possible to use a suitable weighting-function to diminish the effect of such reflections.[104] The electron-density function is calculated at intervals of 25 pm over one asymmetric unit of the unit cell, and is contoured at intervals of 2 e/(100 pm)^3 (see Fig. 41a, which shows a fragment of the structure, namely, part of the base that is recognizable). It was not clear at this stage whether the imidazole ring would develop on the side of O-6 or N-9. The six atoms shown in Fig. 41a were included in a second set of structure-factor calculations in which the R value was decreased to 0.49. By using the newly derived phases, a

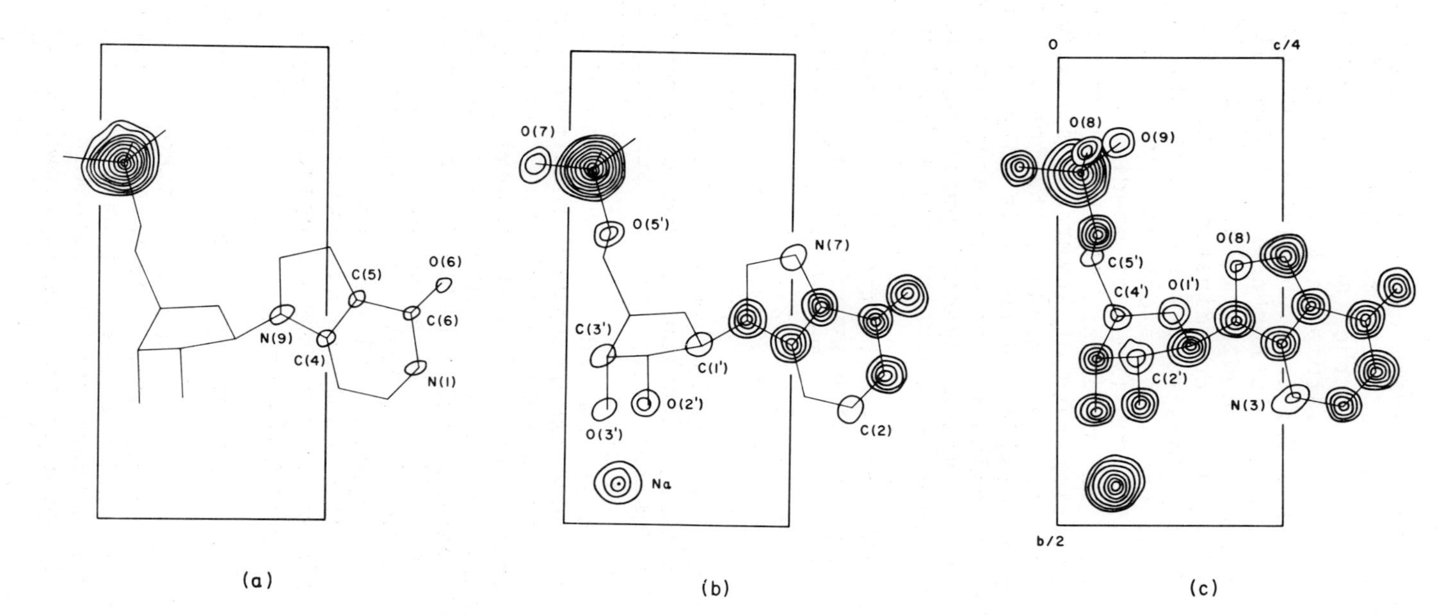

Fig. 41. The Development of the Structure of the Sodium Salt of Inosine 5'-Phosphate.[57] [The atoms that were identified at each step of the structure-factor-electron-density cycle are numbered. The contours are at intervals of 2 $e/(100\ \text{pm})^3$, starting from 2 $e/(100\ \text{pm})^3$.]

second electron-density distribution was calculated (see Fig. 41b). Seven additional atomic sites could now be recognized, and a further set of calculations, which included these seven atoms in the phasing calculation, gave an R value of 0.41; the subsequent electron-density map revealed the entire structure (see Fig. 41c). The sites corresponding to the nine molecules of water were then identified, and the coordinates of all of the atoms were subjected to a least-squares refinement.

Sometimes, the heavy atoms occupy special positions in the unit cell. Two types of special position should be considered. In the first, the heavy atoms occupying special positions are of lower multiplicity. For example, for $HgCl_2 \cdot (uracil)_2$, the heavy atom is at the center of inversion and there is only one such atom in tl cell that contains two chlorine atoms and two uracil rin In the second type of special position, the heavy atoms have one or more coordinates at cardinal points, such as 0, 0.25, 0.5, etc., and the multiplicity of the heavy atoms is the same as for the other atoms. Such a situation occurs for two heavy atoms in a cell belonging to the space group $P2_1$. The coordinates of the two heavy atoms can be written as $x,0.25,z$ and $-x,-0.25,-z$, and these coordinates are related by an inversion at the origin (see Fig. 42a). Thus, the electron-density distribution calculated by use of the phases from the heavy atoms contains a spurious center of symmetry; this produces the desired structure and its antipode, usually called "the ghost." As more atoms belonging to one enantiomorph are included in the phasing, the ghost-structure fades progressively into the background.

For the barium salt of uridine 5'-phosphate, there are in the cell eight barium atoms having eight general positions, but the eight atoms are distributed over two sets of special positions on two-fold axes. The relativ position of the eight barium atoms is shown in Fig. 42b. As a rule, when the heavy atoms occupy special positions they have a symmetry higher than that of the space-group, and the resulting electron-density map also contains this extra symmetry; consequently, interpretation is a little complicated.

4. Isomorphous-replacement Method

The isomorphous-replacement method is important in the solution both of centrosymmetric and noncentrosymmetric structures, and has been effectively employed in solving the crystal structures of macromolecules.[105–107] Indeed all of the protein structures known to date have been

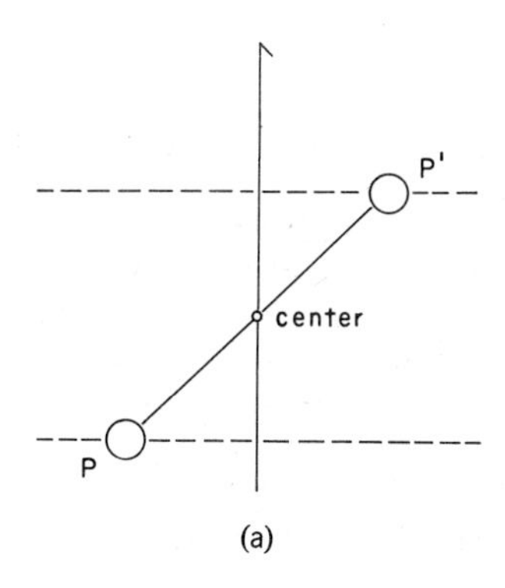

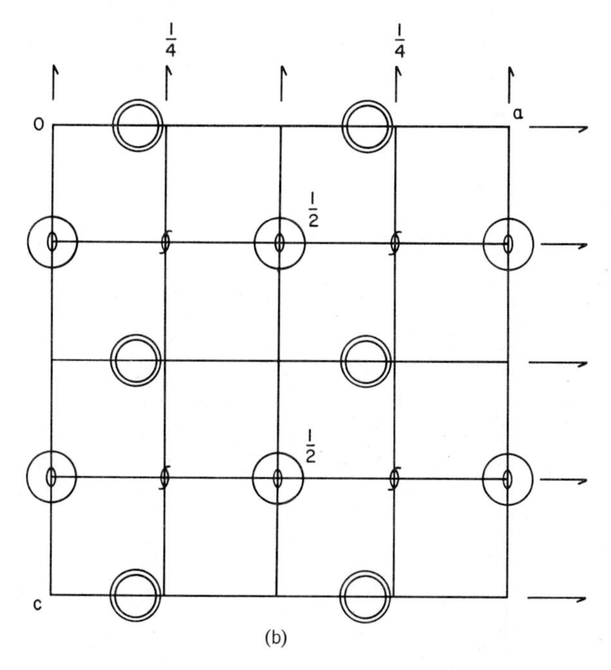

Fig. 42. Special Positions of the Heavy Atoms. [(a) Two heavy atoms in the unit cell in the space group $P2_1$. (b) The view of the eight barium atoms in the cell, distributed over two sets of special positions on two-fold axes in the barium salt of uridine 5'-phosphate.55]

solved by this method. Hence, a brief description of the method is given here.

Two crystals are isomorphous if the atoms common to the two structures occupy the same positions in the cryst lattice. Isomorphous crystals are of two types. In one type, a particular atom in a crystal can be replaced by an atom having a different atomic number, to produce a pair of isomorphous crystals. The alums studied by Cork[107a] are examples of such crystals. In the second type, an atom is *added to* the structure without disturbing the atoms in the parent crystal, and it is this type of isomorphous situation that is of interest in the study of macromolecules.

Consider a reflection from a protein crystal (the parent) of structure amplitude $|F|$. The phase of this reflection may be defined by any point on the circumference of a circle having its center at the origin and having a radius $|F|$. If the exact position of the point on the circle is found, the phase of the reflection is determined. This circle is called the phase circle and

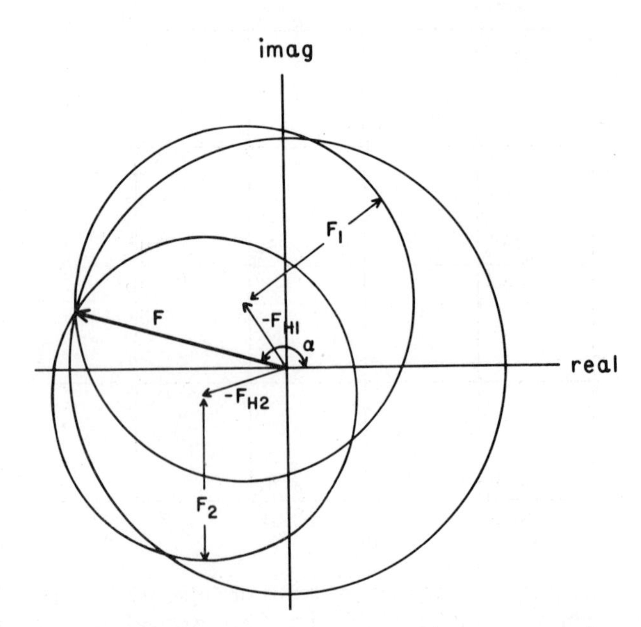

Fig. 43. The Phase Circles for the Parent and Two Hea atom Derivatives Isomorphous with the Parent. [The poi at which the three circles intersect provides a unique solution to the phase problem.]

s shown in Fig. 43. If a heavy-atom derivative of the rotein is formed without affecting the position of the arent molecule in the crystal lattice, that is,it is crystal isomorphous with the parent, the structure actor from this derivative may be written as

$$F_1 = F + F_{H_1} \ ,$$

r

$$F = F_1 + (F_{H_1})$$

n principle, the position of the heavy atom can be found rom the Patterson map, and, hence, the structure factor $_{H_1}$ for the heavy atom can be calculated; therefore, the ector $-F_{H_1}$ can be constructed as shown in Fig. 43. With he center at the tip of this vector, and radius equal o $|F_1|$, a second circle is drawn. The intersections of hese two circles provide the two possible values for the hase of the parent reflection. The intersections are sually at two points that are symmetrical about the vec- or $-F_{H_1}$ (see Fig. 44a). This "ambiguity" is referred to s the horizontal ambiguity. To resolve the ambiguity, t is necessary to prepare a second heavy-atom derivative hat is isomorphous and in which the heavy atom is ituated at a different site. The structure-factor or the second derivative can be written as

$$F_2 = F + F_{H_2} \ ,$$

r

$$F = F_2 - F_{H_2} \ .$$

y following the same procedure, a second circle is drawn hat has its center at the tip of the vector $-F_{H_2}$ nd a radius of $|F_2|$; this circle also intersects the riginal circle at two points. One of the points is ommon to all three circles in an ideal situation, such s that shown in Fig. 43. The common point defines the hase of the reflection of the parent, and provides a

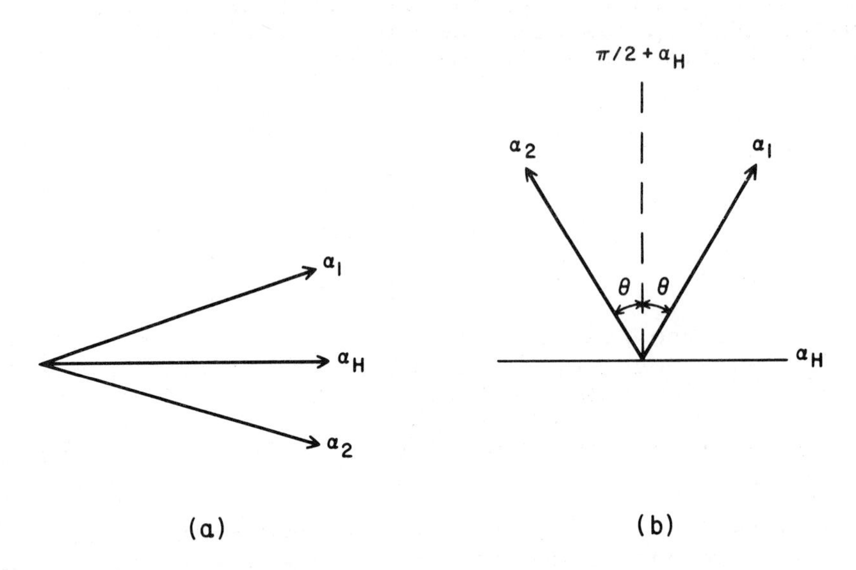

Fig. 44. Ambiguities in the Phase Determination in (a) the Isomorphous-replacement Method, and (b) the Anomalous dispersion Method.

unique solution to the phase problem. However, the ambiguity in the phase could not have been resolved had the second heavy atom been situated at the *same* site, because the phase-circles of the heavy-atom derivatives would hav intersected the phase-circle of the parent at the same two points. Therefore, in principle, a unique solution to the phase problem is obtainable if two heavy-atom derivatives can be prepared in which the heavy atoms occupy two different sites without affecting the original structure of the crystal. In the heavy-atom derivatives of macromolecules, small perturbations in the structure o the parent invariably occur, and the phase circles do not intersect at precisely one point; consequently, it is desirable to prepare additional heavy-atom derivatives in order that a better estimate of the phases may be made. This procedure is known as the multiple, isomorphous-replacement method.

5. Anomalous-dispersion Method

When the wavelength of the X-rays is shorter than the absorption-edge of an atom, anomalous scattering results.

consequence of this scattering in a noncentrosymmetric
stal is that the intensities of the inverse reflections
ℓ) and ($\bar{h}\bar{k}\bar{\ell}$) are unequal, and this results in the break-
n of Friedel's law. Because the difference in the in-
sities for such a pair is usually small, it is desirable
t the intensities be measured as accurately as possible.
difference between the intensities for inverse pairs of
lections provides an elegant way of obtaining the phases
the reflections.[108-110] Fig. 45 shows the construction of

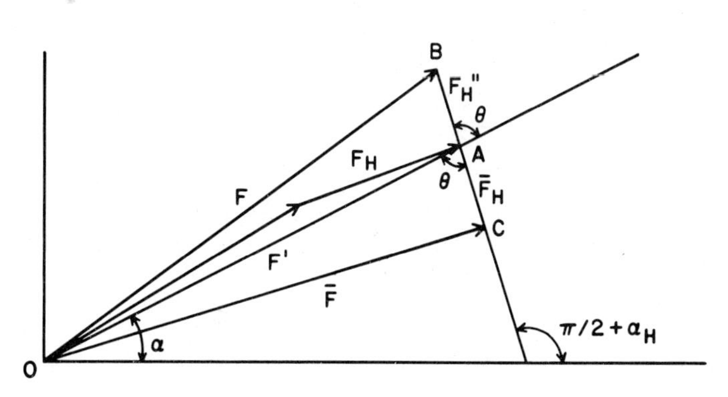

. 45. Construction of the Vector Diagram for a
edel Pair of Reflections when Anomalous Scattering
urs. [The vectors corresponding to the inverse re-
ction are reflected about the real axis, for clarity.]

ucture-factors for inverse reflections. For clarity,
vectors corresponding to the inverse reflections
e been reflected about the real axis, and it is
umed that there is only one kind of anomalous scatterer
the cell. The following relations are obtained from
triangles *OAB* and *OAC*.

$$|F|^2 = |F'|^2 + |F_H''|^2 + 2|F'|\ |F_H''| \cos\theta$$

$$|\bar{F}|^2 = |F'|^2 + |F_H''|^2 - 2|F'|\ |F_H''| \cos\theta$$

Addition of these two expressions yields

$$|F|^2 + |\bar{F}|^2 = 2(|F'|^2 + |F_H''|^2 ,$$

and subtraction of the former from the latter yields

$$\cos\theta = \frac{|F|^2 - |\bar{F}|^2}{4|F'|\ |F_H''|} .$$

Because anomalous scatterers are also heavy atoms, their positions can be found from the Patterson synthesis, and, hence, $|F_H''|$ can be calculated. The value of θ can be calculated by using the difference between the intensitie of a pair of inverse reflections and, consequently, the phase of the reflection is given by

$$\alpha = \frac{\pi}{2} + \alpha_H - \theta .$$

However, the value of θ obtained is ambiguous in sign, because $\cos\theta = \cos(-\theta)$. Thus, there is a two-fold ambiguity in phase, and this is written as

$$\alpha = \frac{\pi}{2} + \alpha_H \pm |\theta| .$$

Here, the ambiguity in phase is about a line at right angles to the heavy-atom phase; it is referred to as the vertical ambiguity (see Fig. 44b). There are several ways of resolving this ambiguity, but only two will be considered here.

Unlike the situation for the isomorphous-replacement method, in the anomalous-dispersion method, the phases can be sorted out in a more or less direct way, as the phase of a reflection is likely to be nearer to than farther from the heavy-atom phase. The number $N(90)$ defined earlier (see p. 482) gives the number of reflections for which the phase-errors are less than 90°. If the phase closer to the heavy atom is chosen, the correct phases are obtained for the majority of the reflections. When the value of θ is small, the phase chosen may be the wrong one; fortunately, in this case, the difference between the correct and the incorrect solution is small.

The most impressive application of this method thus far reported is the solution of the structure of the aquocyanide of the natural vitamin B_{12} nucleus, called[99] Factor V la. The molecule contains one cobalt atom and

77 other non-hydrogen atoms, and the crystal belongs to the space group $P2_1$. Cobalt is a strong, anomalous scatterer for CuKα radiation. The intensities of about 4,000 Friedel pairs of reflections were visually estimated, the position of the cobalt atom was found from the Patterson map, and the phases of almost 2,000 reflections that showed appreciable differences in the intensities of the inverse pairs were then deduced by the method just described. The resulting electron-density map clearly revealed the positions of almost all of the non-hydrogen atoms in the molecule.

It is also possible to solve such structures by the heavy-atom technique, but, for the foregoing example, this would have entailed several structure-factor—Fourier cycles, because of the small value of σ^2, as shown in Table XVI (see p. 481) and the problem of higher symmetry (see p. 486).

A combination of the horizontal ambiguity in phase in the isomorphous-replacement method with the vertical ambiguity in phase in the anomalous-dispersion method leads to a unique solution of the phase problem that has been employed[111] in solving the structure of the protein rubredoxin at a resolution of 3 Å.

Use of the differences in intensity for calculating an antisymmetric Patterson function of the structure was proposed by Pepinsky and co-workers; the method has been employed in solving several crystal-structures.[112]

It would appear that the anomalous-dispersion method is ideally suited for medium-sized structures having ~100 non-hydrogen atoms in the molecule.

6. Direct Method

a. Centrosymmetric Crystal. —In the direct method of determination of structure, the phases, or signs of the reflections, are directly deduced from the intensities observed.[113] That this procedure is possible is implied by the fact that a solution of the structure can be obtained from the Patterson methods, which employ only the observed intensities. Moreover, the electron density expressed by $\rho(xyz)$ is not negative anywhere in the unit cell. Thus, certain restraints exist on the structure-factors that are used as coefficients in computation of the electron density, or, more simply, relationships exist between the phases of the reflections. The problem with centrosymmetric crystals is simple, because the phase-ambiguity is merely one of signs, namely, + or -. The method of establishing signs in a centrosymmetric crystal is well established and extensively used. In this Section, two terms related to the structure-factor are used, namely, the unitary structure-factor U, given by

$$U = F/\Sigma f_j \ ,$$

and the normalized structure-factor E, given by

$$E = F/\sqrt{\Sigma f_j{}^2} \ . \qquad (2)$$

Both U and E represent the scattering from the crystal, in which the actual atoms having a spatial spread are point atoms.

The first breakthrough in this direction came when Harker and Kasper[114] discovered unequal relationships between unitary structure-factors. One such expression that determines the signs of certain reflections is

$$|U_H|^2 \leq \frac{1}{2}\ (1 + U_{2H})\ ,$$

where H and $2H$ stand for (h,k,ℓ) and $(2h,2k,2\ell)$, respectively.

The most important relation between the structure-factors can be written as

$$E_H \sim \sum_K E_K E_{H-K}, \qquad (3)$$

where H stands for (h,k,ℓ) and, similarly, K and $H-K$ represent three corresponding indices. The summation extends over the entire reciprocal space, and is called the *Sayre relation*[115] or the Σ_2 formula.[116] The relation is general, applies both to centrosymmetric and noncentrosymmetric crystals, and means that the phase of a reflection is determined by the phases of all pairs of reflections that have a specific relation to it. Such pairs for a reflection are listed in Table XVIII.

In order to apply this method, a few initial signs must be known. By substituting the known signs on the right-hand side of equation *3*, more signs can be obtained. At the beginning of the determination of sign, only a few terms in the summation would be known exactly. The probability that the sign of E_H is the same as that of the product of one pair, $E_H E_{H-K}$, is given[117] by

$$P_+ = \frac{1}{2} + \frac{1}{2} \tanh \left(\frac{\sigma_3}{\sigma_2{}^{3/2}} \ |E_H| \ |\Sigma E_K E_{H-K}| \right), \ \sigma_n = \Sigma f_j{}^n .$$

TABLE XVIII

Some Pairs of Reflections that Interact with the Reflections 2, 3, 6 through the Σ_2 Formula

H	K	$H-K$
2 3 6	1 1 2	1 2 4
	3 5 4	-1 -2 2
	6 7 4	-4 -4 2
	5 4 3	-3 -1 3
	-1 -1 2	3 4 4

/hen more terms are introduced, the products are added to-
;ether, and, for an equal-atom structure, the expression
:educes to

$$P_+ = \frac{1}{2} + \frac{1}{2} \tanh \left(|E_H| \quad |\Sigma E_K E_{H-K}| / \sqrt{N} \right) ,$$

'here N is the total number of atoms in the unit cell. The
uccess of the method depends on the fact that P_+ can be
s high as 0.99 for even one term involving strong reflec-
ions. The dependence of P_+ on the E values and on the
umber of atoms in the unit cell is given in Table XIX.
he indications provided by such relationships are accept-
d as being certainties, and the sign of the reflection so
etermined is used on the right-hand side of equation (3)
n ascertaining the signs of more reflections by a process
f iteration. It is important that the first few signs
e correctly determined as, otherwise, errors tend to
ropagate, resulting in a large number of incorrect signs.
Because the structure could be described with reference
o any one of several equivalent origins in the space-
roup, a reference-origin has to be established in order
o describe the structure. Referred to different origins,
he structure will produce the same intensities, but,
n general, the signs of the reflections are different.
hus, for the one-dimensional view of one atom in a cen-
rosymmetric crystal (see Fig. 46), the structure could be
escribed by a coordinate x when referred to the origin
t A, or by the coordinate $1/2-x$ when referred to origin
. The structure factors F_A and F_C referred to the two
rigins can be written as

TABLE XIX

Probability that the Product of the Signs of Three Reflections H, K, and H-K Is Positive for Different Values of $\bar{E}$ and Numbers of Atoms in the Unit Cell (N)

$\bar{E}$	Value of N 20	40	60	80	100	200
3.0	1.00	1.00	1.00	1.00	0.99	0.98
2.8	1.00	1.00	1.00	0.99	0.99	0.96
2.6	1.00	1.00	0.99	0.98	0.97	0.92
2.0	0.97	0.93	0.89	0.86	0.83	0.75
1.5	0.82	0.74	0.71	0.68	0.66	0.62

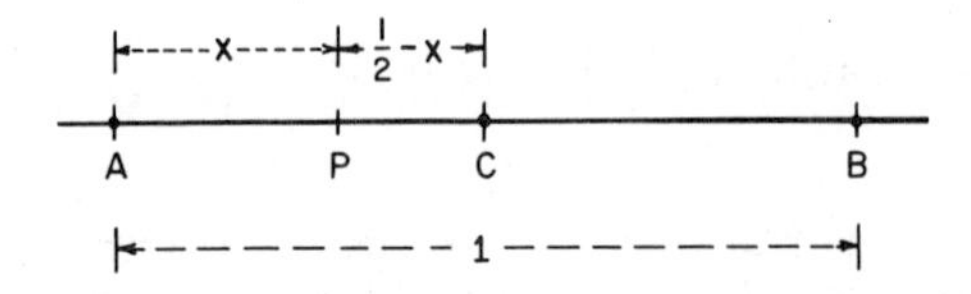

Fig. 46. Two Equivalent Origins that can be Chosen to Describe the Structure.

$$F_A = 2 \cos 2\pi hx$$

and $$F_C = 2 \cos 2\pi h(x-1/2) = (-1)^h F_A$$

When h is even, F_A and F_C have the same sign; but when h is odd, they have opposite signs. The signs of reflections when h is even are independent of the origin chosen and are called *structure invariants*. The reflections when h is odd are called *semi-invariants*. The sign of one reflection in the group of reflections having h odd can arbitrarily be assigned as + or −, and this fixes the origin at A or C; the signs of all other reflections in this group are then referred to this origin. This principle can be generalized for the three-dimensional situation[116]. In the space-group $P2_1/c$, three linearly independent reflections, suitably chosen, can be assigned arbitrary signs in order to define the origin. The reflections may be considered to be distributable among eight parity groups that depend on whether the indices are even (E) or odd (O). The eight groups can be written as (EEE), (EEO), (EOE), (EOO), (OEE), (OEO), (OOE), and (OOO). The reflections belonging to group (EEE) form the invariant set, and these cannot be used to define the origin. Two origin-specifying reflections can be chosen, one from each of the seven other groups, for example, (OOE) and (OOO). The reflections in the two groups interact and thereby produce a reflection belonging to the group (EEO). The third origin-specifying reflection should, therefore, be chosen from any one of the four groups (EOE), (EOO), (OEE), or (OEO).

The practical details of application of this method will now be illustrated for crystals of hydrouracil;[118] these are monoclinic, and belong to the space group $P2_1/c$. The E values of all of the 800 reflections from this crystal were calculated, and 150 reflections having $E > 1.25$ were sorted out in descending magnitude of the value of E. With the aid of a computer, the relations between the reflections in this set were found; these relations were then studied, and the three reflections defining the origin were chosen (see Table XX). A few other signs were determined by using these reflections and, by assigning a symbolic phase to a reflection that interacted with the reflections having known signs, the symbolic signs of more reflections were determined. The signs of all of the 150 reflections could then be determined in terms of three symbolic phases, A, B, and C (see Table XX).

This procedure is called the *symbolic addition method*.[119] Because the sign of the symbols can be either + or -, there are $2^3 = 8$ solutions to consider. It is possible to determine the most probable solution by determining the consistency index, c, defined by

$$c = \Sigma(E_H \Sigma E_K E_{H-K}) \Sigma(E \ \Sigma|E_K E_{H-K}||),$$

where the summation is over all of the Σ_2 terms of a reflection, and over all of the reflections. The values of c for the eight solutions are listed in Table XXI; that corresponding to $A = +$, $B = -$, and $C = +$ has the highest con sistency index, and represents the most probable solution. The signs obtained from this set were combined with the 150 E values, and an E-map, a Fourier synthesis in which the E values, not the F values, are used as coefficients, was calculated.[120] The composite E-map shown in Fig. 47

TABLE XX

Reflections for Hydrouracil that are Used to Specify the Origin and the Symbolic Phases

h	k	ℓ	|E|	Sign	Remark
1	5	2	2.98	+	origin
-1	1	3	2.72	+	origin
2	1	12	2.72	+	origin
-2	2	6	2.71	*A*[a]	
-2	4	1	2.50	*B*[a]	
2	5	8	2.48	*C*[a]	

[a]*A*, *B*, and *C* are three symbolic signs, and each may assume either a positive or a negative value.

TABLE XXI

Consistency Index for the Eight Possible Solutions for the Crystal Structure of Hydrouracil

Combination				
No.	*A*[a]	*B*[a]	*C*[a]	Consistency index
1	+	+	+	0.812
2	+	+	–	0.511
3	+	–	+	0.895[b]
4	+	–	–	0.637
5	–	+	+	0.586
6	–	+	–	0.521
7	–	–	+	0.588
8	–	–	–	0.520

[a]*A*, *B*, and *C* are three symbolic signs, and each may assume either a positive or negative value. [b]Correct solution.

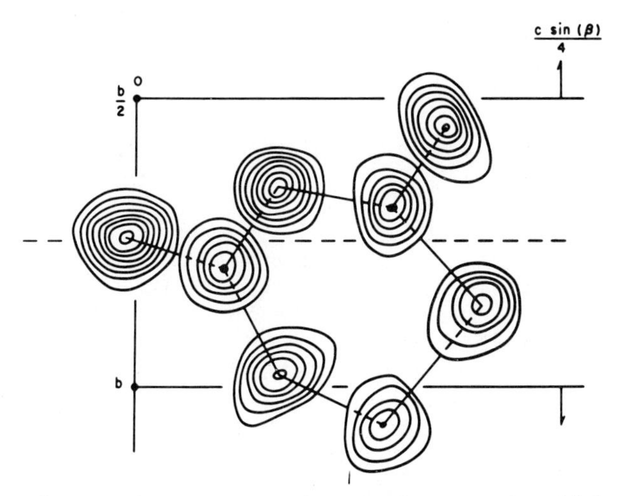

Fig. 47. Composite *E*-Map for Hydrouracil.[118] [Contours are at equal, arbitrary intervals.]

clearly reveals the structure, and the coordinates that were obtained from this map for the atoms were subsequently subjected to a least-squares refinement.

Occasionally, the map corresponding to the most probable solution having the highest value of *e* fails to reveal the structure. In such an event, the solution having the next highest value of *e* is considered. Also, when several *E*-maps are calculated for a structure, it is sometimes found that, in some of these maps, the molecule is placed in a position related by a translation-component to the correct position in the cell. In other maps, the distribution of the peaks has no chemical meaning. In the former example, if there is more than one solution possible, the correct solution is that which satisfies the packing requirements of the lattice.

It may be seen from Table XIX that, as the number of atoms in the unit cell increases, the probability P_+ decreases; this result implies that the method would increasingly fail in such instances. The practical limit at which this method fails is not yet known, but structures containing up to ~150 non-hydrogen atoms in the unit cell have been solved.[120]

b. Noncentrosymmetric Crystal. — Here, the problem is more complex, because the phase-angles can assume values anywhere in the range of 0–360°; however, the method shows considerable promise. Equation *3* (see p. 494) is also valid for noncentrosymmetric crystals, as

$$E = |E| \exp (i\alpha).$$

Substitution of this value in equation *3* gives

$$\tan \alpha_H = \Sigma|E_K E_{H-K}| \sin (\alpha_K+\alpha_{H-K})/\Sigma|E_K E_{H-K}| \cos (\alpha_K+\alpha_{H-K}). \quad (4)$$

This expression, called the *tangent formula*,[121] can be used not only in determining phases but also in refining the approximate phases already obtained. The first appl cation of this method was made in a determination of the structure of L-arginine dihydrate.[122] The structures of a number of nucleic acid components have been solved by this method, which may be illustrated by its application to the structure of 3'-*O*-acetyladenosine.[123,124]

Crystals of 3'-*O*-acetyladenosine belong to the orthorhombic space group $P2_12_12_1$. The intensities were reduce to E values by using equation (*1*), and the Σ_2 relations from 285 reflections having E>1.2 were generated. Phase were assigned to three reflections in order to define the origin (see Table XXII), and, in addition, the phase of a fourth reflection was assigned as $\pi/2$ to define the mirr image with reference to the chosen axes. Initially, the phase extension was conducted by use of a simplified version of equation *4*, namely,

$$\alpha_H = \langle\alpha_K + \alpha_{H-K}\rangle,$$

where the average is taken over the reflections whose phases are defined. It was found that a knowledge of the phases of the four starting reflections was adequate to define the phases of most of the strong reflections with E>1.7. Values for these phases were then inserted in the tangent formula *4* to determine new phases, as well as to refine the phases already obtained. The phases of a total of 285 reflections with E>1.2 were thus determined and were used to calculate an E-map (see Fig. 48). The entire structure was obtained from this map, the 22 heav

TABLE XXII

Reflections Used to Define the Origin and Enantiomorph for 3'-*O*-Acetyladenosine

h	k	ℓ	$\|E\|$	α	
12	3	0	2.79	0	origin
7	0	2	1.99	0	origin
0	10	1	2.06	0	origin
3	7	0	2.18	$\pi/2$	enantiomorph

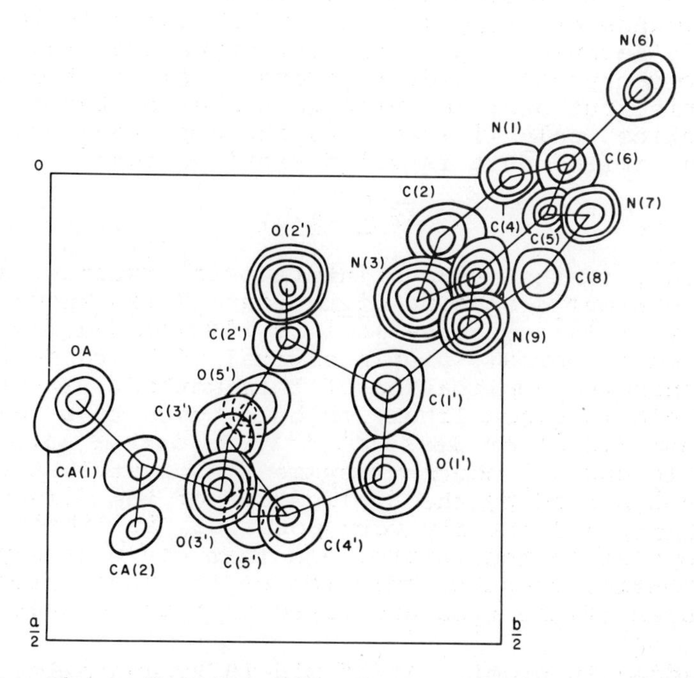

ig. 48. The *E*-Map for 3'-*O*-Acetyladenosine. [Computed ith 285 reflections. Contours are at equal, but arbitrary, ntervals.]

atoms being found from among the 23 highest peaks in the map.

In some instances, it is found that the origin and mirror image-defining phases are not adequate to propagate through the Σ_2 relation and generate phases of other strong reflections. It is then necessary to assign symbolic phases to a few other reflections, as in the centrosymmetric case, and to examine the various solutions corresponding to all possible combinations of phases for the symbols.

Sometimes, only a portion of the structure is visible in the initial E map. All of the high peaks that define a reasonable geometry are then selected for a calculation of the structure factor, and the phases from this calculation are refined by the tangent formula. A new *E*-map, which usually reveals additional atoms of the structure, is then computed, and the process is repeated until the entire structure is obtained.[125]

If a fragment of the structure can be located from the Patterson map as described on p. 476, or the position of the heavy atom can be determined from the Patterson map, the phases from these atoms can be refined by the tangent formula before an electron-density map is calculated. It is found that such a method leads to the structure in fewer rounds of structure-factor-Fourier calculations.[12

The Σ_2 formula *3* for both centrosymmetric and noncentrosymmetric crystals holds rigorously only if the summation is carried out over an infinite number of terms. However in practice, only limited data are employed, but the use of such limited data is valid provided that

$$\overline{\Delta r} \geq d_{\min},$$

where $|\overline{\Delta r}|$ is the average interatomic distance, and $d_{\min}$ is the minimum *d*-spacing of the data (also known as the resolution).[127] Because $|\overline{\Delta r}|$ is ∿100 pm (∿1 Å), the data collected in normal, single-crystal work are sufficient for meaningful application of this method. In practice, these relations are likely to hold, even to a resolution of 250 pm (2.5 Å) or less.[128,128a] To date,* ∿100 centrosymmetric and ∿20 noncentrosymmetric crystal-structures have been solved by the application of the direct method The method, undoubtedly very powerful, is likely to become popular in the future, and that may possibly lead to an automatic, computer solution of the phase problem for structures of the type discussed in this Chapter.

*Note added in proof. As of mid-1972, approximately 200 centrosymmetric and 80 noncentrosymmetric structures have been solved by the direct method.

VIII. REFINEMENT OF THE STRUCTURE

1. Least-squares Procedure and Determination of Positions of Hydrogen Atoms

The approximate structure obtained by any one of the methods described should now be refined, in order to provide the best solution. In earlier work, the difference-synthesis method was extensively used, whereas the method now most commonly employed is the method of least squares.[129] This method is applicable, because the number of reflections from a single crystal is 1000–2000 and is always much larger than the number of atomic parameters that have to be determined. Moreover, the method is readily amenable to automatic cycling of data on high-speed computers.

In the least-squares method, the quantity minimized is the weighted difference between the observed and the calculated structure-amplitudes; this difference is given by

$$\Sigma w(|F| - |F_c|)^2,$$

where w is the weight of the observation; $\sqrt{w}$ is inversely proportional to the standard error in the measurement of F, that is,

$$w = 1/\sigma^2(F).$$

The summation is carried over all of the reflections, and the minimization is performed with reference to the parameters that define the structure.

The weighting scheme has to be so chosen that it reflects the errors in the measurement of the structure amplitudes, $|F|$. Several weighting schemes have been employed, and some of them are:

$$\sigma(F) = F^* \qquad |F| \leq F^*$$

$$= |F|/F^* \qquad |F| > F^* \quad \text{(Hughes[129])},$$

$$(F) = [a+|F| + c|F|^2]^{0.5} \quad \text{(Cruickshank[130])},$$

$$\text{and } (F) = [a - |F| + c|F|^2]^{0.5} \quad \text{(Modified Cruickshank)}.$$

These curves, plotted as a function of $|F|$, are schematically shown in Fig. 49. When the data are collected by diffractometric methods, an estimate of $\sigma(F)$ can be obtained from the counting statistics.[131] The general shape of the curve obtained therefrom, for small crystals, is given in Fig. 49, and, even though the accuracy of obser-

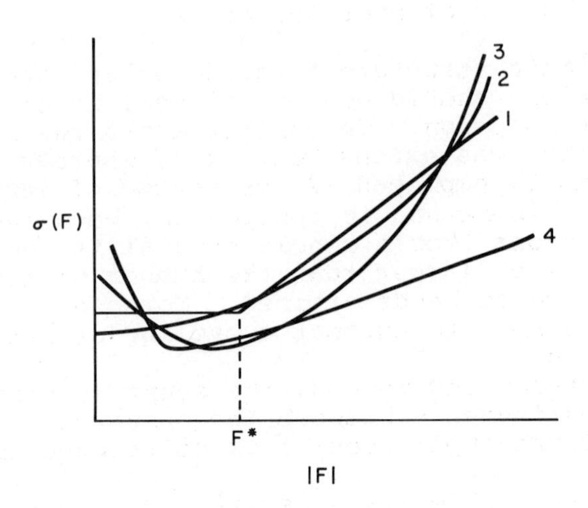

Fig. 49. Schematic Representation of the Error-schemes used in the Least-squares Refinement. [(1) Hughes, (2) Cruickshank, (3) modified Cruickshank, and (4) counting statistics.]

vation for the strong reflections is high, the intensitie embody systematic errors, because of secondary extinction and absorption. To circumvent this problem, the errors for such reflections are increased by adding a linear term in $|F|$.

Let the structure be determined by n atomic parameters, $p_1, p_2, \ldots p_n$. Then the set of equations that determines the shifts Δp_i in the parameters is given by the normal equation

$$A\Delta p = X,$$

where A is the normal matrix of order n, having coefficients

$$A_{ij} = \Sigma\, w \frac{\partial |F_c|}{\partial p_i} \cdot \frac{\partial |F_c|}{\partial p_j},$$

and X is a column-vector of order n, having coefficients

$$X_i = \Sigma w\,(|F| - |F_c|)\, \frac{\partial |F_c|}{\partial p_i}.$$

The shifts in the parameters are obtained by inverting the normal matrix A, and then multiplying it by the column vector X.

$$\Delta p = A^{-1}X.$$

These shifts are added to the old parameters, and a second round of calculation proceeds. The iterative process is deemed to have converged when the shifts in the parameters are much smaller than the corresponding standard errors that are given by

$$\sigma_i = \left[\frac{w(\Delta|F|)^2}{m-n}\right]^{0.5} A^{-1}_{ii},$$

where m is the number of reflections, and A^{-1}_{ii} is the ith diagonal element of the inverse, normal matrix. As the refinement progresses, the differences between the observed and calculated structure-amplitudes decrease, and the R value drops steadily.

In problems involving a larger number of parameters, that is, ∿200 or more, the size of the normal matrix becomes large, and the core size in the high-speed computers currently available becomes inadequate. Here, the parameters are grouped into several blocks, and refinement is performed in stages.

The refinement process usually starts with the variation of a scale-factor, three positional parameters, and one thermal parameter describing the isotropic thermal motion for each nonhydrogen atom. At the end of such a refinement, a difference-map, a Fourier synthesis in which values of $|F| - |F_c|$ are used as coefficients with the phase of F_c, is computed.[132] The difference-map also represents the electron density $\rho_o - \rho_c$, where ρ_o is the actual electron-density and ρ_c is the electron density of the model assumed. In general, if the intensity data are good, all of the hydrogen atoms are locatable on such a difference-map. For cytidine 3'-phosphate, a composite map from which all hydrogen atoms were located is given in Fig. 50. If an atom is undergoing anisotropic motion, the site of the corresponding atom in the difference-map is flanked by characteristic regions of positive and negative density, as shown in Fig. 51. The hydrogen atoms are included in the refinement process, and the heavier atoms are now allowed to undergo anisotropic, thermal motion described by six thermal parameters (in contrast to one for isotropic, thermal motion). The refinement is terminated when the shifts in the parameters are much smaller

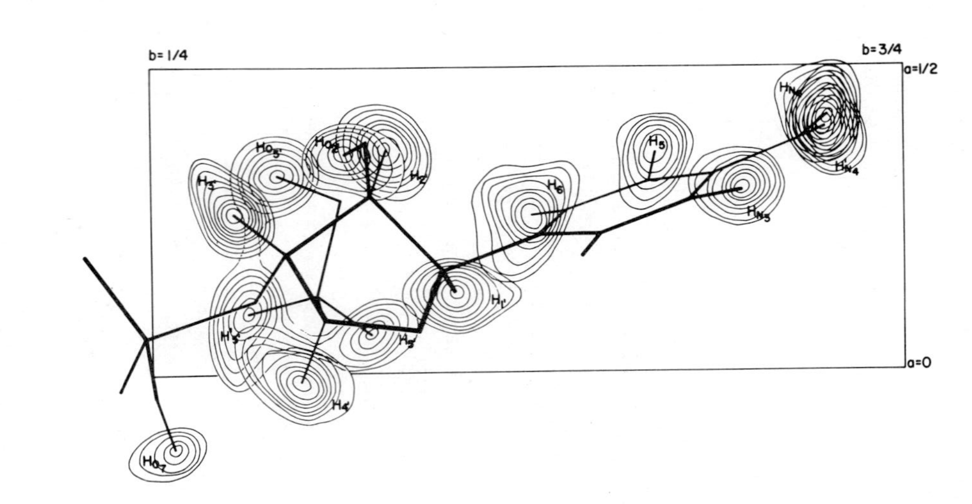

Fig. 50. Composite Diagram Showing the Hydrogen Atoms in Cytidine 3'-Phosphate. [Contours start from 0.1 $e/(100pm)^3$, and are at intervals of 0.05 $e/(100pm)^3$. Reproduced from M. Sundaralingam and L. H. Jensen, *J. Mol. Biol.*, 13, 914 (1965).]

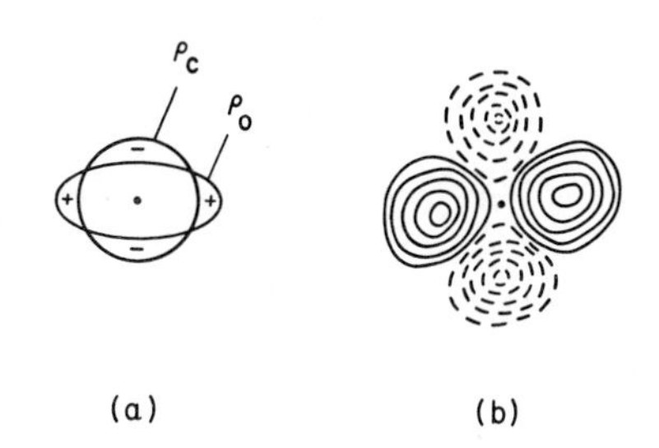

'ig. 51. (a) The Anisotropic Thermal Motion of an Atom when it is Approximated by Isotropic Thermal Motion. (b) The Corresponding Region in the Difference Map.

than their estimated standard errors, usually $\leq$ 1/3. At the end of such a procedure, a set of parameters best fitting the experimental data is obtained, along with their estimated standard errors. By using these parameters, the bond lengths and bond angles of the molecule may be calculated, and their estimated standard errors computed. For a well determined structure, the *R* value is usually ∿0.05, and the average errors in the bond lengths are ∿0.5 pm (∿0.005 Å) and in the bond angles, 0.2°. These values for nucleic acid components, that is, the bases, nucleosides, nucleotides, paired bases, and paired nucleosides are listed in Tables X–XIII, respectively.

At the end of the analysis, it is customary to compute an electron-density distribution by using the final phases obtained from the refinement, and a map so obtained for cytidine 3'-phosphate is shown in Fig. 52.

2. Disorder in Crystals

It is possible for the molecules to occur in the crystal in more than one orientation ("configuration"), giving configurational* disorder, or conformation, giving conformational disorder, and then the "configuration" or conformation is not the same throughout the entire crystal. This phenomenon, termed disorder, can be diagnosed by the presence of residual electron-densities in the difference-map, large thermal motions for the atoms, and/or abnormal geometry in bond lengths and bond angles involving these atoms.

*The meaning of the word "configuration" differs from that of the word as used in carbohydrate chemistry.

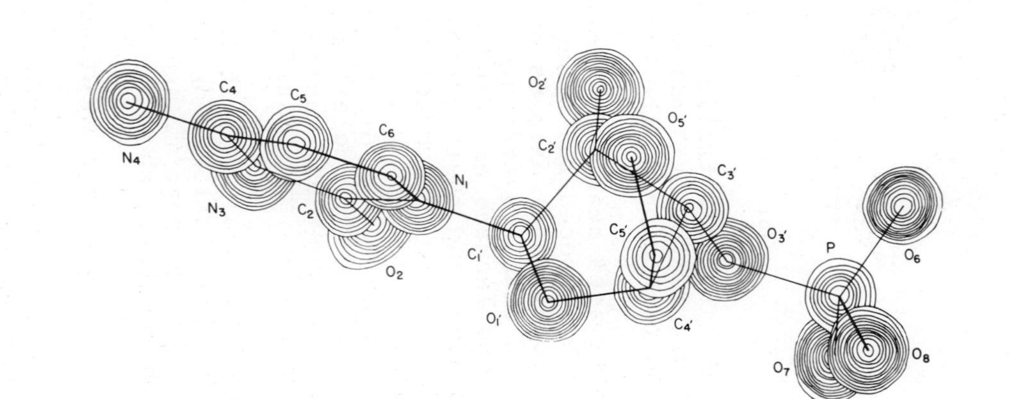

Fig. 52. Composite Fourier Synthesis for Cytidine 3'-Phosphate. [Contours at every 1 $e/(100\ \mathrm{pm})^3$, beginning at 2 $e/(100\ \mathrm{pm})^3$, except near the phosphorus atom, where it is at intervals of 5 $e/(100\ \mathrm{pm})^3$ beginning at 5 $e/(100\ \mathrm{pm})^3$. Reproduced, by permission of Academic Press, Inc., from M. Sundaralingam and L. H. Jensen, *J. Mol. Biol.*, 13, 914 (1965).]

For 5-bromo-2'-deoxycytidine,[75] the position of the O-5' atom was found to be distributed over two positions, corresponding to two different orientations about the C-4'—C-5' bond, as shown in Fig. 53. For dihydrothymine (5-

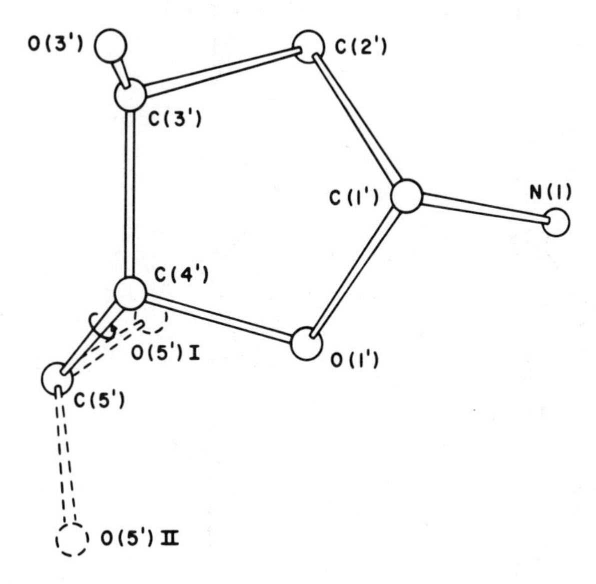

ig. 53. Conformational Disorder[75] in 5-Bromo-2'-deoxy-ytidine about the C-4'—C-5' Bond, giving Two Possible ositions for O-5'.

ethylhydrouracil)[133] (see Fig. 54a), the space-group is entrosymmetric, so that molecules occur in mirror-image airs in the lattice, as shown in an edgewise view of the olecule in Fig. 54b. It was found that the thermal paraeters of C-5 and C-6 are quite high, and that the ond lengths involving these atoms are abnormal; this disovery led to postulation of the coexistence, in the latice, of the mirror image obtained by mirroring the moleule on the mean plane (see Fig. 54c). It is clear that ie only atoms that would not exactly superpose are C-5, -6, and C-7, and the associated hydrogen atoms. In a ibsequent difference-synthesis, the hydrogen atoms asso-

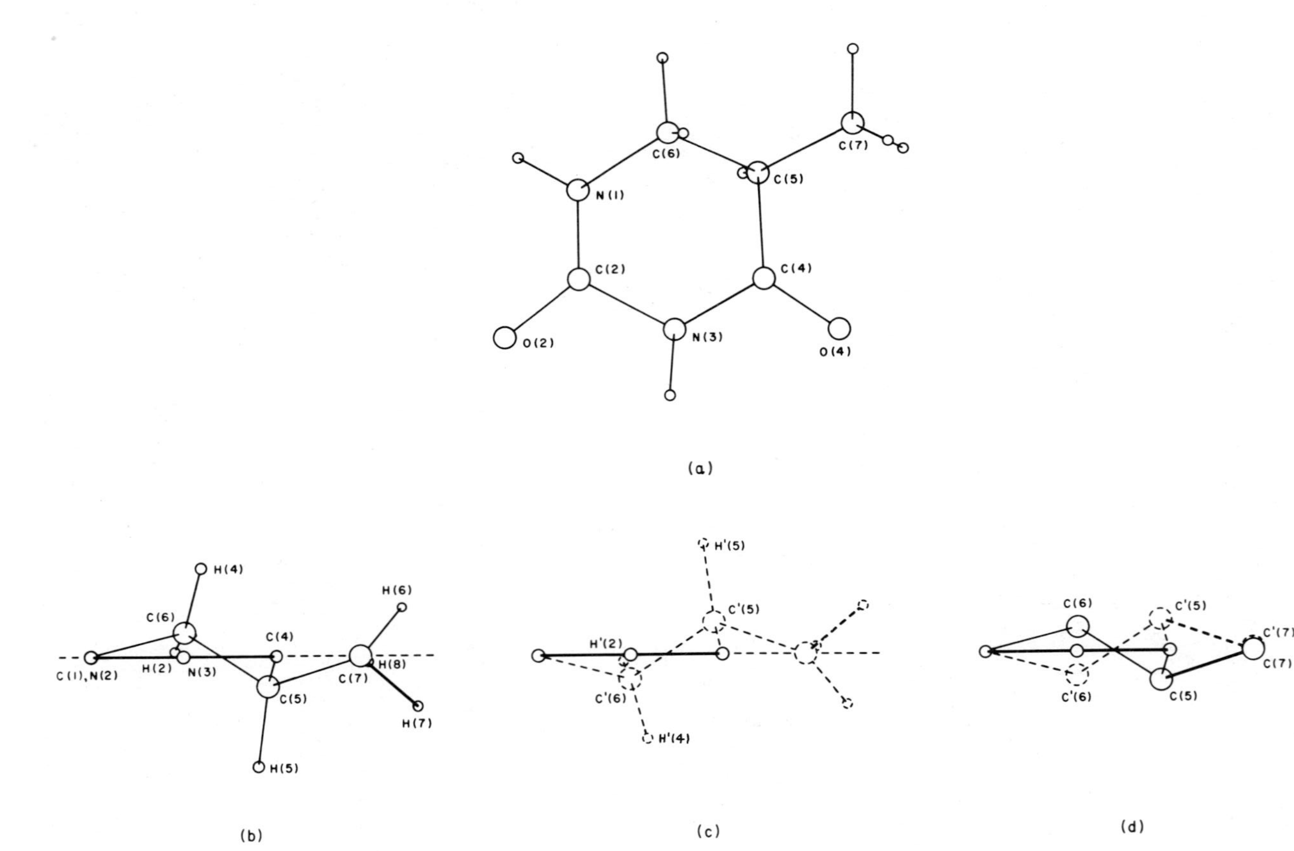

Fig. 54. Disorder in 5-Methylhydrouracil.[133] [(a) The basic molecule, (b) the molecule viewed along its edge, (c) the antipode of (b) mirrored on the mean plane of the molecule, and (d) a composite diagram giving both enantiomorphs. Hydrogen atoms are omitted for the sake of clarity.]

ciated with the two different "configurations" were also detected, thus lending support to the foregoing postulation. By considering the heights of the peaks in electron-density maps, and also the refinement, it was concluded that, at a given site in every two of five unit-cells, the mirror image occurs instead of the original (see Fig. 54d).

An interesting situation involving disordered molecules of water occurs for the sodium salt of inosine 5'-phosphate,[57] as shown in Fig. 55. The water molecules in the structure are distributed over nine sites, W(1)—W(9). Of these, the water molecules W(4), W(6), W(7), and W(8) are disordered. One water molecule is distributed over the two positions W(6) and W(8), ∿200 nm (2 Å) apart, and another molecule is distributed over the positions W(4) and W(7), as indicated by the corresponding occupancy parameters. In every two of five unit-cells, one water molecule is at W(6), where it coordinates with the Na^+ ion; but, when the same water molecule is present at W(8), it takes part in hydrogen-bond formation with other water molecules. The second water molecule, distributed over the sites W(4) and W(7), takes part in hydrogen-bond formation with other water molecules in the crystal. In crystals containing a large proportion of solvent molecules, especially in those of biological macromolecules, the molecules of solvent are invariably disordered.

IX. DETERMINATION OF ABSOLUTE CONFIGURATION

The method of determining the absolute configuration of asymmetric molecules was first worked out by Bijvoet and co-workers.[134] If a molecule in a certain configuration is described by the atomic coordinates x_j, y_j, z_j in the unit cell, the antipode is described by the coordinates $-x_j, -y_j, -z_j$. Where Friedel's law holds, the intensities from these two structures, namely, the intensities from the planes $(hk\ell)$ and $(\bar{h}\bar{k}\bar{\ell})$, are the same, and so, the X-ray structure analysis cannot determine the absolute configuration of the molecule.

However, when an anomalous scatterer is present in the structure, $I(hk\ell) \neq I(\bar{h}\bar{k}\bar{\ell})$. This difference in the intensities of the Friedel pair (also known as the Bijvoet pair) may be used to distinguish between the two configurations. The method currently employed is to calculate the intensity of a reflection for both configurations; then, the calculated value that best agrees with the intensity observed represents the correct configuration. In this way, the absolute configuration was established for the barium salt of uridine 5'-phosphate.[55] As may be

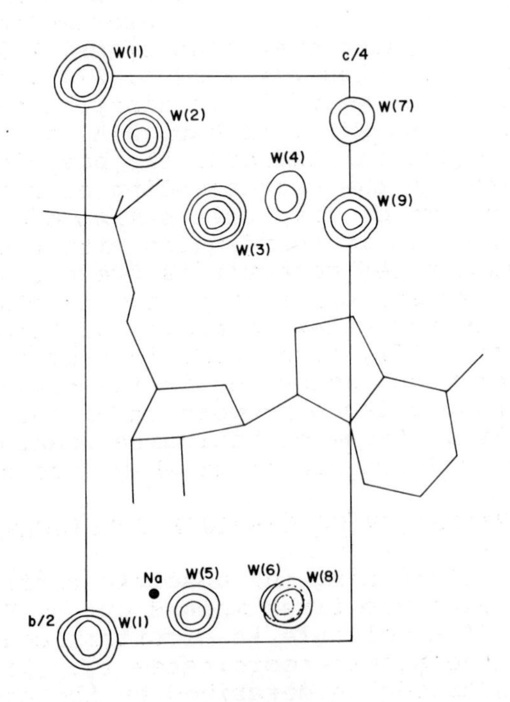

Fig. 55. The Difference Map for the Sodium Salt of Inosine 5'-Phosphate, Showing the Sites of the Water Molecules.[57] [The sites W(4), W(6), W(7), and W(8) are disordered and have heights lower than those at the other sites.] [Reprinted from S. T. Rao and M. Sundaralingam, *J. Amer. Chem. Soc.*, 91, 1210 (1969), published by the American Chemical Society. Reprinted by permission of the copyright owner.]

seen from Table V, barium is a strong, anomalous scatterer for CuK_{α} radiation. Some calculated and observed values of $|F|$ for the β-D-ribosyl nucleotide and its enantiomorph [1-(5-*O*-phosphono-β-L-ribofuranosyl)uracil]

TABLE XXIII

Values of Structure Amplitudes Observed for Various Reflections, Compared with Those Calculated for the D and L Forms of the Barium Salt of 1-D-Ribofuranosyluracil 5'-Phosphate

			Observed Structure amplitude	Calculated structure amplitudes for D-configuration	Calculated structure amplitudes for L-configuration
h	k	ℓ	$\lvert F_o \rvert$	$\lvert F_c \rvert_D$	$\lvert F_c \rvert_L$
1	3	$\bar{1}$	156	141	112
4	2	1	122	112	153
4	2	$\bar{1}$	159	153	112
1	5	1	47	49	89
1	5	$\bar{1}$	86	89	49
3	5	1	213	197	175
3	5	$\bar{1}$	154	175	197
9	3	2	141	153	191
5	1	$\bar{3}$	180	179	135

are compared in Table XXIII. It may be seen that the values for the D-ribosyl nucleotide agree better with the experimental values, thus establishing that the nucleotide contains a D-ribose residue.

Instead of choosing a few reflections and making individual comparisons, the agreement between all of the observed values can be calculated for both configurations, and the one that gives a significantly lower *R* value[135] gives the correct configuration. In the case of the barium salt of 1-(5-*O*-phosphono-D-ribofuranosyl)uracil,[55] the *R* values for the β-D- and β-L-nucleotides were 0.108 and 0.143, respectively. With diffractometric data, it is possible to obtain significant differences in the *R* values, even with lighter atoms, such as chlorine. For 2'-deoxycytidine hydrochloride,[38] with chlorine as the anomalous scatterer, the *R* values for the β-D and β-L isomers were 0.035 and 0.061, respectively. Thus, the correct structure of 2'-deoxycytidine is established as containing a β-D-ribose residue. It has been demonstrated that even oxygen, a light atom and a weak anomalous scatterer may, with CuK_α radiation, be utilized in determining the absolute configuration of a molecule.[136]

X. RESULTS OF X-RAY ANALYSIS

As soon as the atomic coordinates of the molecule in the crystal lattice have been established, the bond distances (r_{ij}), the bond angles (θ_{ijK}), and the torsion angles (ϕ_{ijkl}) of the molecule can be calculated. The torsion angle ϕ about the bond jk is defined as the angle made by the projection of the bond $k\ell$ with respect to the bond ij, when one sights along the bond jk. The clockwise rotation of the $k\ell$ bond is considered to be positive, and so, counterclockwise rotation of $k\ell$ is negative (see Fig. 56).

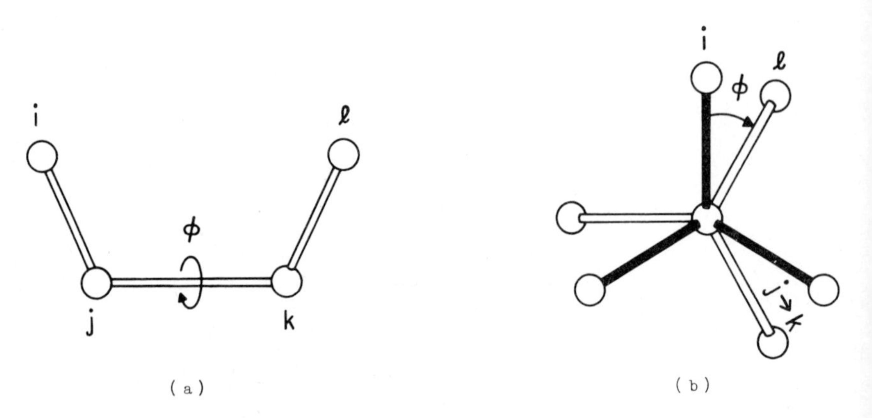

Fig. 56. Definition of the Torsion Angle ϕ Involving Four Atoms, i, j, k, and ℓ.

The bond distances and bond angles for a nucleotide, cytidine 3'-phosphate, are shown in Figs. 57 and 58, and some torsion angles for the nucleotide are listed in Table XXIV. It may be seen that the molecule exists in the "extended conformation," with the cytosine residue twisted at an angle of 60.5° to the mean plane of the D-ribofuranose ring. The D-ribofuranose ring is puckered, C-2' displaced by 610 pm (0.61 Å) from the least-squares plane formed by the remaining ring-atoms. Owing to the

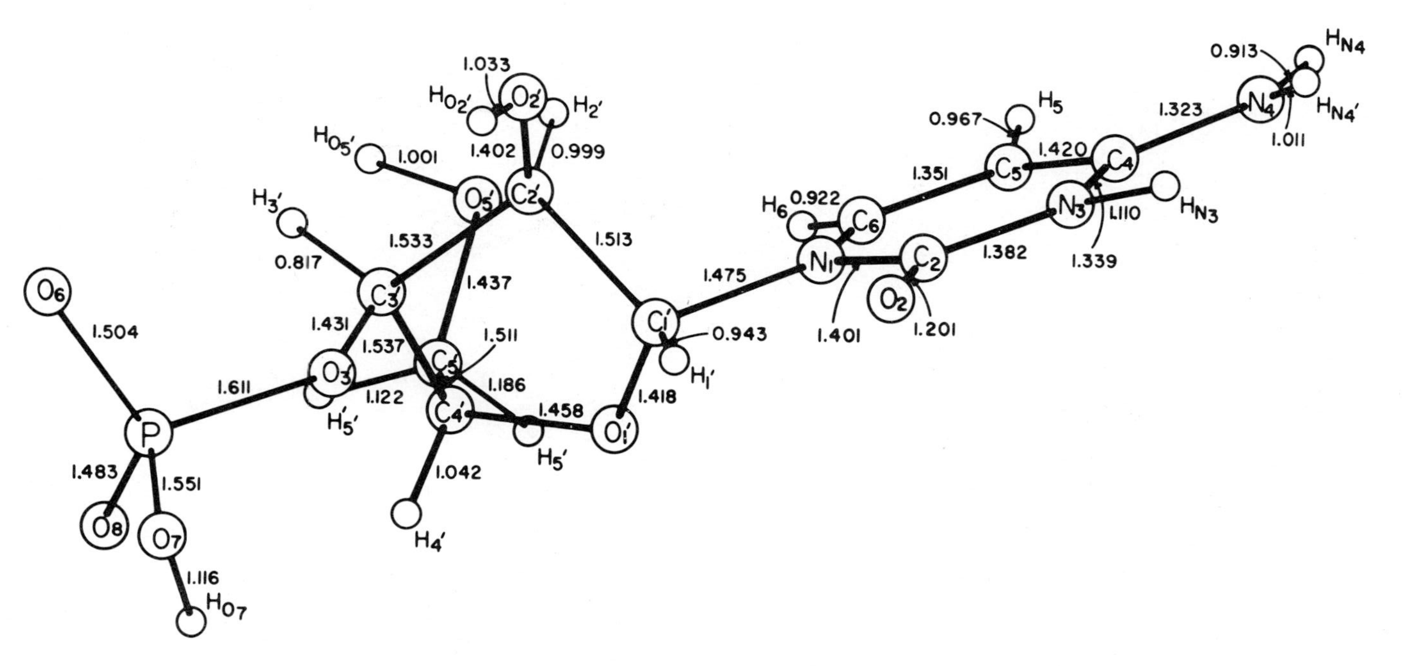

Fig. 57. Bond Lengths in Cytidine 3'-Phosphate. [Reproduced, by permission of Academic Press, Inc., from M. Sundaralingam and L. H. Jensen, *J. Mol. Biol.*, 13, 914 (1965).]

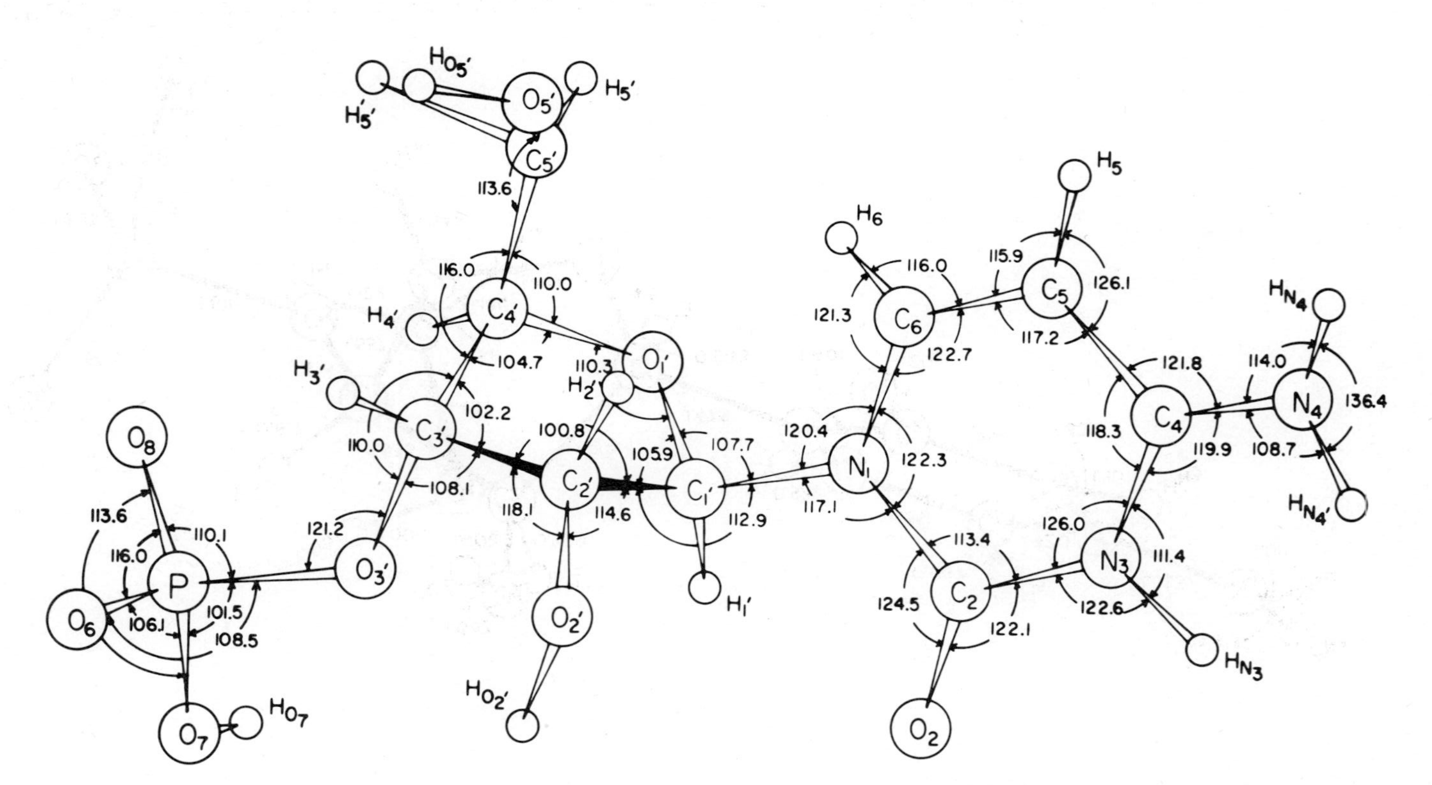

Fig. 58. Bond Angles in Cytidine 5'-Phosphate. [Reproduced, by permission of Academic Press, Inc., from M. Sundaralingam and L. H. Jensen, *J. Mol. Biol.*, 13, 914 (1965).]

TABLE XXIV

Some Torsion Angles for Cytidine 3'-Phosphate

Atoms involved	Torsion angle (degrees)
O-1'-C-1'-N-1-C-6	+41.8
C-4'-O-1'-C-1'-C-2'	-19.8
O-1'-C-1'-C-2'-C-3'	+36.6
C-1'-C-2'-C-3'-C-4'	-38.7
C-2'-C-3'-C-4'-O-1'	+28.1
C-3'-C-4'-O-1'-C-1'	- 5.6
O-3'-P-O-7-H-O-7	-75.3
H-O-5'-C-5'-C-4'	-108.2
O-5'-C-5'-C-4'-C-3'	+43.9
C-5'-C-4'-C-3'-O-3'	+152.1
C-4'-C-3'-O-3'-P	-91.5
C-3'-O-3'-P-O-7	+170.7

puckering of the D-ribose ring, the bond angles about C-1' and C-2' are different from those about C-3' and C-4', respectively. Similarly, the bond distances for C-2'-O-2' and C-3'-O-3' are also different. The bond distances and bond angles of the phosphate group are typical of those for a monophosphoric ester.

Bond distances and bond angles for the common nucleic acid bases, averaged over the more precisely determined structures (estimated standard deviation in bond length, less than 0.01 Å), are given in Figs. 59 through 62. As the precision of the several determinations is comparable, arithmetic (unweighted) mean values were found. In Fig. 59, the dimensions of neutral cytosine and *N*-3-protonated cytosine are compared. The protonated base shows some striking differences from the neutral base, the largest differences occurring in the bond angles C-4-N-3-C-2 and N-3-C-2-N-1. The former bond angle is 6.6° greater than that in protonated cytosine, and the latter angle is

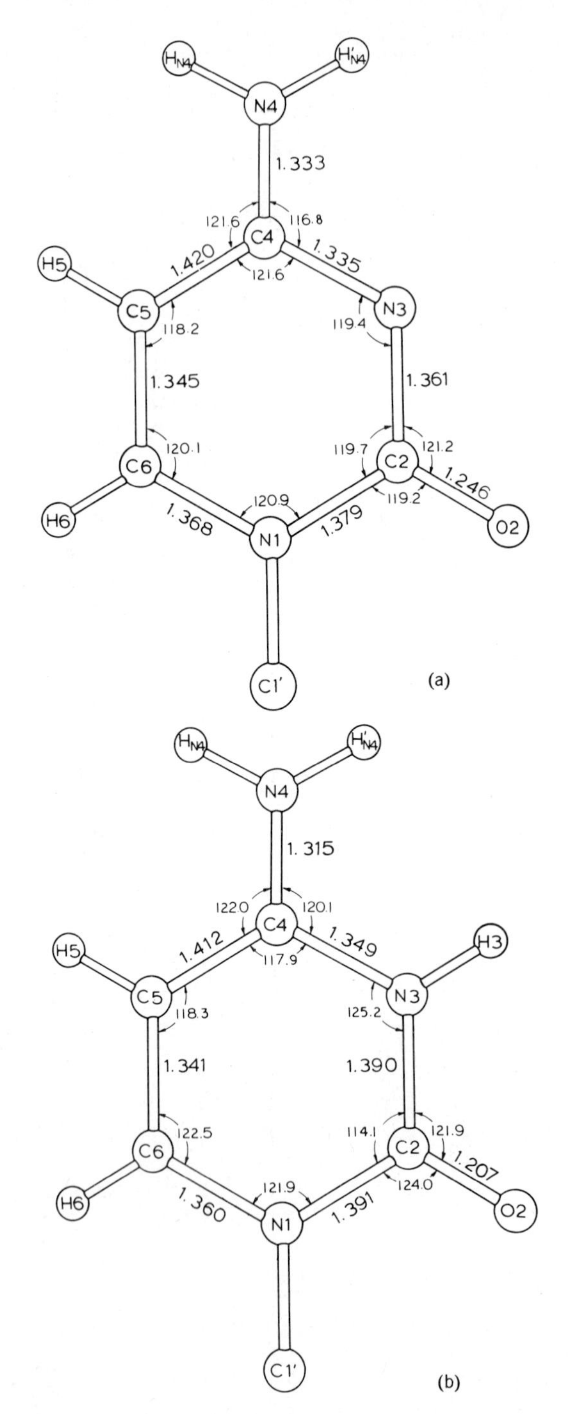

Fig. 59. Bond Lengths and Bond Angles in (a) Neutral and (b) Protonated Cytosine. [The values for the neutral base are from those for cytidine,[37] and those of the protonated base are averaged over those for cytidine 3'-phosphate (orthorhombic form[48]), cytidine 3'-phosphate (monoclinic form[49]), and 2'-deoxycytidine hydrochloride.[3]

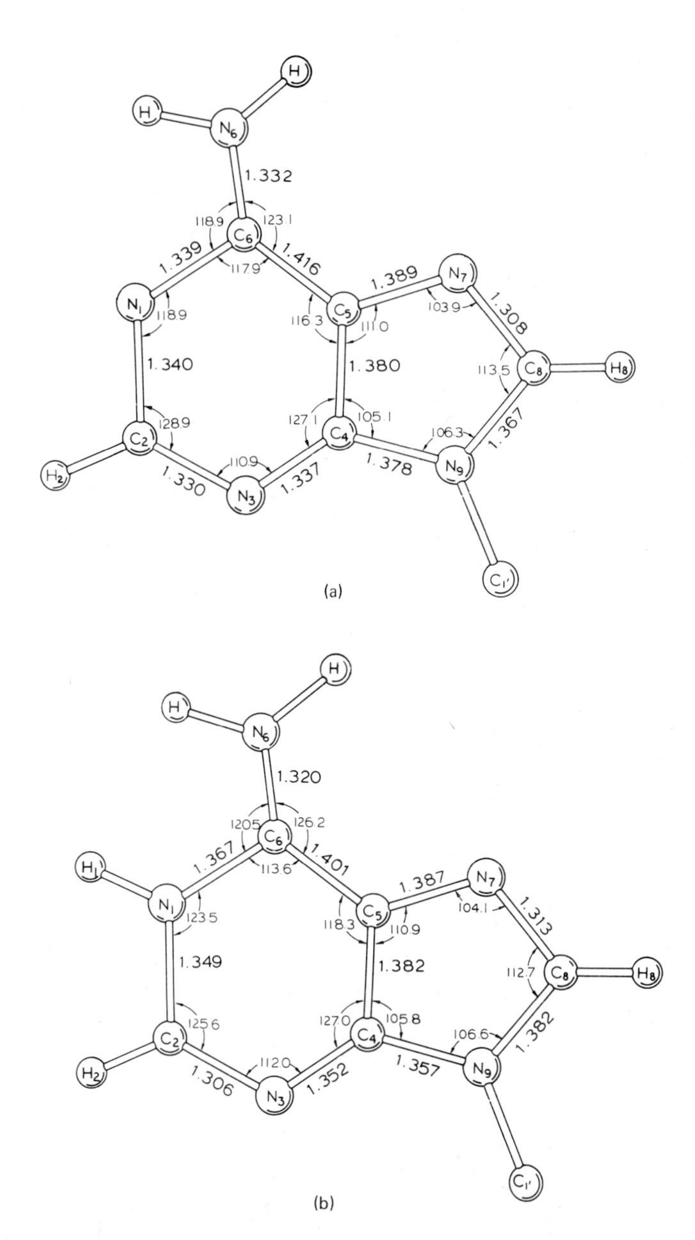

Fig. 60. Bond Lengths and Bond Angles in (a) Neutral and (b) Protonated Adenine. [The values for the neutral base are averaged over those for 2'-deoxyadenosine monohydrate,36 *N*-acetyladenosine,[139] and 2'-amino-2'-deoxyadenosine monohydrate,[139] and those of the protonated base are averaged over those for adenosine 3'-phosphate dihydrate[50] and adenosine 2'-phosphate-(2'→ 5')-uridine 5'-phosphate tetrahydrate.[140]]

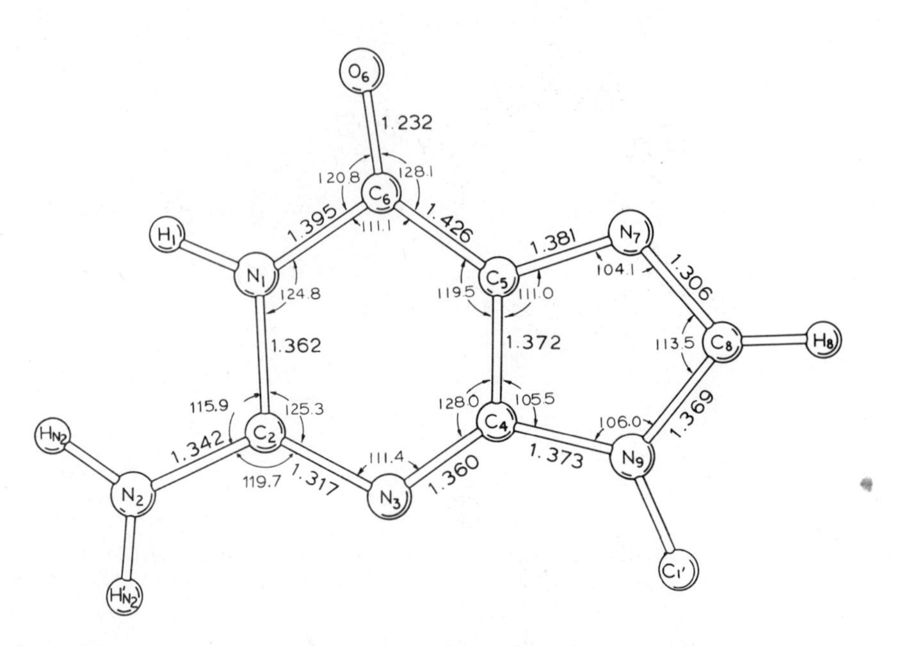

Fig. 61. Bond Lengths and Bond Angles in Guanine and Hypoxanthine, Averaged over Molecules 1 and 2 of Guanosine [141] and Molecules 1 and 2 of Inosine.[141]

6.3° smaller. Smaller differences, 2-3°, are seen for som other bond-angles and there are noticeable differences (∿2 pm; ∿0.02 Å) in the bond distances. The differences in the bond angles and bond distances that involve the non-hydrogen atoms are of importance in deciding whether N-3 is protonated, particularly when the hydrogen atom is not detectable by X-ray data.[137,138]

In neutral and protonated adenine (see Fig. 60), similar differences in the bond angles and bond distances are found. Both in the aminopyrimidine cytosine and the aminopurine adenine, the exocyclic bond angles involving C-6 are unequal; the C-5-C-4-N-4 and the C-5-C-6-N-6 angle are greater than those for N-3-C-4-N-4 and N-1-C-6-N-6, respectively. Interestingly, the differences in these bond angles become larger for protonated adenine, and smaller for protonated cytosine. Furthermore, the internal angle at C-4 in cytosine and at C-6 in adenine is considerably smaller in the protonated bases than in the corresponding neutral bases.

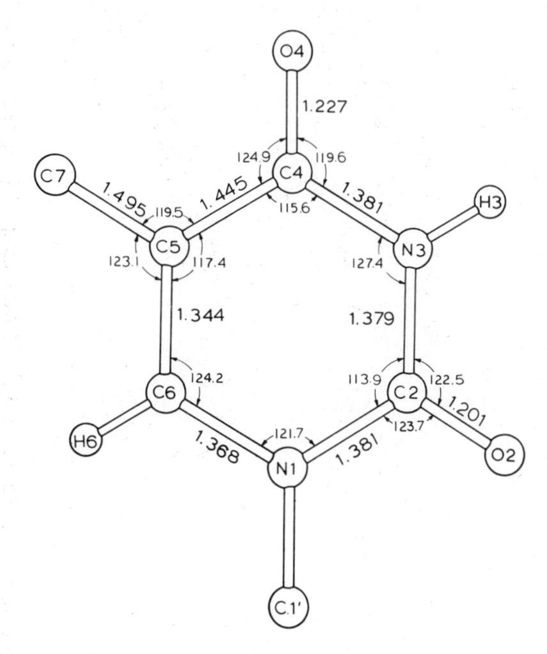

j. 62. Bond Lengths and Bond angles in Thymine, Aver-
ed over Thymidine[142] and 5-Methyluridine.[44]

Che bond distances and bond angles in the oxo bases,
anine and thymine, are given in Figs. 61 and 62, re-
ectively. The geometry of uracil is only slightly
fferent from that of thymine; in uracil, the bond
gle C-4-C-5-C-6 is ∿2° greater than that observed for
mine, whereas, in the 5-halogenated uracil (see Fig.
, this angle is ∿2° greater than that in uracil and
greater than that in thymine. The exocyclic angles in-
ving the halogen atom are unequal, and both the exocyclic
les are about 2° smaller than the exocyclic angles in-
ving C-5 of thymine.
etailed presentations of the geometry of the phosphate
up of sugar phosphates, and the possible conformations
nucleotides have been made elsewhere.[124,137,140–142,145]
n addition to the molecular geometry, knowledge regard-
the environs of the molecule may also be obtained
m the atomic coordinates. For cytidine 3'-phosphate,
molecular packing, viewed down the *c* axis, is shown
Fig. 64. Because the three parts of the nucleotide
se and sugar residues, and phosphate group) have
eral hydrogen-bonding sites, a complex scheme of

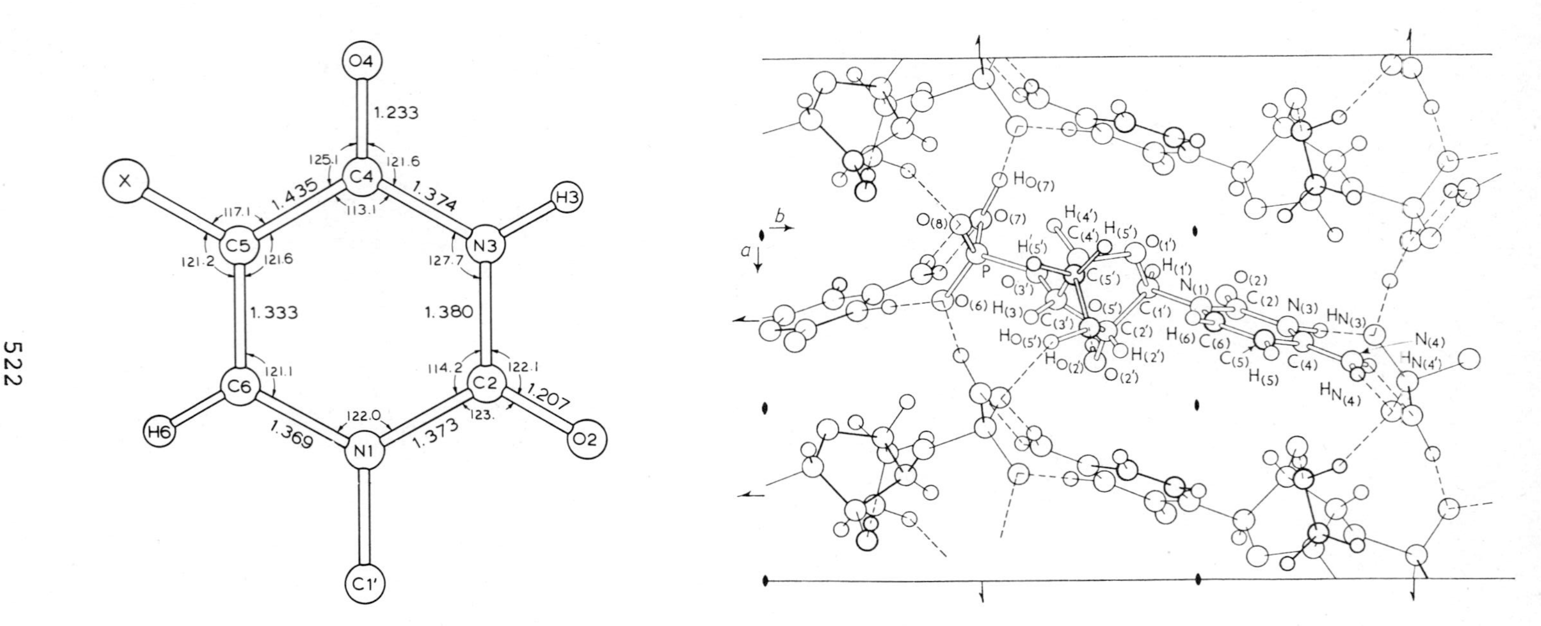

Fig. 63. Bond Lengths and Bond Angles in 5-Halogenated Uridine, Averaged over 2'-Deoxy-5'-fluorouridine[45] and 5-Chlorouridine.[143]

Fig. 64. Packing Diagram and Hydrogen-bonding Scheme in Cytidine 3'-Phosphate, as Viewed Down the *c*-Axis. [Reproduced, by permission of Academic Press, Inc., from M. Sundaralingam and L. H. Jensen, *J. Mol. Biol.*, 13, 914 (1965).]

hydrogen bonding is displayed in the lattice. All of the potential hydrogen-bond donor and acceptor sites are involved in the hydrogen bonding, except the carbonyl oxygen atom on C-2. There are three hydrogen bonds between the protonated cytosine residue and the phosphate group; their distances are N-3-H-N3...O-6 (278.3 pm); N-4-H-N4...O-8,273.4 pm; and N-4-H-N4...O-7,298.3 pm. The D-ribose residue is involved in hydrogen bonding with a neighboring D-ribose residue through O-2', O-2'-H-2'... O-5', 278.2 pm, and with the phosphate group through O-5', O-5'-H'...O-8, 277.1 pm. Thus, O-5' is involved in two hydrogen bonds. The phosphate group shows great versatility in its hydrogen bonding; there are six hydrogen-bonds involving the phosphate group: three to the base, two to the phosphate, and one to the D-ribose residue. The shortest hydrogen bond, of length 253.2 pm (2.532 Å), is between the phosphate groups, PO-7-H...O-8-P. Similar, short hydrogen-bonds are generally observed in crystal structures of phosphates.

The molecular vibration is described by the atomic, thermal parameters, and the thermal ellipsoids for cytidine 3'-phosphate are illustrated in Fig. 65. The major axes of the ellipsoids, which are a measure of the amplitude of thermal vibration in the three directions, are shown by the corresponding envelopes on the surface. The diagram is drawn for the ellipsoids that correspond to 50% probability; that is, they enclose 50% of the scattering material of the atoms. The diagram is shown as a stereoescopic pair, and the three-dimensional effect can be obtained if the diagram is viewed with a pair of stereoscopic glasses. The plot was produced on a Calcomp plotter by use of the program "ORTEP" developed by Johnson.[146]

XI. SUMMARY

The steps involved in a crystal-structure analysis are represented in the flow diagram shown in Fig. 66. When an adequate single crystal has been obtained, the steps leading to intensity-data collection are straightforward. With an automatic diffractometer, it is possible to collect 1,000-3,000 intensities (Tables X-XIII) in 3-7 days. The intensity data are processed to yield a set of structure-amplitudes, used in the structure analysis. The solution of the phase problem may require a few hours to several months, depending on the nature and complexity of the problem. A successful solution of the phase problem yields the approximate atomic coordinates; these are subjected to least-squares refinement to obtain the final

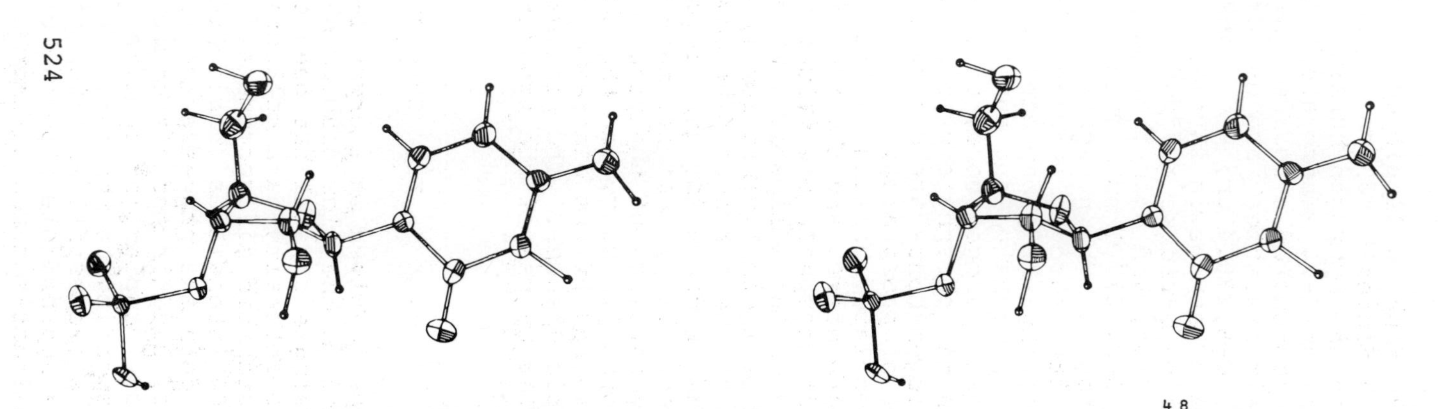

Fig. 65. Stereoscopic Plot of a Molecule of Cytidine 3'-Phosphate[48] (Orthorhombic), with the Thermal Ellipsoids Corresponding to 50% Probability. [Produced with the program ORTEP of Johnson.[146]]

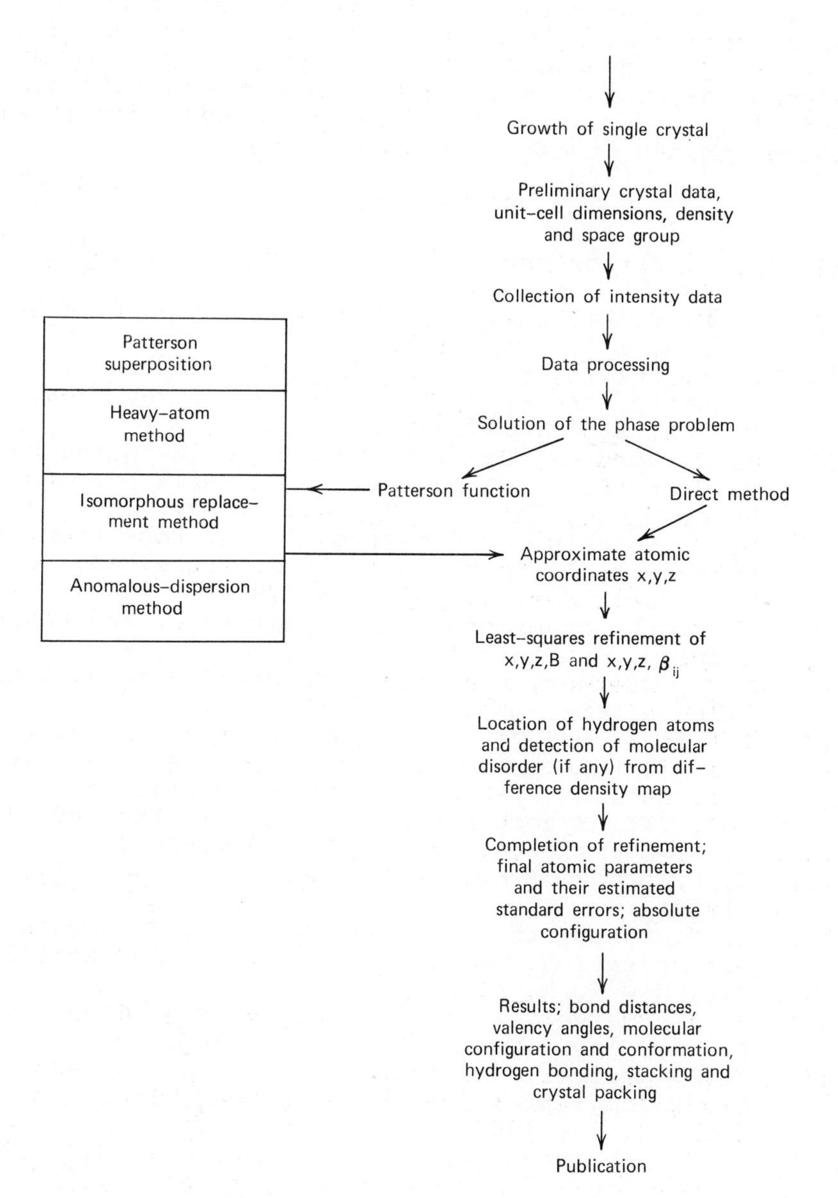

Fig. 66. Flow Diagram for a Single Crystal-structure Analysis.

atomic parameters that define the molecular structure and vibration. Access to unlimited time on a high-speed, electronic computer may yield a refined structure in less than a week, barring complications caused by molecular disorder. In favorable cases, a crystal structure may be analyzed to completion in less than a month, although, amusingly, the preparation of the manuscript for journal publication has often taken much longer.

REFERENCES

(1) J. D. Watson and F. H. C. Crick, *Nature*, 171, 964 (1953).

(2) M. H. F. Wilkins, A. R. Stokes, and H. R. Wilson, *Nature*, 171, 738 (1953).

(3) A. F. Cullis, H. Muirhead, M. F. Perutz, M. G. Rossmann, and A. C. T. North, *Proc. Roy. Soc., Ser. A*, 265, 161 (1962).

(4) J. C. Kendrew, R. E. Dickerson, B. E. Strandberg, R. G. Hart, D. R. Davies, D. C. Phillips, and V. C. Shore, *Nature*, 185, 422 (1960).

(5) D. C. Hodgkin, J. Pickworth, J. H. Robertson, R. J Prosen, R. A. Sparks, and K. N. Trueblood, *Proc. Roy. Soc., Ser. A*, 251, 306 (1959).

(6) P. Groth, *Chemische Krystallographie*, Englemann, Leipzig, Germany, 1906.

(7) M. J. Buerger, *Elementary Crystallography*, John Wiley & Sons, Inc., New York, N.Y., 1963.

(8) *International Tables for X-Ray Crystallography*, Vol. I, Kynoch Press, Birmingham, England, 1965.

(9) *International Tables for X-Ray Crystallography*, Vol. III, Kynoch Press, Birmingham, England, 1962.

(10) R. A. Ross, *Phys. Rev.*, 28, 425 (1926).

(11) W. L. Bragg, *Proc. Cambridge. Phil. Soc.*, 17, 43 (1913).

(12) G. H. Stout and L. H. Jensen, *X-Ray Structure Determination. A Practical Guide*, The Macmillan Co., New York, N. Y., 1968.

(13) M. J. Buerger, *X-Ray Crystallography*, John Wiley & Sons, Inc., New York, N. Y., 1942.

(14) M. J. Buerger, *The Precession Method in X-Ray Crystallography*, John Wiley & Sons, Inc., New York, N. Y., 1964.

(15) J. M. Broomhead, *Acta Crystallogr.*, 1, 324 (1948).

(16) W. Cochran, *Acta Crystallogr.*, 4, 81 (1951).

(17) R. F. Stewart and L. H. Jensen, *J. Chem. Phys.*, 40, 2071 (1964).

(18) R. F. Bryan and K. Tomita, *Acta Crystallogr.*, 15, 1179 (1962).

(19) D. L. Barker and R. E. Marsh, *Acta Crystallogr.*, 17, 1581 (1964).
(20) G. A. Jeffrey and Y. Kinoshita, *Acta Crystallogr.*, 16, 20 (1963).
(21) F. S. Mathews and A. Rich, *Nature*, 201, 179 (1964).
(22) R. E. Marsh, R. Bierstadt, and E. L. Eichhorn, *Acta Crystallogr.*, 15, 310 (1962).
(23) C. E. Bugg, U. T. Thewalt, and R. E. Marsh, *Biochem. Biophys. Res. Commun.*, 33, 436 (1968).
(24) J. M. Broomhead, *Acta Crystallogr.*, 4, 92 (1951).
(25) J. Iball and H. R. Wilson, *Nature*, 198, 1193 (1963).
(26) J. Iball and H. R. Wilson, *Proc. Roy. Soc., Ser. A*, 288, 418 (1965).
(27) H. M. Sobell and K. Tomita, *Acta Crystallogr.*, 17, 126 (1964).
(28) R. Gerdil, *Acta Crystallogr.*, 14, 333 (1961).
(29) K. Hoogsteen, *Acta Crystallogr.*, 16, 28 (1963).
(30) G. S. Parry, *Acta Crystallogr.*, 7, 313 (1954).
(31) R. F. Stewart and L. H. Jensen, *Acta Crystallogr.*, 23, 1102 (1967).
(32) D. W. Green, F. S. Mathews, and A. Rich, *J. Biol. Chem.*, 237, 3573 (1962).
(33) H. M. Sobell and K. Tomita, *Acta Crystallogr.*, 17, 122 (1964).
(34) G. N. Reeke, Jr., and R. E. Marsh, *Acta Crystallogr.*, 20, 703 (1966).
(35) B. M. Craven, *Acta Crystallogr.*, 23, 376 (1967).
(36) D. G. Watson, D. J. Sutor, and P. Tollin, *Acta Crystallogr.*, 19, 111 (1965).
(37) S. Furberg, C. S. Petersen, and C. Romming, *Acta Crystallogr.*, 18, 313 (1965).
(38) E. Subramanian and D. J. Hunt, *Acta Crystallogr.*, B26, 303 (1970).
(39) P. Tollin, H. R. Wilson, and D. W. Young, *Nature*, 217, 1148 (1968).
(40) M. Huber, *Acta Crystallogr.*, 10, 129 (1957).
(41) M. Woolfson, *Acta Crystallogr.*, 9, 974 (1956).
(42) J. Iball, C. H. Morgan, and H. R. Wilson, *Proc. Roy. Soc., Ser. A.*, 295, 320 (1966).
(43) J. Iball, C. H. Morgan, and H. R. Wilson, *Proc. Roy. Soc., Ser. A.*, 302, 225 (1968).
(44) D. J. Hunt and E. Subramanian, *Acta Crystallogr.*, B25, 2144 (1969).
(45) D. R. Harris and W. M. MacIntyre, *Biophys. J.*, 4, 203 (1964).
(46) N. Camerman and J. Trotter, *Acta Crystallogr.*, 18, 203 (1965).
(47) E. Alver and S. Furberg, *Acta Chem. Scand.*, 13, 910 (1959).

(48) M. Sundaralingam and L. H. Jensen, *J. Mol. Biol.*, 13, 914 (1965).
(49) C.E. Bugg and R. E. Marsh, *J. Mol. Biol.*, 25, 67 (1967).
(50) M. Sundaralingam, *Acta Crystallogr.*, 21, 495 (1966).
(51) J. Kraut and L. H. Jensen, *Nature*, 186, 798 (1960).
(52) J. Kraut and L. H. Jensen, *Acta Crystallogr.*, 16, 79 (1963).
(53) P. Horn, V. Luzzati, and K. N. Trueblood, *Nature*, 183, 880 (1959).
(54) K. N. Trueblood, P. Horn, and V. Luzzati, *Acta Crystallogr.*, 14, 965 (1961).
(55) E. Shefter and K. N. Trueblood, *Acta Crystallogr.*, 18, 1067 (1965).
(56) S. T. Rao and M. Sundaralingam, *Chem. Commun.*, 1968, 995.
(57) S. T. Rao and M. Sundaralingam, *J. Amer. Chem. Soc.*, 91, 1210 (1969).
(58) N. Nagashima and Y. Iitaka, *Acta Crystallogr.*, B24, 1136 (1968).
(59) E. Shefter, M. Barlow, R. Sparks, and K. N. Trueblood, *J. Amer. Chem. Soc.*, 86, 1872 (1964).
(60) N. Cameraman and J. Trotter, *Acta Crystallogr.*, 19, 867 (1965).
(61) K. Hoogsteen, *Acta Crystallogr.*, 12, 822 (1959).
(62) K. Hoogsteen, *Acta Crystallogr.*, 16, 907 (1963).
(63) F. S. Mathews and A. Rich, *J. Mol. Biol.*, 8, 89 (1964).
(63a) L. Katz, K. Tomita, and A. Rich, *J. Mol. Biol.*, 13, 340 (1965).
(64) R. F. Bryan and K. Tomita, *Nature*, 192, 812 (1961).
(65) R. F. Bryan and K. Tomita, *Acta Crystallogr.*, 15, 1174 (1962).
(66) E. J. O'Brien, *J. Mol. Biol.*, 7, 107 (1963).
(67) E. J. O'Brien, *Acta Crystallogr.*, 23, 92 (1967).
(68) H. M. Sobell, K. Tomita, and A. Rich, *Proc. Nat. Acad. Sci. U. S.*, 49, 885 (1963).
(69) S. H. Kim and A. Rich, *Science*, 158, 1046 (1967).
(70) L. L. Labana and H. M. Sobell, *Proc. Nat. Acad. Sci. U. S.*, 57, 460 (1967).
(71) H. M. Sobell, *J. Mol. Biol.*, 18, 7 (1966).
(71a) D. Voet and A. Rich, *Abstracts Meeting Amer. Crystallogr. Assoc.*, Tucson, Arizona, Feb. 1968, No. F1.
(72) Yu. G. Baklagina, M. V. Vel'kenshtein, and Yu. D. Krondraskev, *Zh. Strukt. Khim.*, 7, 399 (1966).
(73) K. Tomita, L. Katz, and A. Rich, *J. Mol. Biol.*, 30, 545 (1967).

(74) A. E. V. Haschemeyer and H. M. Sobell, *Nature*, 202, 969 (1964).
(75) A. E. V. Haschemeyer and H. M. Sobell, *Acta Crystallogr.*, 19, 125 (1965).
(76) A. E. V. Haschemeyer and H. M. Sobell, *Proc. Nat. Acad. Sci. U. S.*, 50, 872 (1963).
(77) A. E. V. Haschemeyer and H. M. Sobell, *Acta Crystallogr.*, 18, 525 (1965).
(78) J. M. Robertson, *J. Sci. Instrum.*, 20, 175 (1943).
(79) T. C. Furnas, *Single Crystal Orienter Instruction Manual*, General Electric Company, Milwaukee, Wisconsin, 1957.
(79a) J. de Meulenaer and H. Tompa, *Acta Crystallogr.*, 19, 1014 (1965).
(80) D. C. Phillips, *Acta Crystallogr.*, 7, 746 (1954).
(81) D. C. Phillips, *Acta Crystallogr.*, 9, 819 (1956).
(82) W. C. Hamilton, J. S. Rollett, and R. A. Sparks, *Acta Crystallogr.*, 18, 129 (1965).
(83) A. J. C. Wilson, *Nature*, 150, 152 (1943).
(84) A. J. C. Wilson, *Acta Crystallogr.*, 2, 318 (1949).
(85) F. Hanic, *Acta Crystallogr.*, 21, 332 (1966).
(86) E. R. Howells, D. C. Phillips, and D. Rogers, *Acta Crystallogr.*, 3, 210 (1950).
(87) A. L. Patterson, *Z. Kristallogr.*, A90, 517 (1935).
(88) D. M. Wrinch, *Phil. Mag.*, 27, 98 (1939).
(89) M. J. Buerger, *Vector Space and Its Applications in Crystal Structure Investigation*, John Wiley & Sons, Inc., New York, N. Y., 1959.
(90) J. A. Carrabine and M. Sundaralingam, *J. Amer. Chem. Soc.*, 92, 369 (1970).
(91) D. Harker, *J. Chem. Phys.*, 40, 381 (1936).
(92) J. A. Carrabine and M. Sundaralingam, presented at the Eighth Congress of the International Union of Crystallography, Buffalo, N. Y., Aug. 1969.
(93) P. Tollin and W. Cochran, *Acta Crystallogr.*, 17, 1322 (1964).
(94) H. Lipson and W. Cochran, *The Determination of Crystal Structures*, G. Bell and Sons, Ltd,. London, England, 1957.
(95) C. E. Nordman and K. Nakatsu, *J. Amer. Chem. Soc.*, 85, 353 (1963).
(96) C. E. Nordmann and S. K. Kumara, *J. Amer. Chem. Soc.*, 87, 2059 (1965).
(97) R. A. Jacobson, J. A. Wunderlich, and W. N. Lipscomb, *Acta Crystallogr.*, 14, 598 (1961).
(98) J. M. Robertson and I. Woodward, *J. Chem. Soc.*, 1940, 36.
(99) D. Dale, D. C. Hodgkin, and K. Venkatesan, in G. N. Ramachandran (Ed.), *Crystallography and Crystal Perfection*, Academic Press, Inc., New York, N. Y. , 1963, pp. 237-242.

(100) G. A. Sim, *Acta Crystallogr.*, 10, 177 (1957).
(101) G. A. Sim, *Acta Crystallogr.*, 10, 536 (1957).
(102) S. Parthasarathy, *Acta Crystallogr.*, 18, 1022 (1965).
(103) S. Parthasarathy, *Acta Crystallogr.*, 18, 1028 (1965).
(104) G. A. Sim, in R. Pepinsky, J. M. Robertson and R. A. Speakman (Eds.), *Computing Methods and Phase Problem in X-Ray Crystal Analysis*, Pergamon Press, New York, N. Y., 1961, pp. 227-235.
(105) D. Harker, *Acta Crystallogr.*, 9, 1 (1956).
(106) M. F. Perutz, *Acta Crystallogr.*, 9, 867 (1956).
(107) R. F. Dickerson, in H. Neurath (Ed.), *The Proteins*, Vol. 2, Academic Press, Inc., New York, N. Y., 1964, pp. 603–658.
(107a) J. M. Cork, *Phil. Mag.*, 4, 688 (1927).
(108) A. F. Peerdeman and J. M. Bijvoet, *Acta Crystallogr.*, 9, 1012 (1956).
(109) G. N. Ramachandran and S. Raman, *Curr. Sci.*, 25, 348 (1956).
(110) S. Ramaseshan, in G. N. Ramachandran (Ed.), *Advanced Methods of Crystallography*, Academic Press, Inc., New York, N. Y., 1963, pp. 67–95.
(111) J. R. Herriott, L. C. Sieker, L. H. Jensen, and W. Lovenberg, *J. Mol. Biol.*, 50, 391 (1970).
(112) Y. Okaya and R. Pepinsky, in Ref. 104, pp. 273–299.
(113) I. L. Karle and J. Karle, *Acta Crystallogr.*, 21, 849 (1966).
(114) D. Harker and J. S. Kasper, *Acta Crystallogr.*, 1, 70 (1948).
(115) D. Sayre, *Acta Crystallogr.*, 5, 60 (1952).
(116) H. Hauptman and J. Karle, *The Solution of the Phase Problem. I. The Centrosymmetric Crystal*, ACA Monograph No. 3, 1953.
(117) W. Cochran and M. M. Woolfson, *Acta Crystallogr.*, 8, 1 (1955).
(118) D. Rohrer and M. Sundaralingam, *Chem. Commun.*, 1968, 746.
(119) I. L. Karle and J. Karle, *Acta Crystallogr.*, 16, 969 (1963).
(120) I. L. Karle, H. Hauptman, J. Karle, and A. B. Wing *Acta Crystallogr.*, 11, 257 (1958).
(121) J. Karle and H. Hauptman, *Acta Crystallogr.*, 9, 635 (1956).
(122) I. L. Karle and J. Karle, *Acta Crystallogr.*, 17, 835 (1964).

(123) S. T. Rao, M. Sundaralingam, S. K. Arora, and S. R. Hall, *Biochem. Biophys. Res. Commun.*, 38, 496 (1970).
(124) S. T. Rao and M. Sundaralingam, *J. Amer. Chem. Soc.*, 92, 4963 (1970).
(125) J. Karle, J. A. Estlin, and I. L. Karle, *J. Amer. Chem. Soc.*, 89, 6510 (1967).
(126) J. Karle, J. Flippen, and I. L. Karle, *Z. Kristallogr.*, 125, 201 (1967).
(127) W. Cochran, *Acta Crystallogr.*, 11, 579 (1958).
(128) C. L. Coulter, *J. Mol. Biol.*, 12, 292 (1965).
(128a) J. E. Weinzierl, D. Eisenberg, and R. E. Dickerson, *Acta Crystallogr.*, B25, 380 (1969).
(129) E. W. Hughes, *J. Amer. Chem. Soc.*, 63, 1737 (1941).
(130) D. W. J. Cruickshank, in Ref. 104, pp. 32-78.
(131) H. T. Evans, Jr., *Acta Crystallogr.*, 14, 689 (1961).
(132) W. Cochran, *Acta Crystallogr.*, 4, 408 (1951).
(133) S. Furberg and L. H. Jensen, *J. Amer. Chem. Soc.*, 90, 470 (1968).
(134) J. M. Bijvoet, A. F. Peerdeman, and A. J. van Bommel, *Nature*, 168, 271 (1951); J. M. Bijvoet, *Proc. Kon. Ned. Akad. Wetensch. Amsterdam*, 52, 313 (1949); A. F. Peerdeman, A. J. van Bommel, and J. M. Bijvoet, *Kon. Ned. Akad. Wetensch. Proc., Ser. B*, 54, 16 (1951).
(135) W. C. Hamilton, *Acta Crystallogr.*, 18, 502 (1965).
(136) H. Hope and U. de la Camp, *Abstracts Winter Meeting Amer. Crystallogr. Assoc.* , Seattle, Washington, March 1969, No. H5.
(137) M. Sundaralingam and L. H. Jensen, *J. Mol. Biol.*, 13, 930 (1965).
(138) C. Singh, *Acta Crystallogr.*, 19. 861 (1965).
(139) D. C. Rohrer and M. Sundaralingam, *J. Amer. Chem. Soc.*, 92, 4956 (1970).
(140) E. Shefter, M. Barlow, R. A. Sparks, and K. N. Trueblood, *Acta Crystallogr.*, B25, 895 (1969).
(141) U. T. Thewalt, C. E. Bugg, and R. E. Marsh, *Acta Crystallogr.*, B26, 1089 (1970).
(142) D. W. Young, P. Tollin, and H. R. Wilson, *Acta Crystallogr.*, B25, 1423 (1969).
(143) S. W. Hawkinson and C. L. Coulter, *Acta Crystallogr.*, B27, 34 (1971).
(144) M. Sundaralingam, *J. Amer. Chem. Soc.*, 87, 599 (1965).
(145) M. Sundaralingam, *Biopolymers*, 7, 821 (1969).
(146) C. K. Johnson, *A Fortran Thermal Ellipsoid Plot Program for Crystal Structure Illustrations*, Oak Ridge National Laboratory, Oak Ridge, Tennessee, Report ORNL-3794, 1965.

CHAPTER 9

Chromatography of Nucleic Acid Components

S. ZADRAŽIL*

INSTITUTE OF ORGANIC CHEMISTRY AND BIOCHEMISTRY, CZECHOSLOVAK ACADEMY OF SCIENCES, PRAGUE, CZECHOSLOVAKIA

---

*The author is indebted to Drs. A. Holý, M. Prystaš, and S. Chládek, Institute of Organic Chemistry and Biochemistry, Czechoslovak Academy of Sciences, for submitting a number of the practical chromatographic processes presented, and for their discussions and able collaboration.

I. INTRODUCTION

Progress in any scientific field must be preceded by the achievement of a methodological level that secures the means for practically carrying out the scientist's aims in the laboratory. For organic chemists, such a basic prerequisite is a sufficiently wide range of separation and analytical methods that will permit his laboratory work to be assessed accurately. Nowadays, chromatographic techniques have, in fact, become a separate field of analytical chemistry, and they belong to the basic experimental techniques in any organic-synthesis laboratory.

In the laboratory of an organic chemist working in the field of nucleic acid components, nearly all of the basic chromatographic techniques are extensively used — partition, adsorption, ion-exchange, and gel chromatography (in the latter, compounds are separated according to their size and shape). In the classification according to technical arrangement and the type of carrier used, all of the well known chromatographic methods are again represented, including paper, column, and thin-layer chromatography. Because the following descriptions will be subject to the main principle that this handbook shall have the widest practical applicability in the laboratory, we shall use the latter classification of techniques, as it is the most illustrative from the experimental point of view. For the same reason, no details will be given for theories, mechanisms, or descriptions of elementary findings relating to such matters as carriers and instrumentation, unless they are essential with respect to practical experiments. Nonetheless, we believe that it will be useful to include a brief general section before the detailed description of each individual technique, because of the scattered character of the problems and materials discussed.

II. PAPER CHROMATOGRAPHY

1. Introduction

Probably the oldest and most widely used chromatographic technique employed for preparative, separative, and purification purposes, and for analysis and characterization of compounds, as well as in studying the course of chemical reactions and chemical kinetics, is paper chromatography. Although it was discovered more than 100 years ago, the actual start of its development dates back only to 1944. Separation of nucleic acid components by means of this technique was first mentioned[1] as early as 1947; the method has since achieved such wide popularity that it is now used in all organic and biochemical laboratories as a qualitative technique and, combined with physicochemical techniques, as an important quantitative method. It is similarly important for preparative separations.

Most authors believe that the main advantages of this technique are, primarily, its simplicity and low demands in terms of materials and instrumentation, and its high degree of sensitivity and good capacity for separation, compared, for example, to adsorption chromatography of homologs. For these reasons, paper chromatography is being used increasingly in synthetic organic chemistry, also. Compared to thin-layer chromatography, the resolution achieved by paper chromatography is slightly inferior; thin-layer chromatography appears to be quicker (which considerably increases its popularity), but it is, again, more exacting in terms of equipment and preparation of the material used for the separation.

In terms of the original functional classification, paper chromatography is almost exclusively a partition technique when ordinary cellulose paper is used, or the ion-exchange properties of the carrier may be employed with substituted celluloses. The types of paper most often used are summarized in Table I.

2. Preparation of Paper and Application of Sample

In most cases, the paper does not require modification or preparation before the sample is applied, a fact that speeds up the entire process considerably. For quantitative purposes only, some authors recommend washing the paper with 100 m*M* hydrochloric acid to remove any possible impurities that might absorb u.v. light. After it has been washed (on a special sieve with facilities for removing water by suction from a whole sheaf of papers, or by developing the paper in the same way as in the actual separation), the paper must be thoroughly washed with distilled water until the washings are neutral, and dried.

The sample may be applied by means of a conventional pipet or micropipet, the volume of which depends only on

TABLE 1

Properties of Some Important Chromatographic Papers[a]

Paper	Thick-ness (mm)	Capac-ity (μequiv/cm^2)	Weight per area (g/m2)	Characteristics
Whatman				
No. 1	0.16		85-95	standard paper
No. 2	0.18		95-100	standard paper
No. 3	0.36		185	preparative paper
No. 4	0.19		90-95	fast
No. 3MM	0.31		180	preparative paper
No. 540	0.15		85-90	washed, hardened
DE-81	0.17		85	(2-diethylaminoethyl
CM-82	0.17	2-6	85	(carboxymethyl)-
P-81	0.17	18	85	phosphorylated
Schleicher-Schüll				
No. 2040a	0.18-0.19		85-90	fast
No. 2043a	0.18-0.19		90-95	standard paper
No. 2043 (MG1)	0.22-0.24		120-125	standard paper
No. 2071	0.65-0.70		600-700	preparative paper

[a]According to Ref. 10 and the Whatman information leaflets

the capacity (thickness) of the paper and the spot-shape selected (a point or a line). For the most part, volumes in the order of 10 μℓ are applied (the optimum being 10 μℓ), and the concentration should be such as to provide good visibility on detection (in u.v. light, 10-15 μg of one component of the sample is the optimum); the upper limit of compound is decided by the fact that large spots are usually difficult to separate and are highly irregular. When it is required to detect a compound the concentration of which is slight compared to that of other components of the mixture (*e.g.*, when the quantitative course of a reaction is being monitored), it is recommended that several amounts of sample be applied in parallel, in order to avoid interference by the other components. The process of sample application may be speeded up, especially when large sample-volumes or a preparative technique are involved, by gently heating the origin with a stream of warm air (obtained either through a slit below the paper, when a special application table is used, or by use of a fan heater). The thermolabile character of compounds need be considered in exceptional cases only.[2]

In most cases, no special modification of the sample is required before application, owing to the very high sensitivity of detection in u.v. light (small amounts of reaction mixtures of final samples are applied). Where there is present an excessive amount of salts which would interfere with the regularity of separation and the R_F values, desalting may first be performed. With bases and nucleosides, gel filtration[3,4] is best, especially with Sephadex* G-10 and Bio-Gel** P-2, characterized by strong adsorption of heterocyclic compounds, mainly purines, as compared to low molecular-weight inorganic or organic compounds. With nucleotides, ion-exchange resins[5] or substituted cellulose[6] and activated carbon are preferable. The last-mentioned adsorbent may either be added directly to the sample, or may be used as the filling in small columns.[7,8] Examples of desalting by gel filtration and by ion-exchange resins are described in detail in the Sections dealing with these chromatographic techniques (see pp. 626 and 609, respectively). The process of desalting nucleotide samples by adsorption on carbon in the way in which it is done as a routine procedure in the author's laboratory will next be briefly described.

Activated carbon (Norit or Darco G-60, although other types may also be used) is successively washed with 100 m*M* hydrochloric acid, water, 100 m*M* ammonium hydroxide, water, 60% ethyl alcohol, water, 100 m*M* hydrochloric acid, and water. The final solid is dried at 100° overnight, and, thus prepared, may be used for adsorption. In adsorption from solution, as well as on a column, the nucleotide solution is adjusted to be weakly acid (pH 4-5). The amount of carbon employed is usually in the proportion of 1 g per 4 mg of nucleotide phosphorus (about 40-50 mg of the nucleotide). When adsorption and the other washing and elution operations are conducted in a centrifuge, it is advantageous to cover the liquid in the centrifuge tube with a thin layer ot ethyl alcohol, which keeps the carbon particles from being suspended on the surface, and thus eliminates loss during

---

*Pharmacia AB, Uppsala, Sweden.
**Calbiochem Inc., Los Angeles, California.

decanting operations. To remove salts, washing the charcoal layer of about 10 g with two 50-ml portions of distilled water will generally suffice. Elution of nucleotides is usually performed with a dilute solution of ammonia in ethyl alcohol (0.5% of ammonia in 50% ethyl alcohol). For the above-mentioned amount of nucleotide, about 300 ml of eluant (*i.e.*, for centrifugation, 3 x 100 ml) is suitable. The eluate is evaporated to dryness in a rotary evaporator under diminished pressure at 25-30°. After being dissolved, this material may be applied to paper or to an ion-exchange column for further separation. This treatment affords good results, and usually has no substantial effect either on the qualitative or quantitative composition of the components in the sample.

The distance of the origin from the edge of the paper must be chosen in accordance with the mode of development to be used. In the descending method of operation, this distance should be 7-10 cm, depending on the size of the developing trough and the size of the paper, whereas, with ascending development, it is correspondingly less (4-6 cm). For two-dimensional development, equal distances should be kept from both edges, and the form of the spot to be used is obviously a dot.

3. Development

Development is an important part of the chromatographic process and, besides the modes already discussed, repeated development for substances having low, or similar, R_F values should also be mentioned. With respect to R_F values, the most important influence in development is saturation of the chamber with all of the components of the developing mixture (usually achieved by placing the mixture on the chamber bottom or by means of paper strips, wet with the solvent, affixed to the chamber walls). The temperature is not very important, provided that a standard is chromatographed simultaneously with the sample. The use of standards is recommended for every case of separation that involves identification. The method most often used is descending development; it is also used as the standard method in many laboratories. Individual modifications have no fundamental effect on the quality of the separation process. The differences between R_F values obtained by the descending and ascending methods are very slight (usually negligible), provided that the same conditions are maintained. The descending method is more advantageous in the overflow-development technique and in repeated separations.[9]

With two-dimensional chromatography, the important method of development with the same solvent system in both directions should be mentioned.[10] Fig. 1 shows six cases from which it is possible to decide whether two separate

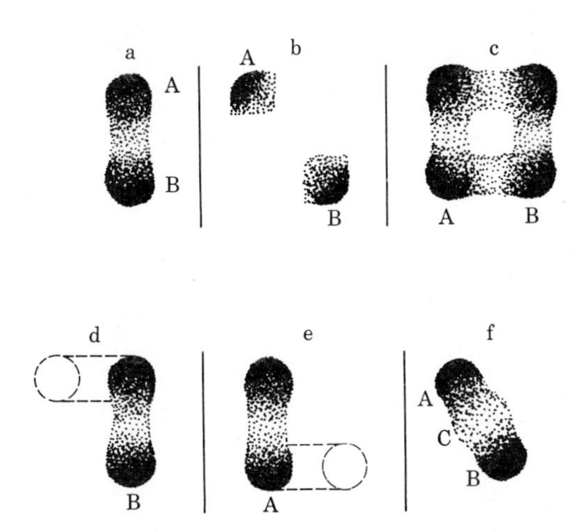

Fig. 1. Diagrammatic Illustration of the Behavior of a Mixture of Two Compounds in Two-dimensional Chromatography with the Same Solvent System.[10] [(a) One-dimensional development; (b) two separate compounds A and B; (c) the reaction A ⇌ B takes place in the course of development; (d) the spot, after the first development, is an artifact, or the reaction A ⟶ B takes place before or during the chromatographic process; (e) the same for the reaction B ⟶ A; and (f) the additional spot C is located between spots A and B. Starting points are omitted; directions of development are downwards, and are from left to right.]

spots of two different compounds or only two forms of the same compound are involved, or whether the compound changes during the chromatographic process.

4. R_F Values[10a]

The R_F value depends on many factors, and so the data usually given are more or less relative, even though R_F values are functions of the partition coefficient and, therefore, are physical constants for specific compounds and specific separation systems. Besides the R_F value, which expresses the ratio of the distance of the spot and of the solvent front from the origin, R_X values are often used, especially in connection with organic synthesis. The R_X value defines the ratio between the distance travelled by a given compound and by a reference compound for which the distance is considered to be equal to unity.

R_X values are more useful than R_F values when systems are compared in terms of suitability for separating a specified mixture, and they are the only possible way of characterizing the position of compounds in repeated development and in the overflow technique, or when the solvent system forms two fronts. Because the use of standard and reference compounds for analytical assessment of the mixture separated is here advocated, a wider use of R_X values is also recommended.

A sufficient number of aids now exist that serve in the determination of R_F values,[10] and there is no need to dwell upon this subject. A suitable ruler for accurate measurement of the distance from the origin of the compound separated and of the solvent front should be available in the laboratory. It should be stressed that distances should be measured to the center of spots. The possibility should be kept in mind that double spots may occur in preparative, as well as in quantitative, application of a whole sample after washing out the vessel (see p. 553).

The most important effects on R_F values are related to saturation of the atmosphere in the developing chamber, the temperature, the content of salts in the sample, the source of the paper, and the purity of the solvent. Highly reproducible R_F values may, therefore, be achieved under the following conditions[11]; (*a*) establishment of equilibrium in the separation mixture for 72 hr at the development temperature and complete saturation of the atmosphere in the chamber, (*b*) use of a constant temperature (within a range of 0.5°), (*c*) use of a reference compound, (*d*) use of a minimum front-to-origin distance of 25 cm (this requirement is reversed in comparison with thin-layer chromatography), and (*e*) adherence to standard operating procedures in application, development, drying, and detection.

5. Selection of the Solvent System

The simplest way of trying to select a solvent for separating a specified mixture of compounds is to study literature data on solvents used for separating similar mixtures. Tradition and preferred systems known from preceding work play a role in this selection. To date over 100 different solvent systems have been published as being suitable for nucleic acid components, including a number of mixtures that differ only slightly in composition; the difference between their separating capacity for the same mixture of compounds is usually small and inconsequential.

In order to select a suitable solvent system, the properties of the compounds to be separated and of the solvents that may be considered as potential components of the system must first be evaluated. The basis for assessing the solvents may be a mixotropic series (see p. 580), defining the solvents in the order of decreasing polarity and indicating the isopartitive properties (the possibility of mutual replacement of vicinal members of the series in the separating system), just as in partition chromatography on columns. Generally, the following demands must be met by the system selected: (*a*) the R_F values of the compounds should be in the range of 0.05 to 0.85, the difference between two neighboring compounds being at least 0.05 (however, compounds having different R_F values in the range of 0.01 to 0.2 can be separated by overflow development or by repeated development), (*b*), spots of the compounds repeated must be equal in size to the sample spot, and their shape should be retained, even when the concentration changes (round spots), (*c*) the system should not react with the compounds analyzed, and it must not interfere with or diminish the detection reaction (*e.g.*, pyridine used as a component of the solvent system interferes with detection by u.v. light, and should therefore be used only to separate radioactive compounds), (*d*) the system should be stable and reproducible (for example, a binary is more stable than a ternary system). When the compounds separated are to be further processed, certain salts should be replaced by more volatile ones (*e.g.*, ammonium sulfate by ammonium carbonate); volatile salts are preferred wherever possible.

The fundamental characteristic of the compounds separated that should mainly be considered is the kind of individual functional groups present in each kind of molecule: *hydrophilic*, such as hydroxyl, carboxyl, amino, amido, aldehyde, and sulfo groups; *fairly hydrophilic*, such as oxo, ether, tertiary amino, nitro, nitrile, and ester groups; and *hydrophobic*, such as methyl, higher alkyl, and aryl groups (the aromatic ring corresponds to about four methylene groups).

The most widely used and most conventional systems for separating nucleic acid components are those that are based on alcohols (mainly propyl alcohol, isopropyl alcohol, butyl alcohol, and isoamyl alcohol). These systems may differ considerably in the content of other components included, and this makes possible the adjustment of the mobility of similar and related compounds; increase in the water content of the mobility phase increases the mobility of polar compounds, for example, nucleotides. Changes in the pH value also influence, to a large extent, the mobility of compounds having different ionizable groups. Addition of ammonia decreases the mobility of compounds having an ionizable enol group (uracil and thymine, as compared to cytosine) and affords a system suitable for separating purine from pyrimidine bases, the system being very sensitive to the concentration of ammonia present (see Table II). Addition of strong acids likewise affects the partition coefficient by decreasing the mobility of both types of base. For guanine and its derivatives, a higher concentration of acid is required,[16,18] owing to the low solubility of these compounds. The incorporation of formic acid speeds up the migration of compounds typified by hypoxanthine, thymine, uracil, and xanthine. For complete separation of bases from D-ribosyl nucleosides, water may be replaced by a saturated aqueous solution of boric acid; borate *cis*-diol complexes then remain at the origin. The presence of salts is important, as these, in general, decrease the mobility of compounds. Salt solutions may be used alone, although a better separation will be achieved in the presence of a small proportion of an organic solvent (see Table III). For example, the system consisting of ammonium sulfate (degree of saturation, 0.8) with 2% of isopropyl alcohol separates nucleoside 2′- and 3′-phosphates[20] (see Table IV). It will be noted that the content of water, or of a component that suppresses the dissociation of strongly polar phosphate groups, has to be increased in order to effect separation of nucleotides containing these groups (see Table V). In this separation, the dissociation of other groups on the heterocyclic ring (which serve to differentiate between different bases and nucleosides in other cases) cannot, in general, participate in the process. In some instances, it is difficult to separate nucleotides by the process, and electrophoretic separation must be resorted to. The most conventional and widely applied solvent systems are described as examples in Table VI.

6. Detection

As in the chromatography of many other compounds, different solutions for detection are used with purines, pyrimidines, and their derivatives, the paper either being sprayed with the solution, or transported gently through a bath containing the solution. Many methods of detection are known[22] that involve the formation of colored salts (mercury bound to both types of base, silver for purine

TABLE II

R_F Values, in Various Chromatographic Systems,[a] of Purine and Pyrimidine Bases of Nucleic Acids

Base	R_F Value (x 100) in system[b]					
	1	2	3	4	5	6
Adenine	28	28	43	64	37	32
Cytosine	19	24	43	52	32	44
Guanine	6	11	23	35	16	22
5-(Hydroxymethyl)cytosine	-	12	-	-	25	44
Hypoxanthine	17	12	37	52	16	29
5-Methylcytosine	-	27	-	-	37	52
Thymine	48	35	66	71	52	76
Uracil	32	19	72	55	38	66
Xanthine	6	5	28	41	11	21

[a]According to the References given in footnote *b*. Paper chromatography was conducted by the decending technique at room temperature on Whatman No. 1 paper.
[b]System, 1, 43:7 butyl alcohol-water, *i.e.*, satd. with water,[1,12,13] 2, as for 1, with 5% (v/v) of concentrated ammonium hydroxide added to the solvent at the bottom of the tank;[13,14] 3 16:4:1 *tert*-butyl alcohol-water-formic acid;[15]

TABLE II (continued)

4, 3:1:1 isopropyl alcohol-concentrated ammonium hydroxide-water;[15] 5, 255:45:4 isopropyl alcohol-water-concentrated ammonium hydroxide;[16] and 6, 170:41:39 isopropyl alcohol-hyrochloric acid-water.[17]

TABLE III

R_F Values of Some Methylated 8-Azaxanthines[a]

Derivative of 8-Azaxanthine (*v*-Triazolo[4,5-*d*]-pyrimidine-5,7(4*H*,6*H*)-dione)	R_F Value (x 100) in system[b]			
	1	2	3	4
8-Azaxanthine	38	32	40	48
1,4-dimethyl-	50	57	70	73
1,6-dimethyl-	67	58	65	73
2,4-dimethyl-	60	60	74	75
2,6-dimethyl-	66	58	68	72
3,4-dimethyl-	39	44	78	80
3,6-dimethyl-	61	74	62	78
4,6-dimethyl-	71	74	55	66
1-methyl-	43	43	56	62
2-methyl-	42	40	59	69
3-methyl-	34	46	54	75
4-methyl-	60	47	41	58
6-methyl-	58	51	41	59

TABLE III (continued)

Derivative of 8-Azaxanthine (*v*-Triazolo[4,5-*d*]-pyrimidine-5,7(4*H*,6*H*)-dione)	R_F Value (x 100) in system[b]			
	1	2	3	4
1,4,6-trimethyl-	76	79	80	82
2,4,6-trimethyl-	76	79	81	83
3,4,6-trimethyl-	62	76	84	89

[a]This is a modified form of that in the original paper,[19] where the descending technique on Schleicher-Schüll 2043 MGl paper was used. [b]1, 2:1 Butyl alcohol-acetic acid; 2, 2:1 propyl alcohol-1% ammonium hydroxide; 3, 4% aqueous sodium citrate; and 4, 3% aqueous ammonium chloride.

TABLE IV

R_F Values[a] of Some Isomeric Nucleotides

Nucleotide	R_F Value (x 100) in system[b]			
	1	2	3	4
Adenosine 2'-phosphate	17	11	56	31
3'-phosphate	17	11	56	20
5'-phosphate	12	7	44	34
3':5'-cyclic phosphate	47	37	56	-
Cytidine 2'(3')-phosphate	15	9	-	-
5'-phosphate	10	5	32	-
3':5'-cyclic phosphate	39	33	33	-
Guanosine 2'-phosphate	7	5	21	56
3'-phosphate	7	5	21	45
5'-phosphate	3	1	14	56
3':5'-cyclic phosphate	26	18	16	-
Uridine 2'(3')-phosphate	18	12	-	-
5'-phosphate	7	4	15	-
3':5'-cyclic phosphate	35	26	20	-

TABLE IV (continued)

[a]According to data of Smith *et al.*,[21] obtained on Whatman No. 40 paper (doubly acid-washed) by the descending technique. [b]1, 7:1:2 isopropyl alcohol-concentrated ammonium hydroxide-water; 2,7:1:2 isopropyl alcohol-concentrated ammonium hydroxide-100 m*M* boric acid; 3, 100:60:1:6 and isobutyric acid-*M* ammonium hydroxide-100 m*M* (ethylenedinitrilo)tetraacetic acid, disodium salt; and 4, 40:9:1 saturated aqueous ammonium sulfate-*M* sodium acetate-propyl alcohol. This is a slight modification of a system used many years ago[20] for the separation of isomeric nucleotides of purines.

TABLE V

R_F Values[a] of Important Nucleotides from Nucleic Acids

Nucleotide	R_F Value (x 100) in system[b]		
	1	2	3
Adenosine 5'-phosphate	17	17	38
2'-deoxy-	26	22	50
Cytidine 5'-phosphate	15	22	36
2'-deoxy-	22	28	48
Guanosine 5'-phosphate	12	10	26
2'-deoxy-	17	13	36
Thymidine 5'-phosphate	36	32	50
Uridine 5'-phosphate	22	26	36

[a]Compiled from data in Ref. 22; chromatography performed on Whatman No. 1 paper.
[b] 1, 3:3:2 Isopropyl ether-butyl alcohol-formic acid; 2, 9:6:3:2 methanol-isopropyl alcohol-concentrated ammonium hydroxide-water; 3, 5:8:4:5:3 80% aqueous formic acid-butyl alcohol-propyl alcohol-acetone-30% (w/v) trichloroacetic acid; ascending technique.

TABLE VI

General Solvent Systems for Paper Chromatography of Nucleic Acid Components

Solvent system	Ratio	Group separation
Butyl alcohol (water-saturated)	43:7	bases, nucleosides
Isopropyl alcohol-hydrochloric acid-water	170:41:39	bases, nucleosides
Butyl alcohol-acetic acid-water	various	bases, protected nucleosides, nucleotides
Propyl alcohol-concentrated ammonium hydroxide-water	6:3:1	nucleosides, nucleotides
Isopropyl alcohol-concentrated ammonium hydroxide-water	7:1:2	nucleosides, mono- and oligo-nucleotides
Saturated aqueous ammonium sulfate-water-isopropyl alcohol	79:19:2	isomeric nucleotides
Isobutyric acid-concentrated ammonium hydroxide-water	various	mono- and oligo-nucleotides
Ethyl alcohol-*M* ammonium acetate (pH 7.5)	7:3 or 5:2	mono- and oligo-nucleotides

bases,[23] reactions of purine mercury complexes with chlorine and eosine or Bromophenol Blue), as well as color reactions of sugar components (the diphenylamine reaction for 2'-deoxy-D-ribonucleosides, for other pyrimidine derivatives after bromination,[24] Schiff-type reactions of *cis*-diol D-ribosyl nucleosides after reaction with periodate), or specific reactions of certain bases (for example, hydrouracil derivatives with *p*-(dimethylamino)benzaldehyde in a mixture of hydrochloric acid and ethyl alcohol[25,26]).

In general, however, absorption of u.v. radiation in the 260-nm range appears to provide the most suitable detection. This includes the fluorescence by some bases under specific conditions [guanine in acid medium,[27] 8-azaguanine (5-amino-*v*-triazolo[4,5-*d*]pyrimidin-7(6*H*)-one) in acid and alkaline medium[28]], permitting the bases to be distinguished from one another. This method of detection is used wherever possible, because it provides evidence that substantially facilitates the further processing of the compounds separated. It is likewise very advantageous for documentation.[29]

7. Elution

Paper is a suitable carrier, permitting rather easy elution of the compounds separated. The most conventional way consists in excision of a spot after the chromatogram has been dried, inserting it in a test tube, and extracting with, most frequently, 5 ml of 100 m*M* hydrochloric acid at room temperature, with occasional agitation. For qualitative and quantitative evaluation in the author's laboratory, this procedure is mainly used in conjunction with u.v. spectrophotometric examination of the extract.

Elution may be speeded up by "developing" the excised spots with the eluant, the edge of the paper being placed between two microscope slides immersed in a trough containing the solution. The first drops obtained in this operation, that is, dropping from a suitably cut tip of a piece of the paper, contain practically all of the compound eluted. If a capillary tube is applied to the tip of the paper eluted (to suck up the eluate), the rate of elution may be raised considerably.[30]

Finally, for preparative separation on paper, an elution process may be used in which a long paper strip is wound onto a test tube to form a column that is readily eluted[31] (see p. 554). This procedure is suitable when a large amount of the substance separated is to be eluted.

The eluants usually employed are water and dilute acids.

For nucleotides separated on ion-exchange paper (for example, DEAE-cellulose paper), volatile buffers (*e.g.*, triethylammonium carbonate, pH 10) are suitable, as they are readily removed afterwards.

8. Some Practical Examples

a. Separation of a Mixture Containing a Nucleoside and Nucleotides by Means of Preparative Paper-chromatography.*

Paper: Whatman No. 3MM (2 sheets, 46 x 57 cm); Whatman No. 1 for analytical control.

Equipment: Chromatographic chamber, application pipet (usually calibrated from 0 to 1 ml, or of the Pasteur type), "polythene" or poly(vinyl chloride) foil (20 x 20 cm), bacteriological test-tube (without the widened edge; dimensions about 1.5 x 15 cm), Whatman cellulose powder.

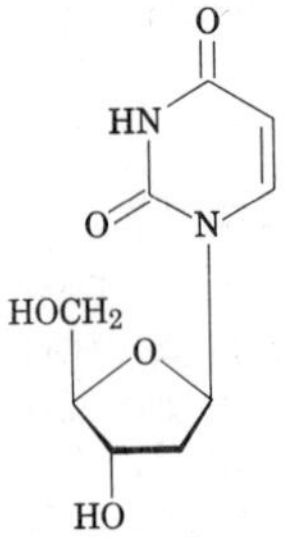

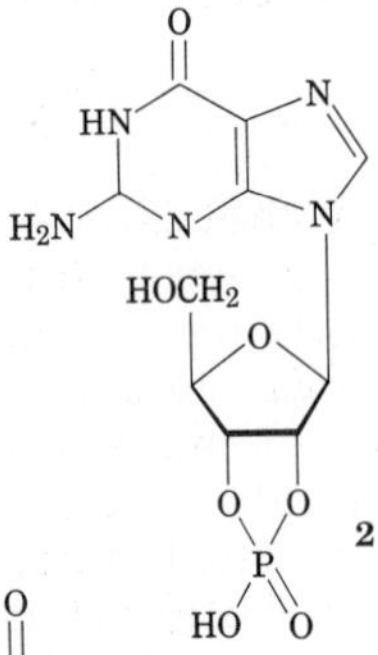

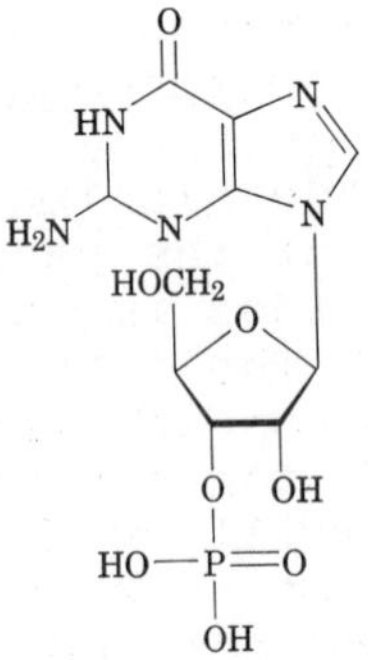

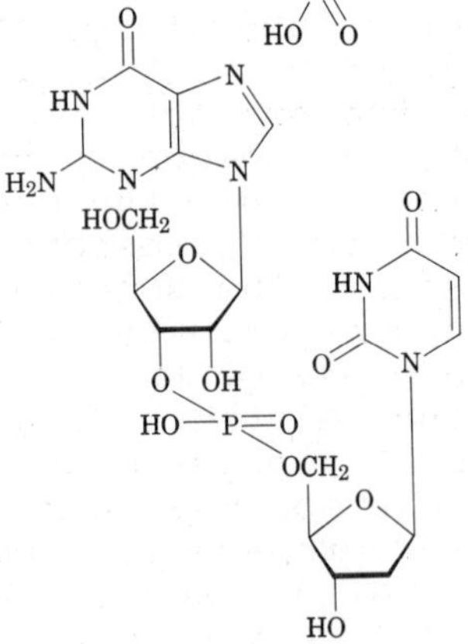

*Details provided by Dr. A. Holy.

Mixture Separated: Incubation mixture from the enzymic synthesis of the dinucleoside phosphate GpdU with ribonuclease T1 (see Ref. 32), containing 2'-deoxyuridine (1, 80 μmoles), guanosine 2':3' cyclic phosphate (2, 60 μmoles), guanosine 3'-phosphate (3, 90 μmoles), and guanylyl-(3'→5')-2'-deoxyuridine (GpdU, 4, 20 μmoles), in 600 μl of 50 m*M* Tris-hydrochloric acid buffer (pH 7) with 45 μg of the enzyme. The total weight of the compounds in the mixture is about 90 mg. The main component, to be further processed, is the ammonium salt of 4.

Eluant: Isopropyl alcohol-water—concentrated ammonium hydroxide (7:2:1); descending treatment. The increasing polarity of the compounds separated is manifested by decreasing R_F values.

Procedure: A starting line is drawn in pencil along the narrow edge of the paper sheets at a distance of 12 cm from the edge. The paper is placed on a suitable chromatographic application table, or over two glass rods, so as not to touch the underlying substrate in the place where the starting line is. Half of the reaction mixture (300 μl) is then taken up in the application pipet, and the solution is allowed to be absorbed in the paper along the entire width of the starting line by use of a continuous motion along the starting line, with the tip of the pipet slightly open and touching the paper gently. It is recommended that spaces, about 1.5 cm wide, be left at each side. The part to which the solution has been applied is then dried with a gentle stream of unheated air, and the procedure is repeated with the other half of the reaction mixture. In general, about 1 ml of solution containing a maximum of 60 mg of the compounds can be applied (the optimal weight is 20 to 40 mg). Care must be taken when applying the dilute solution obtained by washing out the vessel; when this is applied over the more concentrated zone of mixture on the starting line, the zone may become diffuse and will then divide in two on development. It is recommended that the dilute solution be applied close to the *edge* of the zone that has already been applied, parallel to the starting line and between this and the nearer end of the paper. After the two sheets have been dried, they are bent, parallel to the starting line and some 3 cm away from it. The sheets are placed in the chromatographic chamber and immersed in the trough, the shorter part being weighted down with a glass rod. The chromatographic system is poured into the trough and onto the chamber bottom, and the chamber is tightly closed and allowed to stand undisturbed in a thermostated room. When the solvent front has reached a distance of about 5 cm from the lower edge of the chromatographic paper, the sheets are removed, dried in a stream of air, and placed in a suitable holder with the starting line downwards.

Detection: Detection is conducted in u.v. light; because the u.v. absorption of the dry paper is considerable, it is often better to observe the paper in the incident light of the u.v. lamp. With insufficient separation of the zones, development may be repeated when the paper has dried. The edges of the bands are lightly marked with a pencil. The separated mixture contains the following u.v.-absorbing bands (R_F values are calculated for the center of the zone): 0.08 (3), 0.16 (4), 0.26 (2), and 0.60 (1).

Elution: The zones containing the product (R_F 0.16, as determined by the analytical separation) are cut out, and tightly rolled onto a test tube in such a way as to bring the lower end about 1 cm away from the upper, open end of the test tube. Without unrolling of the roll, it is so placed on the plastic foil that the foil extends 1 cm beyond the open end of the test tube, and the test-tube (with the paper roll) is parallel to the edge of the foil. The test tube with the paper roll is now tightly rolled in the foil, which is then fixed with adhesive tape or soft wire at the places where the upper and lower edges of the paper are located, and the free ends of the foil are prevented from unrolling by application of adhesive tape. This tube of foil is now fixed in a holder with the bottom of the test tube upwards, and a 100-ml, round-bottomed flask bearing a funnel wider than the diameter of the elution tube is placed underneath.[31] Whatman cellulose powder (about 5 g) is mixed with water to make a thick suspension which is poured into the elution tube, the water is allowed to be absorbed, and 50 ml of dilute ammonium hydroxide (1:100 concentrated ammonia—water) is carefully added to keep the surface of the cellulose undisturbed. The first drop of effluent, with a high concentration of product, is analyzed by paper electrophoresis and enzymic cleavage.[32] The first portion (∿10 ml) of effluent is collected; it usually contains at least 95% of the total amount of the product. The effluent is transferred to a 25-ml volumetric flask, made to the mark with water, and diluted with 100 m*M* hydrochloric acid, in the ratio of 1:25, for a determination of the amount of the compound by measurement of the optical absorbance at 260 nm (ε_{260} 21.8 x 10^3). The yield is 18 μmoles of the compound.

Isolation: The aqueous solution is evaporated to dryness at 30°/15 torr, the residue (the ammonium salt of the product) is dissolved in 1 ml of water, and the solution is transferred quantitatively with a Pasteur pipet into a 10-ml ampoule, where it is frozen onto the walls by slow rotation in an oblique position, while it is being cooled with a mixture of Dry Ice and ethyl alcohol. After being dried on the outside, the ampoule is quickly transferred to an Abderhalden dryer (containing phosphorus pentaoxide), which is then evacuated with an oil pump to 0.1–0.2 torr; the lyophilization is complete after evacuation overnight.

b. Some Other Applications. – A similar procedure, employing the same solvent system under the same conditions, serves to separate a mixture of 2',3'-*O*-(ethoxymethylene)-uridine,[32a] 2'-*O*-(1-ethoxyethyl)cytidylyl-(3'→5')-2',3'-*O*-(ethoxymethylene)uridine,[32b] cytidylyl-(3'→5')-uridine,[32b] and uridylyl-(3'→5')-cytidylyl-(3'→5')-uridine;[32b] these components may be studied analytically in the same way, the respective R_F values being[33] 0.72, 0.55, 0.18, and 0.04. The same procedure is applicable to components of a mixture of adenosine, adenosine 2',(3')-phosphite, adenosine 2'(3'),5'-diphosphite, and adenosine 2',3',5'-triphosphite[34]; these have R_F values of 0.48, 0.37, 0.22 and 0.10, respectively. The final product, namely, adenosine 2'(3'),5'-diphosphate (for which other compounds are intermediates in the synthesis), has an R_F value of 0.02.

This system is likewise suitable for studying the course of individual, partial reactions, for example, the cyclization of uridine 2'(3')-phosphate to uridine 2':3'-cyclic phosphate (R_F values, 0.1 and 0.32), and cyclization of 6-azuradine [2-β-D-ribofuranosyl-*as*-triazine-3,5(2*H*,4*H*)-dione] 2'(3')-phosphate to 6-azuridine 2':3'-cyclic phosphate, having R_F values of 0.06 and 0.40, respectively.[35] The system may also be used for separating nucleosides from nucleosides, nucleotides from nucleotides, or nucleosides from nucleotides (*e.g.*, protected derivatives), as may be seen from the R_F values given in Tables IV and VII–XI.

c. Separation of *O*-Methylated Nucleosides by Preparative, Paper Chromatography

Paper: Whatman No. 3MM; Whatman No. 1 for analytical evaluation.

Equipment: The same as in the preceding example.

Mixture Separated: Products of the methylation of adenosine[47] (3.74 μmoles) with diazomethane (131 μmoles), after evaporation from ethyl alcohol and dissolution in a small volume of distilled water or dilute ethyl alcohol. The individual components are adenosine (5), 2'-*O*-methyladenosine (6), 3'-*O*-methyladenosine (7), and 2',3'-di-*O*-methyladenosine (8).

TABLE VII

R_F Values[a] of Some Nucleosides and Their 2',3'-*O*-Isopropylidene Derivatives

Compound	R_F Value[b] (x 100) in system[c]		
	1	2	3
Adenosine	57	60	58
2',3'-*O*-isopropylidene-	75	80	78
8-Azaguanosine (5-amino-3,6-dihydro-3-β-D-ribofuranosyl-7*H*-*v*-triazolo[4,5-*d*]pyrimidin-7-one)	40	47	-
2',3'-*O*-isopropylidene-	67	81	-
Cytidine	55	51	55
2',3'-*O*-isopropylidene-	75	78	80
Guanosine	37	46	49
2',3'-*O*-isopropylidene-	68	78	75
Inosine	50	46	52
2',3'-*O*-isopropylidene-	70	77	78
6-Thioguanosine (2-amino-9-β-D-ribofuranosyl-9*H*-purine-6-thiol)	36	46	-
2',3'-*O*-isopropylidene-	61	78	-
6-Thioinosine (9-β-D-ribofuranosyl-9*H*-purine-6-thiol)	45	48	57
2',3'-*O*-isopropylidene-	70	79	79
Uridine	47	55	63
2',3'-*O*-isopropylidene-	70	79	82
Bis(*p*-nitrophenyl) hydrogen phosphate	83	84	85

[a]Data obtained from Ref. 36.

[b]Ascending technique on Schleicher—Schüll 597 paper. Values are relative to that (100) of bis(*p*-nitrophenyl) hydrogen phosphate, having absolute values (given in the Table) that are averages from all chromatographic runs where the compound is used as a standard.

[c]1, 14:5:1 Isopropyl alcohol—water—concentrated ammonium hydroxide; 2, 5:2:3 butyl alcohol—acetic acid—water; and 3, 7:3 isopropyl alcohol—water.

TABLE VIII

R_F Values of Some Nucleoside Derivatives in Different Chromatographic Systems[a]

Compound	R_F Value (x 100) in system[b]				
	1	2	3	4	5
Adenosine	27	61	29	54	35
2'-deoxy-	36	66	28	63	44
N-methyl-	47	76	45	70	57
2'-*O*-methyl-	47	75	48	67	57
1-methyl-	22	51	30	53	3
Guanosine	5	42	25	26	13
2'-deoxy-	9	59	20	39	24
N,*N*-dimethyl-	24	62	37	48	25
N-methyl-	10	60	46	43	21
2'-*O*-methyl-	15	67	46	49	30
1-methyl-	16	57	34	58	18
7-methyl-	4	36	32	35	5
Inosine	5	55	25	42	14
1-methyl-	20	63	41	59	22
Pseudouridine (5-β-D-ribofuranosyluracil)	4	53	50	34	11
2'-*O*-methyl-	18	-	56	74	18
Uridine	10	60	63	43	27
2'-deoxy-	19	71	78	55	45
2'-*O*-methyl-	23	74	83	57	58
3-methyl-	43	78	89	77	59
5-methyl-	25	69	72	57	42

[a]Data obtained from Ref. 36. The nucleosides are found in hydrolyzates of soluble D-ribonucleic acid (sRNA), and the solvent systems are those used for the estimation of nucleosides.[37,38]

[b]1, 86:14:5 Butyl alcohol-water-concentrated ammonium hydroxide; 2, 2:1 isopropyl alcohol–1% (w/v) ammonium sulfate; 3, 340:85:72 isopropyl alcohol–hydrochloric acid–water; 4, 7:2:1 isopropyl alcohol–water–concentrated ammonium hydroxide; and 5, 4:1:2 ethyl acetate–propyl alcohol–water.

TABLE IX

R_F Values of Some Protected Derivatives of Nucleotides[a]

Compound	R_F Value (x 100) in system[b]	
	1	2
N-Acetylguanosine 3'-phosphate	6	31
5'-*O*-(*p*-methoxyphenyl)diphenyl-methyl-	44	64
2'-*O*-acetyl-5'-*O*-(*p*-methoxyphenyl)-diphenylmethyl-	-	70
Adenosine 3'-phosphate	10	21
N-benzoyl-2'-*O*-benzoyl-5'-*O*-(*p*-methoxyphenyl)diphenylmethyl-	-	75
5'-*O*-bis(*p*-methoxyphenyl)phenyl-methyl-	55	-
N-benzoyl-2'-*O*-benzoyl-5'-*O*-bis(*p*-methoxyphenyl)phenylmethyl-	75	75
N-acetyl-2',5'-di-*O*-acetyl-	-	67
Cytidine 3'-phosphate	10	27
5'-*O*-(*p*-methoxyphenyl)diphenyl-methyl-	49	-
N-acetyl-2'-*O*-acetyl-5'-*O*-(*p*-methoxyphenyl)diphenylmethyl-	-	74
5'-*O*-bis(*p*-methoxyphenyl)phenyl-methyl-	47	-
N-acetyl-2'-*O*-acetyl-5'-*O*-bis(*p*-methoxyphenyl)phenylmethyl-	-	76
N-benzoyl-2'-*O*-benzoyl-5'-*O*-bis(*p*-methoxyphenyl)phenylmethyl-	69	80
N-acetyl-2',5'-di-*O*-acetyl-	-	70
Cytidine 2':3'-cyclic phosphate	30	51
5'-*O*-bis(*p*-methoxyphenyl)phenyl-methyl-	78	-
N-Acetyl-2',5'-di-*O*-acetyladenosine 3'-(acetylphosphate)	-	78
Uridine 3'-phosphate	9	30
5'-*O*-(*p*-methoxyphenyl)diphenyl-methyl-	48	-
2'-*O*-benzoyl-5'-*O*-(*p*-methoxyphenyl)-diphenylmethyl-	60	82
5'-*O*-bis(*p*-methoxyphenyl)phenyl-methyl-	49	-

TABLE IX (continued)

2',5'-di-*O*-bis(*p*-methoxyphenyl)-phenylmethyl-	68	-
5'-*C*-pyridinium-	5	26
2'-*O*-acetyl-5'-*O*-bis(*p*-methoxyphenyl)-phenylmethyl-	-	75
2'-*O*-benzoyl-5'-*O*-bis(*p*-methoxy-phenyl)phenylmethyl-	62	84
Uridine 2':3'-cyclic phosphate	25	56
5'-*O*-(*p*-methoxyphenyl)diphenymethyl-	81	-
5'-*O*-bis(*p*-methoxyphenyl)phenyl-methyl-	83	-

[a]Compiled according to data of Khorana,[39,40] obtained on Whatman No. 1 paper by the descending technique.
[b]1, 7:1:2 Isopropyl alcohol—concentrated ammonium hydroxide—water; and 2, 7:3 ethyl alcohol-*M* ammonium acetate (pH 7.5).

TABLE X

R_F Values of Some Derivatives of 6-Azauridine [2-β-D-Ribofuranosyl-*as*-triazine-3,5(2*H*,4*H*-dione][a]

Compound	R_F Value (x 100) in system[b]				
	1	2	3	4	5
6-Azauridine					
2':3'-cyclic phosphate	-	58	47	-	40
5'-(dibenzylphosphate)	88	-	-	-	-
5'-(dibenzylpyrophosphate)	72	-	-	-	-
2'(3')-phosphate	-	77	20	-	30
5'-phosphate	-	80	20	-	30
2',3'-di-*O*-acetyl-	-	64	-	61	-
2',3'-*O*-isopropylidene-	-	60	-	60	-
2',3'-di-*O*-propionyl-	-	65	-	73	-
5'-pyrophosphate	-	-	-	-	16
2',3'-di-*O*-acetyl-	69	-	-	-	-
2',3'-*O*-isopropylidene-	78	-	-	-	-
6-Azauridylyl-(5'→3')-uridine	-	-	10[c]	-	-
6-Azauridylyl-(5'→3')-2'-*O*-(tetrahydropyran-2-yl)uridine	-	-	25[c]	-	-
2',3'-*O*-Benzylidene-6-azauridylyl-(5'→3')-uridine	-	-	55[c]	-	-
2',3'-*O*-Benzylidene-6-azauridylyl-(5'→3')-2'-*O*-(tetrahydropyran-2-yl)uridine	-	-	60[c]	-	-

[a]Compiled from data, published in Refs. 41–44, obtained by the descending technique on Whatman No. 1 paper.

[b]1, 43:7 Butyl alcohol–water, *i.e.*, water-saturated;[44] 2, 79:19:2 saturated aqueous ammonium sulfate–*M* ammonium acetate–isopropyl alcohol;[42,43] 3, 7:1:2 isopropyl alcohol–concentrated ammonium hydroxide–water;[41,42] 4, 5:2 ethyl alcohol–*M* ammonium acetate (pH 7.5);[43] and 5, 2:1 isopropyl alcohol–1% (w/v) aqueous ammonium sulfate.[42,44]

[c]On Whatman No. 3 paper.[41]

TABLE XI

R_F Values of Some (3'→5')-Dinucleotides[a]

(3'→5')-Dinucleotide	R_F Value (x 100) in system[b]			
	1	2	3	4
Adenylyluridine	21	36	-	49
Cytidylyladenosine	21	40	73	57
Cytidylylcytidine	18	41	65	49
Cytidylylguanosine	8.7	34	51	34
Cytidylyluridine	17	42	52	36
Guanylyladenosine	10	33	56	47
Guanylylcytidine	10	29	51	37
Guanylylguanosine	4	24	34	25
Guanylyluridine	10	36	39	25
Uridylyluridine	16	46	-	26

[a]Compiled according to data of Khorana,[45,46] obtained on Whatman No. 1, 40, or 3MM paper by the descending technique.

[b]1, 7:1:2 Isopropyl alcohol—concentrated ammonium hydroxide—water;[45] 2, 7·3 ethyl alcohol—*M* ammonium acetate (pH 7.5);[45] 3, 57:4:39 isobutyric acid—concentrated ammonium hydroxide—water;[45] and 4, 60:1:33 isobutyric acid—concentrated ammonium hydroxide—water (pH 3.1).[46]

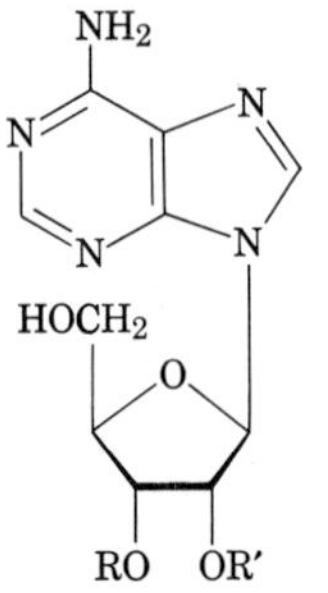

5, R=R'=H
6, R=H, R'=Me
7, R=Me, R'=H
8, R=R'=Me

Eluant: Butyl alcohol—water—concentrated ammonium hydroxide (86:14:5) in the descending arrangement. The R_F values increase with increasing hydrophobic character of the compounds separated ($OH \rightarrow OCH_3$).

Procedure: Because the practical technique is already described in detail in the preceding example, only the separation of the mixture will be mentioned. After the paper, on which a maximum of 75 mg of the mixture is applied, has been dried, the chromatogram is developed in the solvent system for ∿18 hr, depending on the temperature of the room (∿23°). The solvent is removed from the paper by placing it in a drying box provided with circulating air (maximum temperature, 35°). Three absorbing zones may be detected under u.v. light, the R_F values of which are 0.27 (5), 0.47 (6 and 7), and 0.65 (8). The zones are excised, and eluted with water as described on p. 554. The eluate is evaporated, and the residue is dissolved in 3 ml of 30% methanol (aqueous). Because the derivatives (6 and 7) are not separated in the paper-chromatographic process, they are separated on an ion-exchange column[48] (see p. 601).

d. Study of the Acetylation of 2-Amino-6-chloropurine by Paper Chromatography

Paper: Whatman No. 1.

Equipment: Two chromatographic chambers and several application micropipets.

Mixture Separated: An aliquot of the mixture formed

after the acetylation of 2-amino-6-chloropurine (94 mmoles) with acetic anhydride.[49] The final product of the two-stage synthesis is 2-acetamido-6-chloropurine (a versatile intermediate in the synthesis of purine nucleosides).

Composition of the Mixture: 2-Amino-6-chloropurine (9), 2-acetamido-9-acetyl-6-chloro-9*H*-purine (10), 9-acetyl-2-(*N*-acetylacetamido)-6-chloro-9*H*-purine (11), 2-acetamido-6-chloropurine (12), 9-acetyl-2-amino-6-chloro-9*H*-purine (13), and 2-(*N* -acetylacetamido)-6-chloropurine (14). Some of the compounds named are decomposition products of labile components that decompose before, or during, the chromatographic separation process (13 → 9 or 14 → 12). Compound 13 interferes with obtaining a quantitative yield of 12, and its presence in or absence from the mixture is, therefore, first determined chromatographically.

9, R=R′=R″=H
10, R=H, R′=R″=Ac
11, R=R′=R″=Ac
12, R=Ac, R′=R″=H
13, R=R′=H, R″=Ac
14, R=R′=Ac, R″=H

where Ac is acetyl.

Eluants: (*1*) Butyl alcohol-water (43:7); this does not separate 10 from 11, or 12 from 13. (*2*) Aqueous disoduim hydrogen phosphate (5%); this does not separate 11 from 12, or 9 from 13. The two systems are used in the descending arrangement.

Procedure: Because the individual components are represented in the mixture in variable concentrations, it is advantageous to apply, at the origin, several separate spots of relative concentrations of 1:2:5 of the mixture. The application and development techniques are the same as those described on p. 553, the individual samples being applied at a single starting-line (each sample forms a band 1-1.5 cm wide), with gaps between the samples of 2-2.5 cm. Two equal chromatograms are developed in parallel in each solvent system. After the chromatograms have been dried, detection in u.v. light reveals the following R_F values in systems *1* and *2*, respectively: 9, 0.40 and 0.26; 10, 0.70 and 0.68; 11, 0.70 and 0.41; 12, 0.30 and 0.41; 13, 0.30 and 0.26. The main products, which are to be processed further, are 10 and 11, which form compound 12 on partial deacetylation; this final product is chromatographed in system *1* in order to allow the extent of deacetylation of 10 and 11 to be assessed.

Another application involves butyl alcohol saturated with water, sometimes in combination with ammonia in the gaseous phase; this is the system most often used for the group of bases and nucleosides; and its separative ability is very good. Cytosines, methylated in different ways, are separated in the system with ammonia[50] (see Table XII).

The system lacking ammonia serves for determination of the purity of (*a*) 2'-deoxy-*N*-[(dimethylamino)methylene]-adenosine (R_F 0.41), after its synthesis from 2'-deoxy-adenosine[52] (R_F 0.30); (*b*) individual components in the synthesis of isocytidine[53] (R_F 0.12), 2',3'-*O*-iso-propylideneisocytidine (R_F 0.58), and 2,5'-anhydro-2',3'-*O*-isopropylideneuridine (R_F 0.62); and (*c*) 4-thiouridine[53a] (R_F 0.14), eluted from a cellulose column, by chromatographing the fractions on paper.[54] The R_F values of some other compounds in the same or similar solvent systems are shown in Tables II, VIII, X, and XIII-XV. In contrast, for the separation of different hydroxy and methyl derivatives of purine, an acid solution of ethyl alcohol appears to be very suitable (see Table XVI).

e. Separation of Pyrimidine Nucleosides and their Phosphates on DEAE-cellulose Paper[65]

Paper: DEAE-cellulose paper (Whatman DE-20), capacity 400 µequiv/g, with no special modification.

Equipment: Chromatographic chamber and application micropipets.

Mixture Separated: A mixture (50 µl) of cytidine, cytidine 2'(3')-phosphate, cytidine 5'-phosphate, and cytidine 5'-pyrophosphate and triphosphate in a concentration of 1 mg/ml for each component. The same procedure is employed to chromatograph mixtures (in the same concentration) derived from uridine and from thymidine.

TABLE XII

R_F Values of Some Derivatives of Pyrimidine Bases in Two Chromatographic Systems[a]

Compound	R_F Value (x 100) in system[b]	
	1	2
Cytosine		
N,*N*-dimethyl-	48	15
N,1-dimethyl-	58	13
N-methyl-	45	6
1-methyl-	34	6
N,*N*,1-trimethyl-	63	46
4-Ethoxy-2-pyrimidinol	-	55
1-methyl-	80	89
2-Ethoxy-4-pyrimidinol	-	83
Uracil	54	2

[a]Data obtained by paper chromatography on Whatman No. 1 paper according to Ref. 50. [b]1, 43:7 Butyl alcohol-water, *i.e.*, water-saturated, with ammonia in the vapor phase[51] (descending technique); and 2, upper phase of 169:45:15 benzene-ethyl alcohol-water (ascending technique).

TABLE XIII

R_F Values of Some Derivatives of Purines and Pyrimidines[a]

Compound	R_F Value (x 100) in system[b]			
	1	2	3	4
Adenine	41	60	21	32
N-methyl-	53	75	36	54
2-methyl-	50	67	23	50
Guanine	30	43	10	8
N,*N*-dimethyl-	43	69	28	20
N-ethyl-	61	82	-	-
N-methyl-	50	66	25	18
1-methyl-	26	62	19	13
6-Methylpurine	54	75	56	40
8-Methylpurine	56	76	38	46
Thymine	78	70	42	42
Uracil	69	72	38	20

TABLE XIII (continued)

Compound	R_F Value (x 100) in system[b]			
	1	2	3	4
5-amino-	27	45	15	16
1-methyl-	53	70	-	-
4-methyl-	82	68	37	33
5-bromo-	75	65	44	21
5-chloro-	71	63	43	18
5-(dimethylamino)-	52	71	-	-
5-iodo-	76	65	47	26
5-(methylamino)-	44	61	-	-

[a]According to data, published in Refs. 55-57, obtained on Whatman No. 1 paper by the descending technique. [b]1, 85:22:18 Isopropyl alcohol—water; 2, 7:3 isopropyl alcohol—water, with ammonia in the vapor phase; 3, 77:13:10 butyl alcohol—water—formic acid; and 4, 43:7 butyl alcohol—water, *i.e.*, water—saturated, with ammonia in the vapor phase.

TABLE XIV

R_F Values of Important Nucleosides of Nucleic Acids, in Various Chromatographic Systems[a]

Nucleoside	R_F Value (x 100) in system[b]				
	1	2	3	4	5
Adenosine	34	17	54	59	56
2'-deoxy-	36	31	64	66	66
Cytidine	15	14	65	58	47
2'-deoxy-	23	22	73	66	61
Guanosine	10	10	53	41	36
2'-deoxy-	22	18	55	48	50
Inosine	-	12	66	36	39
2'-deoxy-	23	20	69	47	49
Thymidine	50	42	77	67	68
Uridine	-	18	72	46	46
2'-deoxy-	38	29	73	51	58
Xanthosine	-	35	62	33	34

TABLE XIV (continued)

[a] According to data, published in Refs. 22, 58, and 59, obtained on Whatman No. 1, paper. [b] 1, 43:7 Butyl alcohol-water, *i.e.*, water saturated;[58] 2, 3:3:2 isopropyl ether-butyl alcohol-formic acid;[22] 3, 9:6:3:2 methanol-isopropyl alcohol-concentrated ammonium hydroxide-water;[22] 4, 5:8:4:5:3 80% aqueous formic acid-butyl alcohol-propyl alcohol-acetone-30% (w/v) aqueous trichloroacetic acid by the ascending technique;[22] and 5, 2:1:1 butyl alcohol-acetic acid-water.[59]

TABLE XV

R_F Values of Some Pyrimidines and Nucleosides, and Analogs, in Free and Protected Forms[a]

Compound	R_F Value (x 100) in system[b]	
	1	2
6-Azacytidine [5-Amino-2-β-D-ribofuranosyl-*as*-triazin-3(2*H*)-one]	10	32
2',3'-di-*O*-acetyl-	57	71
5'-*O*-trityl-	82	-
N-acetyl-	86	90
2',3',5'-tri-*O*-acetyl-	71	80
N-acetyl-	80	85
6-Azathymine [6-Methyl-*as*-triazine-3,5(2*H*,4*H*)-dione]	59[c]	-
6-Azauracil [*as*-Triazine-3,5(2*H*,4*H*)-dione]	50[c]	-
6-Azauridine [2-β-D-Ribofuranosyl-*as*-triazine-3,5(2*H*,4*H*)-dione]	15	-
5'-*O*-acetyl-	43	-
2',3'-di-*O*-acetyl-	66	75
5'-*O*-trityl-	94	-

TABLE XV (continued)

Compound	R_F Value (x 100) in system[b] 1	2
2',3',5'-tri-*O*-acetyl-	80	-
2',3'-*O*-isopropylidene-	78	-
Cytidine	13	-
2',3',5'-tri-*O*-acetyl-	71	80
N-acetyl-	82	87
2-Thio-6-azathymine [6-Methyl-3-thio-*as*-triazine-3,5(2*H*,4*H*)-dione]	75[d]	-
2-Thio-6-azauracil [3-Thio-*as*-triazine-3,5(2*H*,4*H*)-dione]	61[c]	-
4-Thio-6-azauridine [2-β-D-Ribofuranosyl-5-thio-*as*-triazine-3,5(2*H*,4*H*)-dione]	36[d]	-
Thymidine	49	-
Uridine	15	52

[a] Compiled according to data published in Refs. 43, 44, and 60-62. Paper chromatography was conducted by the descending technique, in most cases on Whatman No. 1 paper. [b] 1, 43:7 Butyl alcohol-water, *i.e.*, water-saturated; and 2, 5:2:3 butyl alcohol-acetic acid-water. [c] On Whatman No. 4 paper. [d] On Whatman No. 3 paper.

TABLE XVI

R_F Values[a] of Some Derivatives of Purine

Compound	R_F (x 100)[b]
2,6-Dihydroxypurine (Xanthine)	27
3-methyl-	42
2,8-Dihydroxypurine (Purine-2,8-diol)	22
3-methyl-	39
6,8-Dihydroxypurine (Purine-6,8-diol)	30
3-methyl-	20
2-Hydroxypurine (Purin-2-ol)	25
3-methyl-	48
6-mercapto-	42
6-Hydroxypurine (Hypoxanthine)	47
3-methyl	35
8-Hydroxypurine (Purin-8-ol)	55
3-methyl-	42
6-mercapto-	48

TABLE XVI (continued)

Compound	R_F (x 100)
6-(methylthio)-	73
6-(benzylthio)-	81
3-Methyluric acid	40
6-Thiopurine (Purine-6-thiol)	52
8-hydroxy-	42
3-methyl-	40
6-Thiouric acid	8

[a]According to data published in Refs. 17,63, and 64. [b]Obtained by the descending technique on Whatman No. 1 paper with 17:1:2 95% ethyl alcohol-acetic acid-water.

Eluant: Ammonium hydrogen carbonate (200 m*M*) - sodium tetraborate (5m*M*); a solution devoid of borate was used in comparing the influence of borate on the mobility of compounds having a *cis*-diol group.

Procedure: Samples are applied to one sheet of DEAE-cellose paper on a common starting-line, the sample taking up 2 cm and the gaps between samples being 3 cm. Care must be exercised in the application and later treatment, because DEAE-cellose paper, when wet, is not very tear-resistant and is readily damaged by the tip of a pipet, and during removal of the paper from the trough and the chromatographic chamber after development. It is, therefore, recommended that excess of solvent be removed from the trough by suction, and that the paper be allowed to dry without being moved, or that it be transferred, together with the trough, to a stand in the drying chamber. Development is conducted at a constant temperature of 24° (the front migrates 40 cm in 6 hr), and, after the paper is dried, detection is readily accomplished with a u.v. lamp. Each mixture is separated into 4 (for thymidine derivatives) or 5 more-or-less well-separated spots having the R_F values listed in Table XVII.

When the borate component of the solvent system is omitted, the R_F values remain unchanged for most of the compounds, with the exception of isomeric monophosphates (2',3', or 5'), which are not separated; this separation is simpler than the electrophoretic method of separating polyphosphates, which is more commonly employed.

Elution: The spots are cut out, and are most readily eluted by immersion of an ∿3-mm wide part of the paper, held between two microscope slides, placed in a chromatographic trough containing aqueous triethylammonium carbonate of pH 10. The apparatus is covered with a lid (preferably of organic, plastic glass) to permit the elution to be performed in a saturated environment. The eluant (transported by capillary forces) is absorbed by the paper spot, so cut as to form a tip along which the eluate drops into a small beaker, or from which it may be taken up by a capillary tube. Before further processing, the eluant is removed by evaporating the eluate to dryness in a desiccator and adding and evaporating water several (usually 2 or 3) times.

TABLE XVII

R_F Values[a] of Some Pyrimidine Nucleosides and Their Mono-, Pyro-, and Tri-phosphates on DEAE-cellulose Paper and on Cellulose Paper

Compound	R_F Value (x 100) in system[b]		
	1	2	3
Cytidine	83	90	67
2'(3')-phosphate	48	58	53
5'-phosphate	36	60	50
5'-pyrophosphate	26	2	36
5'-triphosphate	21	0	24
Thymidine	76	89	68
5'-phosphate	48	4	53
5'-pyrophosphate	33	0	39
5'-triphosphate	25	0	29
Uridine	59	90	58
2'(3')-phosphate	40	4	44
5'-phosphate	30	4	39
5'-pyrophosphate	19	0	24
5'-triphosphate	15	0	20

[a]From data in Ref. 65. [b] 1, Ammonium hydrogen carbonate (200 m*M*)–sodium tetraborate (5 m*M*) on Whatman DE-81 (formerly DE-20) paper; 2, formic acid (50 m*M*) on DE-81 paper (converted into the formate form by *M* formic acid, and washed with water); and 3, 577:38:385 isobutyric acid–concentrated ammonium hydroxide–water on Whatman No. 1 or 3MM paper.

III. PARTITION AND ADSORPTION CHROMATOGRAPHY ON COLUMNS

1. Introduction

Column chromatography, which came into use at the beginning of the present century, is equally useful in many functional fields of analytical chromatography and in preparative organic chemistry, where it considerably extends the range of separation techniques.

According to the carrier and the elution system selected, two methods of separation are used. One method is that known as partition chromatography, although it is almost exclusively concerned with paper and, in part, thin-layer, microanalytical chromatography. Another method of column chromatography is classified as adsorption chromatography, and equally embraces column and thin-layer chromatography. Finally, ion-exchange chromatography is almost exclusively performed by the column technique, as is gel filtration on molecular sieves, although some workers also use this type of carrier for thin-layer chromatography. The column technique is, however, for all preparative separations, the method of choice, because the variations possible in the dimensions of the column are practically unlimited.

Because the same, or similar, types of carriers are used in partition and adsorption chromatography on columns, it is difficult to decide which mechanism is mainly responsible for the separation of a specific mixture; they will therefore be discussed together, and the two will be compared with the somewhat different techniques encountered in ion-exchange chromatography and gel filtration.

2. Practical Details

a. Carriers.—The main types of carrier are kieselguhr,[66,67] (Celite 545), cellulose[68] (used almost exclusively in partition chromatography), silica gel,[69] and aluminum oxide[70] (used for both methods, depending on the activity of the carrier; sometimes,the adsorptive type has some advantages). Most carriers are supplied commercially, in modifications, suited to column chromatography, that have a specified degree of activity (see Thin-layer Chromatography, p. 636) and particle size (*e.g.*, for aluminum oxide and silica gel) and that require no purifcation. Only commercial kieselguhr must usually be purified before use, by washing it with a strong acid (2–3 *M* hydrochloric acid) on a Büchner funnel or sintered-glass plate; it is then washed with distilled water (at least 10 volumes) until the washings are neutral, and then with alcohol (sometimes, finally, with ether). Kieselguhr is usually dried overnight at 100–130°.

b. Preparation of the Column.–It is important to perform all preliminary operations at the temperature at which the column will later be used, in order to avoid rupture and deformation of the column. It is recommended that, in general, the weight of adsorbent should be ∿30-60 times that of the material to be adsorbed. The height-to-diameter ratio of the column usually lies in the range of 10-20:1 (*cf.*, gel filtration, p. 614). The lower part of the column is filled with glass wool, over which a layer of glass beads is so placed as to form a horizontal surface. Sintered-glass plates are readily clogged and are, therefore, not recommended.

Although a number of methods are available for filling the column, the main ones include direct pouring of (*a*) a slurry of the packing with enough of the polar phase to ensure perfect pouring; (*b*) a suspension of the carrier in the stationary phase (the more polar one) or in the first component of the solvent system to be used later (see p.589); or (*c*) the dry carrier into the eluant-filled chromatographic tube through a funnel whose end lies below the liquid level, thus permitting a suspension to be formed immediately. Sedimentation is usually rapid, and the filling of a large column does not generally take more then 30 minutes. The column is then stabilized (at a constant flow-rate) either with the mobile phase or with the respective components. It is advantageous to place a circle of filter-paper on the top of the column to protect the carrier surface from being dispersed, but this is necessary only if the sample is applied in solution.

c. Preparation and Application of the Sample.–A mixture (obtained by evaporation of a solution to dryness) is usually dissolved in the least polar solvent in which its solubility is sufficient to provide a high concentration, necessary for sharp separation, and the solution is applied to the top of the column by means of a solvent of equal polarity, in such a way as not to disturb the carrier surface. The sample may also be applied to the carrier surface in the form of a thick slurry, obtained by dissolving the material in the stationary phase and adding the dry carrier (usually five times the weight of the sample). All of these methods afford a sufficiently sharp separation subsequently.

d. Selection of the Eluant and of the Method of Elution.-It must be stressed that, in general, the solvents employed in adsorption chromatography must be pure in the sense of being free from admixtures of more polar components and of impurities that would interfere in elution and would contaminate any of the compounds later separated.

In partition column-chromatography, a so-called mixotropic solvent series (see Table XVIII) is used to illustrate the similar properties of individual solvents (neighbors in the Table), the properties involved being mainly related to differences in polarity. The systems employed, are therefore, mainly binary and ternary, and are similar to those used for paper chromatography (the differences, usually, lying only in the ratios employed). The solvent systems are mainly based on alcohols and esters, the addition of acids, ammonia, and borate having the same purpose and influence as in paper chromatography (see p. 540). Such systems are suitable, in general, for separating bases, but are mainly used for nucleosides, as these have small structural differences, best exploited by use of the partition method.

The so-called eluotropic series (see Table XIX) is, in some respects, an abbreviated mixotropic series arranged in the reverse order (by ascending polarities); it permits selection of suitable solvents for adsorption chromatography according to their increasing capacity of elution (indicated, besides adsorbent properties, mainly by the solvent's dielectric constant, dipole moment, miscibility with water, and surface tension). This means that the solvent order is not absolute, but may vary slightly, depending on the adsorbent used.[72] The suitability of a solvent or mixture of solvents for a given separation is tested (see p. 640) by starting with a nonpolar solvent and progressing to a more polar one. In general, mixing starts with the addition of a small proportion (1–10%) of a more polar component to the nonpolar solvent, followed by (*a*) successive additions of the more polar component in steps of 10%, up to 50%, or, (*b*) continuous additions thereof, in gradient fashion. There is no point in increasing the concentration of the more polar component beyond 50%, because such a mixture would have, essentially, the eluant properties of the pure solvent itself.

For protected nucleosides, for which adsorption separation is best (as in the separation of nucleoside anomers, position isomers, and *O*- and *N*-glycosyl derivatives), the eluotropic region given by benzene–chloroform–ethyl acetate is used most, benzene and ethyl acetate being preferred because of their low reactivity, especially with respect to protective groups.

Elution may be conducted with a system of constant composition and concentration, or by a stepwise or continuous increase in the polarity of the mixture by

TABLE XVIII

Shortened Mixotropic Solvent Series[a]

Solvent	Solubility in water[b]	Dielectric constant[c]	Solvent	Solubility in water[b]	Dielectric constant[c]
Formamide		84	Acetic acid[d]		6.3
Water		81.1	Ethyl acetate	8.6	6.1
Formic Acid		58.5	Chloroform	1[g]	5.1
Methanol		31.2	Butyl acetate	0.5[i]	5
Isopropyl alcohol[d]		26	Diethyl ether	7.5	4.4
			Trichloroethylene	0.1	3.4
Ethyl alcohol[d]		25.8	*p*-Dioxane[d]		3
Propyl alcohol[d]		22.2	Isobutyric acid	20	2.6
Acetone[d]		21.5	Carbon tetrachloride	0.08	2.25
Butyl alcohol	7.9	19.2	Benzene	0.08[h]	2.24
tert-Butyl alcohol[d]		18.7	Toluene	0.05[i]	2.3
Cyclohexanone	2.4[f]	18.2	Hexane	0.01[g]	1.88
2-Butanone	35.3[e]	18	*s*-Collidine[k]	3.5	
Pentyl alcohol	2.7[h]	16	Dichloromethane[k]	2	

TABLE XVIII (continued)

Shortened Mixotropic Solvent Series[a]

Solvent	Solubility in water[b]	Dielectric constant[c]	Solvent	Solubility in water[b]	Dielectric constant[c]
Cyclohexanol	5.7[g]	15	Isopentyl alcohol[k]	2.7[h]	
1,1-Dichloroethane	0.6	10.65	Di-isopentyl ether[k]		
Tetrahydrofuran[d]		7.6			
Ethyl formate	11.8[i]	7.1			

[a]According to data published in Ref. 10. [b]In g per 100 ml of the solvent. Unless noted, the value given is the solubility at 20°. [c]Measured at room temperature. [d]Miscible with water in all proportions. [e]At 10°. [f]At 31°. [g]At 15°. [h]At 22°. [i]At 25°. [j]At 16°. [k]Values for dielectric constant are not available.

TABLE XIX

Eluotropic Solvent Series[a]

Solvent	Dielectric constant[b]	Shortened series
Acetic acid	-	
Water	81.0	
Methanol	31.2	+
Ethyl alcohol	25.8	
Propyl alcohol	22.2	+
Acetone	21.5	
Dichloromethane	9.08	
Ethyl acetate	6.1	
Chloroform	5.2	+
Diethyl ether	4.5	+
Trichloroethylene	3.4	
Benzene	2.3	+
Toluene	2.3	
Carbon tetrachloride	2.2	
Cyclohexane	2	
Hexane	1.88	+

[a]Compiled from data published in Ref. 71. [b]At room temperature.

changing the ratio of the components (different gradient systems). The progress of the elution is usually monitored by spectrophotometric measurement of the fractions [each, individually, or continuously, for example, by means of a Uvicord (LKB) u.v. absorption meter] which are collected by means of a suitable, automatic device (among the best-known collectors are those sold by LKB, Stockholm, Sweden). The volume of the fraction should generally lie in the range of 2 to 10% of the total volume of the column.

Some examples of the use of different carriers and solvents are described in detail in the following Section. Because of similarities to thin-layer separation, it is suggested that the two techniques be compared by means of these practical examples (see p. 641). For columns, the volume for elution is usually used to describe the separation; thin-layer chromatography is characterized, just as for paper chromatography, by the R_F values of the compounds separated.

3. Some Practical Examples

a. Chromatography* of Protected Nucleosides on a Column of Neutral Aluminum Oxide.[70,73–76]

(*i*) Quantitative Separation of Anomeric 2-Deoxy-D-*erythro*-pentofuranosyl Derivatives.–*Column:* Neutral aluminum oxide (Laborchemie, Apolda, Germany; 500 g) in a column having a ratio of diameter:height of 1:12, is used. To prepare the column, the carrier (supplied commercially in the ignited form) is deactivated by the addition of 4.5% of water, and benzene is added after the carrier has been kept for one day at room temperature. The homogeneous suspension thus obtained is poured into the column in one lot, and the column is washed with about three volumes of dry benzene.

Mixture Separated: A mixture[70] formed from the reaction of 6.0 mmoles of 2-deoxy-3,5-di-*O*-*p*-toluoyl-D-*erythro*-pentosyl chloride[76a] (15) with 6.0 mmoles of 5-chloro-2,4-dimethoxypyrimidine (16). The reaction yielded 5-chloro-1-(2-deoxy-3,5-di-*O*-*p*-toluoyl-α-D-*erythro*-pentosyl)-4-methoxy-2(1*H*)-pyrimidinone (17), the β-D anomer (18), the demethylation product 5-chloro-1-(2-deoxy-3,5-di-*O*-*p*-toluoyl-α-D-*erythro*-pentosyl)-uracil (9), and furfuryl *p*-toluate (20) [a degradation product of the halide].

Eluant: Benzene and ethyl acetate are gradually applied in the ratios and exponential gradient to be described.

The commercial solvents are further purified. Benzene is purified by thorough agitation with concentrated sulfuric acid, and then with saturated aqueous potassium hydrogen carbonate. After being dried, the benzene is distilled, and stored over sodium. Ethyl acetate is washed with saturated aqueous sodium hydrogen carbonate, and 50% aqueous calcium chloride, dried with anhydrous calcium chloride, and distilled.

*Details provided by Dr. M. Pryštaš.

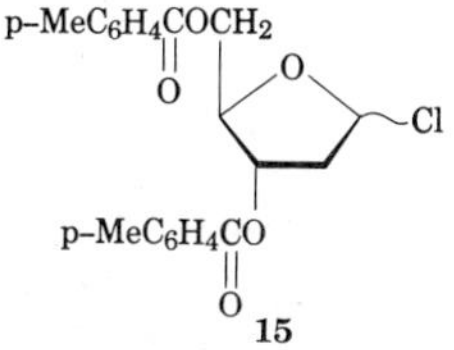

15 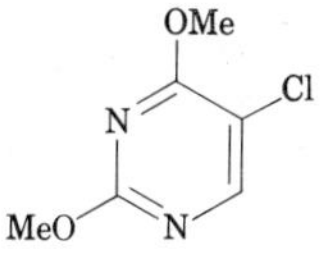 **16**

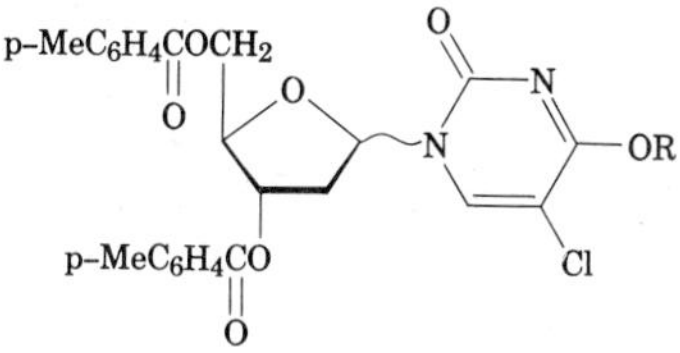

17, R=Me (α-anomer)
18, R=Me (β-anomer)
19, R=H (α-anomer)

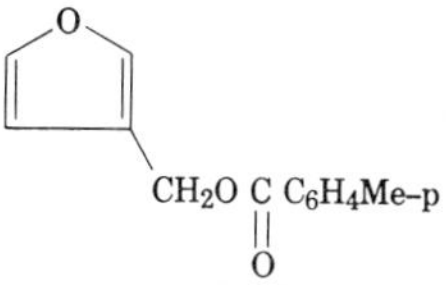

20

The gradient-elution equipment consists of a 2-liter, three-necked bottle containing 1.35 liters of benzene, and a 2-liter flask, turned upside down, that keeps the level in the 3-necked bottle at the original height by means of a ground glass joint (diameter 29 mm) drawn out to form a tube 6 mm in diameter. A total of 1.4 liters of 2:7 ethyl acetate-benzene in the flask will gradually flow through this tube, while the mixture is being agitated with a vibrating stirrer. The solvent mixture is removed continuously through an overflow tube that is connected to the column.

Flow Rate: This is set at 120 ml/hr, fractions of 20 ml being collected. Elution is monitored by the quenching of the fluorescence of paper in u.v. light, after an aliquot of the fraction has been applied dropwise to the paper.[a]

[a]The individual fractions are applied to a strip of chromatographic paper, to determine which of the fractions quench the fluorescence of the paper under a u.v. lamp. The character of the fractions is checked by chromatography on a loose, thin layer of the same aluminum oxide.

Procedure: A solution of the reaction mixture in 15 ml of benzene is carefully applied to the column. The column is eluted with 400 ml of benzene (fractions 1–20), an exponential gradient of 1.4 liters (fractions 21–90) of 7:2 benzene-ethyl acetate (1.4 liters) and benzene (1.35 liters), 600 ml of 2:7 ethyl acetate-benzene (fractions 91–120), 300 ml of 1:1 benzene-ethyl acetate (fractions 121–135), and 500 ml of ethyl acetate containing 0.5% of water (fractions 136–160). Fractions 5–9 yield 42% of the *p*-toluate 20, m.p. 49%; and fractions 11-15 yield 17% of the starting pyrimidine[a] 16, m.p. 71–72°. Fractions 55–81 are evaporated, and yield, after crystallization from ether, 3% of the β-D anomer (18), m.p. 161-162°. Fractions 83–115, processed in a similar way, yield the α-D anomer 17 (17%), m.p. 177–178 . Three recrystallizations (from ethanol) of the residue from the evaporation of fractions 143–148, yield 7% of the demethylated ester 19, m.p. 181–182°.

Column: Neutral aluminum oxide (500 g; activity grade II-III, Brockmann), in the form of a column (10:1), with benzene as the eluant. Preparation of the column is performed as in experiment *(i)*.

Mixture Separated: A mixture[76] of 6-methyl-1-(2,3,5-tri-*O*-benzoyl-β-D-ribosyl)uracil (22) and its anomer (23), obtained after the reaction of 7.7 g of the mercury(II) salt of 6-methyluracil with 260 ml of 110 m*M* 2,3,5-tri-*O*-benzoyl-D-ribosyl chloride[76b] (21) in acetonitrile.[77]

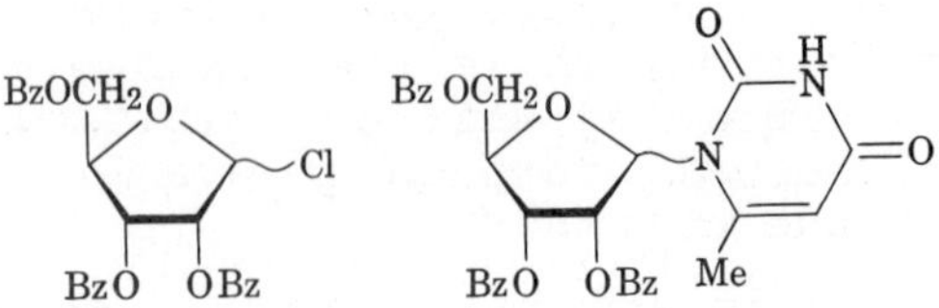

21

22 (β-anomer)
23 (α-anomer)

where Bz is benzoyl.

[b]Fractions 55–81 have an R_F value of 0.6 when they are chromatographed on a loose, thin layer of aluminum oxide of equal activity, with 3:2 benzene-ethyl acetate. Fractions 83–115 afford a single spot (R_F 0.35) under the same conditions.

Eluant: Benzene and ethyl acetate in the ratios discussed. The solvents are purified as described in procedure (*i*).

Flow Rate: This is set at 500 ml/h, and 500-ml fractions are collected.

Procedure: The reaction mixture containing compounds 22 and 23 is processed as usual, and is applied (in 50 ml of benzene) to the column, which is then successively washed with 1.5 liters of benzene, 1.5 liters of 3:17 ethyl acetate-benzene (fractions 1-3), 2.0 liters of 1:3 ethyl acetate-benzene (fractions 4-7), and 2.0 liters of 1:2 benzene-ethyl acetate (fractions 8-11). Fractions 3-5 are combined, evaporated to dryness, and the residue co-evaporated with toluene, to afford 65% of amorphous 23; fraction 9 (homogeneous) yields 12% of 6-methyluridine tribenzoate (22), m.p. 125–128°. The purity of the fractions is checked by thin-layer chromatography on dry-poured, neutral aluminum oxide of the same activity, with 1:1 benzene–ethyl acetate.

b. Chromatography* of Nucleosides and their Derivatives on a Column of Silica Gel

(*i*) Quantitative Separation of Protected Adenosine and its Aminoacyl Derivatives.—*Column:* Silica gel (250 g; 60-120um from the service laboratory of the author's Institute). Column (3.2 x 75 cm) prepared by filling the tube with a thin suspension in dichloromethane.

Mixture Separated: *N*-(Benzyloxycarbonyl)phenyl-L-alanine (24), 2',3'-di-*O*-[*N*-(benzyloxycarbonyl)phenyl-L-alanyl]-5'-*O*-trityladenosine (25), 2',3'-di-*O*-[*N*-benzyloxycarbonyl)phenyl-L-alanyl]adenosine (26), 2'(3')-*O*-[(*N*-benzyloxycarbonyl)phenyl-L-alanyl]-adenosine (27), 5'-*O*-trityladenosine (28), and adenosine (29) in the crude mixture obtained by the reaction of 3 mmoles of *N*-[(dimethylamino)methylene]-5'-*O*-trityl-adenosine with 4.1 mmoles of *N*-(benzyloxycarbonyl)-phenyl-L-alanine anhydride,[69,78] after removal of the (dimethylamino)methylene and 5'-*O*-trityl groups.

Eluant: A linear gradient of dichloromethane and dichloromethane containing 5% of methanol (total volume, 3 liters; volume of the gradient vessels, 2 liters) is used. The gradient-elution equipment consists of two cylindrical vessels, connected near the bottom by a ground-glass joint and a tube. The mixing vessel contains a magnetic stirring-bar, and the solvent is delivered to the top of the column by gravity.

*Details provided by Dr. S. Chládek.

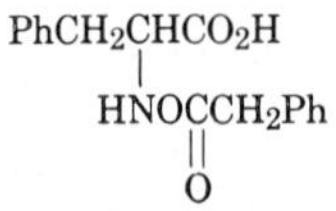

$PhCH_2CHCO_2H$ | $HNOCCH_2Ph$ (C=O)

24

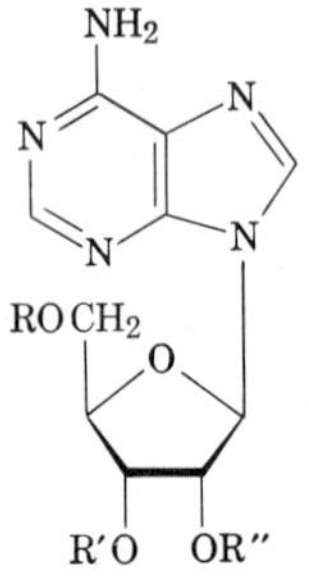

25, R = Tr, R′ = R″ = $PhCH_2CHC(=O)$–$HNOCCH_2Ph$ (C=O)

26, R = H, R′ = R″ = $PhCH_2CHC(=O)$–$HNOCCH_2Ph$ (C=O)

27 (2′,3′-O-linked: H, $PhCH_2CHC(=O)$–$HNOCCH_2Ph$ (C=O))

28, R = Tr, R′ = R″ = H

29, R = R′ = R″ = H

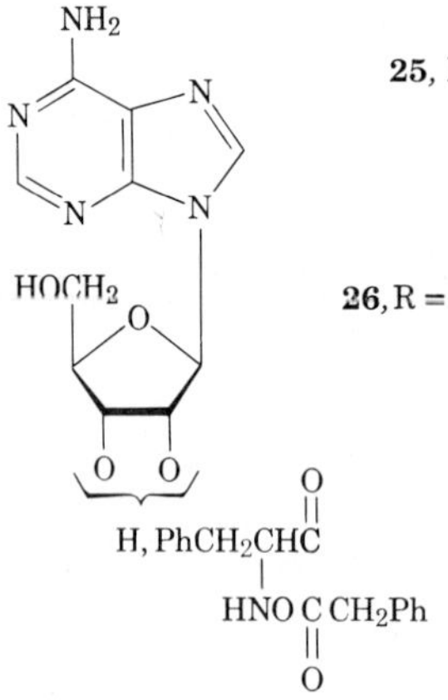

where Ph is phenyl and Tr is trityl.

Commercial dichloromethane is purified by distillation.

Flow Rate: This is set at 3 ml/min, and 15-ml fractions are collected. The elution is monitored continuously at 254 nm with a Uvicord absorption meter (LKB, Stockholm, Sweden).

Procedure: The crude reaction product is dissolved in 20 ml of dichloromethane, and the solution is carefully poured onto the column, so as to keep the adsorbent from being disturbed. The mixing vessel of the gradient equipment is connected to the column and, after the solvents constituting the linear gradient have passed through the column, chromatography is continued with 1 liter of dichloromethane containing 5% of methanol. The flow rate is controlled by the outflow valve of the column. Two main fractions (3 and 4) are detected from the u.v. absorption record for the eluate, besides small fractions that contain the aforementioned products having intact protecting groups. After the corresponding fractions have been combined, the solutions are evaporated in a rotary evaporator (30°/15 torr). The yield of compound 26 (eluted first) is 10%, and that of compound 27 (eluted later) is 52%. The two products are identified by elementary analysis and by i.r., u.v., and n.m.r. spectroscopy.[69]

(*ii*) Quantitative Separation* of the Oxo and Thio Derivatives of "6-Azauridine."—*Column:* Silica gel (125 g, 60-120 μm) as in section (*i*); column dimensions, 12:1. The silica gel is deactivated with 9% of water, and the column is prepared in benzene as described for aluminum oxide (see p. 582).

Mixture Separated: 2',3',5'-Tri-*O*-benzoyl-5-methyl-4-thio-"6-azauridine" (30), the product of the thiation[79] of 6.67 mmoles of 2',3',5'-tri-*O*-benzoyl-5-methyl-"6-azauridine" (31) with phosphorus pentasulfide (4.65 mmoles).

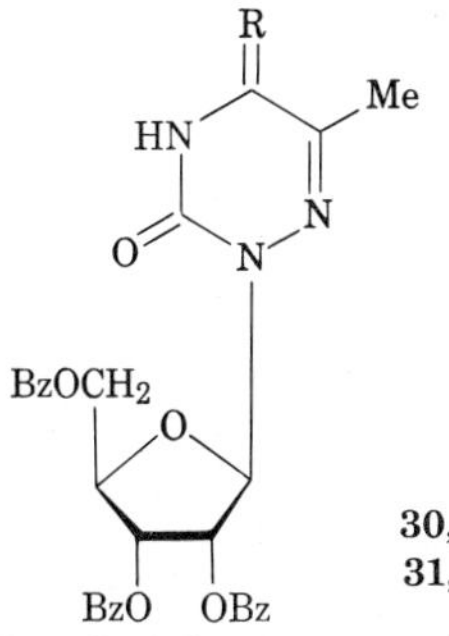

30, R = S
31, R = O

where Bz is benzoyl.

Eluant: Ethyl acetate-benzene in the ratios of 1:50, 1:19, and 1:9. The solvents are purified as before (see p. 583).

Flow Rate: This is set at 150 ml/hr, and fractions of 20-25 ml are collected. The elution is monitored by the orange color of the thio derivative and the quenching of fluorescence of paper in u.v. light by drops of the compound.

Procedure: The reaction mixture, conventionally treated,[79] is evaporated, and a solution of the residue in 20 ml of benzene is applied to the column, which is then successively washed with benzene (250 ml),

*Details provided by Dr. M. Prystaš.

1:50 ethyl acetate-benzene (250 ml), 1:19 ethyl acetate-benzene (750 ml; fractions 1-42) and 1:9 ethyl acetate-benzene (500 ml; fractions 43-63). The orange-colored fractions (20-43) yield 80% of compound 30, m.p. 135-138°. Fractions 52-54 contain the original nucleoside 31 (6%).

c. Partition Chromatography of Purine and Pyrimidine Bases on a Column of Kieselguhr.[80] — *Column:* Celite 545 and Microcel E (9:1 ratio) are purified by washing with a maximum of 5 volumes of 3 *M* hydrochloric acid, either by decantation or on the column (the eluate is yellow at first), followed by washing with distilled water until the eluate is neutral, and drying overnight at 100°. The column is prepared by placing in the tube (1.9 x 50 cm) 50 g of a 9:1 (w/w) mixture of the carriers by use of 28 ml of the lower phase of solvent *1)* (total height of column, 42 cm).

Mixture Separated: A solution (2 ml) of a mixture of adenine, cytosine, guanine, thymine, and uracil (0.25% of each component) in the lower phase of solvent 1.

Eluants: The upper phase (Solvent *1*) of 4:1:2 ethyl acetate-2-ethoxyethanol-10% formic acid, and (Solvent *2*) of 4:1:2 ethyl acetate-2-ethoxyethanol-water in the form of a linear gradient (total volume, 400 ml; decrease in concentration of formic acid, 10-0%); the linear gradient is arranged as with ion-exchange chromatography (see p. 605). This is followed by elution with 600 ml of the upper phase of solvent *3* (1:1:1 ethyl acetate-butyl alcohol-water).

Flow Rate: This is set at 60 ml/hr, 10-ml fractions being collected. Elution is monitored by measuring the u.v. absorption at 270 nm or by the drop-test on chromatographic paper (see p.584).

Procedure: The sample is applied to the top of the column as a mixture with 4 g of the carrier mixture. Several ml of the upper phase of solvent *1* are then added, and the column is connected to the mixing vessel of the gradient system. In the course of the elution by the linear gradient (400 ml), the following bases are eluted in succession: thymine (fractions 5-8; 40 ml), uracil (11-18; 80 ml), and adenine (21-36; 160 ml) On elution with the upper phase (600 ml) of solvent *3*, the bases eluted are: guanine (fractions 43-56; 140 ml) and cytosine (71-91; 210 ml). The yield of bases is usually greater than 95%.

IV. ION-EXCHANGE CHROMATOGRAPHY ON COLUMNS

Whereas separation of compounds by partition and adsorption chromatography is mainly based on the solubility, or on the reversible capacity of adsorption of the compounds to be separated when they are in contact with the solvent system and the carrier, separation by the ion-exchange chromatographic technique is based on the differences in the pK_a values of the respective bases, nucleosides, and nucleotides. The pK_a value gives a practical indication of the relative affinity of each compound for an exchanger.

1. Carriers

Because all nucleic acid components contain at least one group capable of ionization, these compounds in solution form cations and anions under suitable conditions. It is, therefore, possible to separate them with anion-, as well as with cation-exchange resins, the most commonly used of which are listed in Table XX. Besides synthetic ion-exchange resins (such as Dowex and Amberlite), which are used most frequently, weak ion-exchangers of the substituted-cellulose and dextran types are becoming increasingly popular (see Table XXI).

2. Preparation of the Column

The preparation of the ion-exchanger starts with its being converted into a suitable form. Most cation-exchange resins are supplied either in the H^+ or Na^+ form; and most anion-exchangers, in the Cl^- form. Even where the ion-exchange resin is to be used in the form supplied commercially, it is recommended that it be passed through at least one cycle, as in Cl→OH→Cl-, or Na^+→H^+→Na^+. When a different ionic form is needed, it is essential that this operation be performed. The procedure involves decanting, or washing the material in a column or on a large Büchner funnel, with 1-2 *M* hydrochloric acid or sodium hydroxide; with substituted celluloses and dextrans, the concentration should be 1/5th to 1/10th of these values. The resin is then washed with distilled water until the washings are neutral. The respective ionic form is then prepared by washing the OH- or H^+ form with a concentrated aqueous solution of a salt of the respective acid or hydroxide, and by rewashing with water. The final suspension in water is de-aerated and, by pouring down the wall of the

TABLE XX
Types of Ion-exchanger Commonly Used on the Column Chromatography of Components of Nucleic Acids[a]

Trade name	Form supplied	Cross linking	Capacity (mequiv/g)	Range of pH
Polystyrene sulfonic acids				
Dowex-50[b]	Na^+, H^+	1-16	5.1-5.4	0-14
Amberlite IR-120[c]	Na^+	8-10	4.25	1-14
BioRad AG-50 (w)[d]	H^+	1-12	4.9-5.2	1-14
Polystyrene quaternary ammonium derivatives				
Dowex-1[b]	Cl^-	1-16	2.0-3.6	0-14
Dowex-2[b]	Cl^-	4-10	2.7-3.7	0-14
Amberlite IRA-400[c]	Cl^-		3.3	0-12
BioRad AG-1[d]	Cl^-	1-10	3.0-3.5	0-14
Polymethacrylic acids (having carboxylic groups)				
Amberlite IRC-50[c]	H^+	2-3	10	1-14

TABLE XX (continued)

Trade name	Form supplied	Cross linking	Capacity (mequiv/g)	Range of pH
BioRex-70[d]	Na^+		10.2	4-14
Polyamine polymer				
Amberlite IR-45[c]	OH^-		5.6	0-7

[a]According to data published in Ref. 81 and in the respective producers' information leaflets. [b]Dow Chemical Co., Midland, Michigan. [c]Rohm and Haas Co., Philadelphia, Pennsylvania. [d]Bio-Rad Laboratories, Richmond, California.

TABLE XXI

Types of Commercially Available Cellulose and Dextran Ion-Exchangers Used for Separation of Nucleic Acid Components[a]

Trade name[b]	Cellulose (dextran) derivative	Functional group	Capacity (meq/g)
P-1 and P-11	phosphate	$-OPO_3H_2$	3.7
AE-11	*o*-(2-aminoethyl)cellulose	$-OC_2H_4NH_2$	1.0
ET-11	(ECTEOLA)	not fully known	0.5
DE-1 and DE-11	*o*-(2-diethylamino)ethyl-	$-OC_2H_4NEt_2$	1.0
CM-1 and CM-11	*o*-(carboxymethyl)-	$-OCH_2COOH$	0.6
DEAE-Sephadex	*o*-(2-diethylamino)ethyl-	$-C_2H_4NEt_2$	3.5
CM-Sephadex	*o*-(carboxymethyl)-	$-CH_2COOH$	4.5
SE-Sephadex	*o*-(2-sulfoethyl)-	$-C_2H_4SO_2OH$	2.3
QAE-Sephadex	*o*-{2-[diethyl-(2-hydroxypropyl)ammonium]ethyl}-	$-C_2H_4-\overset{+}{N}(C_2H_5)_2-CH_2-CH(OH)-CH_3$	3.0

[a]According to data in Whatman and Pharmacia information leaflets. [b]Nos. 1 and 11 denote the floc and powder forms, respectively.

tube, is poured into a column fitted, ordinarily, with a sintered-glass plate at the bottom to support the resin filling. The diameter-to-height dimensions of the column are usually in the ratio of 1:10-20. The adsorptive capacity of ion-exchange columns is generally considerable, and, because their dimensions may be varied at will, application of laboratory results to large-scale preparations is readily made.

3. Preparation of the Sample

For separation on a column, preparation of the sample consists mainly in adjusting the concentration of the salts and the pH of the solution to suitable values, as these are the main factors affecting the affinities of compounds for the exchanger. Tables XXII and XXIII list the pK_a values of various derivatives of some purines, pyrimidines, nucleosides, and nucleotides; on the basis of these values, conditions for application of the sample and its elution may usually be evaluated. For example, it is clear that, for a mixture of nucleotides, it is best to apply the mixture to the column at pH 6 and at a rather low ionic strength (total dissociation of phosphate groups). The process of application, to the column, of a sample thus modified is the same as with other column techniques, including gel filtration.

4. Eluants and Elution

Aqueous solutions of acids (formic and hydrochloric), hydroxides (ammonium and sodium), volatile ammonium salts (acetate, carbonate, formate, and hydrogen carbonate) are mostly employed, elution being conducted stepwise or by use of a solvent gradient. The advantage of such compounds is their relative ease of removal in further processing of the fractions collected. In principle, two types of chromatography are possible: *(a)* exchange or replacement chromatography, in which the ionic form of the ion-exchange resin is changed in the elution process (exchange of ions of the salt-forms of compounds separated), and *(b)* elution chromatography, the eluant containing, as the main ions, the ions of the equilibrium state of the resin in the column (this method is mainly employed for separation).

The aforementioned example of a nucleotide mixture (after adsorption at pH>6) requires a considerable decrease in the pH value and an increase in the ionic strength for successive elution, in order to decrease the dissociation of phosphate groups; for example, linear gradient of 0-4 *M* formic acid. Mixtures of

TABLE XXII

pK_a Values of Some Purines and Pyrimidines[a]

Compound	pK_a Value	
	Basic	Acid
Adenine	4.22	9.8
N,*N*-dimethyl-	3.87	10.5
N-methyl-	4.18	9.99
Adenosine	3.45	-
Cytidine	4.22	-
5-methyl-	4.28	-
Cytosine	4.45	12.2
5-methyl-	4.6	12.4
Guanine	3.3	9.2, 12.3
Guanosine	1.6	9.16
Hypoxanthine	1.98	8.94, 12.1
Inosine	-	8.75
Thymine	-	9.8

TABLE XXII (continued)

Uracil	-	9.5
Uridine	-	9.17
Xanthine	-	7.44, 11.2
Xanthosine	-	5.75

[a]Compiled from data in Ref. 82 (see Chapter I of this book).

TABLE XXIII

pKa Values of Some Important Nucleotides Related to Nucleic Acids[a]

Nucleotide	Primary phosphate	Amino group	Secondary phosphate	Enol group
Adenosine				
2'-phosphate	0.89	3.80	6.15	
3'-phosphate	0.89	3.65	5.88	–
5'-phosphate		3.74	6.05	–
5'-pyrophosphate		3.95	6.25	–
5'-triphosphate		4.0	6.48	
Cytidine				
2'-phosphate	0.8	4.36	6.17	
3'-phosphate	0.8	4.28	6.0	
5'-phosphate		4.5	6.3	
5'-pyrophosphate		4.6	6.4	
5'-triphosphate		4.8	6.6	

Guanosine				
2'-phosphate	0.7	2.3	5.9	9.7
3'-phosphate	0.7	2.3	5.9	9.7
5'-phosphate		2.4	6.1	9.4
5'-pyrophosphate		2.9	6.3	
5'-triphosphate		3.3	6.5	
Inosine				
5'-phosphate	1.54	-	6.0	8.9
Thymidine				
5'-phosphate	1.6	-	6.5	10.0
Uridine				
2'-phosphate	1.0		5.9	9.4
3'-phosphate	1.0	-	5.9	9.4
5'-phosphate		-	6.4	9.5
5'-pyrophosphate		-	6.5	9.4
5'-triphosphate		-	6.6	9.5

[a] Compiled from data in Ref. 82 (see Chapter I of this book).

dilute aqueous solutions of alcohols and salts are also used as eluants, especially for the separation of very similar nucleosides (see Table XXIV).

The relative positions of peaks on elution curves are determined by the different degrees of ionizability of the compounds separated; their sharpness and width are also influenced by the length of the column, the particle-size of the resin, and the flow-rate of the eluant. Elution of the individual peaks is monitored by continuous examination of the eluate (by use of, for example, a Uvicord u.v. absorption meter), or by measurement of the u.v. absorption of individual fractions collected, a method which readily provides quantitative evaluation.

5. Some Selected Examples

a. Chromatography* of Isomeric Dinucleotides on a Dowex-1 Column.[83] - *Column:* Dowex-1 X2 ($HCOO^{\ominus}$; 200-400 mesh); 1.5 x 25 cm. The corresponding amount of the ion-exchange resin is suspended in water, and successively washed with water, 2 *M* sodium hydroxide, and water, until the effluent is neutral. The resin is suspended in 2 volumes of 4 *M* formic acid, and then in 50 ml of 85% formic acid, and is finally washed with water until the effluent is neutral. The suspension is now poured into the column, or the second washing with formic acid and the final one with water may be done in the column.

Mixture Separated: A mixture of cytidylyl-(3'→5')-adenosine (32) and cytidylyl-(2'→5')-adenosine (33) (a total of 140 mg, including 9% of compound 33) in 5 ml of water. The mixture is obtained as a result of rigorous removal of protecting groups from the components before they are separated.[84]

Eluant: A linear gradient of 20-100 m*M* ammonium formate (total, 2 liters) in the arrangement described on p. 605, the volume of the mixing vessel and reservoir being 1 liter).

Flow Rate: This is set at 1.5 ml/min, 15-ml fractions being collected. Elution is followed by measuring the optical absorbance at 260 nm of fractions, with, for example, a Beckman DU spectrophotometer.

Procedure: The pH of the sample is adjusted to 8.0 with dilute ammonium hydroxide, and the sample is applied to the column, the top of the column is washed down with 2 ml of 20 m*M* ammonium formate, and the mixing

*Details provided by Dr. S. Chládek.

TABLE XXIV

Separation[a] of Nucleoside Mixtures on Columns of Dowex-1

Mixture No.	Components of mixture	Eluant[b]	Volume of effluent (ml)	Notes
1	2'-Deoxycytidine	A	140	
	Cytidine	A	725	
2	9-(2-Deoxy-α-D-*erythro*-pentofuranosyl)adenine	B	570	
	2'-Deoxyadenosine	B + A	600 + 120	
3	Cytidine	A	695	partial separation, only displaced by HCO_3^-
	1-β-D-Xylofuranosylcytosine	A	800	
	1-β-D-Arabinofuranosylcytosine	A + C	1000 + 350	
4	Adenosine	D	600	
	9-β-D-Psicofuranosyladenine	D	750	
5	Cytidine	A	710	
	Adenosine	A + E	1185 + 250	

TABLE XXIV (continued)

Mixture No.	Components of mixture	Eluant[b]	Volume of effluent (ml)	Notes
	Uridine	A + E + C	1185 + 275 + 285	displaced by HCO_3^-
	Guanosine	A + F + C	1185 + 275 + 430	
6	2'-*O*-Methyladenosine	A	170	
	3'-*O*-Methyladenosine	A	350	

[a]According to data published in Refs. 47 and 48. [b]A, aqueous methanol, 30%; B, water; C, 100 m*M* ammonium hydrogen carbonate; D, aqueous methanol, 55%; and E, aqueous methanol, 60%.

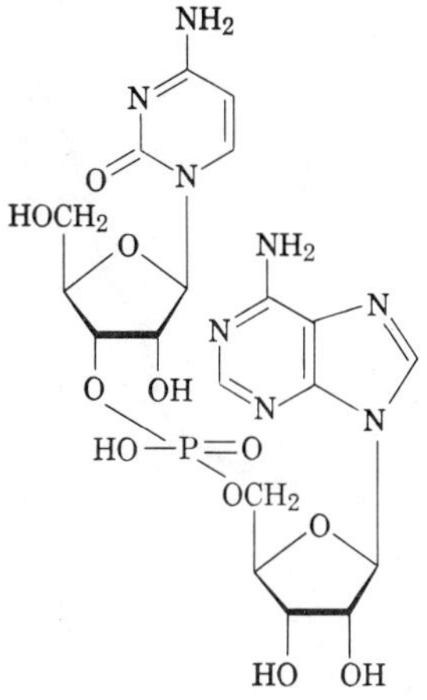

32 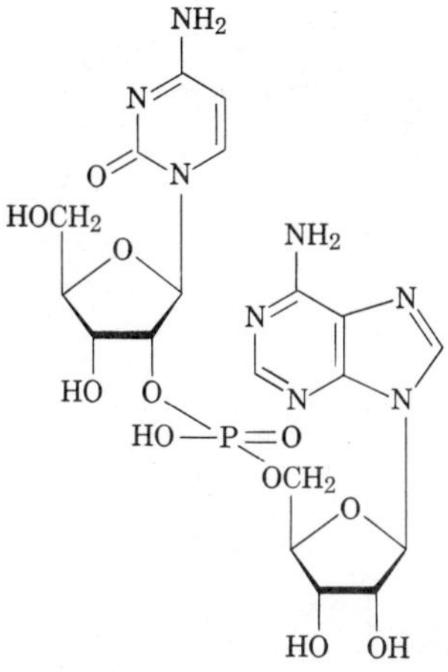 33

vessel of the gradient device is connected to the column. The elution profile is shown in Fig. 2. Fractions 52-57 contain the 2'→5' isomer (compound 33), which is stable to ribonuclease[32] and is not processed further. By combining fractions 62-85, 360 ml of a solution of compound 32 is obtained; this is concentrated in a rotary evaporator at 30°/15 torr, and the pH is adjusted to 8.0 with dilute ammonium hydroxide. Desalting is conducted on a column (2.5 x 70 cm) of Bio-Gel P-2 (200-400 mesh; 100 g) in water, with elution with water at a rate of 1 ml/min (30-ml fractions). The u.v.-absorbing fractions are combined, and lyophilized. Because the resulting residue still contains a small proportion of salts, a solution of the entire material in 3-5 ml of water may be applied to 3 sheets of Whatman 3MM paper (see p. 553) and developed with 7:1:2 isopropyl alcohol-concentrated ammonium hydroxide-water. The zones of dinucleoside phosphates are eluted as described on p.554, the eluate is concentrated in an evaporator, and the concentrate is lyophilized. The yield is 109 mg of chromatographically pure 32, containing less than 1% of isomer 33.

b. Separation of Nucleosides on a Dowex-1 Column.[48]
Column: Bio-Rad AG-1 X2 ($OH^{\ominus}$, 200-400 mesh); 1.6 x 20 cm. The resin is washed with 30% methanol (aqueous). Preparation of the column involves washing the original

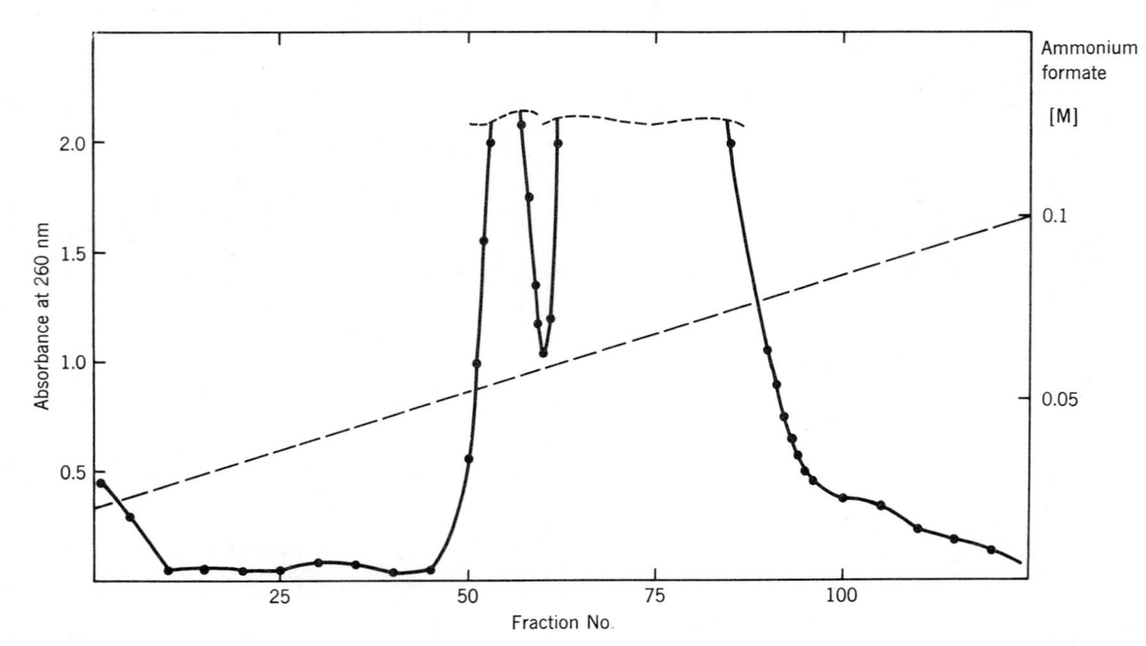

Fig. 2. Elution Curve Obtained on Separating Isomeric Dinucleoside (2'→5') and (3'→5') Monophosphates on a Column of Dowex-1. [The first peak corresponds to the (2'→5') isomer and the second one to the (3'→5') isomer. For experimental conditions, see text.]

$Cl^{\ominus}$ form of the commercial material with five volumes of 2 *M* sodium hydroxide, and then with water. The resin is washed 2-3 times with water by decantation and is poured into the column as a thick, aqueous slurry. Th resin in the column is washed with 2-3 volumes of 3:7 (v/v) methanol-water; the water must be free from carbon dioxide. The capacity of the column allows for a maximum of 100 mg of the sample.

Mixture Separated: A mixture prepared by dissolving adenosine (5), 2', 3'-*O*-isopropylidenadenosine (34), 2'-deoxyadenosine[84a]

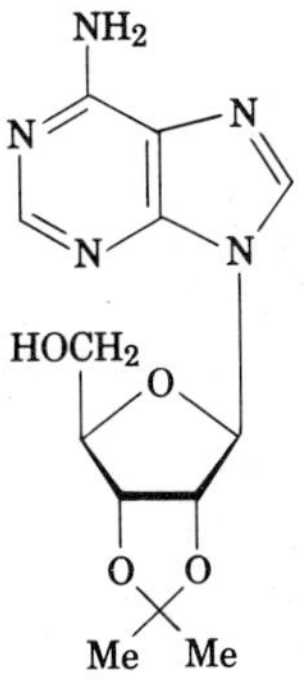

34

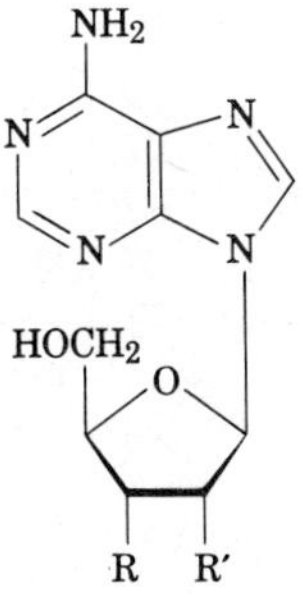

35, R = OH, R' = H
36, R = H, R' = OH

(35), and 3'-deoxyadenosine[84b] (36) (maximum of 3 mg of each component) in 5 ml of 30% methanol.

Eluant: This is 30% methanol (aqueous), followed by 46% methanol.

Flow Rate: This is set at 30 ml/hr (in some cases, up to 1 ml/min), 10-ml fractions being collected.

Elution: This is monitored by measuring the optical absorbance *A* (at 260 *n*m) of the fractions with a u.v. spectrophotometer, the contents of the various fractions being assessed from the A_{280}/A_{260} ratio.

Procedure: A sample (5 ml) is applied to the top of the column, and washed down with a small volume of the eluant; elution is then immediately continued. The volume of eluant required for quantitative elution of the components, while maintaining the order of elution from the column, is: 90 ml (34), 190 ml (35), 375 ml (36), and 830 ml of 46% methanol (5). With some of the mixtures mentioned in Table XXIV, eluant solutions (or their concentrations) must be combined, as with compound 5. The components may be recovered by evaporation of the eluate in a rotary, vacuum evaporator, after which the individual components can be processed further.

The same system, that is, 30% methanol, or combinations with more concentrated aqueous methanol or with salt solutions, may be used for separating other mixtures listed in Table XXIV.

c. Preparative Separation* of a Mixture of a Nucleoside, a Nucleotide, and a Dinucleoside Phosphate by Chromatography on a Column of DEAE-cellulose. – *Column:* DEAE-cellulose (4 x 100 cm) in the hydrogen carbonate form. The column is prepared from 500 g of DEAE-cellulose (Cellex D, standard capacity, from Calbiochem Inc., Los Angles, California, U. S. A.), which is stirred for 5 minutes with 5 liters of 100 m*M* sodium hydroxide. The suspension is filtered by suction on a large sintered-glass funnel, and washed with about 3 liters of water. The cellulose is resuspended in 5 liters of 100 m*M* hydrochloric acid, refiltered after 5-10 minutes, and washed with about 3 liters of water. The step of suspension in sodium hydroxide and washing with water is repeated. A vigorous stream of carbon dioxide (generated from Dry Ice) is passed overnight into the final suspension. The suspension is then filtered by suction, washed with about 10 liters of water, and washed by decantation with two 5-liter portions of water, the suspension being kept for 20 minutes before decantation. The resulting suspension of DEAE-cellulose is poured into a chromatographic tube (110 x 4 cm) fitted with a sintered-glass plate, a slight overpressure of air being used to give a flow rate of about 6-10 ml/min. The tube is thus filled to about 10 cm from the top. When correctly filled, the column should be not quite homogeneous; it should be composed of fine, horizontal bands caused by the fibrous structure of the material (because the cellulose employed is rather coarse). To stabilize the column, it is washed with 1-2 volumes of water under increased pressure to give a flow rate of 10-15 ml/min. The column is washed with 250 ml of 2 *M* triethylammonium hydrogen carbonate and 3-4 liters of water before use; this procedure is also used for regeneration after every separation. Thus prepared, the column is ready for use for some 50 separations without complications, provided that it is not damaged by the penetration of air or by mechanical contamination of the upper part of the column, or that its flow rate is not diminished as a

*Details provided by Dr. A. Holý.

result of excessive compression of the column.

Equipment: A Uvicord absorption meter and a fraction collector.

Mixture Separated: The reaction mixture obtained after condensation of *N*-acetyl-2',5'-di-*O*-acetylcytidine 3'-phosphate with 2',3'-*O*-(ethoxymethylene)-inosine.[85] Composition of the mixture: 2',3'-*O*-(ethoxymethylene)inosine (about 3.5 mmoles), cytidine 3'-phosphate (500 μmoles), and cytidylyl-(3'→5')-2',3'-*O*-(ethoxymethylene)inosine (about 500 μmoles) in 25 ml of water. The separation is conducted in order to isolate cytidylyl-(3'→5')-2',3'-*O*-(ethoxymethylene)inosine [an intermediate in the synthesis of cytidylyl-(3'→5')-inosine].

Eluant: A linear gradient of 0-200 m*M* triethylammonium hydrogen carbonate (total, 4 liters). A stock solution of 2 *M* triethylammonium hydrogen carbonate (pH 7.5) is prepared by cooling 3 liters of water in a 6-liter, Erlenmeyer flask in an ice-bath. During 3-4 hr, 1 kg of triethylamine is added dropwise from a separating funnel, carbon dioxide being simultaneously introduced through a sintered plate (the gas is generated in a 5-liter, round-bottomed flask that is half-filled with crushed Dry Ice). The introduction of carbon dioxide is continued overnight, and the solution is then filtered by gravity through a plate of low-density sintered-glass and made up to 5 liters with water. The pH of the solution should be tested before filtering; it should be 7.5, and, if it is more alkaline, it is adjusted by brief introduction of carbon dioxide. The 2 *M* stock solution thus obtained is stable for several months, if kept in a closed bottle at room temperature.

The linear-gradient system is provided by two 2-liter, two-necked Woulf bottles, the lower tubes of which are fitted with tight-fitting, one-holed rubber stoppers. A thick-walled capillary tube (1 mm i.d.) so bent as almost to reach the bottom at a distance of 3-5 cm from the bottle wall is pushed through the hole in each stopper. The outer ends of the two capillary tubes are connected by a tube of rubber or poly(vinyl chloride). It is advantageous to calibrate the two bottles by volumes of ∿500 ml. The solution in the mixing chamber is agitated by a magnetic stirrer or similar device (*e.g.*, a vibrating stirrer), and the eluant is fed from this bottle to the column by gravity or by use of a pressure pump. In the mixing chamber is placed 2 liters of water, and in the second bottle (the reservoir) is placed 2 liters of 200 m*M* triethylammonium hydrogen carbonate, prepared by diluting the stock solution. Air is removed from the tubing connecting the two bottles by elevating the reservoir, and the two bottles are placed on the same level. The stirrer in the mixing chamber is started, and the system is then ready for use.[c]

Flow Rate: This is initially set at 5 ml/min (water) and then at 3 ml/min (gradient); 30-ml fractions are collected.

Elution: This is monitored by continuous measurement of the u.v. absorption at 254 nm with a Uvicord absorption meter (LKB, Stockholm, Sweden).

Procedure: The sample solution is filtered through a layer (5 mm thick) of Kieselguhr on a dense, sintered-glass plate prewashed with water,[d] the filter is washed with 10 ml of water, and the filtrate is transferred with a pipet to the column, without disturbing the adsorbent. The solution is allowed to become gradually absorbed, and the walls are then washed down with a small volume of water, which is likewise allowed to be absorbed; finally, a layer of 50 ml of water is carefully formed above the top of the column. The inflow valve for buffer from the mixing chamber is opened, and the entire column is sealed air-tight and washed with water. The u.v-absorbing fractions (neutral, protected-nucleoside derivative) are collected in one flask during this washing, until the absorption decreases to zero. Elution is then continued by so opening the inflow valve for buffer from the mixing vessel as to give a flow rate of 3 ml/min. By consulting the u.v.-absorption record, fractions corresponding to the two separate absorbing peaks are combined, and the two solutions are separately evaporated to dryness in a rotary evaporator at 35°/15 torr. Each residue is redissolved in 50 ml of water, and the solution is evaporated to dryness under the same conditions. This procedure, which serves to remove traces of the volatile buffer, is repeated once more, the residue is dissolved in 20 ml of water and made to 50 ml with water in a volumetric flask, and the concentration is determined spectrophotometrically at 260 nm by diluting an aliquot

---

[c]A simple variant involves use of two beakers, or bottles of square cross-section, connected by a tube.

[d]Filtration through kieselguhr or some other neutral filtering material is recommended before every separation, even though, to the naked eye, the solution may appear to be clear. This procedure removes fine particles that would eventually clog the upper part of the column. Likewise, the eluant applied to the column should be prefiltered through a low-density, sintered-glass plate.

with 100 *mM* hydrochloric acid. These stock solutions are also tested for homogeneity by paper-chromatographic and electrophoretic means.[85] The first fraction (80-110 *mM* buffer) contains 470 μmoles of cytidylyl-(3'→5')-2',3'-*o*-(ethoxymethylene) inosine, and the second fraction (150-180 *mM* buffer), 320 μmoles of cytidine 3'-phosphate.[e] The stock solution of the first fraction is re-evaporated to dryness under the same conditions, for use in the preparation of cytidylyl-(3'→5')-inosine.[85]

d. Separation* of a Mixture of Isomeric Monophosphites of 1-β-D-Lyxofuranosyluracil by Preparative Column-chromatography on DEAE-cellulose in the Borate Cycle.- *Introduction:* The principle of the method consists in the formation of complexes of boric acid with the sugar moiety; in these complexes, vicinal hydroxyl groups are oriented *cis* to each other. The complexes formed are highly acidic and their pK_a values depend on the structure of the complex. In the example under discussion, separation of the 5'-phosphite (37), having a 2',3'-*cis*-diol group, from the 2'-isomer (38) is involved (with the possibility of formation of a complex between the 3'-and 5'-hydroxyl groups), as well as of separation from the 3'-phosphite (39) with no possibility of complex formation with boric acid at all. The difference between the respective pK_a values is utilized for separation on a weak ion-exchanger, namely, DEAE-cellulose.

Column: DEAE-cellulose (4 x 100 cm) in the hydrogen carbonate form. The column, prepared as described on p. 604, is washed with 500 ml of 2 *M* triethylammonium borate solution (for preparation, see the eluant), and then with water at a rate of 5-6 ml/min. The column is then equilibrated with 1 liter of 50 m*M* triethylammonium borate (prepared from the stock solution).

Mixture Separated: The reaction mixture from the preparation of monophosphites from 1-β-D-lyxofuranosyluracil,[86] containing 300 μmoles of 37, 200 μmoles of 38, and 450 μmoles of 39 in 25 ml of 50 m*M* triethylammonium borate (pH 7.5).

---

[e] The initial and final concentration of the buffer, which characterizes the course of the chromatographic process in the respective arrangement, is calculated from the elution record by plotting a straight line for the linear increase in concentration of the buffer from zero to the final value; the respective concentrations of buffer are obtained from the ordinates of the points for the first and last fractions.

---

*Details provided by Dr. A. Holý.

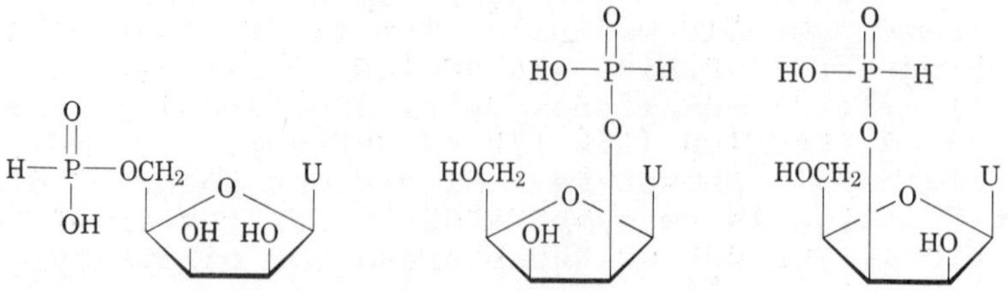

37 38 39

where U is uracil-1-yl.

Eluant: A linear gradient of 50-500 m*M* triethylammonium borate (total, 4 liters). A stock 2 *M* buffer solution is prepared as follows. A suspension of 318.5 g of boric acid in 3 liters of water is heated to 50° with agitation. Triethylamine (∿800 g) is added dropwise with stirring, until the solid dissolves, the solution is cooled to room temperature, and its pH is adjusted to 7.5 by adding more triethylamine. The solution is then made to 5 liters with water; it is stable for several months at room temperature, if kept in a closed bottle.

Flow Rate: This is set a 3 ml/min, 30-ml fractions being collected.

Elution: This is monitored with a u.v. absorption meter (*e.g.*, Uvicord).

Procedure: The pH of the solution is checked (if it is too low, it is adjusted with trietyhlamine) and the solution is applied quantitatively to the column. After it has been absorbed, about 50 ml of 50 m*M* buffer is carefully added, the buffer inflow-valve from the mixing vessel is opened, and elution is performed with a linear gradient of 2 liters of 50 m*M* triethylammonium borate (in the mixing vessel) and 2 liters of the 500 m*M* buffer (in the reservoir); the technique is described on p. 605. Three completely separated fractions are isolated (see Fig. 3); these are separately evaporated to dryness in a rotary evaporator at 35°/15 torr, and the separate residues are co-evaporated, under the same conditions, with five 50-ml portions of methanol. (Triethylammonium borate decomposes in the process, and volatilizes *in vacuo* in the form of triethylamine and trimethyl borate.[f])

---

[f] As a rule, five co-evaporations with methanol suffice to remove the buffer completely. If this is not achieved, it is recommended that the material be redissolved in 50 ml of methanol, 10 ml of water be added, and the evaporation and co-evaporation be repeated until the buffer has been completely removed.

The residues of the triethylammonium salts from the foregoing three fractions are transferred with methanol to 10-ml volumetric flasks, and each is made to volume with methanol. The composition and purity of the fraction are determined by electrophoresis in 200 m*M* triethylammonium borate, and by paper chromatography.[86] The content of each compound is determined spectrophotometrically by diluting aliquots with 10 m*M* hydrochloric acid at 260 nm (log ε_{260}^{pH}=4.0). Yields are 410 μmoles of 39 (220-310 m*M* buffer), 150 μmoles of 38 (420-450 m*M* buffer), and 280 μmoles of 37 (460-500 m*M* buffer), in the order of elution.

The column is regenerated by washing it with 250 ml of 2 *M* triethylammonium borate, 3 liters of water, and, 1 liter of 50 m*M* triethylammonium borate (pH 7.5).

e. Desalting of Monocleotides on a Column of DEAE-cellulose.[6] — *Column:* DEAE-cellulose carbonate (20 g; capacity 700 μequiv/g) in a column (4 x 17 cm). In preparing the column, the aqueous suspension of ion-exchanger is washed three times with water by decantation, to remove the fine particles. This treatment is followed by washing with 100 m*M* hydrochloric acid and 100 m*M* sodium hydroxide, as described in the preceding examples (see p. 604). After the $Cl^{\ominus}$ form has been washed with water, the aqueous suspension is poured into the chromatographic tube (fitted with a sintered-glass plate), washed with two volumes of *M* ammonium carbonate, and equilibrated with 100 m*M* ammonium carbonate buffer (pH 8.6).

Mixture Separated: A 20-ml sample, containing a mixture of monocleotides (derived from adenosine, cytidine, guanosine, and uridine; concentrations of each component 100-500 μg/ml) in 500 m*M* ammonium sulfate. In parallel samples, ammonium sulfate is replaced with an equal concentration of sodium chloride or 100 m*M* phosphate buffer of pH 7.

Eluant: 100 m*M* and 700 m*M* ammonium carbonate, pH 8.6.

Flow Rate: 180 ml/hr while the sample is being applied to the column; 100 ml/hr during elution of the components.

Elution: This is monitored by measuring the optical absorbance of individual fractions at 260 nm and by conductivity measurements.

Procedure: The nucleotide sample is diluted 1:10 with the 100 m*M* eluant, and the resulting solution is allowed to flow through the prepared column. Sample application is followed by elution with 100 m*M* ammonium carbonate until the conductivity values of the fractions collected are comparable to those of the eluant; the column is then eluted with 700 m*M* buffer. The

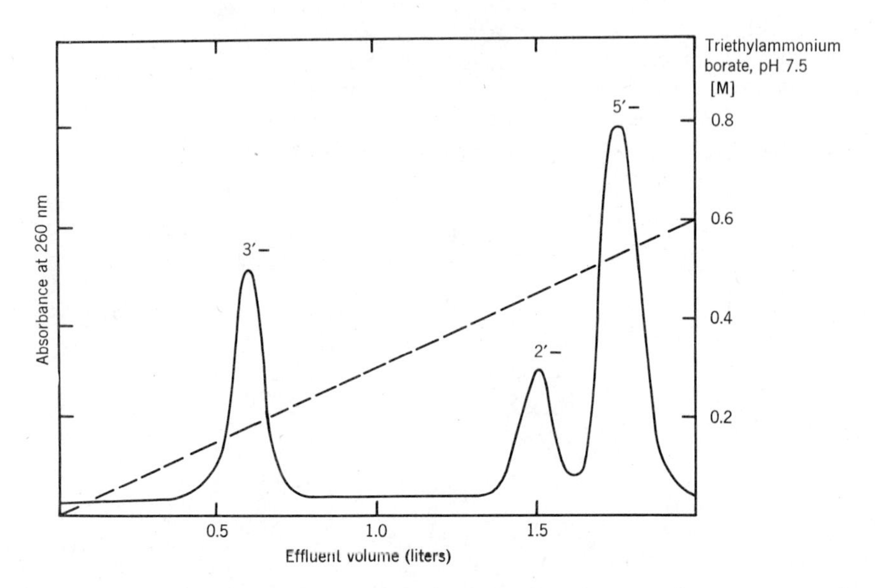

Fig. 3. Elution Curve Obtained on Separating Isomeric Uridine Monophosphites on a Column of DEAE-cellulose in the Borate Cycle. [Individual peaks are marked with the numbers corresponding to those of the positions of the phosphite group. For experimental conditions, see text.]

yield of nucleotides exceeds 98% under these conditions. Volatile ammonium salts may be removed by repeated evaporation to dryness in a rotary evaporator, or by sublimation

V. GEL FILTRATION

1. Introduction

Compared to other chromatographic techniques, gel filtration (the separation, on molecular sieves, of compounds, based on the size and shape of their molecules) is the most recent branch of chromatography, specifically, of column chromatography. The beginning of laboratory use of gel filtration dates back to the time when Sephadex (Pharmacia Fine Chemicals, Uppsala, Sweden), a modified dextran cross-linked to a three-dimensional network of polysaccharide chains, became commercially available.

The density of the cross-links between the individual chains determines the pore size and, therefore, the molecular-weight range in which the respective type of dextran may be used. The original application of this type of material was in the field of purification, where it replaced dialysis, as well as for the concentration and isolation of polymeric substances of high molecular weight. This use was then extended to include the range of compounds of low molecular weight, for which Sephadex exhibits excellent separative ability. This property does not depend solely on the slight differences in the molecular weight of otherwise closely related substances, but also on the ability of aromatic and heterocyclic compounds to be more or less strongly adsorbed on the carrier in the chromatographic column; the large number of hydroxyl groups per molecule of the carrier is apparently responsible for this interaction. Because of this property (even when it does not follow the main gel-filtration principle), gel filtration is a technique very suitable for the separation of nucleic acid components, especially bases and nucleosides, by use of eluants containing a low concentration of salts and having neutral pH. The carrier types applicable in this field are listed in Table XXV. This Table includes suitable types of another molecular sieve, namely, Bio-gel (Bio-Rad Laboratories, U.S.A.), based on acrylamide. Bio-gel is a cross-linked copolymer of acrylamide and methylenebis-(acrylamide) having gel properties similar to those of the dextrans.

Because, in gel filtration, some terms not commonly used in describing other chromatographic techniques are employed, as they serve to characterize the separation columns used, as well as the compounds to be separated, these terms will first be discussed.

The elution volume (V_e), the simplest characteristic of any compound, is the volume of eluant needed to elute the compound, from the application of it to the column to the point at which its maximum concentration in the respective fraction (the top of the relevant peak on the elution record) is reached; the volume of sample applied must be sufficiently small compared to the volume of the column (one-tenth of this value at most). If a large volume of the sample is applied, a plateau is formed on the elution curve, and the elution volume is then calculated from the point of application of the sample to the point of inflection on the rising part of the elution peak. In the majority of cases, in contrast to adsorption chromatography or ion-exchange chromatography with other carriers, the elution volume does not depend on the eluant employed.

Depending on the size of the molecules or particles of the compound studied, that is, on its molecular weight and, in part, on its conformation (molecular shape), the volume, V_o, of the external water (between

TABLE XXV

Types of Molecular Sieve Used in Gel Filtration of Compounds of Low Molecular Weight[a]

Trade name[b]	Particle size, μm	Water regain, ml/g	Bed volume, ml/g	Operating range of molecular weight
Sephadex G-10	40-120	1.0	2.0-3.0	700
Sephadex G-15	40-120	1.5	2.5-3.5	1500
Sephadex G-25, coarse	100-300	2.5	4.0-6.0	100-5000
medium	50-150	2.5	4.0-6.0	100-5000
fine	20-80	2.5	4.0-6.0	100-5000
Sephadex G-50, coarse	100-300	5.0	9-11.0	500-10000
medium	50-150	5.0	9-11.0	500-10000
fine	20-80	5.0	9-11.0	500-10000
Bio-Gel P-2	50-100 mesh	1.5	3.8	200-2600
	100-200 mesh	1.5	3.8	200-2600
	200-400 mesh	1.5	3.8	200-2600

TABLE XXV (continued)

Trade name[b]	Particule size, μm	Water regain, ml/g	Bed volume, ml/g	Operating range of molecular weight
Bio-Gel P-4	50-100 mesh	2.4	5.8	500-4000
	100-200 mesh	2.4	5.8	500-4000
	200-400 mesh	2.4	5.8	500-4000

[a]According to information leaflets of Pharmacia AB, and Calbiochem Inc.
[b]Chemical nature of Sephadex is dextran, and of Bio-Gel, is poly(acrylamide).

the individual particles of carrier) and the volume, V_i, of the inner water may both influence the elution volume in accordance with the equation: $V_e = V_o + K_D \cdot V_i$, where K_D is a factor to be discussed next. All compounds of high molecular weight that are larger than the pores in the gel employed are excluded from the column immediately after the volume V_o, whereas compounds of low molecular weight penetrate into the inner space of the gel particles, and should be eluted successively in the range V_i.

The aforementioned factor K_D is, therefore, the partition coefficient of the partition of the respective compound between the two aqueous media mentioned (the outer and inner volume), and it is defined as:

$$K_D = (V_e - V_o)/V_i = (V_e - V_o)/(V_t - V_o - m_g.V_g),$$

where V_t is the overall filling volume (which may be obtained by calibrating the column), and m_g is the weight and V_g is the partial specific volume (~0.6) of the gel employed. Most authors use the K_D value to characterize, in a sufficient manner, the behavior of the compound to be studied on the gel column. Consequently, it is used as the equivalent of the R_F value employed in paper and thin-layer chromatography. The factor K_D is, for non-adsorbing substances (those that do not interact with the carrier; $0 < KK_D < 1$), independent of the dimensions of the column, and it does not depend on the buffer and on the pH used.[87] With compounds that interact with the carrier (thus, for the K_D value, disturbing the independence of the factors mentioned[88]), $K_D > 1$ always applies (see Tables XXVI–XXIX); with Sephadex G-10, K_D may achieve values[3] of up to 14.

Gel filtration may be recommended for the separation of compounds of low molecular weight, especially because of the relative simplicity and ease of application of all operations involved. In contrast to ion-exchange chromatography on columns, gel filtration does not require gradient elution and cycling of the filler. Its application is not limited solely to qualitative, rapid separation of mixtures; it also satisfies the demands of precise, quantitative analysis of small amounts of such components as nucleoside bases and nucleosides.[88]

2. Preparation of the Column

The preparation of a gel column is similar to that for ion-exchange chromatography, but the regularity of the separation process is more sensitive to homogeneity of the column in terms of the particle size and the layers in the chromatographic tube.

Because high height-to-diameter ratios (10–100:1 and above) for the tube are usually used in gel filtration, the influence of the walls should be eliminated by applying a film of a silicone to the inner surface of the tube. For this purpose, the use of a 1% solution of dichlorodimethylsilane in benzene or carbon tetrachloride is recommended. The column, thoroughly cleaned and degreased, is almost filled with the solution; it is allowed to stand for some time, and then the solution is poured out and the tube (bearing the resulting inner film) is dried at 60-100°. The procedure may well be repeated in order to thicken the hydrophobic layer. As with all column separations, it is advisable that the "dead space" below the column filling be as small as possible, in order to avoid mixing of the separated zones during outflow from the column.

The preparation of the gel suspension involves addition of the dry gel in small amounts to distilled water, while the slurry is agitated constantly (use of a magnetic stirrer is recommended). A low ionic strength hinders gel aggregation in the suspension; such aggregation is seldom observed when pure distilled water, alone, is used. The suspension should, preferably, be washed by decantation with several portions of fresh solution, in order to remove the slowly sedimenting gel portion, and thereby ensure good permeability of the column. To decrease the content of u.v.-absorbing impurities that might be present, the suspension may be washed several times with 100 m*M* sodium hydroxide and 100 m*M* hydrochloric acid (more-concentrated solutions, that might damage the dextran polymer, should not be used). The final suspension in the elution buffer is de-aerated by keeping it under vacuum for 30 minutes, with occasional agitation; it is then poured into a column (preferably, down a glass rod), the column having first been one-quarter filled with de-aerated, elution buffer. When a column is fitted with a sintered-glass septum, a layer (2 cm deep) of glass spheres (0.5–1 mm diameter) may be placed over the septum to diminish the possibility of its becoming clogged. The use of columns produced by Pharmacia Co. (Uppsala, Sweden) is recommended; these satisfy all of the necessary conditions, namely, only a slight "dead space", a polyamide fiber-network as support for the filling, and a sample applicator that avoids the possibility that the carrier might become disturbed. It is advisable to pour the entire suspension into the column in one portion, in order to avoid formation of unwanted layers in the carrier when the suspension is added portionwise. The filling operation should be slow; this may be achieved by suitable control of the flow by means of hydrostatic pressure. When the filling has settled completely, the flow rate may be gradually raised to the level required for the experiment; the column is then washed at this rate with several volumes of the eluant. If a Pharmacia chromatographic tube and applicator are not used, it is recommended that a circle of filter paper be placed on the top of the filling to keep it from being disturbed.

In the practical determination of the column constants V_o and V_i, small samples of Blue Dextran 2000 (0.4% solution), hemoglobin, RNA, and DNA (for V_o, all of these compounds are completely excluded by the gel matrix), and of such low molecular-weight compounds as acetone (0.6% solution), potassium chloride, and glycine are used. The latter penetrate into the carrier particles, and their elution volume is equal to the sum (V_o + V_i). In some instances, V_i may be calculated as the product $mg.V_r$, where V_r is the water-regain/g of dry gel.

3. Application and Elution of the Sample

An application of sample is made to the surface of the gel filling while it is in the process of drying out; either the entire volume above the filling is allowed to flow out, or the excess liquid is carefully removed by means of a pipet to avoid disturbing the gel. The entire sample is then washed down with a small volume of the eluant, and the column may then be connected to the reservoir for elution. When low molecular-weight compounds are used, the differences between the viscosity and density of the sample and of the eluant are usually not great, so that there is no need to modify the sample in this respect.

A suitable volume of sample corresponds to ∿2–10% of the volume of the column filling; but, the smaller the sample volume, the more distinct is the separation. The important relationship between the separation effect and the sample volume follows from the small capacity of the gel, as compared to that of ion-exchangers. In theory, two compounds having, respectively, the partition coefficients K_D' and K_D'' should not be capable of being separated if the sample volume is greater than the product $(K_D'-K_D'')V_i$. However, this is only the fundamental condition, whereas separation depends, practically, as already stated, on many factors, such as flow rate, sample viscosity, size of gel particles, and homogeneity of the filling in the column.

Elution is usually conducted at a constant flow-rate, which should not differ too much from the self-imposed rate maintained by the maximum difference of levels and the permeability of the column. All types of laboratory pump may be employed. An increase in the flow rate will cause a decrease in the sharpness of the elution peak and, therefore, of the separation. Increase in the size of the gel particles (superfine → coarse) has a similar influence.

Often, the eluant is distilled water, or buffer solutions, or solutions of salts whose concentration should not exceed 100 m*M*. The pH of these solutions usually lies in the neutral region, from pH 6–8 (phosphate buffer, ammonium salts of weak acids, or sodium chloride). Extreme values of pH, especially towards the alkaline region (pH 9-11), mainly influence the magnitude of the elution volume for the compound involved,[3] and decrease the absorption thereof.

In the organic chemistry of nucleic acid components, gel filtration is especially suitable for desalting and for separation of groups of related compounds. When specific adsorption, mainly of purine derivatives, is utilized, bases, nucleosides (as a group or as individual compounds), and nucleotides, as well as the lower oligonucleotides, may be separated from each other.[3, 87-92,96,97]

4. Some Practical Examples

a. Separation of Adenine, Adenosine, and Adenosine 3'-Phosphate on Sephadex,[90] as an Example of Group Separation of Nucleic Acid Components.-*Column:* Sephadex G-25 (50-150 μm); 2.2 x 90 cm.

Mixture Separated: A mixture of 2 mg of each component in a total volume of 2 ml (in distilled water, or elution buffer).

Eluant: 0.005% aqueous ammonium carbonate, pH 7.6.

Flow Rate: This is set at 1 ml/cm^2/min, 5-ml fractions being collected.

Elution: The process is monitored by measurement of the optical absorbance at 260 nm of the fractions, with a Beckman DU spectrophotometer. The total time for elution is 7 hr. To determine the V_o value, ribosomal RNA is used, the V_i value being calculated from the weight of gel used and the water-regain. The K_D values corresponding to the individual elution-peaks in Fig. 4

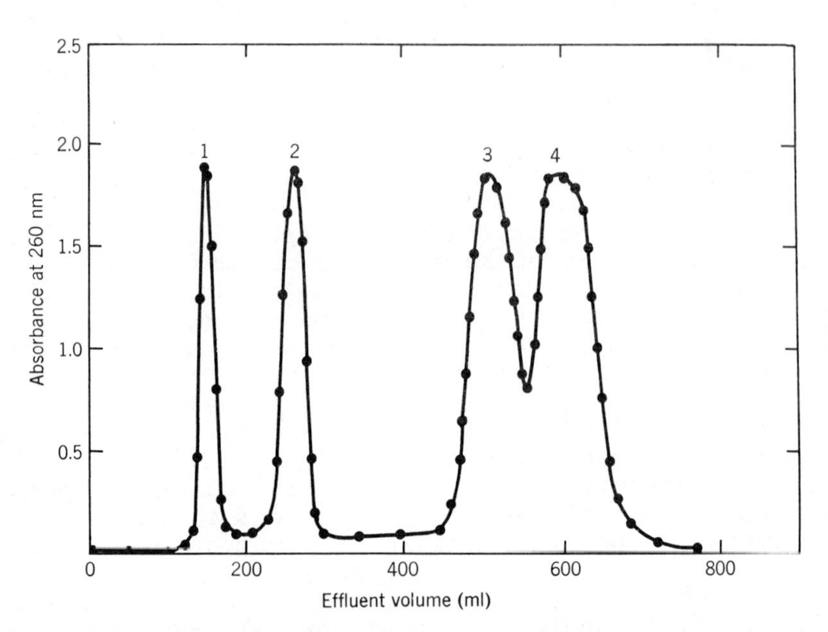

Fig. 4. Separation of a Mixture of Adenine, Adenosine, and Adenosine 3'-Phosphate on a Sephadex G-25 Column, [The order of elution peaks corresponds to D-ribonucleic acid (peak 1; used for estimations of V_o), adenosine 3'-phosphate (2), adenosine (3), and adenine (4), respectively. For experimental conditions, see text.]

are 0 (RNA, V_o), 0.85 (adenosine 3'-phosphate), 2.59 (adenosine), and 3.38 (adenine).

A similar separation may be achieved under the same conditions with guanine derivatives, whereas pyrimidine derivatives are not separated in this way, owing to the small differences in their K_D values[90] (cytidine 1.66, cytosine 1.80, uracil 1.82, and uridine 1.66). When the diameter-to-height ratio of the column is increased to 3.5 x 180 cm and the pH of the eluant is increased to 9, satisfactory conditions are obtained for separating[91] thymine (1.64), thymidine (1.23), and thymidine 5'-phosphate (0.89), for which a solvent system for other types of chromatography is usually difficult to find.

In all these examples, the influence of adsorption in the separation process is involved, because the molecular-weight limit, at which the influence of the molecular sieve starts, is about 1000 with this type of Sephadex.[87] The applicability of this type and other types of Sephadex for similar separations may be judged from Tables XXVI–XXVIII, where the K_D values are given, together with the eluant most frequently employed and

TABLE XXVI

K_D Values of Some Purines and Pyrimidines on Columns of Various Sephadexes[a]

Compound	K_D Value on column of									
	Sephadex G-10			Sephadex G-15		Sephadex G-25				
	1[b]	2[c]	3[d]	1[b]	3[d]	1[b]	3[d]	4[e]	5[f]	6[g]
Adenine	6.00	7.66	6.56	4.26	4.62	2.42	3.51	2.2	3.38	3.62
Cytosine	1.43	1.84	1.50	1.26	1.50	1.29	1.55	1.6	1.80	1.69
Guanine	3.35	5.84	3.93	2.80	3.22	2.02	2.70	-	3.40	3.23
Hypoxanthine	2.17	2.83	2.31	1.89	2.20	1.56	-	1.6	-	-
Orotic acid	1.31	2.05	-	1.22	-	1.22	-	-	-	-
Thymine	1.86	2.23	1.56	1.63	1.45	1.24	1.30	-	-	1.54
Uracil	1.56	1.91	1.12	1.50	1.18	1.18	1.20	1.1	1.82	1.54
Xanthine	3.15	4.34	2.31	2.89	2.20	2.00	-	1.8	-	-

[a]According to data published in Refs. 3 and 87–91. [b]Column dimensions, 1.5 x 25 cm; elution with 130 m*M* ammonium formate (pH 6).[87] [c]Column dimensions, 1 x 100 cm; elution with 50 m*M* sodium dihydrogen phosphate (pH 7).[3] [d]Column dimensions, 1.3 x 150 cm; elution with 10 m*M* ammonium carbonate (pH 9).[88] [e]Column dimensions, 3.5 x 35 cm; elution with water.[89] [f]Column dimensions, 2.2 x 90 cm; elution with 0.005% aqueous ammonium carbonate (pH 7.6).[90] [g]Column dimensions, 3.5 x 180 cm; elution with 10 m*M* ammonium carbonate (pH 9).[91]

TABLE XXVII

K_D Values* of Some Nucleosides on Columns of Various Sephadexes[a]

Compound	K_D Value on column of									
	Sephadex G-10			Sephadex G-15		Sephadex G-25				
	1[b]	2[c]	3[d]	1[b]	3[d]	1[b]	3[d]	4[e]	5[f]	6[g]
Adenosine	3.36	4.23	3.37	2.92	2.79	1.87	2.15	1.7	2.59	2.50
2'-deoxy-	3.23	4.23	-	2.81	-	1.81	-	-	-	-
Cytidine	0.99	1.32	1.06	1.06	1.18	1.13	1.25	1.2	1.60	1.42
2'-deoxy-	1.09	-	-	1.16	-	1.20	-	-	-	-
Guanosine	2.43	3.14	2.21	2.29	2.09	1.77	1.90	1.6	2.60	2.30
2'-deoxy-	2.66	3.28	-	2.40	-	1.73	-	-	-	-
Inosine	1.26	1.56	1.95	1.31	1.50	1.25	-	1.2	-	-
2'-deoxy-	1.26	1.60	-	1.30	-	1.14	-	-	-	-
Orotidine	0.62	-	-	0.70	-	0.93	-	-	-	-
1-β-D-Ribofuranosylthymine	1.00	-	-	1.15	-	1.09	-	-	-	-
Thymidine	1.24	2.23	0.75	1.29	1.02	1.04	1.10	-	-	1.23
Uridine	1.07	1.28	0.75	1.21	0.85	1.12	1.00	1.0	1.66	1.27
2'-deoxy-	1.09	-	-	1.15	-	1.05	-	-	-	-
Xanthosine	1.83	2.51	1.94	1.95	1.50	1.56	-	-	-	-

TABLE XXVIII

K_D Values* of Some Nucleotides on Columns of Various Sephadexes[a]

Compound	K_D Value on column of									
	Sephadex G-10			Sephadex G-15		Sephadex G-25				
	1[b]	2[c]	3[d]	1[b]	3[d]	1[b]	3[d]	4[e]	5[f]	6[g]
Adenosine										
5'-phosphate	0.91	0.93	0.25	1.37	0.58	1.35	0.80	0.1	0.85	1.18(2')
5'-pyrophosphate	0.41	-	-	0.74	-	1.03	-	-	-	1.27(3')
5'-triphosphate	0.29	-	-	0.50	-	0.80	-	-	-	-
Cytidine 5'-phosphate	0.4	0.40	0.12	0.54	0.37	0.87	0.70	0.1	0.66	0.69
Guanosine 5'-phosphate	0.8	0.82	0.18	1.07	0.43	1.25	0.80	0.4	0.91	1.04
Inosine 5'-phosphate	0.45	-	-	0.71	-	0.85	-	-	-	-
Thymidine										
5'-phosphate	0.52	-	-	0.73	-	0.92	-	-	-	0.89
5'-pyrophosphate	0.28	-	-	0.44	-	0.60	-	-	-	-
5'-triphosphate	0.19	-	-	0.32	-	0.56	-	-	-	-
Uridine 5'-phosphate	0.45	0.44	0.12	0.74	0.10	0.86	0.70	0.1	0.66	0.69
Xanthosine 5'-phosphate	0.60	-	-	0.79	-	1.09	-	-	-	-

*Footnotes *a–g* as in Table XXVI.

the column dimensions. The possibility of desalting solutions of similar compounds may also be judged on this basis, because most of the simple cations and anions have K_D values close to unity[98] (see p. 626). Compounds of similar types are compared in Fig. 5 by elution volumes (V_e) for columns of different types of Sephadex.

Sephadex G-10, having the highest cross-linkage density, is the most suitable carrier for separating bases from nucleosides and their derivatives, because, besides having increased adsorption, the molecular sieve may also participate in the separation process. Purine derivatives, which are adsorbed most strongly[3,88–90] do not, however, always have reproducible elution-volumes with this type of Sephadex—neither with different pH values of the eluant, nor when applied to the column in different concentrations, volumes, and combinations with other compounds[96] (this situation applies mainly to the bases). On the other hand, pyrimidine derivatives have very stable elution-volumes.[96] The best separation of thymine, thymidine, and thymidine 5'-phosphate may be achieved on a column (1.5 x 25 cm) of this type of Sephadex with 130 m*M* ammonium formate at pH 6 at 30 ml/hr, the K_D values obtained being 1.86, 1.24, and 0.52, respectively.[87] In the field of thymidine analogs, thymidine may be readily separated[97] from 5-bromo-2'-deoxy-uridine[98a] and 2'-deoxy-5-iodouridine (1.5 x 90 cm; elution with citric acid–phosphate buffer of pH 3.5 at a rate of 25 ml/hr), but it cannot be separated from 2'-deoxy-5-fluorouridine.

It may therefore be stated that, in general, Sephadex G-10 and, similarly, Bio-Gel P-2 are suitable mainly for separating pyrimidine bases from other compounds and from each other, and for separating pyrimidine and purine nucleosides from nucleotides[96] (which, however, are not separated from each other).

b. Influence of the Position of the Terminal Phosphate Group, and Separation of Oligonucleotides on a Sephadex Column.[92]—*Column:* Sephadex G-25 (20-80 μm), 1 x 175 cm; Sephadex G-50 (20-80 μm), 1 x 170 cm.

Mixture Separated: A synthetic mixture of oligonucleotides containing a known number of monomer components in which the position of the phosphate group is known (for comparison with products obtained by the reaction of thymidine 5'-phosphate with ethyl phosphate followed by pyrophosphate cleavage with acetic anhydride[99]).

Eluant: 0.005% aqueous triethylammonium hydrogen carbonate.

Flow Rates: 0.7 ml/min for G-25 Sephadex, and 0.3 ml/ for G-50 Sephadex. Elution is monitored at 254 nm by

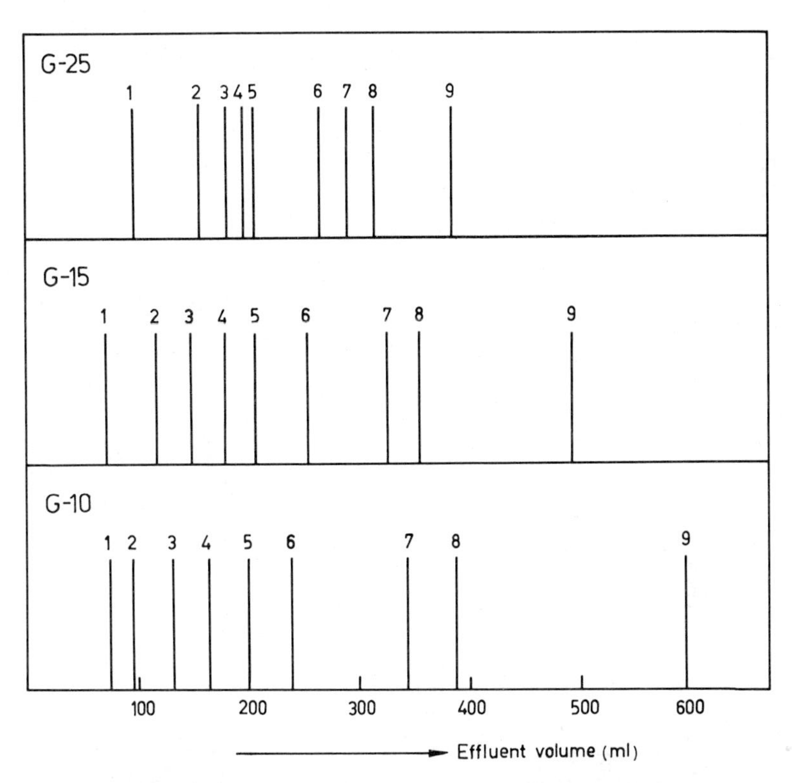

Fig. 5. Diagrammatic Representation of Elution Volumes of Some Purines, Pyrimidines, and Nucleosides on Sephadex G-10, G-15, and G-25 Columns.[88] [(1) D-Ribonucleic acid, (2) adenosine 5'-phosphate (both used for characterization of a column), (3) uridine, (4) uracil, (5) thymine, (6) guanosine, (7) adenosine, (8) guanine, and (9) adenine.]

means of a Uvicord absorption meter. The V_o value is determined with a soluble RNA preparation, and the V_i value, from the elution volume of tritiated water. The K_D values obtained in this way are summarized in Table XXIX.

Assessment: The number of phosphate groups and their different positions determine the behavior of the respective compound on any one type of Sephadex. The K_D value of a specific compound decreases in relation to the position of the terminal phosphate group in the series: zero

TABLE XXIX

K_D Values of Thymidine and Some of its Oligonucleotides having Differences in the Position of Their Terminal Phosphate Group[a]

Compound	K_D Value on column of		
	Sephadex G-25		Sephadex G-50
	1[b]	2[c]	2[c]
Thymidine	1.23	0.99	0.98
TpT[d]	-	0.51	0.61
$T(pT)_2$[d]	-	0.34	0.54
$T(pT)_3$[d]	-	0.12	0.30
$T(pT)_5$[d]	-	0.04	0.14
$T(pT)_9$[d]	-	0.01	0.05
pT[e]	0.89	0.31	0.50
$(pT)_2$[e]	0.48	0.13	0.30
$(pT)_3$[e]	0.24	0.06	0.22
$(pT)_4$[e]	0.12	0.02	0.13
$(pT)_5$[e]	0.06	-	-
$(pT)_6$[e]	0.03	-	0.06
Tp[f]	-	0.32	0.47
$(Tp)_2$[f]	-	0.12	0.30
pTp[f]	-	0.07	0.22
$p(Tp)_2$[f]	-	0.04	0.18

[a]According to data in Refs. 91 and 92. [b]Column dimensions, 3.5 x 180 cm; elution with 10 m*M* ammonium carbonate (pH 9).[91] [c]Column dimensions 1 x 170 cm; elution with 0.005% aqueous triethylammonium hydrogen carbonate.[92] [d]Prepared from 5'-derivatives by treatment with alkaline phosphatase from *Escherichia coli*. [e]Synthesized according to Ref. 93. [f]Synthesized according to Ref. 94.

>5'(3' or 2')>3',5'-diphosphate. Thus, thymidine dinucleotides having phosphate groups at different terminal positions have the following K_D values under the conditions specified: 0.51 and 0.61 [thymidylyl-(3'→5')-thymidine on G-25 and G-50, respectively], 0.13 and 0.30 [thymidylyl-(5'→3')-thymidine 5'-phosphate], 0.12 and 0.30 [thymidylyl-(3'→5')-thymidine 3'-phosphate], and 0.04 and 0.18 [5'-*O*-phosphonothymidylyl-(3'→5')-thymidine 3'-phosphate]. Sephadex G-25 and G-50 are very suitable carriers for separating oligonucleotides up the tetranucleotide level.[90,92]

For terminal phosphate groups, simulation of molecular-weight variations are involved, rather than the influence of adsorption. Terminal phosphate groups probably bind neighboring water molecules, and these may be bound with a strength sufficient to give the effect of a higher molecular weight when they are in solution. This influence may be utilized for ascertaining the presence of a terminal phosphate group.[92]

c. The Influence of Substituents on the Separation of Compounds of Purine Bases (Nucleosides and Nucleotides) by Means of Gel Filtration on Sephadex G-10 and on Predicting the V_e Value for a Compound of Specified Structure.[3]—*Column:* Sephadex G-10, 1 x 100 cm.

Mixture Separated: Mostly mixtures of individiual compounds that are used to determine their V_e values, in amounts sufficient for detection by measurement of u.v. absorption.

Eluant: 50 m*M* Sodium dihydrogen phosphate pH 7 (adjusted with sodium hydroxide). Regeneration of the column is accomplished by washing with several volumes of 50 m*M* sodium hydroxide and the elution buffer.

Flow Rate: This is set at 24.4 $ml/cm^2/hr$.

Results: The V_o value is determined with a 0.04% aqueous solution (32.5 ml) of Blue Dextran 2000, and the V_i value, from the elution volume of acetone ($V_o + V_i$ = 61 ml). Because, with this type of carrier, the V_i value depends on the compound used for determining it, the corrected V_e^o value[3] and other data[3] are given in Table XXX.

$$V_e^o = (V_e - V_o)/V_o = K_D\ (V_i/V_o),$$

which, in logarithmic form, is

$$\log V_e^o = \log K_D + \log\ (V_i/V_o) = \log K_D + \text{a constant}.$$

The value $\Delta \log V_e^o$ denotes the difference caused in the elution volume of the compound by one substituent, assuming that the $\log V_e^o$ values are additive. Therefore,

Table XXX is a record of the results achieved under the aforementioned conditions, the individual compounds, differing from each other by one substituent ($\Delta \log v_e^o$), being compared each time with a single "mother" compound. The elution diagrams of different compounds grouped according to their elution volumes are given in Figs. 6 and 7.

The possibility of predicting V_e values on the basis of $\Delta \log v_e^o$ values for different chemical groups is illustrated by means of an example taken from Ref.3; with the specific column already mentioned, the V_e value for 6-thioguanine (2-aminopurine-6-thiol) may be determined as follows.

$$\log v_e^o \text{ (6-thioguanine)} = \log v_e^o \text{ (purine)} + \Delta \log v_e^o \text{ (6-SH)} + \Delta \log v_e^o \text{ (2-NH}_2\text{)}$$

$$\log v_e^o \text{ (purine)} = +0.326$$
$$\Delta \log v_e^o \text{ (6-SH)} = +0.525 \qquad V_o = 32.5 \text{ ml}$$
$$\Delta \log v_e^o \text{ (2-NH}_2\text{)} = +0.353 \qquad V_e = (v_e^o + 1).V_o$$

$\log v_e^o$ (6-thioguanine) = **+ 1.204**,
and hence v_e^o = **16.0**.
Calculated V_e = 552 ml; V_e found experimentally = 555 ml. Thus, this procedure not only allows prediction of the behavior and possibility of separation of a specified compound on the column, but also permits preliminary consideration of the structure of very simple purine derivatives. From the results, it appears that the adsorption of purines depends mainly on the *free* nitrogen atoms of the heterocycle. If they are substituted, the adsorption strength decreases gradually, in the series $N^9 > N^3 > N^7 > N^1$.

d. <u>Desalting of Nucleotides, Nucleosides, and Bases by Gel Filtration on a Column[4] of Bio-Gel P-2</u>.—*Column:* Bio-Gel P-2, 0.9 x 55 cm (similar results are obtained with an 800-ml column.

Mixture Separated: A mixture of nucleotides and other components of nucleic acids obtained after separation of ion-exchange columns with salt gradients, the salts being used mostly in 1-molar concentration (sodium chloride, ammonium chloride, sodium bromide, potassium bromide, ammonium bromide, urea, formic acid, acetic acid, ammonium sulfate, ammonium hydrogen carbonate, formate and acetate anions, and hydrogen and dihydrogen orthophosphate). The volume of the sample is generally not more than 2% of the volume of the column filling, and should not exceed 10% thereof.

Eluant: Distilled water.

Flow Rate: This is set at $2 \text{ml/cm}^2\text{/min}$; an increase in this rate has no substantial influence on the desalting.

TABLE XXX

Values[a] of log V_e^o for Different Substituents on the Purine Ring

Compound	V_e	V_e^o	log V_e^o	Δlog V_e^o
Adenine	251	6.38	+0.805	-
2-amino-	439	12.52	+1.098	+0.293
9-(2-deoxy-β-D-*erythro*-pentofuranosyl)-	153	3.50	+0.544	+0.261
2-hydroxy- (Isoguanine)	135	3.17	+0.501	-0.304
1-methyl-	75	1.31	+0.116	-0.689[b]
N^1-oxide	91	1.80	+0.256	-0.549[b]
9-β-D-ribofuranosyl- (Adenosine)	153	3.50	+0.544	-0.261
2':3'-cyclic phosphate	59	0.87	-0.093	-0.898
2-Aminopurine	187	4.77	+0.679	-
6-amino-	439	12.52	+1.098	+0.419
6-hydroxy- (Guanine)	199	4.85	+0.686	+0.007
6-mercapto-	555	16.07	+1.206	+0.527
Guanine	199	4.85	+0.686	-

TABLE XXX (continued)

Compound	V_e	V_e^o	log V_e^o	Δlog V_e^o
9-(2-deoxy-β-D-*erythro*-pentofuranosyl)-	126	2.70	+0.432	-0.254
8-hydroxy-	225	5.92	+0.772	+0.086
1-methyl-	166	3.88	+0.589	-0.097
7-methyl-	153	3.70	+0.568	-0.118
9-β-D-ribofuranosyl- (Guanosine)	122	2.60	+0.415	-0.271
5'-phosphate	56	0.66	-0.179	-0.865
Hypoxanthine	113	2.57	+0.366	-
2-amino- (Guanine)	199	4.85	+0.686	+0.320
9-(2-deoxy-β-D-*erythro*-pentofuranosyl)-	78	1.29	+0.112	-0.254
2-(dimethylamino)-				+0.170
2-hydroxy- (Xanthine)	156	3.59	+0.555	+0.189
8-hydroxy-	122	2.76	+0.442	+0.076
1-methyl-	92	1.85	+0.266	-0.100
7-methyl-	93	1.74	+0.240	-0.126

TABLE XXX (continued)

9-methyl-	83	1.56	+0.194	-0.172
2-(methylamino)-	184	4.41	+0.645	+0.279
9-β-D-ribofuranosyl- (Inosine)	77	1.265	+0.102	-0.264
6-Mercaptopurine	275	7.09	+0.851	-
2-amino-	555	16.07	+1.206	+0.355
Purine	106	2.12	+0.326	-
2-amino	187	4.77	+0.678	+0.353
6-amino- (Adenine)	251	6.38	+0.805	+0.479
6-bromo-	278	7.57	+0.880	+0.554
6-carboxy-	82	1.52	+0.183	-0.143
6-chloro-	206	5.06	+0.704	+0.378
6-cyano-	284	7.73	+0.889	+0.563
6-(dimethylamino)-	281	7.65	+0.884	+0.558
6-(furfurylamino)-	788	23.24	+1.367	+1.041
2-hydroxy-	82	1.52	+0.183	-0.143
6-hydroxy- (Hypoxanthine)	113	2.57	+0.366	+0.040

TABLE XXX (continued)

Compound	V_e	V_e^o	log V_e^o	Δlog V_e^o
6-iodo-	400	11.34	+1.054	+0.728
6-mercapto-	275	7.09	+0.851	+0.525
6-methoxy-	159	3.60	+0.556	+0.230
6-methyl-	116	2.32	+0.366	+0.084
6-(methylamino)-	250	6.69	+0.826	+0.500
6-(methylthio)-	319	8.82	+0.946	+0.620
6-(phenylamino)-	1754	52.97	+1.724	+1.398
6-(trimethylammonium)-	65	1.00	00.000	-0.326[b]
Xanthine	156	3.59	+0.555	-
8-hydroxy- (Uric acid)	171	4.09	+0.612	+0.057
1-methyl-	139	3.09	+0.490	-0.065
7-methyl-	120	2.54	+0.406	-0.149
9-β-D-ribofuranosyl-	104	2.07	+0.316	-0.239

[a] According to data of Sweetman and Nyhan,[3] for a Sephadex G-10 column (1 x 100 cm) with elution with 50 m*M* sodium dihydrogen phosphate (pH 7). [b] Compound has a positive charge on the ring.

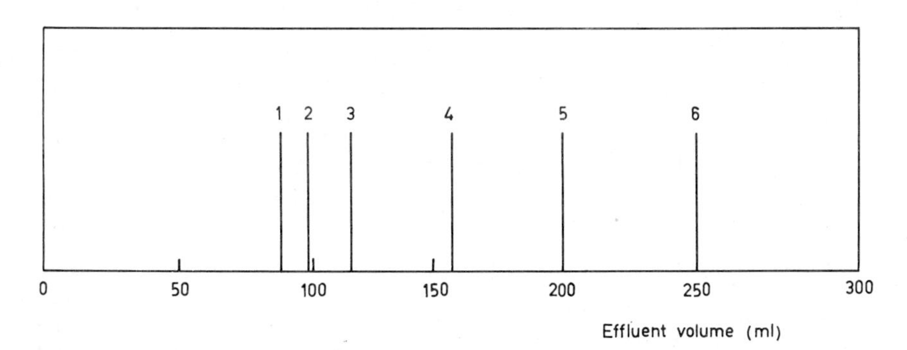

Fig. 6. Diagrammatic Representation of Elution Volumes of Purines and Pyrimidines on a Sephadex G-10 Column.[3] [(*1*) Uracil + cytosine, (*2*) thymine, (*3*) hypoxanthine, (*4*) xanthine, (*5*) guanine, and (*6*) adenine. Column dimensions are 1 x 100 cm; the eluant is 50 m*M* sodium dihydrogen phosphate (pH 7.0).]

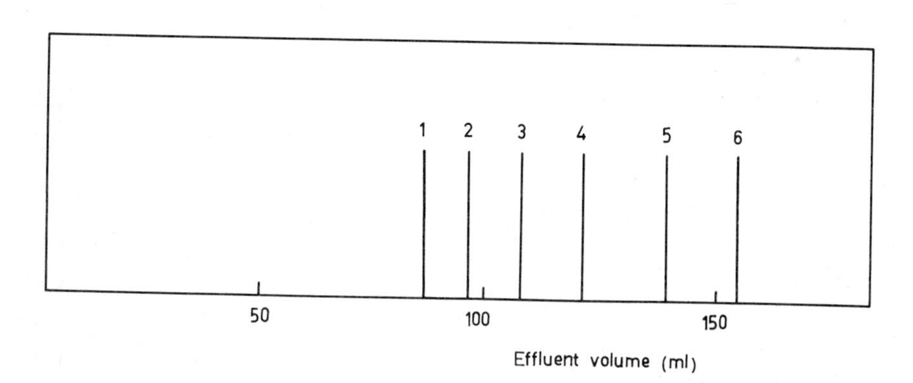

Fig. 7. Diagrammatic Representation of Elution Volumes of Methylxanthines on a Sephadex G-10 Column.[3] [(*1*) 1,3,7-Trimethylxanthine, (*2*) 3,7-dimethylxanthine, (*3*) 1,7- and 1,3-dimethylxanthine, (*4*) 7-methylxanthine, (*5*) 1-methylxanthine, and (*6*) xanthine. For experimental conditions, see Fig. 6 and text.]

Elution: This is monitored by measuring the optical absorbance at 260 and 220 nm ($Br^{\ominus}$, urea) or by titration ($Cl^{\ominus}$, $Br^{\ominus}$).

The V_o value is determined with a serum albumin solution (40% water, by volume). The K_D values obtained in this way are summarized in Table XXXI. Besides the volume of the sample, the pH, temperature, and gel particle-size are factors that influence the desalting process. When the salt zone overlaps the peak of the compound eluted, it is recommended that the desalting operation be repeated with this fraction on the same column.

VI. THIN-LAYER CHROMATOGRAPHY

1. Introduction

Thin-layer chromatography, first described in 1938, came into general use in laboratories in the nineteen-fifties. It rapidly became popular, mainly because of its great advantages in comparison with adsorption chromatography on columns, to which it is rather similar. Two basic techniques may be distinguished, depending on the preparation of the adsorbent layer and its mode of application to a solid support (usually glass, although polyester and similar films are also used): (*1*) layers poured as a wet slurry, and (*2*) loose layers poured as a dry, finely-dispersed powder.

The advantages and disadvantages of the two procedures are as follows. In general, loose layers are simpler to prepare and can be used immediately; this facilitates rapid selection of a suitable material. Loose layers are cheaper as regards apparatus because the simple equipment required is readily available in every laboratory, may consist of carriers of equal activities and grain size (as in column chromatography), are developed more quickly, and are more readily converted to application on a preparative scale, and the capacity of their layers is, in general, higher. On the other hand, the separative capacity of loose layers is usually inferior to that of compacted layers, and, therefore, they can only be used for simple mixtures. The handling of a loose layer demands more care, detection by spraying requires more experimental experience, and the layers cannot be prepared in advance as a supply for future use. Furthermore, storage of the plates and their documentation require much space.

This means, therefore, that the loose-layer technique is only fully satisfactory in those laboratories where it is used occasionally, for the analysis of simple reaction-mixtures. Obviously, more accurate results are obtainable with high-precision, wet-poured layers; these have excellent separative capacity, and are strongly recommended, despite the fact that the preparation of the layer is more laborious, and requires considerably more costly instrumentation.

TABLE XXXI

Desalting[a] of Components of Nucleic Acids on a Column[b] of Bio-Gel P-2

Compound	K_D Value
Mono- and oligo-nucleotides	0.9
Thymidine and ribonucleosides	1.4-2.0[c]
Purine and pyrimidine bases	2.0
Ammonium sulfate and disodium hydrogen phosphate	0.75
Disodium hydrogen phosphate, formate, and acetate	0.80
Ammonium hydrogen carbonate	0.85
Sodium chloride and ammonium chloride	1.0
Formic acid and acetic acid	1.1
Sodium bromide, potassium bromide, ammonium bromide, and urea	1.4

[a]According to data in Ref. 4. [b]Column dimensions, 0.9 x 55 cm; elution with water. [c]The region of K_D value for all of the naturally occurring nucleosides.

All of the fundamental types of chromatography may be used with the thin-layer technique: partition (mainly applicable in classical, paper chromatography), adsorption (characteristic mainly of the column method), and ion-exchange chromatography (applicable, according to the carrier employed, in both column and paper chromatography). Materials of the molecular-sieve type may be used as carriers in thin-layer chromatography, permitting a combination of gel filtration with adsorption and ion-exchange effects. The theoretical and practical similarities between the separations achieved in column and paper chromatography do not limit the difference between these techniques and the thin-layer technique to one of technicalities only. The main advantages of the thin-layer technique are its enhanced sensitivity (as, by use of short-wave u.v. light, amounts down to 0.5 μg may be detected), the considerably faster development, an increased separative capacity, and sharper delineation of the individual spots when, for example, a paper chromatogram is compared with a chromatogram on a thin layer of cellulose.

Every chemical laboratory conventionally equipped is a suitable environment for the use of thin-layer chromatography. The only condition to be met is that it shall have sufficient horizontal space for preparation of the layers, and sufficient space for storing a stock of glass plates and pre-prepared layers. Plates must be dried in a well ventilated fume-hood. A conventional infrared lamp suffices for drying, and, in work with nucleic acid components and their analogs, a source of short-wave u.v. light is essential (for example, a Chromatolite, sold by Hanovia, of Slough, Buckinghamshire, England), as, in this region, it is preferable to almost all other methods of detection.

2. Carriers

The adsorbents and carriers most frequently used in separating nucleic acid components are reviewed in Table XXXII. The basic applicability of an adsorbent is characterized by the following conditions:[100] (_1_) its particle size should lie in the range of 5-10 μm (otherwise, diffusion, slow migration, or smudging of spots will result), (_2_) the material must have a sufficiently large surface area (a minimum of 10 m^2/g), and (_3_) it must lend itself to easy detection of spots (a white carrier is best; charcoal, the separative properties of which are good in all other respects, is not suitable).

Commercial materials for use in column chromatography are also suitable for loose layers, after they have been "classified" to separate particles of a suitable range

TABLE XXXII

Some Commercially Available Carriers for Thin-layer Chromatography[a]

Trade name	Producer	Chemical nature
Kieselgel D5	Fluka A.G.	Silica gel with gypsum
Kieselgel D5-F	Fluka A.G.	with gypsum and fluorescent indicator
Silica gel sheet	Kodak Ltd.	with fluorescent indicator
Kieselgel HF_{254}	Merck A.G.	with fluorescent indicator
Aluminum oxide	Merck A.G.	Aluminum oxide with gypsum as binder
Kieselguhr G	Merck A.G.	Celite with gypsum as binder
Celite TLC	Res.Spec. Co.	with gypsum as a binder
Cellulosepulver	Macherey, Nagel & Co.	Cellulose
MN-300 (F_{254})		with or without fluorescent indicator
MN-300 G		with gypsum as binder
MN-300 GF_{254}		with gypsum and fluorescent indicator
MN-300/DEAE		DEAE-cellulose
MN-300 G/DEAE		with gypsum as binder
MN-300 G/Ecteola		with gypsum as binder
PEI-Cellulose		Poly(ethyleneimine)-cellulose
Polyamid für DC	Merck A.G.	Polyamide
Polyamide layer	Chen-Chin Trading Co.	Polyamide
MN-Polyamidpulver für DC	Macherey, Nagel & Co.	with or without starch as binder

TABLE XXXII (continued)

Sephadex (superfine)	Pharmacia AB	Dextran molecular sieve

[a]Partial reproduction of a Table in Ref. 100.

of sizes. With wet-poured layers, finely powdered materials are used; a number of these are now supplied that also contain an inorganic binder (for example, gypsum with Kieselgel D5, cellulose powder MN 300G, or Kieselguhr G) or an organic binder (for example, starch with MN-polyamide powder "für DC"). Such binders are added during the process of preparation of the carrier by grinding the rough commercial preparation; besides gypsum, such materials as starch,[101] collodion,[102] gelatin,[103] or poly (vinyl alcohol)[104] may be used. Certain commercial preparations contain a fluorescent indicator, which facilitates detection of certain compounds in U.V. light. The binder only improves the mechanical properties of the layer during handling and elution; it seldom affects a separation.

3. Preparation of the Layer

Where there is no need for adjusting the particle size of commercial carriers, preparation of slurries is very simple, and the following may be considered a standard procedure. A mixture of one part of the carrier and two parts of distilled water is thoroughly agitated in a closed flask for 0.5–1 minute. The slurry is then quickly applied to a well degreased, washed plate, especially rapidly if a fast-hardening binder, such as gypsum, is used. Preparation of a slurry in organic solvents (for example, 1:1 methanol–water, 9:1 ethyl alcohol–water, 2:1 chloroform–methanol, or ethyl acetate) usually causes prolongation of the hardening period and gives a layer having a smoother surface, but, generally, less cohesion is obtained.[105 107]

A loose layer is usually prepared with equipment made in the laboratory.[108 110] A glass rod fitted at the ends with rubber rings, or with adhesive tape wound around it, serves as a roller for adjusting the thickness of the layer; a glass plate having rubber spacers may be used similarly. Depending on the pourability and the cohesion of the material, the layer must be adjusted either by a sliding or a rolling motion. This type of plate should be prepared immediately before use, in order to avoid damage to the surface.

Wet-poured layers prepared from thick, aqueous slurries, or from slurries in organic solvents or mixtures thereof, may also be adjusted by means of a glass rod or spatula.

Considerably improved results are obtained by the use of special spreaders having adjustable slit-widths (the best known are those sold by Desaga, Research Specialities Co., Camag, and Shandon); the layers thus prepared give more reproducible R_F values, as these values depend on such factors as the regularity of thickness of the layer, and its homogeneity. All of these spreaders have a reservoir for the slurry, an adjustable slit for modifying the layer thickness, and a smoothing plane. The plates are then dried for several hours in a dust-free atmosphere at or slightly above room temperature.

4. Activation of the Layer

With some carriers used as wet-poured layers, activation is required (for example, silica gel and aluminum oxide). Activation is achieved by heating the plate for a sufficient time (30–60 minutes) at 100° (for silica gel) and at 400-450° (for aluminum oxide, to achieve "activity II," corresponding to a water content of ∿3%). The activity is best determined, immediately, on the plate by use of a mixture of azo dyes (0.06% of azobenzene or 0.04% of *p*-methoxyazobenzene, with Sudan Yellow, Sudan Red, or *p*-aminoazobenzene in carbon tetrachloride); 20 ml of such a solution is developed on the plate with carbon tetrachloride, and the activity is determined by comparison with tabulated R_F values[111] (see Table XXXIII). Preparations having activities II and III are used most often in the study of nucleic acid components by thin-layer chromatography.

TABLE XXXIII

R_F Values of Azo Dyes on Layers of Aluminum Oxide of Different Activities[112]

Dye		R_F Values (x 100)			
	Activity[a]	II	III	IV	V
	Water content (%)	3	6	10	15
Azobenzene		59	74	85	95
p-Methoxyazobenzene		16	49	69	89
Sudan Yellow		1	25	57	78
Sudan Red		0	10	33	56
p-Aminoazobenzene		0	3	8	19

[a]According to Brockmann and Schoder.[113]

5. Application of the Sample

Difficulties with sample application are not normally encountered, especially when wet-poured layers are used. The procedure is essentially the same as in paper chromatography. The necessary volume of sample is applied at the origin (marked with a sharp pencil, usually 2 cm from the edge of the plate, according to the plate size) by touching it with the tip of the micropipet. Individual points may be marked with a pencil, as on paper. The diameter of the spot should not exceed 0.5 cm. Drying of the sample may be speeded by use of a gentle stream of hot air (provided, for example, by an electric fan-heater).

Application of the sample to loose layers is somewhat more difficult and laborious. It is not permissible to touch the surface with the tip of a micropipet, or to let a drop fall upon the surface of the layer, because of the likelihood of damaging the surface. One useful device is a combination of a micropipet with a syringe; this allows small drops of the sample solution to be forced out one by one, and permits touching the surface with these drops, as the layer will absorb the drops with no danger of damage. Experienced workers can perform this operation by hand without difficulty. However, drying cannot be speeded with a stream of air. Marking of the position of the sample must be done on paper below the plate.

The amount of sample to be applied for analytical purposes depends on the sensitivity of the detection. Detection is most frequently achieved by absorption of u.v. light, which, in general, permits detection of 0.5-10 μg of a compound. It is, therefore, recommended that a maximum of 100 μg be applied for analytical separations. When comparing R_F values, their dependence on the amount of sample applied to wet-poured layers must be borne in mind; it is, therefore, important to apply equal concentrations of compounds for an identification process. With loose layers, this dependence is small (<2%) over a wide range of concentration.

6. Development

Development is similar to that employed in paper chromatography in closed vessels, the atmosphere of which is sufficiently saturated with solvent vapors.[114,115] Again, the procedure differs as between loose and wet-poured layers. With loose layers, the inclination of the plate to the horizontal plane should not exceed 20%, and the plate may be carefully placed in the solvent immediately. Also, with this technique, the sensitivity to saturation of the atmosphere in the chamber is less. Wet-poured layers may be developed in an almost vertical position by immersion in a layer of solvent about 0.5 cm deep. The length of the development path should be shortened as much as possible, in order to decrease the influence of unfavorable factors (the optimal length is about 10 cm). Plates having sides about 20 cm long are, therefore, ideal.

With both types of layer, descending development may be used, as well as overflow,[116,117] gradient, [118] repeated, successive, and other procedures; these will not be discussed in detail, because they are not widely used and the results do not differ much from the fundamental, ascending method.

7. Detection and R_F Values

Detection is best accomplished by spraying the layer with a solution of a detection agent (compare paper chromatography). Most of the compounds studied in the author's laboratory may, however, be readily detected by observing the plate under u.v. light; to increase the sensitivity of detection, the fluorescent background of an added fluorescent component may also be used. The two procedures do not require special skill or experience.

Measurement of R_F values is performed as in paper chromatography. The solvent front and the spots should be carefully marked with a sharp pencil, as their intensity may decrease with time. When the reproducibility of R_F values is important, their dependence on a large number of factors must be borne in mind, including the quality of the adsorbent, the activity of the carrier, the thickness and homogeneity of the layer, the quality of the solvents, the development technique used, the concentration of the compound, the path length, and the distance between the solvent level and the origin. These matters must all receive careful consideration, and they should be used only as orientative data when a decision is to be made as to the possibility of using a particular layer and solvent system for separation. Such considerations also apply to meaningful use of the Tables of R_F values given in this Chapter. When compounds are to be identified, standards should be used, and these should migrate parallel to the compound studied.

Elution of the material separated on preparative thin-layers is very simple, especially with loose layers. The marked part of the layer containing the adsorbed compound is removed by gentle suction into a suitable wash-tube, and the mixture obtained is extracted in the usual way, or is eluted after being placed in a column. With wet-poured layers, the zones to be eluted must be separated by means of a spatula, as the carrier adheres strongly to the support.

8. Selection of Solvent Systems

It is very difficult to write general instructions on how to select the adsorbent and solvent for any one group of compounds. In most cases, selection is quite empirical, and, because a large amount of information on chromatography is available, data for the separation of a similar compound should be sought in the literature. Because thin-layer chromatography is very similar to column or paper chromatography, the reader is referred to the Section dealing with the ways and principles of selecting systems for these two techniques (see pp. 540 and 577), where the problem is discussed in detail.

One very practical and rapid way of ascertaining the suitability of a selected solvent or mixture of solvents is reported by Stahl[119]: 2-3 drops of the solvent selected for separation are dropped onto the sample that has been applied directly to a prepared, micro-scale, thin layer, the sample being a part of the mixture to be separated; the suitability of the selection may be readily assessed from the rings formed. In this way, different solvents in the series from nonpolar to highly polar ones may be tested. A general rule, developed by Stahl[120] and illustrated diagrammatically in Fig. 8, may be mentioned.

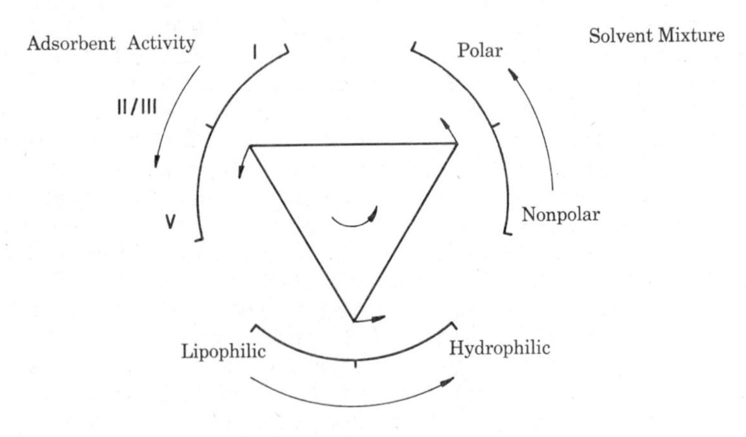

Fig. 8. Diagrammatic Representation of the General Selection of Adsorbent Activity and of a Suitable Solvent-system for Compounds of Different Polarity.[120] [The apexes of a rotatable, equilateral triangle determine the suitable adsorbent activity and polarity of the solvent mixture, when one of them is set to the estimated polarity of the compound chromatographed.]

Separation of the more polar compounds requires an adsorbent of higher activity and a more polar solvent mixture, as compared with mixtures of less polar compounds.

It is recommended that thin layers be used for testing specific adsorbents and solvent mixtures selected for use in column chromatography. For example, compounds that have very low R_F values (<0.2) cannot be separated on a column by use of the same solvent or, possibly, even on the same adsorbent, because, on elution, they will afford zones that are too wide. On the other hand, thin layers often serve as an analytical technique for studying the elution from columns and the efficiency of separation. In the following section, some practical examples of the separation of individual groups of compounds will be discussed, with special reference to different types of carrier and to the summaries of R_F values that are tabulated.

9. Some Practical Examples

a. Chromatography of Protected Nucleosides on Loose, Thin Layers of Aluminum Oxide. -*Introduction:* This type of chromatography is utilized for both preparative and analytical purposes for protected nucleosides. The following examples deal with anomeric mixtures of protected "2-deoxy-D-ribofuranosyl" derivatives[70,73–75] and D-ribofuranosyl derivatives,[76] and $N^1$and O^6 (Refs. 121 and 122), as well as N^1 and N^3 (Ref. 77), isomeric mixtures of nucleosides. Similarly, this technique is very useful in the separation of glycosyl derivatives prepared by both the mercury(II) and silylation methods.[121–123]

Thin Layer: Neutral aluminum oxide (Brockmann, activity II–III) is poured, in a sufficient amount, onto a glass plate (18 x 48 cm, carefully degreased and washed) and smoothed several times with a glass rod so fitted with rubber rings at the ends as to afford a layer-thickness of 1–2 mm (the thickness of the wall of the rubber rings) and the desired layer-width (the distance between the rubber rings). The origin is 4–5 cm from the narrower end of the layer; usually, 2-ml portions of the sample solution are quantitatively applied, by means of a pipet, in a narrow, straight line. Care must be taken not to disturb the layer with the tip of the pipet, and individual drops should be applied side by side. The plate is developed in a glass vessel (45 x 25 x 22 cm) that contains 400-450 ml of the solvent mixture chosen.

Solvents: Benzene and ethyl acetate (for purification, see p. 583).

(*i*) Preparative Separation* of an Anomeric Mixture of 1-(2-Deoxy-3,5-di-*O*-*p*-toluoyl-D-*erythro-pentosyl*)-4-methoxy-2(1*H*)-pyrimidinones [40 (α) and 41 (β)].

Reaction: 2-Deoxy-3,5-di-*O*-*p*-toluoyl-D-*erythro*-pentosyl chloride[76a] (15) (250 μmoles) with 2,4-dimethoxypyrimidine[123a] (250 μmoles) in benzene.[74]

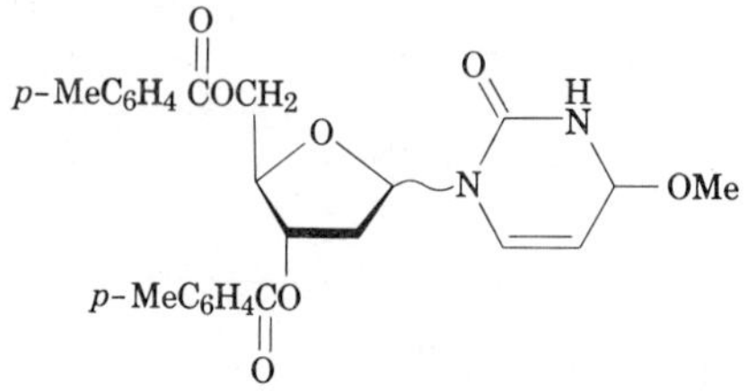

40 (α-anomer)
41 (β-anomer)

Procedure: The reaction mixture containing compounds 40 and 41 is subjected to chromatography on the plate (18 x 48 cm) in 1:1 benzene—ethyl acetate. The separation process is interrupted at the moment when the solvent front reaches the top of the layer. The thin layer is dried at room temperature in a fume hood, and is then inspected under a u.v. lamp. The quenching zones are marked (mechanical means), separated one at a time with a stainless-steel knife, and then eluted individually, in a column, with ethyl acetate. The eluates are then evaporated *in vacuo*, and the residue is crystallized from ether. The quenching zone having R_F 0.5 gives 9% of the β-D anomer (41), m.p. 174—176°, and the less-mobile zone (R_F 0.25) gives 28% of the α-D anomer (40), m.p. 198—199°.

(*ii*) Preparative Separation of an Anomeric Mixture of 5-(Benzyloxymethyl)-4-methoxy-1-(2,3,5-tri-*O*-benzoyl-D-ribosyl)-2(1*H*)-pyrimidinone [42 (α) and 43 (β)].

Reaction: The two components, 2,3,5-tri-*O*-benzoyl-D-ribosyl chloride[76b] (44) and 5-(benzyloxymethyl)-2,4-dimethoxypyrimidine (45) (350 μmoles of each), in acetonitrile were allowed to react by the Hilbert—Johnson method.[76]

Procedure: The reaction mixture is separated by chromatography with 7:3 benzene—ethyl acetate, and affords two zones, R_F 0.3 and 0.5. The faster-moving zone, treated as in the preceding example, yields 37% of the β-D anomer (43), m.p. 73—77°, and the slower-moving zone yields 8% of the amorphous α-D anomer (42).

*Details provided by Dr. M. Pryštaš.

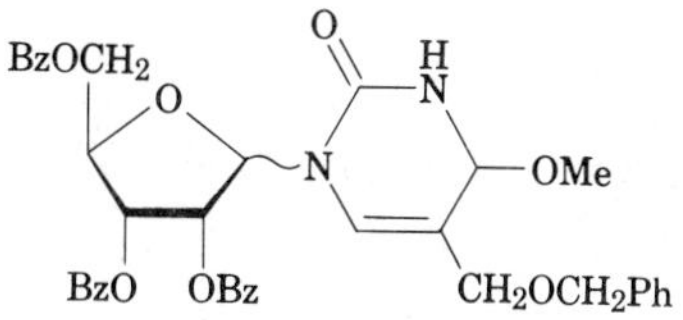

42 (α-anomer)
43 (β-anomer)

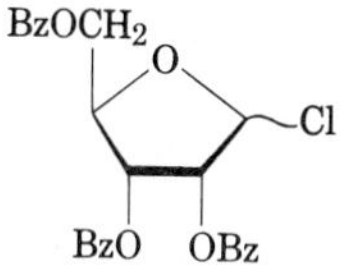

44

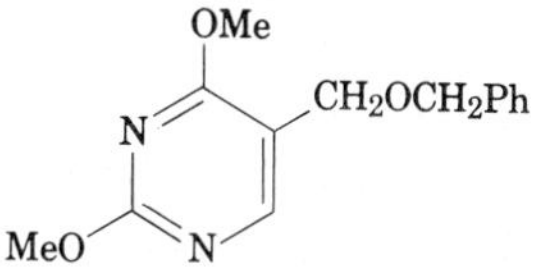

45

where Bz is benzoyl and Ph is phenyl.

(*iii*) Preparative Separation of Mixture of 1-(2,3,5-Tri-*O*-benzoyl-β-D-ribosyl-6(1*H*)-pyrimidinone (46) and 6-(2,3,5-Tri-*O*-benzoyl-β-D-ribosyloxy)pyrimidine (47).

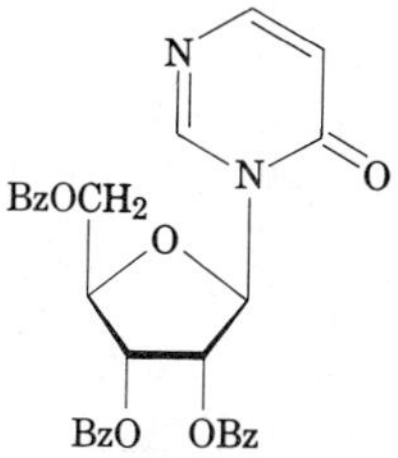

46

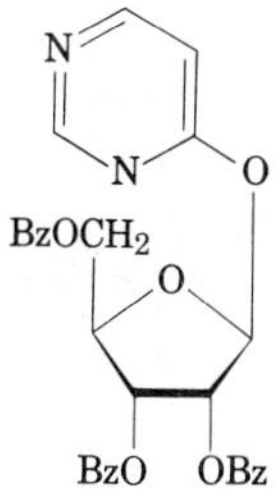

47

where Bz is benzoyl.

Reaction: 1.08 μmoles of 44 with 540 μmoles of 6(1*H*)-pyrimidinone by the mercury(II) method, in toluene.[122]

Procedure: Chromatography of the reaction mixture containing 46 and 47 on two plates (18 x 48 cm) with 3:50

ethyl acetate–benzene gives quantitative separation.[g] The zones having R_F 0.2 yield 57% of 46, m.p. 159°, and the more-mobile zone (R_F 0.4) yields 3.5% of 47, m.p. 136–137°. The weakly quenching zone (R_F 0.7) yields 11% of 2,3,5-tri-*O*-benzoyl-D-*erythro*-pent-1-enofuranose (the degradation product of halide 44), m.p. 121-122°.

(iv) Preparative Separation of a Mixture of 2',3',5'-Tri-*O*-benzoyl-6-methyluridine (49) and 6-Methyl-3-(2,3,5-tri-*O*-benzoyl-β-D-ribosyl)uracil (50).

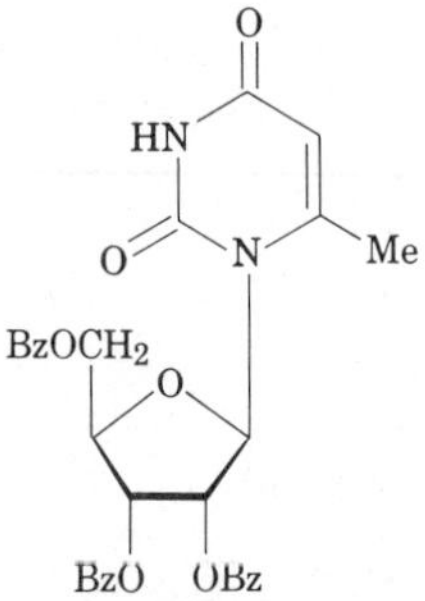

49 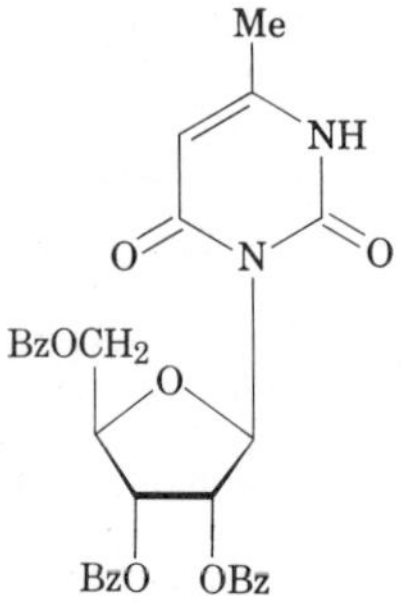 50

where Bz is benzoyl.

Reaction: D-Ribosylation of 6-methyluracil by the mercury(II) method.[77]

Procedure: After conventional treatment, the D-ribosylation mixture of the mercury(II) salt of 6-methyluracil (890 μmoles) and halide[76b] 44 (960 μmoles) is chromatographed on a thin layer (2 mm), dry-poured onto one plate (18 x 48 cm) with 3:5 ethyl acetate-benzene. The strongly quenching zone having R_F 0.65 yields the amorphous N^3-ribofuranosyl derivative (50, 72%), and treatment of the less-mobile zone (R_F 0.1) results in 13% of the N^1-ribosyl derivative (49), m.p. 125-128°.

b. Chromatography* of Adenosine Derivatives on a Loose, Thin Layer of Silica Gel.

Thin Layer: Silica gel, 30-60 μm thick (service laboratory of the author's Institute), containing a fluorescent indicator. The layer is adjusted with a

[g]Before application to the thin layer, the reaction mixture is passed (without separation) through a column of aluminum oxide (Brockmann activity, II-III) by use of 6:1 benzene-ethyl acetate.

*Details provided by Dr. S. Chládek.

rod (as in the preceding example with aluminum oxide). Plate dimensions are 15 x 45 cm, and development is performed in a glass vessel (21 x 40 x 30 cm).

Mixture Separated: After reaction of 600 μmoles of the *p*-nitrophenyl ester of *N*,*N*-bis(benzyloxycarbonyl)-*L*-lysine with 300 μmoles of 2,3-*O*-(aminomethylethoxymethylene)adenosine in *N*,*N*-dimethylformamide.[124]

Chromatographic System: Dichloromethane with 5% of methanol (∿350 ml). The dichloromethane is purified by distillation. Repeated development is used (up to 3 times), each time after the plate has been dried in air at room temperature.

Procedure: The mixture (dissolved in 2 ml of the chromatographic solvent) is applied (in drops forming zones side by side) to the starting-line about 4 cm from the narrower edge. After development, and drying at room temperature, the following bands (in the order of increasing R_F values) are detected on the thin layer with u.v. lamp: (*1*) unreacted orthoester, (*2*) and (*3*) diastereoisomers of 2',3'-*O*-[*N*,*N*-bis(benzyloxycarbonyl)-L-lysyl](aminomethylethoxymethylene)adenosine, from (4) *p*-nitrophenol and (*5*) unreacted *p*-nitrophenyl ester. The bands detected are marked on the thin layer by dotting with a pencil, and bands of product are isolated by removing the other silica gel in the vicinity of the bands with a spatula and brush. The individual, u.v.-absorbing, silica gel bands are then transferred to chromatographic tubes fitted with sintered-glass plates, and are eluted with the chromatographic solvent (with about five volumes of silica gel in the tube). The resulting residues are evaporated in a rotary vacuum evaporator at 30°. The resulting residues are chromatographically homogeneous products, and are further identified by their i.r. and u.v. spectra, and by elementary analysis.[124]

c. Chromatography of Purines and Pyrimidines, Nucleosides, and Nucleotides on Thin Layers of Cellulose.[125]

Thin Layer: Cellulose MN 300 (Macherey, Nagel, and Co., Duren, Germany) (10 g) is quickly suspended (20-30 seconds) in 50-60 ml of acetone with a high-speed shaker. The slurry is applied with a Stahl type of spreader (Desaga, Heidelberg, Germany). As the edges serve as guide edges for the applicator, this method requires glass plates (10x20 cm) having absolutely even edges. The plates (5 to 7), well degreased and washed, must be of equal thickness (to allow the layers to be applied all at once). The finished layers are air-dried; 3-5 minutes suffices when they are dried in a stream of warm air. Immediately after being dried, the plates are ready for use in separation procedures. If plates are prepared from an aqueous slurry, they are air-dried overnight at room temperature. The layers may also be prepared with the use of a glass rod[126] (as with loose layers).

(*i*) Analytical Separation of Purine and Pyrimidine Bases and of Their Nucleosides.[125]

Mixtures Separated: A mixture of bases (adenine, guanine, hypoxanthine, and uracil) and their nucleosides; 2–5 μg of each component. The volume of the sample is ∿5 μl. Among those of other bases, the mobilities of 6-chloropurine and 2,6-diaminopurine were studied.

Chromatographic Solvent: Distilled water.

Procedure: The samples are applied by micropipet to the layer, starting about 2.5 cm from the narrow end of the plate, with gaps of 1–1.5 cm in between. The diameter of the spot should not exceed 3–4 mm (a larger spot would be smudged, resulting in an imperfect separation). The plate is transferred to the developing chamber, which contains a 1-cm layer of distilled water at the bottom, and the chamber is closed. Development takes ∿30 minutes and the front migrates 10 cm from the origin. After the plate has been dried at 100° in a drying chamber, the individual components are detected with a u.v. lamp. Their R_F values are: adenine, 0.30; adenosine, 0.53; 6-chloropurine 0.64; 2,6-diaminopurine, 0.21; guanine, 0.37; guanosine, 0.58; hypoxanthine, 0.55; inosine, 0.70; uracil 0.72; and uridine, 0.81.

The same thin-layer material is suitable for the separation of derivatives of methylated purines[127] (see Table XXXIV). Iodo derivatives of pyrimidine (and their nucleosides) are chromatographable in many solvent systems, most of which are based on butyl alcohol, with cellulose or Silica Gel G plates[128] (see Table XXXV). Table XXXVI summarizes results obtained in the separation of substituted purines on cellulose and substituted-cellulose layers.[129] Bases and nucleosides may be further separated by thin-layer chromatography on a variety of other carriers, including those having absorption or partition and ion-exchange characteristics[129-132] (see Table XXXVII).

(*ii*) Analytical Separation of Mononucleotides and Their Di- and Tri-phosphates.[125]

Mixture Separated: Mixtures of three phosphates (AMP, ADP, and ATP) derived from adenosine, and similar mixture derived from cytidine and uridine; maximum volume of sample, 5 μl; concentration, 1 mg/ml for each component.

Chromatographic System: Butyl alcohol-acetone-acetic acid-5% aqueous ammonium hydroxide-water (9:3:2:2:4). If the components are insufficiently separated, development may be repeated after the layer has been dried.

Procedure: Similar to the procedure for separating the base–nucleoside mixture. However, because of the greater volatility of the solvents, the developing chamber

TABLE XXXIV

R_F Values of Some Purine Derivatives on Thin Layers of Cellulose[a]

Compound	R_F Value (x 100) in system		
	1[b]	2[c]	3[d]
2'-Deoxyinosine	80	75	40
1-methyl-	87	80	66
Guanine	42	25	17
1,7-dimethyl-	59	43	49
1-methyl-	49	33	37
7-methyl-	48	58	39
Guanosine	61	65	20
1,7-dimethyl- (Iodide)	77	86	-
1-methyl-	43	33	37
2'-deoxy-	64	65	53
1-methyl-	69	69	62
Hypoxanthine	63	61	34
1-methyl-	73	68	48

[a]Compiled from Ref. 127, where chromatography was conducted on MN-Cellulose powder 300-G (250 µm). [b]5% Aqueous ammonium hydrogen carbonate. [c]5% Aqueous sodium dihydrogen phosphate, saturated with isoamyl alcohol. [d]2:5:13 Concentrated ammonium hydroxide-*N,N*-dimethylformamide-isopropyl alcohol.

must be well sealed, to ensure saturation of the inner atmosphere with all of the components of the solvent mixture. Development requires ∿60 minutes (the front is then 10 cm from the origin). The following R_F values are obtained (these apply to one developing operation only): 5'-AMP, 0.38; ADP, 0.26; ATP, 0.16; 5'-CMP, 0.34; CDP, 0.22; CTP, 0.13; 5'-UMP, 0.37; UDP, 0.25; and UTP, 0.17.

The same and similar types of compounds are separable on other carriers by use of the techniques of adsorption and partition in thin-layer chromatography[130,133,134] (see Table XXXVIII). Because the ion-exchange carriers, as in paper chromatography and column chromatography of nucleotides on different ion-exchangers, have a higher resolving power for related compounds, they are used preferentially in this field, also [129,130,135–137] (see Table XXXIX).

TABLE XXXV

R_F Values of Some Iodo Derivatives Separated by Thin-layer Chromatography[a]

5-Iodo derivative of	Thin layer[b] of	R_F Value (x 100) in system[c] 1	2	3	4	5	6	7	8
Cytosine	A	47	48	48	52	45	61	57	45
	B	58	49	49	52	58	57	53	40
Cytidine	A	66	30	13	14	24	62	32	30
	B	68	27	27	29	73	64	42	42
Uracil	A	83	77	80	79	71	85	84	83
	B	59	70	69	74	62	56	68	45
Uridine	A	86	54	59	58	48	85	61	57
	B	72	47	47	50	52	65	47	45
2'-deoxy-	A	90	74	78	75	65	74	81	80
	B	57	65	65	70	60	50	63	42

[a]Compiled from Ref. 128; compounds radioactively labeled. [b]A, Cellulose G; B, Silica Gel G. [c]1, Water; 2, 43:7 butyl alcohol-water, *i.e.*, water-saturated; 3, butyl alcohol, saturated with 33 mM boric acid; 4, butyl alcohol, saturated with 330 mM boric acid; 5, saturated aqueous boric acid; 6, 100 m*M* formic acid; 7, butyl alcohol, saturated with 100 m*M* formic acid; and 8, butyl alcohol, saturated with 100 m*M* ammonium hydroxide.

Table XXXVI

R_F Values of Some Base Analogs on Thin Layers of Cellulose and Substituted Cellulose[a]

Compound	R_F Value (x 100) on thin layer[b] of		
	1	2	3
6-Chloro-9-ethylpurine	79	85	86
9-{*p*-[*N*,*N*-Bis(2-chloroethyl)-amino]phenyl}-6-chloropurine	26	47	45
9-{*p*-[*N*,*N*-Bis(2-chloroethyl)-amino]phenyl}hypoxanthine	54	58	42
9-{*p*-[*N*,*N*-Bis(2-hydroxyethyl)-amino]phenyl}adenine	64	69	70
9-{*p*-[*N*,*N*-Bis(2-hydroxyethyl)-amino]phenyl}hypoxanthine	74	78	51
9-{*p*-[*N*,*N*-Bis(2-hydroxyethyl)-amino]phenyl}-9*H*-purine-6-thiol	62	25	17
5-Amino-9-[*p*-(2-hydroxyethyl)-anilino]-6-chloropyrimidine	76	76	63
6-(*p*-Chloroanilino)-9-[*p*-(2-hydroxyethyl)phenyl]purine	19	36	35

[a]According to Ref. 129; data obtained in 25% aqueous *N*,*N*-dimethylformamide (pH 7.5). [b]1, MN Cellulose powder (250 μm); 2, MN 300-G/ECTEOLA-cellulose; and 3, MN 300-G/DEAE-cellulose.

TABLE XXXVII

R_F Values[a] of Some Purines and Pyrimidines on Thin Layers of Various Materials

Compound	R_F Value (x 100) on thin layer[b] of						
	1	2	3	4	5	6	7
Adenine	96	75	85	54	51	38	29
Adenosine	77	65	80	50	65	56	56
Cytidine	55	30	37	23	90	77	82
Guanine	-	0	0	0	44	38	33
Guanosine	-	22	17	17	74	57	50
Hypoxanthine	-	-	-	-	-	57	46
Inosine	21	32	30	23	85	73	61
Uracil	-	58	72	54	83	75	73
Uridine	39	39	50	37	90	84	84
Xanthine	-	48	52	37	74	-	-

[a]According to data in Refs. 102 and 130–132. [b]1, Celite 535 (with starch as a binder), with 9:1 isopropyl alcohol-water;[131] 2, polyamide, on poly(ethylene terephthalate) film, with 4:4:1 heptane-butyl alcohol-acetic acid; 3, as for 2, with 4:1:4 carbon tetrachloride-acetic acid-acetone;[132] 4, as for 2, with 5:1:5:3 toluene-pyridine-2-chloroethanol-800 µM ammonium hydroxide,[132] 5, as for 2, with 500 µm sodium chloride; 6, Kieselgel-G with water;[102,130] and 7, ECTEOLA-cellulose (capacity 260 µequiv/g), with water.[102,130]

TABLE XXXVIII

R_F Values[a] of Some Nucleotides on Thin Layers of Various Materials

Compound	R_F Value (x 100) on thin layer[b] of			
	1	2	3	4
Adenosine				
2'(3')-phosphate	84	82	82	72
5'-phosphate	70	50	56	81
3':5'-cyclic phosphate	87	-	-	-
5'-pyrophosphate	60	-	-	-
5'-triphosphate	34	8	10	5
Cytidine				
2'(3')-phosphate	86	front	front	front
5'-phosphate	72	97	94	front
5'-pyrophate	66	-	-	-
5'-triphosphate	33	-	-	-
2'-Deoxycytidine 5'-phosphate	-	97	94	front
2'-Deoxyguanosine 5'-phosphate	-	51	68	56
Guanosine				
2'(3')-phosphate	80	30	51	28
5'-phosphate	70	50	66	28
5'-pyrophosphate	54	-	-	-
5'-triphosphate	28	-	-	-
Uridine				
2'(3')-phosphate	86(80)	41	48	36
5'-phosphate	74	-	-	-
5'-pyrophosphate	61	-	-	-
5'-triphosphate	45	-	-	-

[a]According to data in Refs. 133 and 134. [b]1, Silica Gel TLC plates (Merck-254, fluorescent) with 6:2:1 methanol–water–concentrated ammonium hydroxide;[133] 2, polyamide, on poly(ethylene terephthalate) film, with 20:1 water–acetic acid;[134] 3, as in 2, with 2:2:1 isopropyl alcohol–water–acetic acid;[134] and 4, as in 2, with 2:2:1 acetone–water–acetic acid.[134]

TABLE XXXIX

R_F Values[a] of Some Nucleotides on Thin Layers of Ion-Exchangers

Compound	R_F Value (of 100) on thin layer[b]					
	1	2	3	4	5	6
Adenosine 5'-phosphate	57	26	-	52	80	74
3'-phosphate	48	-	-	-	-	-
5'-pyrophophate	26	8	48	26	29	52
5'-triphosphate	21	-	11	6	4	12
Cytidine 5'-phosphate	74	31	-	64	80	80
3'-phosphate	71	-	-	-	-	-
5'-pyrophosphate	51	11	53	33	35	56
5'-triphosphate	34	-	13	11	4	16
Guanosine 5'-phosphate	55	14	-	40	50	62
3'-phosphate	44	-	-	-	-	-
5'-pyrophosphate	37	3	27	17	13	27
5'-triphosphate	17	-	7	5	2	5
Inosine 5'-phosphate	74	-	-	59	53	72
5'-pyrophosphate	-	-	-	30	8	30
5'-triphosphate	-	-	-	9	2	5
Uridine 5'-phosphate	80	13	-	74	64	77
3'-phosphate	75	-	-	-	-	-
5'-phophosphate	63	0	15	41	11	40
5'-triphosphate	44	-	4	14	2	7

[a]According to data in Refs. 102, 130, and 135-137.
[b]1, ECTEOLA-cellulose (capacity 410 μequiv/g), with 150 *mM* sodium chloride;[102,130] 2, as for 1, with 10 *mM* hydrochloric acid;[102,130] 3, DEAE-cellulose, with 20 *mM* hydro-

chloric acid;[135] 4, poly(ethyleneimine)cellulose, with *M* lithium chloride;[136,137] 5, as for 4, with 1:1 500 mM lithium chloride,[136,137] and 6, as for 4, with *M* acetic acid-4*M* lithium chloride.[136,137]

REFERENCES

(1) E. Vischer and E. Chargaff, *J. Biol. Chem.*, 168, 781 (1947).

(2) J. Smrt and S. Chládek, *Collect. Czech. Chem. Commun.*, 31, 2978 (1966).

(3) L. Sweetman and W. L. Nyhan, *J. Chromatogr.*, 32, 662 (1968).

(4) M. Uziel and W. E. Cohn, *Biochim. Biophys. Acta*, 103, 539 (1965).

(5) W. E. Cohn and F. J. Bollum, *Biochim. Biophys. Acta*, 48, 588 (1961).

(6) G. W. Rushizky and H. A. Sober, *Biochim. Biophys. Acta*, 55, 217 (1962).

(7) I. G. Walker and G. C. Butler, *Can. J. Chem.*, 34, 1168 (1956).

(8) R. O. Hurst and G. C. Becking, *Can. J. Biochem. Physiol.*, 41, 469 (1963).

(9) R. Rüdiger and H. Rüdiger, *J. Chromatogr.*, 17, 186 (1965).

(10) I. Hais and K. Macek (Eds.), *Paper Chromatography* (in Czech), Publishing House of Czechoslovak Academy of Sciences, Prague, Czechoslovakia, 1959, pp. 140–141 and 148–156.

(10a) R_F values for purine and pyrimidine bases and their nucleosides and nucleotides are given in *Specifications and Criteria for Biochemical Compounds*, R. S. Tipson (Ed.), National Academy of Sciences-National Research Council, Washington, D.C., 1972, pp. 153-154/157-183.

(11) E. C. Bate-Smith and R. G. Westall, *Biochim. Biophys. Acta*, 4, 427 (1950).

(12) R. D. Hotchkiss, *J. Biol. Chem.*, 175, 315 (1948).

(13) R. Markham and J. D. Smith, *Biochem. J.*, 45, 294 (1949).

(14) G. R. Wyatt, in E. Chargaff and J. N. Davidson (Eds.) *The Nucleic Acids*, Vol. I, Academic Press, Inc., New York, N. Y., 1955, p. 252.

(15) D. A. W. Roberts, *J. Chromatogr.*, 6, D7 (Table 12) (1961).

(16) G. R. Wyatt, *Biochem. J.*, 48, 584 (1951).

(17) F. Bergmann and H. Ungar, *J. Amer. Chem. Soc.*, 82, 3957 (1960).

(18) G. R. Wyatt, *Biochem. J.*, 48, 581 (1951).

(19) G. Nübel and W. Pfleiderer, *Chem. Ber.*, 98, 1063 (1965).
(20) R. Markham and J. D. Smith, *Biochem. J.*, 49, 401 (1951).
(21) M. Smith, G. I. Drummond, and H. G. Khorana, *J. Amer. Chem. Soc.*, 83, 698 (1961).
(22) E. Chargaff and J. N. Davidson (Eds.), *The Nucleic Acids*, Vol. I, Academic Press, Inc., New York, N. Y., 1955, pp. 245-248.
(23) D. S. Letham, *J. Chromatogr.*, 20, 184 (1955).
(24) J. Jonsen, L. Haavaldsen, and S. Laland, *J. Chromatogr.*, 1, 291 (1958).
(25) R. M. Fink, R. E. Cline, C. McGaughey, and K. Fink, *Anal. Chem.*, 28, 4 (1956).
(26) C. Janion and D. Shugar, *Acta Biochem. Pol.*, 7, 309 (1960).
(27) J. D. Smith and R. Markham, *Biochem. J.*, 46, 509 (1950).
(28) J. Kream and E. Chargaff, *J. Amer. Chem. Soc.*, 74, 4274 (1952).
(29) A. S. Milton, *J. Chromatogr.*, 8, 417 (1962).
(30) F. Sanger, G. G. Brownlee, and B. G. Barrell, *J. Mol. Biol.*, 13, 373 (1965).
(31) S. Chládek, *Syn. Proc. Nucliec Acid Chem.*, 1, 456 (1968).
(32) A. Holý and G. Kowollik, *Collect. Czech. Chem. Commun.*, 35, 1013 (1970).
(32a) J. Žemlička, *Syn. Proc. Nucleic Acid Chem.*, 1, 422 (1968).
(32b) A. Holý, *Syn. Proc. Nucleic Acid Chem.*, 1, 506 (1968).
(33) A. Holý and J. Smrt, *Collect. Czech. Chem. Commun.*, 31, 3800 (1966).
(34) A. Holý and F. Šorm, *Collect. Czech. Chem. Commun.*, 31, 1544 (1966); A. Holý, *Syn. Proc. Nucleic Acid Chem.*, 1, 452 (1968)
(35) A. Holý and J. Smrt, *Collect. Czech. Chem. Commun.*, 31, 1528 (1966).
(36) A. Hampton, *J. Amer. Chem. Soc.*, 83, 3640 (1961).
(37) R. H. Hall, *Biochemistry*, 4, 661 (1965).
(38) R. H. Hall, *Methods Enzymol.*, 12A, 312 (1967).
(39) D. Söll and H. G. Khorana, *J. Amer. Chem. Soc.*, 87, 352 (1965).
(40) R. Lohrmann, D. Söll, H. Hayatsu, E. Ohtsuka, and H. G. Khorana, *J. Amer. Chem. Soc.*, 88, 819 (1966).
(41) J. Smrt and F. Šorm, *Collect. Czech. Chem. Commun.*, 27, 73 (1962).
(42) J. Beránek and J. Smrt, *Collect. Czech. Chem. Commun.*, 25, 2029 (1960).
(43) J. Beránek and J. Pitha, *Collect. Czech. Chem. Commun.*, 29, 625 (1964).

(44) J. Smrt, J. Beránek, and F. Šorm, *Collect. Czech. Chem. Commun.*, 25, 130 (1960).
(45) R. Lohrmann and H. G. Khorana, *J. Amer. Chem. Soc.*, 86, 4188 (1964).
(46) R. Lohrmann and H. G. Khorana, *J. Amer. Chem. Soc.*, 88, 829 (1966).
(47) J. B. Gin and C. A. Dekker, *Syn. Proc. Nucleic Acid Chem.*, 1, 208 (1968).
(48) C. A. Dekker, *J. Amer. Chem. Soc.*, 87, 4027 (1965).
(49) R. H. Iwamoto, E. M. Acton, and L. Goodman, *J. Med. Chem.*, 6, 684 (1963); E. M. Acton and R. H. Iwamoto, *Syn. Proc. Nucleic Acid Chem.*, 1, 25 (1968).
(50) W. Szer and D. Shugar, *Syn. Proc. Nucleic Acid Chem.*, 1, 58 (1968).
(51) R. Markham and J. D. Smith, *Biochem. J.*, 52, 552 (1952).
(52) J. Žemlička and A. Holý, *Collect. Czech. Chem. Commun.*, 32, 3159 (1967); A. Holý, *Syn. Proc. Nucleic Acid Chem.*, 1, 172 (1968).
(53) D. M. Brown, A. R. Todd, and S. V. Varadarajan, *J. Chem. Soc.*, 1957, 868; B. F. West, *Syn. Proc. Nucleic Acid Chem.*, 1, 313 (1968).
(53a) V. N. Shibaev, M. A. Grachev, and S. M. Spiridonova, *Syn. Proc. Nucleic Acid Chem.*, 1, 503 (1968).
(54) N. K. Kochetkov, E. I. Budowsky, V. N. Shibaev, and M. A. Grachev, *Izv. Akad. Nauk SSSR, Ser. Khim.*, 1963, 1592.
(55) D. B. Dunn and J. D. Smith, *Biochem. J.*, 67, 494 (1957).
(56) D. B. Dunn and J. D. Smith, *Biochem, J.*, 68, 627 (1958).
(57) J. D. Smith and D. B. Dunn, *Biochem. J.*, 72, 294 (1959).
(58) J. G. Buchanan, *Nature*, 168, 1091 (1951).
(59) K. Fink and W. S. Adams, *J. Chromatogr.*, 22, 118 (1966).
(60) J. Morávek, *Collect. Czech. Chem. Commun.*, 24, 2571 (1959).
(61) V. Černěckij, S. Chládek, and F. Šorm, *Collect. Czech. Chem. Commun.*, 27, 87 (1962).
(62) A. Holý, J. Smrt, and F. Šorm, *Collect. Czech. Chem. Commun.*, 30, 3309 (1965).
(63) F. Bergmann, G. Levin, A. Kalmus, and H. Kwietny-Govrin, *J. Org. Chem.*, 26, 1504 (1961).
(64) F. Bergmann and A. Kalmus, *J. Org. Chem.*, 26, 1660 (1961).
(65) K. B. Jacobson, *J. Chromatogr.*, 14, 542 (1964).
(66) R. H. Hall, *J. Biol. Chem.*, 237, 2283 (1962).

(67) R. H. Hall, *Biochemistry*, 3, 876 (1964).
(68) K. Randerath and H. Struck, *J. Chromatogr.*, 6, 365 (1961).
(69) J. Žemlička, S. Chládek, Z. Haladová, and I. Rychlík, *Collect. Czech. Chem. Commun.*, 34, 3755 (1969).
(70) M. Prystaš and F. Šorm, *Collect. Czech. Chem. Commun.*, 29, 131 (1964).
(71) O. Motl and L. Novotný, in O. Mikeš (Ed.), *Laboratory Handbook of Chromatographic Methods*, D. Van Nostrand Co., Ltd., London, England, 1966, p. 205.
(72) W. Trappe, *Biochem. Z.*, 305, 150 (1940).
(73) M. Prystaš and F. Šorm, *Collect. Czech. Chem. Commun.*, 29, 121 (1964).
(74) M. Prystaš and F. Šorm, *Collect. Czech. Chem. Commun.*, 30, 2960 (1956).
(75) M. Prystaš, J. Farkaš, and F. Šorm, *Collect. Czech. Chem. Commun.*, 30, 3123 (1965).
(76) M. Prystaš and F. Šorm, *Collect. Czech. Chem. Commun.*, 31, 1053 (1966).
(76a) C. C. Bhat, *Syn. Proc. Nucleic Acid Chem.*, 1, 521 (1968).
(76b) H. J. Thomas, J. A. Johnson, Jr., W. E. Fitzgibbon, Jr., S. J. Clayton, and B. R. Baker, *Syn. Proc. Nucleic Acid Chem.*, 1, 249 (1968).
(77) M. Prystaš and F. Šorm, *Collect. Czech. Chem. Commun.*, 34, 2316 (1969).
(78) D. H. Rammler and H. G. Khorana, *J. Amer. Chem. Soc.*, 85, 1997 (1963).
(79) M. Prystaš and F. Šorm, *Collect. Czech. Chem. Commun.*, 34, 1104 (1969).
(80) R. H. Hall, *Methods Enzymol.*, 12A, 315 (1967).
(81) O. Mikeš, *Laboratory Handbook of Chromatographic Methods*, D. Van Nostrand Co., Ltd., London, England, 1966, pp. 256-259.
(82) J. D. Smith, *Methods Enzymol.*, 12A, 354, 361 (1967).
(83) P. R. Taylor and R. H. Hall, *J. Org. Chem.*, 29, 1078 (1965).
(84) S. Chladek, unpublished results.
(84a) C. Pedersen and H. G. Fletcher, Jr., *Syn. Proc. Nucleic Acid Chem.*, 1, 132 (1968); R. K. Ness, *ibid.*, 1, 183 (1968); M. Ikehara and H. Tada, *ibid.*, 1, 188 (1968).
(84b) M. Ikehara and H. Tada, *Syn. Proc. Nucleic Acid Chem.*, 1, 188 (1968); D. H. Murray and J. Prokop, *ibid.*, 1, 193 (1968).
(85) A. Holý and J. Žemlička, *Collect. Czech. Chem. Commun.*, 34, 3921 (1969).
(86) A. Holý and F. Šorm, *Collect. Czech Chem.*

Commun., 34, 1929 (1969).

(87) J. de Bersaques, *J. Chromatogr.*, 31, 222 (1967).

(88) G. Gorbach and J. Henke, *J. Chromatogr.*, 37, 225 (1968).

(89) B. Gelotte, *J. Chromatogr.*, 3, 330 (1960).

(90) S. Zadražil, Z. Šormová, and F. Šorm, *Collect. Czech. Chem. Commun.*, 26, 2643 (1961).

(91) T. Hohn and W. Pollman, *Z. Naturforsch., B*, 18, 919 (1963).

(92) F. N. Haynes, E. Hansbury, and W. E. Mitchell, *J. Chromatogr.*, 16, 410 (1964).

(93) H. G. Khorana and J. P. Vizsolyi, *J. Amer. Chem. Soc.*, 83, 675 (1961).

(94) H. G. Khorana, *Some Recent Developments in the Chemistry of Phosphate Esters of Biological Interest*, John Wiley and Sons, Inc., New York, N. Y., 1961.

(95) P. Flodin, *J. Chromatogr.*, 5, 103 (1961).

(96) R. Braun, *Biochim. Biophys. Acta*, 142, 267 (1967).

(97) R. Braun, *Biochim. Biophys. Acta*, 149, 601 (1967).

(98) M. Uziel and W. E. Cohn, *Fed. Proc.*, 24, 668 (1965).

(98a) D. W. Visser, *Syn. Proc. Nucleic Acid Chem.*, 1, 409 (1968).

(99) F. N. Hayes and E. Hansbury, *J. Amer. Chem. Soc.*, 86, 4172 (1964).

(100) L. Lábler and V. Schwarz (Eds.), *Thin-layer Chromatography* (in Czech), Publishing House of Czechoslovak Academy of Sciences, Prague, Czechoslovakia, 1965.

(101) J. G. Kirchner, J. M. Miller, and G. I. Keller, *Anal. Chem.*, 23, 420 (1951).

(102) K. Randerath, *Angew. Chem.*, 73, 674 (1961).

(103) L. Birkofer, C. Kaiser, H. A. Meyer-Stoll, and F. Suppan, *Z. Naturforsch., B*, 17, 352 (1962).

(104) K. Onoe, *Nippon Kagaku Zasshi*, 73, 337 (1952).

(105) P. R. Bhandari, B. Lerch, and G. Wohlleben, *Pharm. Ztg.*, 107, 1618 (1962).

(106) J. J. Peifer, *Mikrochim. Acta*, 1962, 529.

(107) G. R. Duncan, *J. Chromatogr.*, 8, 37 (1962).

(108) V. Černý, J. Joska, and L. Lábler, *Collect. Czech. Chem. Commun.*, 26, 1658 (1961).

(109) S. Heřmánek, V. Schwarz, and Z. Čekan, *Collect. Czech. Chem. Commun.*, 26, 1969 (1961).

(110) M. Mottier and M. Potterat, *Anal. Chim. Acta*, 13, 46 (1955).

(111) S. Heřmánek, V. Schwarz, and Z. Čekan, *Collect. Czech. Chem. Commun.*, 26, 3170 (1961).

(112) J. Pitra, S. Heřmánek, and L. Lábler, in L. Lábler and V. Schwarz (Eds.), *Thin-layer Chromatography*, Publishing House of Czechoslovak Academy of Sciences, Prague, Czechoslovakia, 1965, p. 74.
(113) H. Brockmann and H. Schoder, *Ber.*, 74, 73 (1941).
(114) C. G. Honnegger, *Helv. Chim. Acta*, 46, 1730 (1963).
(115) E. Stahl, *Arch. Pharm.* (Weinheim), 292, 411 (1959).
(116) E. Mistrjukov, *J. Chromatogr.*, 9, 311 (1962).
(117) M. Brenner and A. Niederwieser, *Experientia*, 17, 237 (1961).
(118) S. M. Rybicka, *Chem. Ind.*(London), 1962, 308.
(119) E. Stahl, *Chem.-Ztg. Chem. App.*, 82, 323 (1959).
(120) E. Stahl, *Pharm. Rundsch.*, 1, 2 (1959); L. Lábler and V. Schwarz (Eds.), *Thin-layer Chromatography* (in Czech), Publishing House of Czechoslovak Academy of Sciences, Prague, Czechoslovakia, 1965, p. 117.
(121) M. Prystaš and F. Šorm, *Collect. Czech. Chem. Commun.*, 33, 210 (1968).
(122) M. Prystaš and F. Šorm, *Collect. Czech. Chem. Commun.*, 33, 1813 (1968).
(123) M. W. Winkley and R. L. Robins, *J. Org. Chem.*, 33, 2822 (1968).
(123a) C. C. Bhat and H. R. Munson, *Syn. Proc. Nucleic Acid Chem.*, 1, 83 (1968).
(124) S. Chládek and J. Žemlička, *Collect. Czech. Chem. Commun.*, 33, 4299 (1968).
(125) K. Randerath, *Nature*, 205, 908 (1965).
(126) G. N. Mahapatra and O. M. Friedman, *J. Chromatogr.*, 11, 265 (1963).
(127) A. D. Broom, L. B. Townsend, J. W. Jones, and R. K. Robins, *Biochemistry*, 3, 495 (1964).
(128) A. Massaglia, V. Rosa, and S. Sosi, *J. Chromatogr.*, 17, 316 (1965).
(129) T.-C. Chou and H.-H. Lin, *J. Chromatogr.*, 27, 307 (1967).
(130) K. Randerath, *Biochem. Biophys. Res. Commun.*, 6, 452 (1961/1962).
(131) B. Shasha and R. L. Whistler, *J. Chromatogr.*, 14, 532 (1964).
(132) K.-T. Wang and I. S. Y. Wang, *Biochim. Biophys. Acta*, 142, 280 (1967).
(133) P. Remy, G. Dirheimer, and J.-P. Ebel, *J. Chromatogr.*, 31, 609 (1967).
(134) K.-T. Wang and P.-H. Wu, *J. Chromatogr.*, 38, 153 (1968).

(135) K. Randerath, *Nature*, 194, 768 (1962).
(136) K. Randerath and E. Randerath, *J. Chromatogr.*, 16, 111 (1964).
(137) K. Randerath and E. Randerath, *Methods Enzymol.*, 12A, 323 (1967).

ERRATUM

Volume 1, page 185, paragraph 3, line 2 up. For 69%, read 82.5%.

Subject Index

D

E

F